J	(i)	(ii)	(iii)	(iv)	(v)
0	0.61	61.9	80.5	55.2	64.1
2	1.45	0.8		0.6	
4	2.38	2.5	1.0	3.7	2.2
5	2.15				
6	3.18	6.5	1.7	6.5	4.9
8	3.74	2.6	1.8	4.7	3.3
10	4.41	3.0	1.2	3.5	2.4
12	4.53			1.0	
13	4.07			0.6	0.6
16	4.49			0.7	
18	4.31	1.7		1.3	1.0
28	2.05	1.2		0.9	1.0
33	0.94			0.6	
36	0.66	0.7			
42	0.19	0.6		0.9	0.7
46	0.05	1.0		0.8	0.7
48	0.05	11.5	9.2	12.4	11.7

Tab. 1. Statistics of ground state spins J_0 for $N = 6$ identical particles on a single $j = 21/2$ orbital. Except for column (ii) only statistically significant fractions $f_J > 0.5\%$ are included.
(i) Fraction of states with a given J
(ii) Statistics of ground state spins in random ensemble with Gaussian distribution of V_L with zero mean and dispersion equal to 1.
(iii) The same as (ii) but with $V_0 = -1$.
(iv) The same as in (ii) but with $V_0 = +1$.
(v) The same as in (ii) but without pairing, $V_0 = 0$.

being pushed up by attractive pairing. In cases of an odd number of particles, there is a significant enhancement of the ground state spin $J = j$, in agreement with the prevailing value $J = 0$ in the even neighbor.

We will not review here an avalanche of explanations triggered by the pioneering work.[4] We mention only that the statistical width σ_J of the level density for different values of J was suggested by many authors starting with Ref. 7 as a reason for the high probability of $J_0 = 0$. Even if the class of $J = 0$ states would have a larger width, it would just map the problem to another one of distribution of statistical widths. But in fact σ_J typically grows with J, Fig. 1. The widths σ_J for different J are not independent being defined by the same interaction. The difficulty is increased by the interference of different components V_L in a given set of interaction parameters; their conspiracy makes the results of a direct sampling method [8] unreliable and justified only in simplest cases,[9] such as that of the $j = 7/2$ orbital, where the seniority is conserved and the eigenfunctions are defined by symmetry regardless of the interaction.

More detailed studies [10,11,13,14] brought some clarity into the question of the structure of ground states in systems with random interactions. The distribution of overlaps

$$x = |\langle 0|\text{paired}\rangle|^2 \tag{2}$$

of the empirical ground state wave functions $|0\rangle$ with fully paired states of seniority zero is close to what one would expect for the eigenvectors of the Gaussian

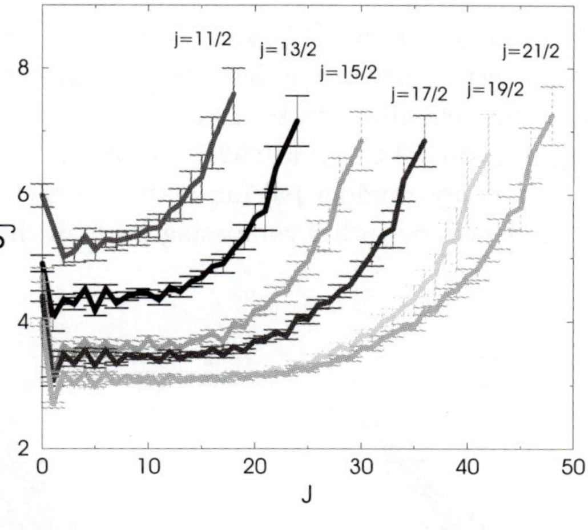

Fig. 1. Widths σ_J for various j-orbitals, $N = 6$

orthogonal ensemble of a corresponding dimension, Fig. 2.

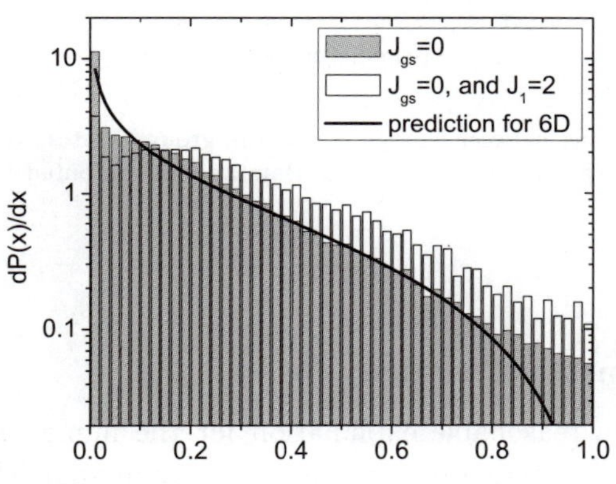

Fig. 2. The distribution of overlaps, eq. (2), for $N = 6$ fermions on the $j = 15/2$ level, shaded histogram, compared to the statistical distribution for a chaotic system of dimension $d = 4$. The unshaded distribution is found for the part of the ensemble that produces the ground state $J = 0$ along with the first excited state $J = 2$.

Fig. 3 shows the lack of pairing correlations in systems driven by random interactions from another angle. Here, despite a selection of states with $J_0 = 0$, energies of the resulting ground states are still noticeably lower that the expectation value of the Hamiltonian for the fully paired state. The shift of the best fit (solid line) from the $x = y$ curve (dashed line) is an indication of correlation energy in random realizations that goes beyond pairing. The left panel corresponds to a half occupied system, where particle-hole symmetry allows for a stronger role of pairing,[12] thus the agreement in energies is slightly better as compared to the plot on the right panel.

As for more realistic models, the average overlap between the realistic shell model solution for the ground state of ^{24}Mg and ground states of randomly chosen Hamiltonian, where one keeps correct non-degenerate single-particle energies and attractive pairing, while the remaining parts of the interaction are random, reaches only $\bar{x} = 0.1$ The largest probability $f_0 = 0.92$ is obtained [11] for the shell model version, where one retains only random pairing matrix elements with degenerate single-particle levels and leaves out other competing parts of the interaction.

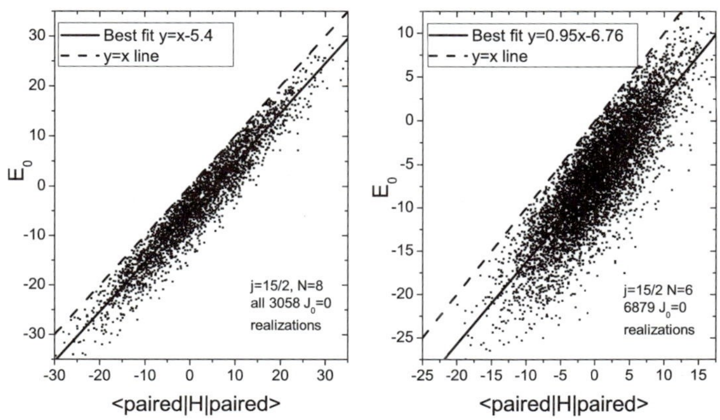

Fig. 3. The scattered plots display correlations between energy E_0 of the ground states with $J_0 = 0$ in random realizations (y-axis), and expectation values of the Hamiltonian computed for the fully paired state (x-axis).

3. Statistical Approach and Geometric Chaoticity

Few general ideas help us in finding a reasonable explanation for the numerical observations. (i) Any realization of the interaction ensemble in a many-particle system leads to the formation of some mean field. In this approximation one can speak about mean occupation numbers of single-particle states. The symmetry of the mean field can be different but rotational modes restore the angular momentum quantum number J, M. The mean field is a shared tool that establishes correlations between classes of states with different quantum numbers. (ii) In each class of state with a given value of J, different schemes of vector coupling of individual spins are possible. The fractional parentage coefficients, for even a small number of intermediate chains, rapidly become almost random numbers. This is what we call *geometric chaoticity*.[1,10] (iii) Averaging over possible coupling chains we come to the effective Hamiltonian $\tilde{H}$ of classes that depends on the constants of motions only, in a simple case on particle number N and total spin $\mathbf{J}$. The same approach can be applied to chaotic quantum dots.[15] In our problem we expect the effective

Hamiltonian of the form

$$\tilde{H} = H_0(N) + H_2(N)\mathbf{J}^2 + H_4(N)(\mathbf{J}^2)^2 + \dots \tag{3}$$

(other conserved quantities as isospin can be added). (iv) Finally, *time-reversal* $\mathcal{T}$-invariance is important for the results. The direct inclusion of a random $\mathcal{T}$-violating interaction [7] may only enhance the fraction of $J_0 = 0$, as discussed in Ref. 5. However, there are situations where $\mathcal{T}$-invariance alone leads to a strong preference of $J_0 = 0$. In a system of many randomly interacting spins 1/2, this invariance ensures the double Kramers degeneracy of $\mathcal{T}$-conjugated states. Then, for an even N, the orbitals in the ground state are always occupied pairwise that automatically leads to $J = 0$; the exception can be only the case of the degeneracy of few pairs. Indeed, such systems are shown [16] to have f_0 close to 1. The difference compared to the earlier mentioned statistics of quantum spin glasses [3] is in the random Hamiltonian that keeps the geometric ratio 3:1 of singlet/triplet amplitudes while in the shell model ensembles these amplitudes are independent. A similar situation with Kramers degeneracy and large f_0 takes place when a set of interactions leads to a strongly deformed mean field, where again the pairwise occupation prefers $J_0 = 0$.

Assuming the axial symmetry of the mean field, we consider an aligned state with $M = J$ and approximate energy of a state through the single-particle occupancies n_m and Clebsch-Gordan coefficients,

$$\langle H \rangle \Rightarrow E(\{n_m\}) = \frac{1}{2}\sum_{mm'} V_{mm'}\langle n_m n_{m'}\rangle, \quad V_{mm'} = 2\sum_{L\Lambda} V_L (C^{L\Lambda}_{jm\,jm'})^2. \tag{4}$$

The next step along the road of approximations is to neglect the correlations of occupancies, $\langle n_m n_{m'}\rangle \approx n_m n_{m'}$ and minimize energy (4) under constraints of total particle number and total spin projection,

$$\sum_m n_m = N, \quad \sum_m m n_m = M. \tag{5}$$

The minimization with the aid of two Lagrange multipliers, chemical potential μ and cranking frequency (or magnetic field) γ defines [5]

$$E(N,M) = \frac{1}{2}[N\mu(N,M) + M\gamma(N,M)]. \tag{6}$$

The calculation of the Lagrange multipliers [5] gives

$$\mu = \frac{2N}{(2j+1)^2}\sum_L (2L+1)V_L,$$

$$\gamma = \frac{3M}{(2j+1)^2 j(j+1)}\sum_L (2L+1)[L(L+1) - 2j(j+1)]V_L. \tag{7}$$

Identifying M with J, we come to the terms H_0 and H_2 of the effective Hamiltonian (3). In this approximation, similar to Hund's rule, the probability $f_0 = 50\%$ depending on the sign of the moment inertia $\propto H_2^{-1}$ that, in turn, is defined by the weight of interactions in aligned, $\mathbf{L}^2 > 2\mathbf{j}^2$, and antialigned, $\mathbf{L}^2 < 2\mathbf{j}^2$, pairs. The statistical approach is not valid for J_{max}, since such a state is in our model unique and the occupation numbers are fully correlated. But for $J = J_{max}$ we know the wave function and can calculate energy directly. The comparison between actual ground state energy for $J_0 = 0$ and energy predicted by the simplest statistical model shows [10,5] their clear correlation plus a constant non-statistical shift. The model can be advanced with the use of more realistic occupation numbers with a nonlinear dependence on total M.[5] A mean field approach was also used for the interacting boson model.[17] Here the simplicity of the model, in particular for p-bosons, and a sharp division of the parameter space into different shape regions,[18] allow for a more complete study.

4. Dynamics and New Questions

The next level of numerical and analytic consideration should investigate the dynamics and low-lying excited states of a system with random interactions. The first steps in this direction [5] lead to unexpected results.

Earlier the researchers claimed that the random interactions do not create rotational spectra. Look however at the unshaded histogram of Fig. 3. It corresponds to the subset of the ensemble that includes random interactions leading to the sequence of spins $J_0 = 0$, $J_1 = 2$. This subset gives an appreciable excess of cases with a large overlap, eq. (2), between actual ground states and fully paired states. From normal nuclear physics we know two generic cases when such a sequence of levels in even-even nuclei takes place: spherical strongly paired (usually near-magic) nuclei with good seniority and well deformed nuclei with the low-lying ground state rotational band.

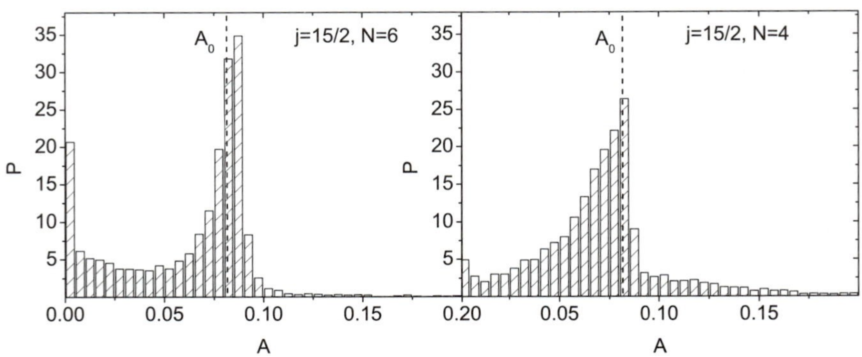

Fig. 4. The histogram for the Alaga ratio of eq. (8) for 4 and 6 particles in a $j = 15/2$ orbital; only the cases with the lowest spin sequence $J_0 = 0$, $J_1 = 2$ are included, 6.7% and 9.2% of all cases for $N = 4$ and $N = 6$, respectively.

The further analysis allows us to distinguish these two cases. An appropriate indicator is the *Alaga ratio*

$$A = \frac{\langle Q \rangle^2_{J=2}}{B(\mathrm{E2}; 0 \to 2)} \tag{8}$$

of the expectation value of the squared quadrupole moment in the $J = 2$ state to the quadrupole transition probability between this and the ground $J = 0$ state. The distribution for this quantity, Fig. 4, reveals two peaks, pronounced stronger at a large particle number, at $A = 0$ and $A = 4/49$. The first value signals a spherical shape, while the second one is characteristic for a rigid axial rotor.

5. Conclusion

The short review above gives an evidence that we routinely underestimate the role of incoherent interactions in many-body physics. Taken in coordination with the symmetry requirements for a finite mesoscopic system, they determine in average quantum characteristics of the ground and low-lying states. The underlying mechanism is related to the formation of a mean field, geometric chaoticity of angular momentum coupling and time-reversal invariance. Moreover, the geometry of a finite system is inclined to single out certain shapes of a system with a probability much larger than expected *a-priori* for a random ensemble. The coherent interactions, such as attractive pairing, enhance the effect up to some extent predetermined by random interactions and symmetries.

Acknowledgments

The authors are thankful to B.A. Brown, M. Horoi and D. Mulhall for collaboration and discussions. The support from the NSF, Grants PHY-0070911 and PHY-0244453, and from the DOE under grant No. DE-FG02-92ER40750 is acknowledged.

References

1. V. Zelevinsky, B.A. Brown, N. Frazier, and M. Horoi, Phys. Rep. **276**, 85 (1996).
2. T.A. Brody, J. Flores, J.B. French, P.A. Mello, A. Pandey, and S.S.M. Wong, Rev. Mod. Phys. **53**, 385 (1981).
3. J. Oitmaa and O.P. Sushkov, Phys. Rev. Lett. **87**, 167206 (2001).
4. C.W. Johnson, G.F. Bertsch, and D.J. Dean, Phys. Rev. Lett. **80**, 2749 (1998).
5. V. Zelevinsky and A. Volya, Phys. Rep. **391**, 311 (2004).
6. Y.M. Zhao, A. Arima, and N. Yoshinaga, nucl-th/0311050 (submitted to Phys. Rep.).
7. R. Bijker, A. Frank, and S. Pittel, Phys. Rev. C **60**, 021302 (1999).
8. Y.M. Zhao, A. Arima, and N. Yoshinaga, Phys. Rev. C **66**, 034302 (2002).
9. P. Chau Huu-Tai, A. Frank, N.A. Smirnova, and P. Van Isacker, Phys. Rev. C **66**. 061302 (2002).
10. D. Mulhall, A. Volya, and V. Zelevinsky, Phys. Rev. Lett. **85**, 4016 (2000).
11. M. Horoi, B.A. Brown, and V. Zelevinsky, Phys. Rev. Lett. **87**, 062501 (2001).
12. A. Volya, Phys. Rev. C **65**, 044311 (2002).
13. V. Zelevinsky, D. Mulhall, and A. Volya, Phys. Atom. Nucl. **64**, 525 (2001).
14. M. Horoi, A. Volya and V. Zelevinsky, Phys. Rev. C **66**, 024319 (2002).

15. J.A. Folk, C.M. Markus, R. Berkowits, I.L. Kurland, I.L. Aleiner, and B.L. Altshuler, Phys. Scr. T **90**, 26 (2001).
16. L. Kaplan, T. Papenbrock, and C.W. Johnson, Phys. Rev. C **63**, 014307 (2001).
17. R. Bijker and A. Frank, Phys. Rev. C **65**, 044316 (2002).
18. P. Cejnar, V. Zelevinsky, and V.V. Sokolov, Phys. Rev. E **63**, 036127 (2001).

Cold Atoms and Quantum Information

SUPERFLUID REGIMES IN DEGENERATE ATOMIC FERMI GASES

G. V. SHLYAPNIKOV

Laboratoire Physique Théorique et Modèles Statistique, Université Paris Sud, Bât. 100, 91405 Orsay, France
Van der Waals-Zeeman Institute, University of Amsterdam, Valckenierstraat 65/67, 1018 XE Amsterdam, The Netherlands

Received 1 December 2004

We give a brief overview of recent studies of quantum degenerate regimes in ultracold Fermi gases. The attention is focused on the regime of Bose-Einstein condensation of weakly bound molecules of fermionic atoms, formed at a large positive scattering length for the interspecies atom-atom interaction. We analyze remarkable collisional stability of these molecules and draw prospects for future studies.

Keywords: Superfluid; fermions; atomic gases.

1. Introduction

The last few years are marked by a remarkable progress and rapidly growing activity in the studies of ultracold trapped Fermi gases. These studies provide links between quantum gases and condensed matter systems, and one of the main directions is a search for novel macroscopic quantum phenomena. In this respect, the situation resembles that in the early studies of Bose-condensed gases after the discovery of Bose-Einstein condensation (BEC) in trapped clouds of alkali atoms.[1,2,3] During a few years these studies found a variety of spectacular macroscopic quantum effects: collective oscillations and their damping, formation of quantized vortices, etc. (see [4] for review).

Quantum degenerate Fermi gases are fundamentally different from Bose gases. Due to the Pauli exclusion principle, in the case of identical fermions not more than one particle can occupy a given quantum state of motion. Therefore, well below the temperature of quantum degeneracy the distribution of particles is characterized by the presence of a Fermi sea. Occupation numbers of quantum states by identical fermions are equal to unity up to energies close to the Fermi energy T_F, and are zero for larger energies. The Fermi energy represents the ratio of the number of particles to the (energy) density of states and in the uniform two-component Fermi gas is given by the relation $T_F = \hbar^2(3\pi^2 n)^{2/3}/2m$, where n is the gas density, and m the atom mass. At a finite temperature $T \ll T_F$, the transition from unity to zero occupation numbers is not step-wise. It occurs in a narrow vicinity of the Fermi surface, i.e. the surface $\varepsilon = T_F$ in the energy space. For such a distribution of fermions, only particles with energies near the Fermi surface provide a response of the system to external perturbations. In this respect, the rest of the Fermi gas becomes a "dead body". This is quite different from Bose-condensed systems where all particles participate in the response.

Macroscopic quantum phenomena in degenerate Fermi gases are commonly expected in relation to the Cooper pairing and superfluid Bardeen-Cooper-Schrieffer (BCS) phase transition. This requires attractive interaction between particles. Then, in the simplest version of the superfluid transition in a two-component Fermi gas, at sufficiently low temperatures fermions belonging to different components and having opposite momenta on the Fermi surface form correlated (Cooper) pairs in the momentum space. This leads to the appearance of a gap in the excitation spectrum which in the uniform case reads (see, e.g., [5])

$$\varepsilon_k = \sqrt{[\hbar^2(k^2 - k_F^2)/2m]^2 + \Delta^2}, \tag{1}$$

where ε_k and k are the excitation energy and momentum, and $k_F = \sqrt{2mT_F/\hbar^2}$ is the Fermi momentum. The gap Δ depends on the interaction parameters, density, and temperature. The property of superfluidity can be understood considering the motion of the Fermi gas as a displacement of the entire Fermi sphere in the momentum space. Then the friction can be viewed as a transfer of particles to the Fermi sphere of the gas at rest, with related creation of holes in the Fermi sphere of the moving gas. In the superfluid phase characterized by the dispersion law (1), this process is forbidden at $T \to 0$ as the creation of a particle-hole pair costs energy 2Δ.

Cooper pairs are really formed only by particles near the Fermi surface and, therefore, the "condensate of Cooper pairs" is small. In a dilute Fermi gas, most efficient is the formation of Cooper pairs due to the attractive binary interaction between particles in the s-wave channel. This requires the presence of at least two fermionic components in the system, with almost equal concentrations and a negative s-wave scattering length a for the intercomponent interaction. Then the superfluid BCS transition occurs at a temperature which in the uniform case is given by:[6]

$$T_c = (2/e)^{7/3} T_F \exp(-\pi/2k_F|a|). \tag{2}$$

In the dilute limit one has the inequality $k_F|a| \ll 1$, and the transition temperature T_c is much smaller than the Fermi energy (temperature of quantum degeneracy).

The small ratio T_c/T_F was the main difficulty in achieving the superfluid BCS transition in the first generation of experiments with alkali atom Fermi gases, which successfully reached the regime of quantum degeneracy. This has been first done in the JILA experiment with two-component clouds of ^{40}K,[7] and then in a mixture of ^{6}Li with ^{7}Li,[8,9] in a two-component gas of ^{6}Li,[10] in a mixture of ^{6}Li with Na [12], and a mixture of ^{40}K with ^{87}Rb.[13] The Fermi degeneracy has been identified through the measurement of the density profile, and densities $n \sim 10^{13}$ cm^{-3} have been reached. The key problem with cooling a mixture of two fermionic isotopes or a mixture of fermions and bosons to well below the temperature of quantum degeneracy was Pauli blocking of elastic collisions and, hence, the reduction of the rate of evaporative cooling.[14] At temperatures $T \ll T_F$, energy states below the Fermi surface are highly occupied and any collisional event resulting in a state with such an energy is suppressed by the Pauli exclusion principle. In other words, there is a reduction of the phase space available for final states of fermions in a collisional event. The "effective" cross section of elastic collisions strongly decreases with temperature as has been measured at JILA,[15] which leads to the suppression of evaporative cooling (see [14]). The ratio $T/T_F \approx 0.2$ achieved in these studies was still not sufficiently small for reaching the superfluid BCS transition. An overview of the first generation of theoretical and experimental studies of trapped Fermi gases is given in Ref. 16.

2. Feshbach Resonances and BCS-BEC Crossover

In the present stage, the studies of macroscopic quantum effects in two-component trapped atomic Fermi gases are focused on tuning and increasing the intercomponent interaction (scattering length a) by using Feshbach resonances. In the vicinity of the resonance, for the two-particle problem one has a strong coupling between (zero-energy) continuum states of colliding atoms and a bound molecular state of another hyperfine domain. The effective coupling strength depends on the detuning from the resonance (the energy difference δ between the bound molecular state and the continuum states), which can be varied by changing the magnetic field. Thus, the scattering length becomes field dependent (see Fig. 1).

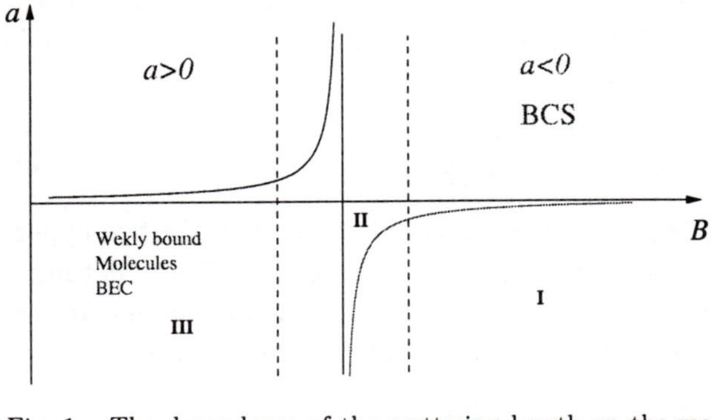

Fig. 1. The dependence of the scattering length on the magnetic field. The symbols I, II, and III stand for the BCS regime, strongly interacting regime, and the BEC regime of weakly bound molecules.

At resonance the scattering length changes from $+\infty$ to $-\infty$, and in the vicinity of the resonance one has the inequality $k_F|a| \gtrsim 1$. The gas then is in the strongly interacting regime. It is still dilute and the mean interparticle separation greatly exceeds the characteristic radius of interparticle interaction R_e. However, the amplitude of binary interactions (scattering length) is larger than the mean separation between particles and, therefore, is no longer a relevant quantity to describe the system.

For a large detuning from resonance the gas is still in the weakly interacting regime, i.e. the inequality $n|a|^3 \ll 1$ is satisfied. On the negative side of the resonance ($a < 0$) corresponding to the region I in Fig. 1, one expects the BCS pairing between distinguishable fermions, well described in literature.[5] On the positive side ($a > 0$) represented by region III in Fig. 1, two fermions belonging to different components can form weakly bound molecules. Therefore, one expects the formation of these composite bosons and encounters the problem of their Bose-Einstein condensation. A subtle question is then to describe a crossover from the BCS to BEC behavior and find the superfluid transition temperature in the crossover strongly interacting regime (region II in Fig. 1). This type of crossover has been discussed in literature in the context of superconductivity [17,18,19,20] and in relation to superfluidity in two-dimensional films of ^{3}He.[21,22] The idea of resonance coupling through a Feshbach resonance for achieving a superfluid phase transition in ultracold two-component Fermi gases has been proposed in Refs. 23, 24

The description of a many-body system near a Feshbach resonance requires a detailed knowledge of the 2-body problem. This is a two-channel problem which can be described in terms of the Breit-Wigner scattering,[25,31] the open channel being the states of colliding atoms and the closed channel the bound molecular state of the other hyperfine domain. Various aspects of this type of problems have been discussed by Feshbach[26] and Fano[27]. In cold atom physics the idea of Feshbach resonances was introduced in Ref. 28, and optically induced resonances have been discussed in Refs. 29, 30.

The 2-body physics is the most transparent if one can omit the (small) background scattering length. Then for low collision energies ε the scattering amplitude is given by [31] :

$$F(\varepsilon) = -\frac{\hbar\gamma/\sqrt{m}}{\varepsilon + \delta + i\gamma\sqrt{\varepsilon}}, \tag{3}$$

where the quantity $\hbar\gamma/\sqrt{m} \equiv W$ characterizes the coupling between the two hyperfine domains. The effective scattering length is $a = -F(0)$. In Eq. (3) the detuning δ is positive if the bound molecular state is below the continuum of colliding atoms. Then for $\delta > 0$ the scattering length is positive, and for $\delta < 0$ it is negative. Introducing a characteristic length

$$R_* = \hbar^2/mW \tag{4}$$

and expressing the scattering amplitude through the relative momentum of particles $k = \sqrt{m\varepsilon}/\hbar$, equation (3) takes the form:

$$F(k) = -\frac{1}{a^{-1} + R_* k^2 + ik}, \tag{5}$$

The validity of Eq. (5) does not require the inequality $kR_* \ll 1$. At the same time, this equation formally coincides with the amplitude of scattering of slow particles by a potential with the same scattering length a and an effective range $R = -2R_*$, obtained under the condition $k|R| \ll 1$.

The length R_* is an intrinsic parameter of the Feshbach resonance problem. It characterizes the width of the resonance. From Eqs. (3) and(4) we see that large W and, hence, small R_* correspond to a wide resonance, whereas small W and large R_* lead to a narrow resonance. The issue of wide and narrow resonances is now actively being discussed in literature.[32,33,34,35,36,37,38,39]

Actually, the use of the terms "wide" and "narrow" depends on the problem under consideration. For example, in the unitarity limit where $a \to \pm\infty$, Eq.(5) shows that the length R_* drops out of the problem under the condition $kR_* \ll 1$. In a quantum degenerate Fermi gas the characteristic momentum of particles is the Fermi momentum $k_F = (3\pi^2 n)^{1/3}$. Thus, for a given R_* the condition of the wide resonance depends on the gas density n and reads [34,35,37,38,39]

$$k_F R_* \ll 1. \tag{6}$$

The two-channel scattering problem is then reduced to a single-channel one, i.e. the scattering properties are the same as in the case of potential scattering at an infinite scattering length.

In the weakly interacting BCS regime, where $a < 0$ and $k_F|a| \ll 1$, the condition (6) is sufficient for the resonance to be wide. The characteristic momentum of atoms is k_F, and both the second and third terms in the denominator of Eq. (5) are small compared to a^{-1}. The 2-particle problem is then equivalent to the potential scattering with a scattering length a.

For $a > 0$ and $na^3 \ll 1$, where one has a weakly interacting gas of diatomic bosonic molecules, the criterion of the wide resonance is different.[36,40] The most important limitation is related to the binding energy and the wavefunction of the molecules. The energy of the weakly bound molecular state (it is certainly assumed that the characteristic radius of interaction $R_e \ll a$) is determined by the pole of the scattering amplitude (5). One then finds [36,40] that this state exists only for $a > 0$ and under the condition

$$R_* \ll a. \tag{7}$$

the binding energy is given by

$$\varepsilon_0 = \hbar^2/ma^2. \tag{8}$$

Then the wavefunction of the weakly bound molecular state has only a small admixture of the closed channel, the size of the molecule is $\sim a$, and atom-molecule and molecule-molecule interactions are determined by a single parameter - the atom-atom scattering length a. In this sense, the problem becomes universal. It is equivalent to the interaction problem for the 2-body potential which is characterized by a large positive scattering length a and has a potential well with a weakly bound molecular state.

The picture becomes somewhat more complicated if the background scattering length $\tilde{a}$ can not be neglected.[41] Nevertheless, the essential physics remains the same: for $|\tilde{a}| \lesssim |a|$ the conditions of the wide resonance given by Eqs. (6) and (7), and Eq. (8) for the binding energy of the molecules are still valid.

Most experiments with two-component atomic Fermi gases are using wide Feshbach resonances. The only exception is the experiment at Rice with ^{6}Li near a narrow Feshbach resonance at 543 G.[42] For this resonance the length R_* is very large and at obtained values of a the condition (7) is not fulfilled. Below we will focus attention on the case of a wide Feshbach resonance.

3. Strongly Interacting Regime

In the BCS regime (region I in Fig. 1) the superfluid transition temperature is given by Eq. (2), and the physics of this regime is well known.[5] Due to a very low value of T_c, this regime is not yet achieved in current experiments.

In the BEC regime of weakly bound molecules (region III), the superfluid transition temperature can be found using a relation for the temperature of Bose-Einstein condensation of an ideal Bose gas. As the density of the gas of molecules is $n/2$, and the mass of a molecule is $2m$, we obtain

$$T_c = \frac{3.31\hbar^2}{2m}\left(\frac{n}{2}\right)^{2/3} \approx 0.2T_F. \tag{9}$$

In the strongly interacting regime (region II), the superfluid transition is strongly influenced by fluctuations,[43,44,45] and the transition temperature is a fraction of T_F. According to Ref. 44, for a spatially homogeneous gas one has $T_c \approx 0.26T_F$.[44] However, the issue of the transition temperature is not completely resolved and calculations are now underway.[46]

The critical temperature T_c for the BCS-BEC crossover (regions from I to III in Fig. 1 and Fig. 2) depends on the magnetic field through the field dependence of the interaction amplitude (scattering length). In Fig. 2 we present a typical field dependence of the ratio T_c/T_F for the coupling through a (wide) Feshbach resonance.

Atomic Fermi gases in the strongly interacting regime are characterized by universal behavior of interactions[47,48,49,50] and universal thermodynamics.[51] The only length scale is the mean interparticle separation $n^{-1/3}$, and the only energy scale is the Fermi energy T_F. Recent Monte Carlo calculations [48,52,53] give the chemical potential

$$\mu = \beta T_F, \tag{10}$$

with the coefficient $\beta \approx 0.4$.

The strongly interacting regime is now actively being studied theoretically in relation to the nature of superfluid pairing [33,34,54,55,56,46,57] and the character of elementary excitations.[58,59,60,61,62,63] This regime has been achieved in experiments

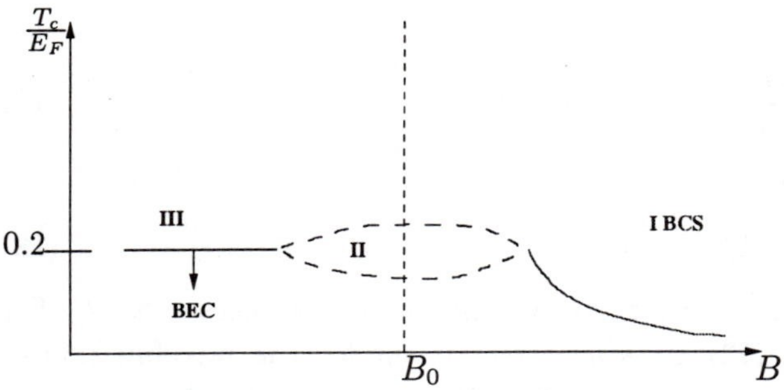

Fig. 2. The ratio of the critical temperature to the Fermi energy, T_c/T_F, versus the magnetic field for the BCS-BEC crossover. In the strongly interacting regime (region II), where the result is not exactly known, the critical temperature is shown by dashed curves.

with two-component atomic gases of ^{40}K [64] and ^{6}Li,[65,11,49,66,67,68,69,70] and experimental data indicate that the coefficient β is close to 0.4. The JILA and MIT experiments [64,65] observed a BEC-type behavior of fermionic atom pairs for $k_F|a|$ on both sides of the resonance. Intriguing experiments at Duke [66] and Innsbruck [67] present an anomalous dependence of frequencies and damping rates of quadrupole excitations on the interaction strength, which may be a signature for a transition from a superfluid to collisionless regime. Strong evidence for achieving the superfluid regime was obtained in the recent Innsbruck experiment 69 through the radio-frequency measurement of the pairing gap for a strongly interacting Fermi gas of ^{6}Li. An argument in support of the presence of superfluid-non-superfluid transition has been found in the Duke experiment[70] from the measurement of the temperature dependence of the heat capacity.

4. BEC Regime of Weakly Bound Molecules

For the BEC regime of the weakly bound bosonic molecules, the key questions are the stability of the condensate with regard to elastic molecule-molecule interactions, and the decay of the gas due to collisional relaxation of the molecules to deep bound states. The elastic interaction should be repulsive, otherwise the Bose-condensed molecular gas undergoes a collapse. The exact solution for the molecule-molecule elastic scattering was found in Ref. 71, and a detailed analysis of this problem is given in Ref. 40. This has been done assuming that the (positive) atom-atom scattering length a for the interspecies interaction greatly exceeds the characteristic radius of interatomic potential, R_e. Then, as in the case of the 3-body problem with fermions (see [72,73,74,75] and references in [75]), the amplitude of elastic interaction is determined only by a and can be found in the zero-range approximation for the interatomic potential.

This approach was introduced in the two-body physics by Bethe and Peierls.[76,77] The leading idea is to solve the equation for the free relative motion of two particles placing a boundary condition on the wavefunction ψ at a vanishing interparticle distance r:

$$\frac{(r\psi)'}{r\psi} = -\frac{1}{a}, \quad r \to 0, \tag{11}$$

which can also be rewritten as

$$\psi \propto (1/r - 1/a), \quad r \to 0. \tag{12}$$

One then gets a correct expression for the wavefunction at distances $r \gg R_e$. For the case where $a \gg R_e$, Eq. (12) gives a correct result for the wavefunction of weakly bound and continuum states even at distances much smaller than a.

The molecule-molecule elastic scattering is a 4-body problem, which at collision energies much smaller than the molecule binding energy ε_0 is dominated by the contribution of the s-wave channel. In the zero range approximation this problem can be analyzed from the solution of the free-particle Schródinger equation in which the total energy is equal to $-2\varepsilon_0$:

$$\hat{P}\Psi \equiv -\left[\nabla^2_{\mathbf{r}_1} + \nabla^2_{\mathbf{r}_2} + \nabla^2_{\mathbf{R}} - 2/a^2\right]\Psi = 0, \tag{13}$$

placing the Bethe-Peierls boundary condition at vanishing distances between fermions in different internal states (labeled by the symbols $\uparrow$ and $\downarrow$). In Eq. (13) $\mathbf{r}_1$ is the distance between two given $\uparrow$ and $\downarrow$ fermions, and $\mathbf{r}_2$ is the distance between the other two; $\mathbf{R}/\sqrt{2}$ is the distance between the centers of mass of these pairs, $\mathbf{r}_\pm = (\mathbf{r}_1 + \mathbf{r}_2 \pm \sqrt{2}\mathbf{R})/2$. The wavefunction Ψ is symmetric with respect to the permutation of the molecules and antisymmetric with respect to permutations of identical fermions:

$$\begin{aligned}\Psi(\mathbf{r}_1, \mathbf{r}_2, \mathbf{R}) &= \Psi(\mathbf{r}_2, \mathbf{r}_1, -\mathbf{R}) \\ &= -\Psi(\mathbf{r}_\pm, \mathbf{r}_\mp, \pm(\mathbf{r}_1 - \mathbf{r}_2)/\sqrt{2}).\end{aligned} \tag{14}$$

For $\mathbf{r}_1 \to 0$ the Bethe-Peierls boundary condition reads:

$$\Psi(\mathbf{r}_1, \mathbf{r}_2, \mathbf{R}) \to f(\mathbf{r}_2, \mathbf{R})(1/4\pi r_1 - 1/4\pi a), \tag{15}$$

and it is readily written for $\mathbf{r}_2 \to 0$ and $\mathbf{r}_\pm \to 0$ by using Eqs. (14). The function f carries the information about the second pair of particles when the first two sit on top of each other. The knowledge of this function is sufficient for describing all of the four boundaries.

For large R the wavefunction is given by

$$\Psi \approx \varphi_0(r_1)\varphi_0(r_2)(1 - \sqrt{2}a_{dd}/R), \tag{16}$$

where a_{dd} is the molecule-molecule scattering length, and

$$\varphi_0(r) = (r\sqrt{2\pi a})^{-1} \exp(-r/a) \tag{17}$$

is the wavefunction of a weakly bound molecule. The value of a_{dd} is deduced from the behavior of f at large R, where from Eqs. (16), (17) and (15) we obtain:

$$f = (2/ra)\exp(-r/a)(1 - \sqrt{2}a_{dd}/R), \quad R \gg a. \tag{18}$$

On the basis of Eqs. (13) and (15) one derives an integral equation for the function f [71,40]:

$$\int_{\mathbf{r}',\mathbf{R}'} \Big\{G(|\bar{S}_1 - S_1|)[f(\mathbf{r}',\mathbf{R}') - f(\mathbf{r},\mathbf{R})] + \big[G(|\bar{S}_1 - S_2|) - \textstyle\sum_{\pm} G(|\bar{S}_1 - S_{\pm}|)\big] f(\mathbf{r}',\mathbf{R}')\Big\} = (\sqrt{2}-1) f(\mathbf{r},\mathbf{R})/4\pi a, \quad (19)$$

where $S_1 = \{0, \mathbf{r}', \mathbf{R}'\}$, $S_2 = \{\mathbf{r}', 0, -\mathbf{R}'\}$, $S_{\pm} = \{\mathbf{r}'/2 \pm \mathbf{R}'/\sqrt{2}, \mathbf{r}'/2 \mp \mathbf{R}'/\sqrt{2}, \mp\mathbf{r}'/\sqrt{2}\}$, $\bar{S}_1 = \{0, \mathbf{r}, \mathbf{R}\}$, and $G(x) = (2\pi)^{-9/2}(xa/\sqrt{2})^{-7/2} K_{7/2}(\sqrt{2}\,x/a)$ is the Green function of Eq. (13), with $K_{7/2}$ being the decaying Bessel function. In contrast to Ψ, the function f depends only on three variables: the absolute values of $\mathbf{r}_2$ and $\mathbf{R}$, and the angle between them. At $R \gg a$, fitting Eq. (18) with $f(\mathbf{r},\mathbf{R})$ obtained numerically from Eq. (19) for all distances $\mathbf{R}$ and $\mathbf{r}$, one finds with 2% accuracy [71,40]:

$$a_{dd} = 0.6a > 0. \quad (20)$$

Calculations show the absence of 4-body weakly bound states, and the behavior of f suggests a soft-core repulsion between molecules, with a range $\sim a$. The result of Eq. (20) is exact, and it indicates the stability of molecular BEC with respect to collapse. Compared to earlier studies which were assuming $a_{dd} = 2a$,[20] Eq. (20) gives almost twice as small a sound velocity of the molecular condensate and a rate of elastic collisions smaller by an order of magnitude. The result of earlier studies [20] was also reconsidered in Ref. [78] by using a diagrammatic approach which leads to $a_{dd} = 0.75a$. However, this approach misses a number of diagrams which give a contribution of the same order of magnitude as those taken into account.

5. Collisional Relaxation of Weakly Bound Molecules

The lifetime of the Bose gas of weakly bound molecules is determined by their collisional relaxation into deep bound states. The released binding energy of a deep state is $\sim \hbar^2/mR_e^2$. It is transformed into the kinetic energy of particles in the outgoing collisional channel and they escape from the sample. The collisional relaxation is a crucial process as these dimers are diatomic molecules in the highest rovibrational state. Several experiments show that such molecules consisting of bosonic ^{87}Rb [79,80] and ^{133}Cs atoms [81] undergo a rapid collisional decay. On the other hand, first observations of weakly bound molecules Li_2 and K_2, consisting of fermionic atoms,[82,83,42,85] showed that they are long-lived at densities $\sim 10^{13}$ cm^{-3}.

The remarkable collisional stability of weakly bound molecules of fermionic atoms was explained in Ref. 71 and analyzed in detail in Ref. 40. The key physical reason is Fermi statistics for the atoms in combination with a large size of the molecular state (small momenta of bound fermionic atoms). The binding energy of the molecules is $\varepsilon_0 = \hbar^2/ma^2$ and their size is close to a. The size of deep bound states is of the order of the characteristic radius of interaction $R_e \ll a$. Hence, the relaxation requires the presence of at least three fermions at distances $\sim R_e$ from each other. As two of them are necessarily identical, due to the Pauli exclusion principle the relaxation probability acquires a small factor proportional to a power of (kR_e), where $k \sim 1/a$ is a characteristic momentum of the atoms in the weakly bound molecular state. The inequality $a \gg R_e$ allows one to obtain the dependence of the atom-molecule and molecule-molecule wavefunctions at short interparticle distances on the two-body scattering length a and thus to establish a strong decrease of the relaxation rate with increasing a.

In Refs. 71, 40 this has been done by using the zero range approximation at interatomic distances much larger than R_e. For the relaxation in atom-molecule collisions the rate constant is $\alpha_{\rm rel} \propto a^{-3.33}$, and it rapidly decreases with an increase in the two-body scattering length a.

This result is easily generalized to the case of the s-wave molecule-molecule scattering. Indeed, the relaxation process requires only three atoms to approach each other to short distances. The fourth particle does not participate. Averaging the relaxation probability over the motion of the fourth particle makes the problem similar to the relaxation in atom-molecule collisions. However, there is a relaxation channel that is more important in the limit of very large a. For the s-wave molecule-molecule scattering, both the fourth particle and the atom bound to this particle can undergo p-wave scattering on the other molecule in such a way that their total angular momentum is equal to zero. This leads to the relaxation rate $\alpha_{\rm rel} \propto a^{-2.55}$.

Assuming that the short-range physics is characterized by the length scale R_e and the energy scale $\hbar^2/mR_e^2$, we can restore the dimensions and write:

$$\alpha_{\rm rel} = C(\hbar R_e/m)(R_e/a)^s, \tag{21}$$

where $s \approx 2.55$ for the molecule-molecule collisional relaxation, and $s \approx 3.33$ for the relaxation in atom-molecule collisions.

Equation (21) shows a slower decrease of the molecule-molecule relaxation rate with increasing a than in the case of atom-molecule collisions. Obviously, in the limit $R_e/a \to 0$ the molecule-molecule relaxation should dominate over the atom-molecule one. The competition between these two relaxation processes can be present if the ratio R_e/a is not too small and the densities of molecules and atoms in the gas are comparable with each other.

Experimental studies of bosonic molecules produced in a Fermi gas by using wide Feshbach resonances at JILA, Innsbruck, MIT, and ENS are well described within the presented theoretical approach. It should be mentioned here that for $a > 0$ and equal concentrations of the two atomic components of the gas, at temperatures well below the Fermi energy practically all atoms should be converted into molecules if the gas density satisfies the inequality $na^3 \ll 1$.[89] In ongoing experiments the imbalance between the atomic components is fairly small, and at sufficiently low temperatures there can only be a small fraction of unpaired fermionic atoms.

The JILA,[84,85,64] Innsbruck,[83,86,67] MIT,[87,65] ENS[82,68], and Rice[88] results show a remarkable collisional stability of weakly bound molecules K_2 and Li_2 consisting of two fermionic atoms. At molecular densities $n \sim 10^{13}$ cm^{-3} the lifetime of the gas ranges from tens of milliseconds to tens of seconds, depending on the value of the scattering length a. A strong decrease of the relaxation rate with increasing a, following from Eq. (21), is consistent with experimental data. The potassium experiment at JILA [84] and lithium experiment at ENS [68] give the relaxation rate constant $\alpha_{dd} \propto a^{-s}$, with $s \approx 2.3$ for K_2, and $s \sim 2.0$ for Li_2. The absolute value of the rate constant for a thermal gas of Li_2 is $\alpha_{dd} \approx 2\times10^{-13}$ cm^3/s for the scattering length $a \approx 800$Å.[82] For K_2 it is by an order of magnitude higher at the same value of a,[84] which can be a consequence of a larger value of the characteristic radius of interaction R_e.

At realistic temperatures the relaxation rate constant α_{dd} is much smaller than the rate constant of elastic collisions $8\pi a_{dd}^2 v_T$, where v_T is the thermal velocity. For example, for the Li_2 weakly bound molecules at a temperature $T \sim 3\mu$K and $a \sim 800$Å, the corresponding ratio is of the order of 10^{-4} or 10^{-5}. This opens wide possibilities for reaching BEC of the molecules and cooling the Bose-condensed gas to temperatures of the order of its chemical potential. Long-lived BEC of weakly bound molecules has been recently observed for $^{40}K_2$ at JILA [85,64] and for $^6Li_2$ at Innsbruck, [86,67] MIT,[87,65] ENS,[68] and Rice.[88] Measurements of the molecule-molecule scattering length confirm the result $a_{dd} = 0.6a$ with accuracy up to 30%. This result is also confirmed by recent Monte Carlo calculations [53] of the ground state energy in the molecular BEC regime.

Long lifetime of weakly bound molecules of fermionic atoms allows interesting manipulations with these molecules. It seems realistic to arrange a deep evaporative cooling of their Bose-condensed gas to temperatures of the order of the chemical potential. Then, converting the molecular BEC into fermionic atoms by adiabatically changing the scattering length to negative values, one provides an additional cooling. The obtained atomic Fermi gas will have extremely low temperatures $T \sim 10^{-2}T_F$ which can be below the BCS transition temperature.[90] At these temperatures one has a very strong Pauli blocking of elastic collisions and expects the collisionless regime for the thermal cloud, which is promising for identifying the BCS-paired state through the observation of collective oscillations or free expansion.[91]

6. Weakly Bound Molecules of Heavy and Light Fermions

The physics of remarkable collisional stability of weakly bound molecules of fermionic atoms is quite transparent in the case of molecules of heavy and light fermions. This is especially important in view of future experiments with mixtures of K and Li fermionic atoms. We will demonstrate the collisional stability of such molecules considering their relaxation to deep bound states in collisions with light fermionic atoms.

The existence of a weakly bound molecular state assumes that the scattering length for the interaction between heavy and light fermionic atoms is positive and it greatly exceeds the characteristic radius of interaction between the atoms, R_e. The size of the weakly bound state is $\sim a$, whereas the size of a deep bound state is $\sim R_e$. Therefore, the relaxation of the weakly bound molecule to a deep state in collisions with light fermions requires that both light atoms (an incident atom and the atom bound to the molecular state) are at a distance $\sim R_e$ from the heavy atom. Assuming for simplicity that the heavy atom is at rest, we then have to analyze the behavior of the system, which depends on the coordinates of the light fermions, $\mathbf{r}_1$ and $\mathbf{r}_2$ (see Fig. 3).

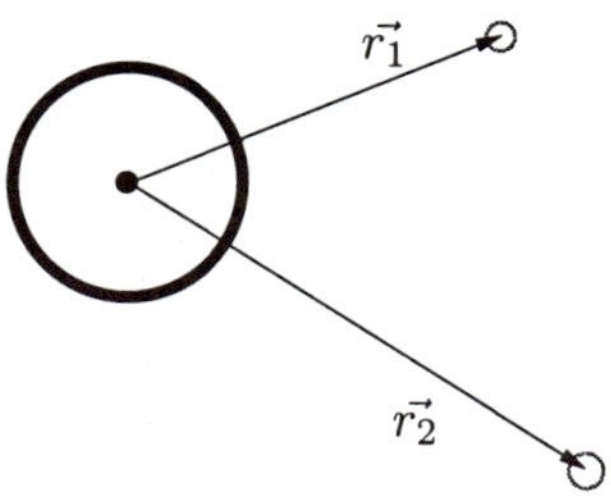

Fig. 3. The collision of a weakly bound molecule of heavy and light fermions with a light fermionic atom. The large circle denotes the heavy atom, and small circles light atoms.

We will establish the dependence of the relaxation loss rate on the scattering length a, without going into a detailed analysis of the behavior at short distances r_1 and r_2. Assuming that the inelastic amplitude of relaxation is much smaller than the amplitude of elastic scattering, the dependence of the relaxation rate on a is related only to the a-dependence of the initial-state wavefunction Ψ. At distances $r_1, r_2 \sim R_e$ where the relaxation occurs, as well as at any distances $r_1, r_2 \ll a$, the wavefunction Ψ is determined by the Schrödinger equation with zero total energy and depends on a only through a normalization coefficient:

$$\tilde{\Psi} = A(a)\psi. \tag{22}$$

In the region where $R_e \ll r_{1,2} \ll a$, the wavefunction Ψ is obtained straightforwardly, which gives the coefficient $A(a)$ and the dependence of the relaxation rate on a. The relaxation rate constant will be $\alpha_{\rm rel} \sim A^2(a)$.

At distances $r_{1,2} \gg R_e$ the zero range approximation gives the wavefunction

$$\Psi = \frac{1}{\sqrt{2}}\{\varphi(r_1)\chi(r_2) - \varphi(r_2)\chi(r_1)\}, \tag{23}$$

where ϕ_0 is the wavefunction of the weakly bound molecular state given by Eq. (17), and χ is the wavefunction of an incident atom. In the ultracold limit, assuming that the momentum of this atom is tending to zero, we have

$$\chi(r) = (1 - a/r). \tag{24}$$

For distances $r_{1,2} \ll a$, Eqs. (17), (24), and (23) give

$$\Psi = \frac{1}{\sqrt{8\pi a^3}}\left\{\frac{r_2}{r_1} - \frac{r_1}{r_2}\right\} \propto \frac{1}{a^{3/2}}. \tag{25}$$

Comparing Eq. (25) with Eq. (22) we immediately find $A \propto 1/a^{3/2}$ and, accordingly, we have

$$\alpha_{\rm rel} \propto \frac{1}{a^3}. \tag{26}$$

Note that the factor $1/a$ is coming from the shape of the wavefunction of the weakly bound state and, hence, from the overlap integral of this function with the wavefunction of a deep bound state (Frank-Condon factor). The factor $1/a^2$ originates from the Pauli principle requiring antisymmetrization of the wavefunction.

We now compare the result of Eq. (26) with the relaxation rate of weakly bound molecules of bosonic light and heavy atoms, caused by their collisions with light bosons. In this case the wavefunction Ψ should be symmetric with regard to interchange of light atoms, and one should replace the sign $-$ by $+$ in Eq. (23). Then at distances $r_{1,2} \ll a$ one has $\Psi \propto a^{1/2}$. Hence, the relaxation rate is $\alpha_{\rm rel} \propto a$ and it rapidly increases with a. Again, the Frank-Condon factor is $\propto 1/a$. However, the resonance enhancement of the wavefunction due to a large value of the scattering length, gives a factor a^2.

This comparison clearly shows why weakly bound molecules of fermionic atoms are long-lived, whereas similar molecules of bosonic atoms undergo a rapid collisional decay.

7. Concluding Remarks

In conclusion, remarkable achievements in the physics of ultracold Fermi gases attracted a lot of attention and draw fascinating prospects for future studies. These prospects are to a large extent related to a very long life-time of weakly bound bosonic molecules of fermionic atoms. One of the ideas is related to transferring weakly bound molecules of fermionic atoms to their ground ro-vibrational state by using two-photon spectroscopy, as proposed in Ref. 92 for molecules of bosonic atoms. Long lifetime of weakly bound molecules of fermionic atoms at densities $\sim 10^{13}$ cm^{-3} should provide a much more efficient production of ground state molecules compared to the case of molecules of bosonic atoms, where one has severe limitations on achievable densities and lifetimes. One can then extensively study the physics of molecular Bose-Einstein condensation. Moreover, heteronuclear molecules which are supposed to be formed in mixtures of different fermionic atoms, can be polarized by an electric field. One then gets a gas of dipoles interacting via anisotropic long-range forces, which drastically changes the physics of Bose-Einstein condensation.[93]

Acknowledgments

This work was supported by the Centre National de la Recherche Scientifique (CNRS) de France, and by the Nederlandse Stichting voor Fundamenteel Onderzoek der Materie (FOM). LPTMS is a mixed research unit of CNRS and Université Paris Sud.

References

1. M.H. Anderson *et al.*, Science **269**, 198 (1995).
2. K.B. Davis *et al.*, Phys. Rev. Lett. **75**, 3969 (1995).
3. C.C. Bradley *et al.*, Phys. Rev. Lett. **75**, 1687 (1995).
4. F. Dalfovo *et al.*, Rev. Mod. Phys. **71**, 463 (1999).
5. E.M. Lifshitz and L.P. Pitaevskii, *Statistical Physics* (Pergamon Press, Oxford, 1980), Part 2.
6. L.P. Gor'kov and T.K. Melik-Barkhudarov, Zh. Eksp. Teor. Fiz. **40**, 1452 (1961) [Sov. Phys. JETP **13**, 1018 (1961)].
7. B. DeMarco and D.S. Jin, Science **285**, 1703 (1999).
8. A.G. Truscott *et al.*, Science **291**, 2570 (2001).
9. F. Schreck *et al.*, Phys. Rev. Lett. **87**, 080403 (2001).
10. S.R. Granade *et al.*, Phys. Rev. Lett. **88**, 120405 (2002).
11. K.M. O'Hara *et al.*, Science **298**, 2179 (2002).
12. Z. Hadzibabic *et al.*, Phys. Rev. Lett. **88**, 160401 (2002).
13. G. Roati *et al.*, Phys. Rev. Lett. **89**, 150403 (2002).
14. M.J. Holland *et al.*, Phys. Rev. A **61**, 053610 (2000).
15. B. DeMarco *et al.*, Phys. Rev. Lett. **86**, 5409 (2001)
16. G.V. Shlyapnikov, in *Proceedings of the 18th International Conference on Atomic Physics*, edited by H.R. Sadeghpour, E.J. Heller, and D.E. Pritchard (World Scientific, Singapore, 2002).
17. D.M. Eagles, Phys. Rev. **186**, 456 (1969).
18. A.J. Leggett, in *Modern Trends in the Theory of Condensed Matter*, edited by A. Pekalski and J. Przystawa (Springer, Berlin,1980).
19. P. Nozieres and S. Schmitt-Rink, J. Low Temp. Phys. **59**, 195 (1985).
20. See for review M. Randeria, in *Bose-Einstein Condensation*, edited by A. Griffin, D.W. Snoke, and S. Stringari (Cambridge University Press, Cambridge,1995).
21. K. Miyake, Progr. Theor. Phys. **69**, 1794 (1983).
22. See for review M.Yu. Kagan, Sov. Physics Uspekhi **37**, 69 (1994).
23. M. Holland *et al.*, Phys. Rev. Lett. **87**, 120406 (2001).
24. E. Timmermans *et al.*, Phys. Lett. A **285**, 228 (2001).
25. G. Breit and E. Wigner, Phys. Rev. **49**, 519 (1936).
26. H. Feshbach, Ann. Phys. **19**, 287 (1962); *Theoretical Nuclear Physics* (Wiley, New York, 1992).
27. U. Fano, Phys. Rev. **124**, 1866 (1961).
28. A.J. Moerdijk, B.J. Verhaar, and A. Axelsson, Phys. Rev. A **51**, 4852 (1995); see also E. Tiesinga, B.J. Verhaar, and H.T.C. Stoof, Phys. Rev. A **47**, 4114 (1993).
29. P.O. Fedichev *et al*, Phys. Rev. Lett. **77**, 2913 (1996).
30. J.L. Bohn and P.S. Julienne, Phys. Rev. A **60**, 414 (1999).
31. L.D. Landau and E.M. Lifshitz, *Quantum Mechanics*, (Butterworth-Heinemann, Oxford, 1999).
32. R. Combescot, Phys. Rev. Lett. **91**, 120401 (2003).
33. G.M. Falco and H.T.C. Stoof, Phys. Rev. Lett. **92**, 130401 (2004).
34. G.M. Bruun and C. Pethik, Phys. Rev. Lett. **92**, 140404 (2004).
35. G.M. Bruun, Phys. Rev. A **70**, 053602 (2004).
36. D.S. Petrov, Phys. Rev. Lett. **93**, 143201 (2004).
37. S. De Palo *et al*, cond-mat/0404672.
38. E.A. Cornell, KITP Conference on Quantum Gases, Santa Barbara, May 10-14, 2004.
39. R. Diener and Tin-Lun Ho, cond-mat/0405174.
40. D.S. Petrov, C. Salomon, and G.V. Shlyapnikov, Phys. Rev. A (2004).

41. P.D. Drummond and K. Kheruntsyan, Phys. Rev. A **70**, 033609 (2004).
42. K.E. Strecker, G.B. Partridge, and R.G. Hulet, Phys. Rev. Lett. **91**, 080406 (2003).
43. M.L. Chiofalo *et al.*, Phys. Rev. Lett. **88**, 090402 (2002); S.J.J.M.F. Kokkelmans *et al.*, Phys. Rev. A, **65**, 053617 (2002).
44. J.N. Milstein *et al.*, Phys. Rev. A **66**, 043604 (2002).
45. Y. Ohashi and A. Griffin, Phys. Rev. Lett. **89**, 130402 (2002); Phys. Rev. A **67**, 033603 (2003).
46. M.J. Holland, C. Menotti, and L. Viverit, cond-mat/0404234.
47. H. Heiselberg, Phys. Rev. A **63**, 043606 (2001).
48. J. Carlson *et al*, Phys. Rev. Lett. **91**, 050401 (2003).
49. M.E. Gehm *et al*, Phys. Rev. A **68** 011401 (2003).
50. T. Bourdel *et al*, Phys. Rev. Lett. **91**, 020402 (2003).
51. Tin-Lun Ho, Phys. Rev. Lett. **92**, 090402 (2004).
52. S.-Y Chang *et al*, Phys. Rev. A. **70**, 043602 (2004).
53. G.E. Astrakharchik *et al*, cond-mat/0406113.
54. A. Bulgac, cond-mat/0309358.
55. A. Perali *et al*, hys. Rev. Lett. **93**, 100404 (2004).
56. O. Chen *et al*, cond-mat/0402274.
57. R. Diener and Tin-Lun Ho, cond-mat/0404517.
58. S. Stringari, Europhys. Lett. **65**, 749 (2004).
59. H. Heiselberg, Phys. Rev. Lett. **93**, 040402 (2004); cond-mat/0406714.
60. H. Hu *et al*, Phys. Rev. Lett. **93**, 190403 (2004).
61. R. Combescot and X. Leyronas, Phys. Rev. Lett. **93**, 138901 (2004).
62. N. Manini and L. Salasnich, cond-mat/0407039.
63. Y. Ohashi and A. Griffin, cond-mat/0410220.
64. C.A. Regal, M. Greiner, and D.S. Jin, Phys. Rev. Lett. **92**, 040403 (2004).
65. M.W. Zwerlein *et al*, Phys. Rev. Lett. **92**, 120403 (2004).
66. J. Kinast *et al.*, Phys. Rev. Lett. **92**, 150402 (2004).
67. M. Bartenstein *et al*, Phys. Rev. Lett. **92**, 203201 (2004).
68. T. Bourdel *et al.*, Phys. Rev. Lett. **93**, 050401 (2004).
69. C. Chin *et al*, Science **305**, 1128 (2004).
70. J. Kinast *et al*, cond-mat/0409283.
71. D.S. Petrov, C. Salomon, and G.V. Shlyapnikov, Phys. Rev. Lett. **93**, 090404 (2004).
72. G.V. Skorniakov and K.A. Ter-Martirosian, Zh. Eksp. Teor. Phys. **31**, 775 (1956) [Sov. Phys. JETP **4**, 648 (1957)].
73. G.S. Danilov, Sov. Phys. JETP **13**, 349 (1961).
74. V.N. Efimov, Yad. Fiz. **12**, 1080 (1970) [Sov. J. Nucl. Phys. **12**, 589 (1971)]; Nucl. Phys. A **210**, 157 (1973).
75. D.S. Petrov, Phys. Rev. A **67**, 010703 (2003).
76. H. Bethe and R. Peierls, Proc. Roy. Soc. (London) **148A**, 146 (1935).
77. K. Huang, *Statistical Mechanics* (Wiley, New York, 1987).
78. P. Pieri and G.C. Strinati, Phys. Rev. B **61**, 15370 (2000).
79. D.J. Heinzen, Workshop on Cold Molecules, Les Houches, 2002.
80. S. Dürr, T. Volz *et al*, Phys. Rev. Lett. **92**, 020406 (2004).
81. J. Herbig *et al*, Science **301**, 1510 (2003).
82. J. Cubizolles *et al.*, Phys. Rev. Lett. **91**, 240401 (2003).
83. S. Jochim *et al.*, Phys. Rev. Lett. **91**, 240402 (2003)
84. C.A. Regal, M. Greiner, and D.S. Jin, Phys. Rev. Lett. **92**, 083201 (2004).
85. M. Greiner, C. Regal, and D.S. Jin, Nature **426**, 537 (2003).
86. S. Jochim *et al.*, Science **302**, 2101 (2003); M. Bartenstein *et al.*, Phys. Rev. Lett. **92**, 120401 (2004).

87. M. W. Zwierlein, *et al.*, Phys. Rev. Lett. **91**, 250401 (2003).
88. R.G. Hulet, KITP Conference on Quantum Gases, Santa Barbara, May 10-14, 2004.
89. S.J.J.M.F. Kokkelmans, G.V. Shlyapnikov, and C. Salomon, Phys. Rev. A **69** 031602 (2004).
90. L.D. Carr, G.V. Shlyapnikov, and Y. Castin, Phys. Rev. Lett. **92**, 150404 (2004).
91. C. Menotti, P. Pedri, and S. Stringari, Phys. Rev. Lett. **89**, 250402 (2002).
92. A.J. Kerman, J.M. Sage, S. Sainis, T. Bergeman, and D. DeMille, Phys. Rev. Lett. **92**, 153001 (2004).
93. L. Santos, G.V. Shlyapnikov, P. Zoller, and M. Lewenstein, Phys. Rev. Lett. **85**, 1791 (2000).

BOSONS IN OPTICAL LATTICES

S. L. ROLSTON

Dept. of Physics, University of Maryland, College Park, MD 20742
rolston@umd.edu

Received 1 December 2004

The advent of coherent matter waves in the form of Bose-Einstein condensates, coupled with periodic potentials in the form of optical lattices, has established a new area of research on the boundary between atomic and condensed matter physics. This article is a brief review of the recent experimental progress in the area of degenerate Bose gases loaded into optical lattices, including strongly correlated systems and the role of dimensionality. Future prospects will also be outlined.

Keywords: Bose-Einstein condensation; Bose-Hubbard model; Mott insulator; optical lattice.

1. Introduction

Observation of the quantization of the energy levels of laser-cooled atoms trapped in optical standing waves in 1992, seen in absorption[1] and emission,[2] can be viewed as the start of research in optical lattices. This early work (see review article by Jessen and Deutsch[3]) was to a large part concerned with issues of laser cooling, an incoherent system with dynamics dominated by spontaneous emission. Although the concepts of condensed matter physics were used, for example in determining the limits of adiabatic cooling in an optical lattice,[4] for the most part the periodic nature of the potential was not very important, as there was little or no coherence from one well to another. In 1995 we saw the first creation of dilute gas Bose-Einstein condensates[5,6] (BECs), which provide a source of deBroglie waves with a coherence length equal to the size of the condensate, typically 10-100μm. In the following years, a number of groups performed experiments loading BECs in optical lattices.[7] Characteristic of these works was that the physics could be described with single-particle physics, i.e. each atom in the condensate is describable by the same Bloch state, or superposition thereof, and interactions between atoms could be neglected or at most treated as a perturbation. To summarize this diverse set of work as a whole, it confirmed that formalism and phenomena of condensed matter physics such as the concepts of quasimomentum, tunneling, and band structure, were all applicable to the optical lattice system. Many clean experiments emphasized the ideal defect- and dislocation-free nature of the optical lattice potential. As an example, Bloch oscillations, very difficult to observe in solid state systems due to scattering, are easily observed in optical lattices. These atomic physics experiments were more ideal condensed matter experiments than many experiments carried out in condensed matter physics. While these results were satisfying, they did not really break much new conceptual ground, as we fully understand single particle motion in periodic potentials.

2. Recent Experiments

The first BEC/optical lattice experiment to begin to probe beyond single-particle physics was carried out at Yale in 2001.[8] A BEC was adiabatically loaded into a one dimensional optical standing wave (1D optical lattice), and the coherence from site to site was analyzed by releasing the atoms from the lattice and looking at the interference pattern formed from the array of overlapping wavepackets. A loss of interference contrast was observed as the adiabatic loading time was increased. This was consistent with a reduction of the fluctuations of the number of atoms in each lattice site corresponding to an increase in the phase uncertainty, due to the number-phase uncertainty relationship. These results were interpreted in the context of the Bose-Hubbard model[9], a canonical model of condensed matter physics that predicts a quantum phase transition between a superfluid state and a Mott insulator state as the ratio of tunneling energy to interaction energy is varied. The Bose-Hubbard Hamiltonian can be written as:

$$H = -J\sum_{\langle i,j\rangle} b_i^\dagger b_j + \sum_i \epsilon_i \hat{n}_i + \frac{1}{2}U\sum_i \hat{n}_i(\hat{n}_i - 1) \tag{1}$$

where J is the tunneling matrix element and $\langle i, j\rangle$ restricts the summation to nearest neighbors, ϵ_i is the individual site energy, U is the interaction energy between two atoms on one site, b_i is the annihilation operator for an atom on site i (in the Wannier state basis), and $\hat{n}_i$ is the number operator for site i. This assumes only nearest-neighbor interactions and is restricted to a single (the lowest) band, both of which are very good approximations for most optical lattice parameters. The quantum phase transition arises from the competition between the kinetic energy(tunneling) term in the Hamiltonian, which minimizes energy with delocalized states, and the atom-atom interaction energy, which minimizes the energy by localizing the individual atoms to avoid each other, evenly distributing the atoms among the available lattice sites. Here was a clear example of physics beyond single-particle physics, and beyond the mean field Gross-Pitaevskii equation, which has been extremely successful in describing much of dilute gas BEC in terms of single-particle wavefunctions of a combined mean field/external confinement potential. The Yale experiment had more than 100 atoms in each potential well, and was not capable of observing a quantum phase transition, but nonetheless observed a 15 dB reduction in particle number fluctuations.

In a 1999 paper, the group of Peter Zoller made the suggestion[10] that the Bose-Hubbard model could be realized in 2 and 3D optical lattices. In a seminal experiment reported in 2002, a group in Munich reported[11] the observation of the superfluid-Mott insulator phase transition in a 3D optical lattice (cubic symmetry), observing the loss of interference as the system reached the insulating state (and its recovery if it is adiabatically returned to the superfluid state) as well as a gap in the spectrum associated with the insulating state in transport with the application of a large gradient, equivalent to a very strong electric field in an electron solid state system[12]. While this experiment realized the Bose-Hubbard model, it was done in a overall harmonic confining potential, so that the homogenous B-H results are not fully applicable. A number of theoretical treatments[10,13,14] have addressed these differences. In this inhomogeneous case, rather than the entire system undergoing a phase transition, different regions may be insulator or superfluid. For example if the atomic density is large enough, the central region will be an N=2 insulator, surrounded by a thin superfluid shell, which is in turn surrounded by a N=1 insulator, surrounded by an outermost superfluid shell. The exact details of this structure depend on the atom density (filling factor), the interaction energy, and the stiffness of the confining potential. Although this inhomogeneity prevents the explicit occurrence of a sharp phase transition, it removes the constraint of the homogenous B-H model on commensurate filling to observe the completely insulating phase. Recent work[15,16] have also experimentally observed signatures of the Mott insulating state in 3D optical lattices.

One of the motivations of studying the B-H model for atoms trapped in optical lattices is that the N=1 Mott insulator is an ideal initialized quantum register for quantum information processing using qubits of single neutral atoms trapped in optical lattices. To fully understand the fidelity of such a state is of both practical interest for quantum computation, as well as fundamental interest for understanding the details of this strongly correlated system[17].

The role of dimensionality is important in strongly correlated quantum systems, as fluctuations tend to become more important in low dimensions. Optical lattices can easily be constructed in 1-,2-, or 3D geometries. A series of recent experiments[15,18,19,20] have been studying 1D Bose gases, by confining them in a 2D optical lattice, which creates an ensemble of tubes where transverse motion is completely frozen out. In these experiments, the number of atoms per tube is in the 20-100 range, while the number of tubes is typically about 10^3. A process of fermionization of the 1D gas is expected as one approaches the Girardeau-Tonks limit[22], where the system is strongly correlated. An experiment at NIST[18] observed a factor of 7 reduction in three-body loss in the tubes over that for a 3D gas. This rate is proportional to the 3rd-order correlation function at r=0, and is a signature that the wavefunction is beginning to have fermionic rather than bosonic character[21]. Recent work at Penn State looking at the total energy of the system[20] is in excellent agreement with the Girardeau-Tonks theory. Very recent results from the NIST group[23] have looked at dipole oscillations of a 1D gas as an optical lattice is turned on. Significant damping is observed at extremely shallow lattice depths (0.25 E_R), and is most likely due to interactions between the superfluid component and the quantum depletion-induced excitations, which are quite large because of the critical sensitivity of 1D systems to fluctuations[24,25].

3. Future Prospects

Research in degnerate gases loaded into periodic potentials is still in the beginning stages. Of course there will be great interest in fermions loaded into optical lattices,[26,27] but I will restrict discussion to bosons. There are many possible topics of interest, most of which have yet to be discovered, so I will mention but a few to display the richness of this research area. Among the future areas of exploration will be a further understanding of the Bose-Hubbard system, including the development of spectroscopic tools to spatially resolve the number distributions in individual lattice sites, of relevance to both a fundamental understanding of many-body physics as well as to the application to quantum registers for quantum computing. Two dimensional gases have yet to be explored, and we can expect a richness of physics, give the unique features 2D systems display, such as the Kosterlitz-Thouless transition. Just as we use deep 2D optical lattices to study 1D gases, we can use deep 1D optical lattices to study 2D gases.

Disorder plays a important role in many areas of condensed matter physics. Optical lattices are inherently disorder-free, making for extremely clean systems, but we can expect to see disorder introduced into optical lattices with a degree of control unheard of in condensed matter physics. While is might be difficult to engineer dislocations, it should be possible to add impurities (different atomic species for example) as well as random optical potentials (generated by light scattering off of rough surfaces). Unlike condensed systems, the optical lattice system will have knobs to be able to change the amount of disorder at will. We can anticipate many possibilities including the study of (Anderson) localization, the Bose glass phase,[9] and perhaps a Bose metal phase[28].

In the BEC/optical lattice systems studied to date, the atom-atom interaction has been the short range van der Waals interaction, well described by a simple contact term. We can expect research in periodic systems where the interaction is modified, either by Feshbach resonances that can greatly increase the scattering length as well as change the interaction from repulsive to attractive, and long range anisotropic dipole-dipole interactions (exciting atoms into Rydberg states for example). So far the lattice work has been confined to single spins, but following on a number of spinor condensate experiments,[29,30] we can also expect two and perhaps three component spinor Bose gases loaded into optical lattices, which will enrich the phases available to the system.

This has been a purposely brief review. I have not attempted to be comprehensive in the list of cited works, but to at least give an idea of the progress and direction of the field, as well as to provide an entry point into the current literature.

Acknowledgments

I would like to acknowledge research support from NIST, NASA, and NSA-ARDA.

References

1. P. Verkerk,it et. al, *Phys. Rev. Lett.* **68**, 1116 (1992).
2. P. S. Jessen, *et. al*, *Phys. Rev. Lett.* **69**, 49 (1992).
3. P. S. Jessen and I. H. Deutsch, *Adv. At. Mol. Opt. Phys.* **37**, 95 (1996).
4. A. Kastberg, *et. al*, *Phys. Rev. Lett.* **74**, 1542 (1995).
5. M. H. Anderson, *et. al*, *Science* **269**, 198 (1995).
6. K. B. Davis, *et. al*, *Phys. Rev. Lett.* **75**, 3969 (1995).
7. as an example, J. Denschlag, *et. al*, *J. Phys. B* **35**, 3095 (2002).
8. C. Orzel, *et. al*, *Science* **291**, 2386 (2001).
9. M. P. A. Fisher, *et. al*, *Phys. Rev. B* **40**, 546 (1989).
10. D. Jaksch, *et. al*, *Phys. Rev. Lett.* **81**, 3108 (1998).
11. M. Greiner, *et. al*, *Nature* **415**, 39 (2002).
12. S. Sachdev, K. Sengupta, and S. M. Girvin, *Phys. Rev. B* **66**, 075128 (2002).
13. G. G. Bartrouni, *et. al.*, *Phys. Rev. Lett.* **89**, 117203 (2002).
14. G. Pupillo, E. Tiesinga, and C. J. Williams, *Phys. Rev. A* **68**, 063604 (2003).
15. T. Storfele, *et. al*, *Phys. Rev. Lett.* **92**, 130403 (2004).
16. J. V. Porto, *et. al*, *Phil. Trans. Roy. Soc. London* **361**, 1417 (2003).
17. G. Pupillo, *et. al*, *cond-mat* **0403052**, (2004).
18. B. L. Tolra, *et. al*, *Phys. Rev. Lett.* **92**, 190401 (2004).
19. B. Paredes, *et. al*, *Nature* **429**, 277 (2004).
20. T. Kinoshita, T. Wenger, and D. S. Weiss, *Science* **305**, 1125 (2004).
21. K. V. Kheruntsyan, *et. al*, *Phys. Rev. Lett.* **91**, 040403 (2003).
22. V. Dunjko, V. Lorent, and M. Olshanii, *Phys. Rev. Lett.* **86** 5413 (2001).
23. C. D. Fertig, *et. al*, *cond-mat* **0410491**, (2004).
24. J. Gea-Banacloche, *et. al*, it cond-mat **0410677**, (2004).
25. A. Polkonikov and D. W. Wang, *Phys. Rev. Lett.* **93**, 070401 (2004).
26. L. Pezze, *et. al*, *Phys. Rev. Lett.* **93**, 120401 (2004).
27. M. Kohl, *et. al*, *cond-mat* **0410389**, (2004).
28. P. Phillips and D. Dalidovich, *Science* **302**, 243 (2003).
29. D. S. Hall, *et. al*, *Phys. Rev. Lett.* **81**, 1543 (1998).
30. H.-J. Meisner, *et. al*, *Phys. Rev. Lett.* **82**, 2228 (1999).

GENERALIZED ENTANGLEMENT AND QUANTUM PHASE TRANSITIONS

ROLANDO SOMMA
Los Alamos National Laboratory, Los Alamos, NM 87545, USA
somma@lanl.gov

HOWARD BARNUM
Los Alamos National Laboratory, Los Alamos, NM 87545, USA
barnum@lanl.gov

EMANUEL KNILL
Mathematical and Computational Sciences Division,
National Institute of Standards and Technology,
325 Boulder, Boulder, CO 80305, USA
knill@boulder.nist.gov

GERARDO ORTIZ
Los Alamos National Laboratory, Los Alamos, NM 87545, USA
ortiz@viking.lanl.gov

LORENZO VIOLA
Department of Physics and Astronomy, Dartmouth College,
6127 Wilder Laboratory, Hanover, NH 03755, USA
Lorenza.Viola@Dartmouth.EDU

Received 1 December 2004

Quantum phase transitions in matter are characterized by qualitative changes in some correlation functions of the system, which are ultimately related to entanglement. In this work, we study the second-order quantum phase transitions present in models of relevance to condensed-matter physics by exploiting the notion of generalized entanglement [Barnum *et al.*, Phys. Rev. A **68**, 032308 (2003)]. In particular, we focus on the illustrative case of a one-dimensional spin-1/2 Ising model in the presence of a transverse magnetic field. Our approach leads to tools useful for distinguishing between the ordered and disordered phases in the case of broken-symmetry quantum phase transitions. Possible extensions to the study of other kinds of phase transitions as well as of the relationship between generalized entanglement and computational efficiency are also discussed.

Keywords: Entanglement; quantum information science; quantum phase transitions.

1. Introduction

During the last decade, the notion of entanglement[1] has acquired a growing significance as one of the key resources in quantum information science. In particular, entanglement helps unlocking the power of quantum computation (QC). The concept of entanglement has been known since the birth of quantum mechanics, and may be regarded as one of the main features that distinguishes quantum from classical mechanics. Entanglement is traditionally understood as a uniquely quantum phenomenon whereby a pure state of a composite quantum system may no longer be determined by the states of its constituent subsystems alone. Thus, entanglement involves the existence of intrinsically quantum correlations, which implies that the measurement outcomes in one subsystem are related to the measurement outcomes in another one.

In quantum many-body physics, quantum correlations between constituent degrees of freedom likewise play a critical role. They are responsible for influencing the behavior of the system at both the microscopic and macroscopic level. Quantum phase transitions (QPTs), in particular, are dominated by purely quantum fluctuations in the correlations present in the many-body ground state.[2] It then becomes natural to apply concepts and tools developed in the context of of entanglement theories to study QPTs from a quantum-information standpoint. However, the lack of a unique measure of entanglement applicable to any physical setting (e.g., fermionic, bosonic, anyonic, etc.) has resulted so far in most of these studies being done using measures of bipartite entanglement,[3] such as concurrence.[4] Obviously, such an approach cannot be expected to always succeed at characterizing criticality, for bipartite measures of entanglement involve only two-body correlations, whereas QPTs might be dominated by three-body or higher-order correlations.

In this work, we apply a subsystem-independent extension of the usual entanglement concept, which we termed *generalized entanglement* (GE),[5,6,7] to the study of some second-order QPTs in models of interest. We also argue how GE naturally allows us to investigate quantum complexity issues, in particular the conditions under which a quantum algorithm (QA) might be efficiently simulated on a classical computer.

The content is organized as follows. After briefly recalling the basics of the GE idea in Sec. 2, in Sec. 3 we present some examples where this notion may be easily applied. In Sec. 4 we show how a simple measure of GE behaves as a disorder parameter for the ferromagnetic-to-paramagnetic second-order QPT present in the spin-1/2 Ising model in a transverse magnetic field. In Sec. 5 we explain why GE is relevant as a resource for performing QC. Finally, Sec. 6 summarizes our results.

2. Generalized Entanglement and the $\mathfrak{h}$-Purity

We begin by briefly revisiting the traditional concept of entanglement. Within the standard framework, given a quantum system S composed of N distinguishable subsystems, a pure state of the total system $|\psi_S\rangle$ is said to be entangled if it cannot be written as a product of states of each subsystem, that is $|\psi_S\rangle \neq |\phi_1\rangle \otimes \cdots \otimes |\phi_N\rangle$, where $|\phi_j\rangle$ refers to an arbitrary state of the j-th subsystem. In the same landmark paper where the term entanglement is introduced,[1] Schrödinger states that "maximal knowledge of a total system does not necessarily include total knowledge of all its parts". This may be equivalently phrased by saying that if one calculates the reduced density operator ρ_j of the jth subsystem by tracing out all the other parties, $\rho_j = \mathsf{Tr}_{1\ldots\hat{j}\ldots N}(\rho)$, with $j = 1, \cdots N$ and $\rho = |\psi_S\rangle\langle\psi_S|$, the state $|\psi_S\rangle$ is entangled if and only if at least one subsystem state is mixed; i.e., $\mathsf{Tr}(\rho_j^2) < 1$.

For instance, let S be composed of N two-level quantum systems such as spin-1/2, in which case the reduced density operator for the j-th spin is given by $\rho_j = \frac{1}{2}[\mathbb{1} + \sum_{\alpha=x,y,z} \langle\sigma_\alpha^j\rangle\sigma_\alpha^j]$, where σ_α^j, $\alpha = x, y, z$, are the spin-1/2 Pauli operators acting on the j-th spin, and $\langle\sigma_\alpha^j\rangle = \langle\psi_S|\sigma_\alpha^j|\psi_S\rangle$ denotes the corresponding expectation value. The vectors $\mathbf{R}_j = (\langle\sigma_x^j\rangle, \langle\sigma_y^j\rangle, \langle\sigma_y^j\rangle)$, $\mathbf{R}_j \in \mathbb{R}^3$, allow a geometric representation of each reduced state in $\mathbb{R}^3$, satisfying $0 \leq |\mathbf{R}_j| \leq 1$. This is the so-called Bloch-sphere representation[8] (Fig. 1). Since $\mathsf{Tr}(\rho_j^2) = \frac{1}{2}(1 + |\mathbf{R}_j|^2)$, the state $|\psi_S\rangle$ is entangled if $|\mathbf{R}_j| < 1$ for at least one j, represented by a point *inside* the corresponding unit sphere. One may therefore consider $|\psi_S\rangle$ to be maximally entangled if $|\mathbf{R}_j| = 0$ for all j. However, the state $|\psi_S\rangle$ is unentangled (i.e., a product state) if $|\mathbf{R}_j| = 1$ for all j, corresponding to points lying on the surface of the unit sphere (Fig. 1).

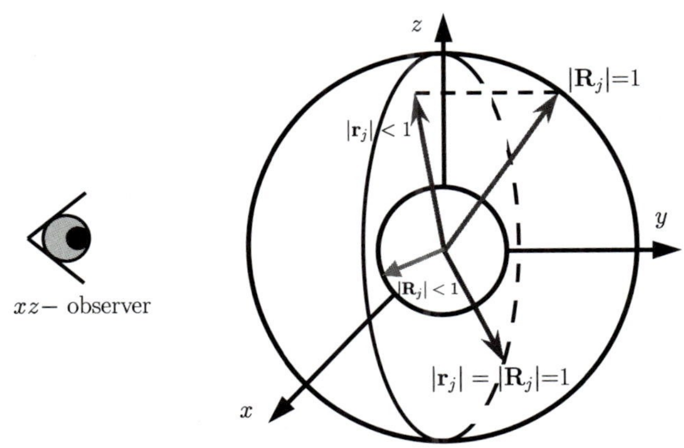

Fig. 1. Bloch-sphere representation of the reduced state of the j-th spin. The red arrow, representing a point on the surface, corresponds to a pure reduced state, hence unentangled with the rest of the spins. However, this state is generalized entangled with respect to the xz-observer (blue arrow). The green arrow represents a mixed reduced state, thus implying entanglement among the spins. Reduced states represented as points on the Equator are neither entangled nor generalized entangled relative to the xz-observer (purple arrow).

Let us assume that for some reason the observer can obtain information about two of the spin components, say $\langle\sigma_x^j\rangle$ and $\langle\sigma_z^j\rangle$, but cannot learn $\langle\sigma_y^j\rangle$. In other words, in the Bloch-sphere representation, such an observer only sees a projection $\mathbf{r}_j$ onto the xz plane of each vector $\mathbf{R}_j$ (Fig. 1). It then becomes natural to ask, What are the entangled states relative to this observer? By analogy with the original setting, where all the *local* spin components are accessible and unentangled states are represented by vectors of extremal length, a straightforward answer is to regard unentangled states–relative to this observer–as those states that correspond to a point on the unit circumference in the xz plane. The remaining states are entangled. Thus, a quantum state $|\psi_S\rangle$ is entangled relative to this observer if there is a subsystem j such that the 2D vector $\mathbf{r}_j = (\langle\sigma_x^j\rangle, \langle\sigma_z^j\rangle)$ satisfies $|\mathbf{r}_j| < 1$.

The basic idea of the GE approach is to extend the intuition underlying the above example to an *arbitrary* system S. Accordingly, a pure quantum state $|\psi_S\rangle$ is called generalized unentangled (entangled) *relative to a preferred set of observables* $\mathfrak{h} = \{O_\ell, \ell = 1, \ldots, M\}$ if it is (is not) extremal in the convex set of reduced states, the reduced state corresponding to $|\psi_S\rangle$ being specified by the expectation values $\langle\psi_S|O_\ell|\psi_S\rangle$. A more detailed explanation of this definition may be found in Ref. [5]. A natural geometric measure of GE is obtained by defining the *purity of a state relative to a set of observables*, $\mathfrak{h} = \{O_\ell, \ell = 1, \ldots, M\}$, or $\mathfrak{h}$-purity,

$$P_h = \mathsf{K} \sum_{\ell=1}^{M} \langle O_\ell \rangle^2 , \tag{1}$$

where K is a global normalization factor chosen in such a way that the maximum purity is equal to 1, and $\langle O_\ell\rangle = \langle\psi_S|O_\ell|\psi_S\rangle$. In the cases studied here, the $\mathfrak{h}$-purity attains its maximum value ($P_h = 1$) whenever the state $|\psi_S\rangle$ of the system is generalized unentangled relative to $\mathfrak{h}$, and is strictly non-maximal otherwise.

Important properties can be established when the observable set $\mathfrak{h}$ is a semi-simple Lie algebra,[5] irreducibly represented in the Hilbert space of the system. In this case, the generalized *un*entangled states relative to $\mathfrak{h}$ turn out to coincide with the *generalized coherent states* (GCSs)[9,10] of $\mathfrak{h}$. Thus, states of the form $|\psi_S\rangle = D(\{\alpha_\ell\})|\mathsf{REF}\rangle$, $\alpha_\ell \in \mathbb{C}$, with $D(\{\alpha_\ell\}) = \exp[i \sum_\ell \alpha_\ell O_\ell]$ being a generalized displacement operator and $|\mathsf{REF}\rangle$ a reference state, are generalized unentangled relative to $\mathfrak{h}$. Remarkably, the $\mathfrak{h}$-purity remains invariant under arbitrary group operations, obtained by exponentiating the algebra. We refer the reader to previous works [5,6,7] for further details.

3. Generalized Entanglement: Examples

3.1. *Spin-1/2 systems*

The standard notion of entanglement for multipartite systems is recovered whenever the algebra $\mathfrak{h}$ is chosen to be the set of *all local observables*,[7] corresponding to each of the parties that compose the actual system. For example, if the system consists of N spin-1/2s, we obtain a measure of conventional entanglement by calculating the purity relative to the set $\mathfrak{h} = \{\sigma_x^1, \sigma_y^1, \sigma_z^1, \ldots, \sigma_x^N, \sigma_y^N, \sigma_y^N\}$, the σ_α^j being the spin-1/2 Pauli operators acting on the j-th spin, as above. That is,

$$P_h = \frac{1}{N} \sum_{j=1}^{N} \sum_{\alpha=x,y,z} \langle\sigma_\alpha^j\rangle^2 , \tag{2}$$

which is equivalent to the measure of global entanglement originally introduced by Meyer and Wallach.[11] Since $\mathfrak{h}$ is a semi-simple Lie algebra, its generalized unentangled states are the GCSs obtained by applying any group operation to a reference state such as $|\mathsf{REF}\rangle = |\downarrow\rangle_1 \otimes \cdots \otimes |\downarrow\rangle_N$. For the algebra $\mathfrak{h}$ of local observables, such group operations are simply local rotations on each spin. In other words, the group orbit describing the GCSs of $\mathfrak{h}$ comprises all the product states of the form $|\psi_S\rangle = |\phi_1\rangle \otimes \cdots \otimes |\phi_N\rangle$, which have maximum purity (i.e., $P_h = 1$). Other states such as $|\psi_S\rangle = |\mathsf{GHZ_N}\rangle = \frac{1}{\sqrt{2}}[|\downarrow\rangle_1 \otimes \cdots \otimes |\downarrow\rangle_N + |\uparrow\rangle_1 \otimes \cdots \otimes |\uparrow\rangle_N]$ are (maximally) generalized entangled relative to the set of local observables (i.e., $P_h = 0$).

Different GE measures are obtained when a set $\mathfrak{h}$ different from the local observables is chosen. An obvious example, in particular, is given by the set of all observables. In this case, the purity takes its maximum value independently of the pure quantum state,[6,7] expressing the fact that any state is a GCS of the Lie algebra of all observables. Physically, choosing this preferred set is equivalent to not making any subsystem decomposition of the total system, from the GE viewpoint.

3.2. *Fermionic systems*

Another class of systems relevant for developing more intuition about the meaning and usefulness of a measure of entanglement such as the relative purity is provided by fermionic systems. Consider a quantum lattice with N sites, where spinless fermions are able to hop between different sites. Remarkably, the purity relative to the set of observables built upon the bilinear (number-conserving) fermionic operators, that is, $\mathfrak{u}(N) = \text{span}\{c_i^\dagger c_j; 1 \le i,j \le N\}$, where $c_i^\dagger$ (c_i) create (annihilate) a fermion at the ith site, takes its maximum value for the fermionic product states, namely states of the form $|\psi_S\rangle = \prod_l c_l^\dagger |\mathsf{VAC}\rangle$ (Slater determinants). Here, the reference state $|\mathsf{VAC}\rangle$ is the no-fermion state and the index l labels a particular set of modes (i.e., sites, wave vectors, etc.). Although this representation of $\mathfrak{u}(N)$ in the N-spin state space is reducible, and the equivalence between maximal purity of the reduced state and generalized unentanglement defined as *extremality* of the reduced state no longer holds in general, maximal purity is still a *sufficient* condition. This is true for reducible representations in general. For this particular representation, the condition is also necessary.

Upon appropriately redefining the $\mathfrak{u}(N)$-generators so that they form an orthonormal Hermitian basis,[7] the $\mathfrak{u}(N)$-purity may be evaluated as

$$P_{\mathfrak{u}(N)} = \frac{2}{N} \sum_{j<j'=1}^{N} \left[\langle c_j^\dagger c_{j'} + H.c. \rangle^2 - \langle c_j^\dagger c_{j'} - H.c. \rangle^2 \right] + \frac{4}{N} \sum_{j=1}^{N} \langle c_j^\dagger c_j - 1/2 \rangle^2 . \quad (3)$$

Then, $P_{\mathfrak{u}(N)}$ has the property of being maximum, $P_{\mathfrak{u}(N)} = 1$, for single Slater determinants (with any number of fermions) corresponding to the generalized unentangled states relative to this algebra, while giving $P_{\mathfrak{u}(N)} < 1$ for any generalized entangled state. Note that, due to the invariance of $P_{\mathfrak{u}(N)}$ under group operations, generalized unentangled states are independent of the specific set of modes chosen.

4. Entanglement and QPTs: The Ising Model in a Transverse Magnetic Field

QPTs are changes in the state of matter occurring at zero temperature in the limit of infinite system size due to modifications of the interactions between the underlying microscopic constituents.[2] Basically, a QPT is driven by the quantum fluctuations in the ground state. Since different quantum phases are distinguished by the behavior of some correlation functions, a measure of entanglement capable of characterizing the QPT should be sensitive to the behavior of such correlations. Here, we focus on the study of the paramagnetic-to-ferromagnetic QPT present in the Ising model in a transverse magnetic field,[12] using the relative purity as our measure of entanglement. A study of other QPTs based on the same GE approach may be found in an earlier work.[7]

The model Hamiltonian for a chain of N spin-1/2s is

$$H = -2g\sum_{j=1}^{N}\sigma_x^j\sigma_x^{j+1} + \sum_{j=1}^{N}\sigma_z^j\,, \tag{4}$$

where periodic boundary conditions are understood. The idea is to drive the system across the QPT by slowly changing the parameter g, which quantifies the ratio between the spin-spin coupling and the applied magnetic field strength. When $g \gg 1$, the spin-spin interactions are dominant and the spins align in the x direction, leading to ferromagnetic order. The ground state is then doubly degenerate in the thermodynamic limit ($N \to \infty$): $|G\rangle \to \frac{1}{\sqrt{2}}[|+\rangle_1 \otimes \cdots \otimes |+\rangle_N \pm |-\rangle_1 \otimes \cdots \otimes |-\rangle_N]$, $|+\rangle_j$ and $|-\rangle_j$ denoting the eigenstates of the σ_x^j operator. In the opposite regime, where $g \to 0$, the spins align along the direction z of the external magnetic field, the ferromagnetic order in the x direction is lost, and the quantum phase is paramagnetic. Thanks to the exact solvability of this model,[12] it is possible to calculate the spin-spin correlation function $\langle\sigma_x^1\sigma_x^{N/2}\rangle$ in the ground state analytically for all g. This yields to $\langle\sigma_x^1\sigma_x^{N/2}\rangle = 0$ when $g \leq 1/2$ and $\langle\sigma_x^1\sigma_x^{N/2}\rangle > 0$ when $g > 1/2$, so the critical point is $g_c = 1/2$ (Fig. 2).

Solving this model requires that we to map the Pauli spin-1/2 algebra onto the fermionic algebra via the Jordan-Wigner transformation.[13] Equation 4 is then rewritten as

$$H + N = -2\sum_k(-1 + 2g\cos k)c_k^\dagger c_k + ig\sin k\,(c_k^\dagger c_{-k}^\dagger - H.c.), \tag{5}$$

where the canonical operators $c_k^\dagger$ (c_k) create (annihilate) a spinless fermion in the k-th mode, and $k \in V = \pm\left[\frac{\pi}{N}, \frac{3\pi}{N}, \cdots, \frac{(N-1)\pi}{N}\right]$. In the fermionic language, H is an element of the $\mathfrak{so}(2N)$ Lie algebra generated by the number-conserving operators of the Lie subalgebra $\mathfrak{u}(N) = \text{span}\{c_k^\dagger c_{k'}; k, k' \in V\}$, together with the operators $\mathfrak{r}(N) = \{c_k^\dagger c_{k'}^\dagger, c_k c_{k'}; k, k' \in V\}$; that is, $\mathfrak{so}(2N) = \mathfrak{u}(N) \oplus \mathfrak{r}(N)$. The exact-solvability of this model is due ultimately to the *polynomial* scaling of the dimension of $\mathfrak{so}(2N)$ with N. In fact, the ground state may be obtained by applying

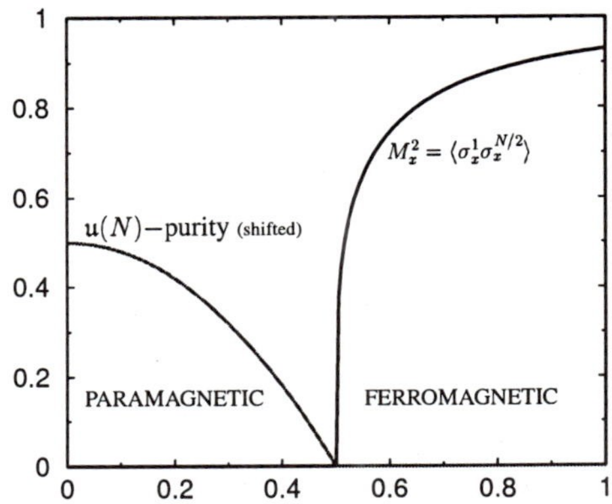

Fig. 2. Magnetization along the x direction (order-parameter) and (shifted) $\mathfrak{u}(N)$-purity in the ground state of the Ising model in a transverse magnetic field. As a function of g, the latter behaves as a disorder parameter, undergoing a sharp change at the critical point $g_c = 1/2$.

a displacement operator of SO$(2N)$ (effecting a mean-field Bogoliubov transformation) to a reference state (the Bogoliubov quasi-particle vacuum). Thus, the ground state is a GCS of $\mathfrak{so}(2N)$, being generalized unentangled relative to this algebra, and leading to a maximum $\mathfrak{so}(2N)$-purity independently of g. Notice that this is a general result: The nondegenerate ground state of a Hamiltonian built from the elements of a Lie algebra $\mathfrak{h}$ is always a GCS of that algebra.[5] Then, the $\mathfrak{so}(2N)$-purity will remain constant when changing the coupling and, in the limit $N \to \infty$, will not detect the QPT.

However, restriction to subalgebras of $\mathfrak{so}(2N)$ will in general cause the ground state to become generalized entangled and the relative purity to depend on g. A natural Lie subalgebra to consider is the $\mathfrak{u}(N)$ algebra, which describes the Hamiltonian in Eq. 5 in the fully polarized limit $g = 0$. By exploiting the symmetries of the model,[7] the relative purity in the ground state takes the simple form

$$P_{\mathfrak{u}(N)} = \frac{4}{N} \sum_{k \in V} \langle c_k^\dagger c_k - 1/2 \rangle^2. \tag{6}$$

In the thermodynamic limit, the sum transforms into an integral that may be evaluated exactly, leading to[7]

$$P_{\mathfrak{u}(N)} = \begin{cases} 1 - 2g^2 & g < g_c \\ 1/2 & g \geq g_c \end{cases}. \tag{7}$$

Remarkably, $P_{\mathfrak{u}(N)}$ sharply distinguishes the paramagnetic-to-ferromagnetic QPT in this model, remaining constant in the ordered (ferromagnetic) phase. Thus, the shifted purity $P'_{\mathfrak{u}(N)} = P_{\mathfrak{u}(N)} - 1/2$ behaves as a *disorder parameter*, being different from zero only in the disordered (paramagnetic) phase (see Fig. 2). In addition, an expansion close to the critical point $(g \to g_c^-)$ shows that $P'_{\mathfrak{u}(N)} \sim 2(g_c - g)^\nu$, $\nu = 1$ being the critical exponent of the Ising universality class. Therefore, the $\mathfrak{u}(N)$-purity fully characterizes this QPT.

5. Generalized Entanglement as a Resource for Efficient Information Processing

Entanglement is known to be an enabling resource for a variety of tasks within both quantum communication and quantum information processing. In the context of QC, in particular, the occurrence of multipartite entanglement is a necessary condition for an algorithm to achieve exponential speedup over classical computation, as least as long as the underlying quantum states are restricted to be pure.[14] It is then natural to ask whether and in what sense GE may be similarly understood and used as a resource for efficient computation. The main goal of this section is to argue that the ability to generate GE relative to *small* (i.e., of polynomial dimension) Lie algebras is a distinctive feature of any QC that is efficient (i.e., has polynomial complexity) compared to its classical counterpart, and to suggest how the same ability may generally provide a necessary condition for allowing efficient simulation of universal QC.

The QAs we are interested in involve three fundamental steps[15]: First, prepare a desired initial state. Second, evolve it by applying controlled unitary operations, such as arbitrary single-qubit rotations and two-qubit gates in a quantum processor based on qubits. Finally, measure some observable; the result yields the solution of the problem we simulated. This algorithm is performed efficiently on a quantum computer if the total number of operations in every step scales polynomially with the degrees of freedom of our problem (e.g., the number of lattice sites when simulating a physical problem, etc.).

Although certain algorithms can be performed efficiently on a quantum computer, their classical simulations are in general not efficient. The reason is that the dimension of the Hilbert space associated with the algorithm increases exponentially with the degrees of freedom of the problem. Thus, the classical simulation must deal with exponentially large algebraic objects (e.g., matrices, etc.).

However, some algorithms can be efficiently simulated on classical computers. For example, if the QA is performed by applying only a polynomially large number of single-qubit rotations to an initial product state, such an evolution can be efficiently mimicked on a classical computer by recording the phases and directions of each rotation, which are drawn from a set consisting of a polynomially large number of elements. This efficiency is lost, in general, when the overall evolution interlaces single qubit rotations with two-qubit gates such as Controlled-NOT gates,[8] which are responsible for generating entanglement. Therefore, there is an extremely important connection between efficiency and entanglement.

Let us now focus on efficient QAs described by a circuit involving a polynomial number of elementary gates obtained by exponentiating a polynomially large Lie algebra $\mathfrak{h}$, with the initial state being a GCS of $\mathfrak{h}$. As explained in Sec. 2, the initial state is generalized unentangled relative to $\mathfrak{h}$. Moreover, the evolved state will remain generalized unentangled with respect to $\mathfrak{h}$ during the whole evolution, since this property is invariant under group operations obtained by exponentiating

$\mathfrak{h}$ (group operations map GCSs into GCSs). We argue that if the dimension of $\mathfrak{h}$ scales polynomially with the degrees of freedom of our problem, such an evolution can also be simulated efficiently on a classical computer. The basic idea is that the latter can be represented by means of the adjoint representation of the Lie algebra $\mathfrak{h}$, in which the Lie algebra acts on itself via the Lie bracket, and thus only a polynomially large number of parameters need to be recorded. Equivalently, if the control Hamiltonian of a quantum computer is an element of a small Lie algebra (i.e., a generalized mean-field Hamiltonian [7]), such device has no universal control, with no advantages over its classical analogue.

Instead, if the QA involves operations that are obtained by exponentiating the elements belonging to an exponentially large-dimensional Lie algebra, straightforward classical simulation of this evolution is, in general, not efficient. Such operations are responsible for the creation of GE relative to polynomially-large-dimensional Lie subalgebras. However, we conjecture that the ability to do gate operations belonging to such a Lie group (generated by a Lie algebra of exponential dimension) allows *efficient simulation of the standard model of QC.* If the Lie algebra is polynomially generated as a Lie algebra (even though it has an exponential dimension), this would also provide an efficient computational model with the power of standard quantum computing. If these conjectures prove valid, that would provide a sense in which the ability to create GE at will efficiently unlocks the power of QC. Another sense in which GE might determine the power of QC would be if gates generated by a low-dimensional algebra, plus the availability of some class of input states that are generalized entangled relative to the algebra generating the gate-set, enabled universal QC. Notably, the Lie-algebraic GE framework appears especially promising for analyzing QC in the adiabatic setting, whose equivalence with the standard QC model has recently been established. [16]

6. Conclusions

The notion of GE plays a crucial role in quantum information science in that it can not only be used to characterize QPTs in matter, but it also proves insightful for understanding the computational efficiency of QAs. The basic idea is that entanglement is a relative notion that depends directly on a preferred set of observables to which we have operational access. This expands both physically and mathematically the traditional notion, which rests on the existence of a preferred decomposition into subsystems.

In Sec. 4 we illustrated how the relative purity fully characterizes the QPT present in the Ising model in a transverse magnetic field. The steps followed for the diagonalization of the Hamiltonian of Eq. 4 naturally single out a preferred Lie algebra of observables for this model. We then showed that the $\mathfrak{u}(N)$-purity undergoes a sharp change at the critical point, behaving as a disorder parameter and therefore capturing the most relevant quantum correlations that drive this broken-symmetry QPT. We also mentioned how a similar behavior is found in other models of interest, such as the anisotropic (spin-1/2) XY model in the presence of a transverse magnetic field. All these models share the property of having a natural preferred set of observables relative to which the purity is calculated. However, such a set does not exist or might be hard to compute in the most general (e.g., the $2D$ Heisenberg model).

Finally, in Sec. 5 we compared quantum and classical efficiency when simulating QAs. We conjectured that GE may be used as a resource for performing faster computation, because a quantum simulation becomes more efficient than the corresponding classical simulation. When GE is involved, the classical simulation needs in general to record an exponentially large number of parameters due to the exponentially increasing Hilbert space describing the problem, while the quantum simulation may remain efficient. We plan to expand these considerations and present more rigorous results elsewhere.

Acknowledgments

Contributions to this work by NIST, an agency of the US government, are not subject to copyright laws.

References

1. E. Schrödinger, Naturwissenschaften **23**, 807 (1935).
2. S. Sachdev, *Quantum Phase Transitions* (Cambridge University Press, Cambridge, 1999).
3. T. J. Osborne and M. A. Nielsen, Phys. Rev. A **66**, 032110 (2002). A. Osterloh, L. Amico, G. Falci, and R. Fazio, Nature **416** 608 (2002). G. Vidal, J. I. Latorre, E. Rico, and A. Kitaev, Phys. Rev. Lett. **90**, 227902 (2003).
4. W. Wootters, Phys. Rev. Lett. **80**, 2245 (1998).
5. H. Barnum, E. Knill, G. Ortiz, and L. Viola, Phys. Rev. A **68**, 032308 (2003).
6. H. Barnum, E. Knill, G. Ortiz, R. Somma, and L. Viola, Phys. Rev. Lett. **92**, 107902 (2004).
7. R. Somma, G. Ortiz, H. Barnum, E. Knill, and L. Viola, Phys. Rev. A **70**, 042311 (2004).
8. M. Nielsen and I. Chuang, *Quantum Computation and Quantum Information* (Cambridge University Press, Cambridge, 2000).
9. R. Gilmore, Rev. Mex. Fis. **23**, 143 (1974).
10. A. Perelomov, *Generalized Coherent States and their Applications* (Berlin, 1985).
11. D. A. Meyer and N. R. Wallach, J. Math. Phys. **43**, 4273 (2002).
12. E. Lieb, T. Schultz, and D. Mattis, Ann. Phys. (N.Y.) **16**, 406 (1961). P. Pfeuty, Ann. Phys. (N.Y.) **57**, 79 (1970). E. Barouch and B. McCoy, Phys. Rev. A **3**, 786 (1971).
13. P. Jordan and E. Wigner, Z. Phys. **47**, 631 (1928).
14. R. Jozsa and N. Linden, Proc. R. Soc. London A **459**, 2001 (2003).
15. R. Somma, G. Ortiz, E. Knill, and J. Gubernatis, Int. Journ. Quant. Inf. **1**, 189 (2003).
16. D. Aharonov, W. van Dam, J. Kempe, Z. Landau, and S. Lloyd, quant-ph/0405098.

GROUND STATE OF MANY-BODY LATTICE SYSTEMS VIA A CENTRAL LIMIT THEOREM

CARLO PRESILLA

Dipartimento di Fisica, Università di Roma "La Sapienza", Piazzale A. Moro 2, Roma 00185, Italy
Center for Statistical Mechanics and Complexity, Istituto Nazionale per la Fisica della Materia, Unità di Roma 1, Roma 00185, Italy
Istituto Nazionale di Fisica Nucleare, Sezione di Roma 1, Roma 00185, Italy

MASSIMO OSTILLI

Dipartimento di Fisica, Università di Roma "La Sapienza", Piazzale A. Moro 2, Roma 00185, Italy
Center for Statistical Mechanics and Complexity, Istituto Nazionale per la Fisica della Materia, Unità di Roma 1, Roma 00185, Italy

Received 1 December 2004

We review a novel approach to evaluate the ground-state properties of many-body lattice systems based on an exact probabilistic representation of the dynamics and its long time approximation via a central limit theorem. The choice of the asymptotic density probability used in the calculation is discussed in detail.

Keywords: Lattice quantum models; probability theory; stochastic processes.

1. Introduction

The real- or imaginary-time dynamics of systems described by a finite Hamiltonian matrix, representing bosonic or fermionic degrees of freedom, admits an exact probabilistic representation in terms of a proper collection of independent Poisson processes.[1,2,3] For a lattice system, the Poisson processes are associated to the links of the lattice and the probabilistic representation leads to an optimal algorithm [3] which coincides with the Green Function Quantum Monte Carlo method in the limit when the latter becomes exact.[4]

In the recent Ref. 5 we have exploited the above probabilistic representation to derive analytical expressions for the matrix elements of the evolution operator in the long time limit. In this way, the ground-state energy as well as the expectation of a generic operator in the ground state of a lattice system without sign problem are obtained as the solution of a simple scalar equation. The result is based on the application of a central limit theorem to the rescaled multiplicities of the values assumed by the potential and hopping energies in the configurations dynamically visited by the system. As a consequence, the probabilistic expectations can be calculated by using a Gaussian-like probability density. In this paper, we briefly review the approach developed in Ref. 5 and discuss in detail the choice of the asymptotic probability density used in the calculation.

2. Exact Probabilistic Representation of Lattice Dynamics

We illustrate our approach in the case of imaginary-time dynamics for a system of hard-core bosons described by the Hamiltonian

$$\hat{H} = - \sum_{i \neq j \in \Lambda} \sum_{\sigma = \uparrow\downarrow} \eta_{ij} c^{\dagger}_{i\sigma} c_{j\sigma} + \hat{V}, \tag{1}$$

where $\Lambda \subset Z^d$ is a finite d-dimensional lattice with $|\Lambda|$ sites and $c_{i\sigma}$ the commuting destruction operators at site i and spin index σ with the property $c^2_{i\sigma} = 0$. The potential operator $\hat{V}$ is arbitrary, *e.g.* for the Hubbard model $\hat{V} = \sum_{i \in \Lambda} \gamma_i c^{\dagger}_{i\uparrow} c_{i\uparrow} c^{\dagger}_{i\downarrow} c_{i\downarrow}$. For simplicity, we assume $\eta_{ij} = \epsilon$ if i and j are first neighbors and $\eta_{ij} = 0$ otherwise.

In order to study the ground-state properties of the Hamiltonian $\hat{H}$ it is sufficient to evaluate the long time behavior of $\sum_{\boldsymbol{n}} \langle \boldsymbol{n} | e^{-\hat{H}t} | \boldsymbol{n}_0 \rangle$, where $\boldsymbol{n} = (n_{1\uparrow}, n_{1\downarrow}, \ldots, n_{|\Lambda|\uparrow}, n_{|\Lambda|\downarrow})$ are the lattice occupation numbers taking the values 0 or 1. In fact, the ground-state energy is given by

$$E_0 = \lim_{t \to \infty} -\partial_t \log \sum_{\boldsymbol{n}} \langle \boldsymbol{n} | e^{-\hat{H}t} | \boldsymbol{n}_0 \rangle, \tag{2}$$

while the quantum expectation of a generic operator $\hat{O}$ in the ground state of $\hat{H}$ can be obtained via the Hellman-Feynman theorem [5] by evaluating the ground-state energy $E_0(\xi)$ of the modified Hamiltonian $\hat{H} + \xi \hat{O}$.

At any finite time t, the matrix elements of the evolution operator considered above admit the exact probabilistic representation

$$\sum_{\boldsymbol{n}} \langle \boldsymbol{n} | e^{-\hat{H}t} | \boldsymbol{n}_0 \rangle = \mathsf{E} \left(\mathcal{M}^t_{\boldsymbol{n}_0} \right), \tag{3}$$

where $\mathcal{M}^t_{\boldsymbol{n}_0}$ is a stochastic functional defined in terms of independent Poisson processes associated to the links of the lattice, see Ref. 3 for a detailed description. At each jump of a Poisson process relating sites i and j with spin σ and taking place at a given configuration $\boldsymbol{n}$, a particle of spin σ moves from site i to site j or vice versa if the mod 2 sum of the occupations of these two sites is $\lambda_{ij\sigma}(\boldsymbol{n}) = 1$, while the lattice configuration $\boldsymbol{n}$ remains unchanged if $\lambda_{ij\sigma}(\boldsymbol{n}) = 0$. Hereafter, links with $\lambda_{ij\sigma} = 1$ will be called active. By ordering the jumps according to the times s_k, $k = 1, \ldots, N_t$, at which they take place in the interval $[0, t)$, we define a trajectory as the Markov chain $\boldsymbol{n}_1, \boldsymbol{n}_2, \ldots, \boldsymbol{n}_{N_t}$ generated from the initial configuration $\boldsymbol{n}_0$. The number of jumps N_t is, of course, a random integer associated to each trajectory. We associate to each trajectory also two sequences, $A_0, A_1, \ldots, A_{N_t - 1}$ and $V_0, V_1 \ldots, V_{N_t}$, representing the number of active links and the potential energy of the visited configurations

$$A_k = \sum_{(i,j) \in \Gamma} \sum_{\sigma = \uparrow\downarrow} \lambda_{ij\sigma}(\boldsymbol{n}_k), \tag{4}$$

$$V_k = \langle \boldsymbol{n}_k | \hat{H} | \boldsymbol{n}_k \rangle = \langle \boldsymbol{n}_k | \hat{V} | \boldsymbol{n}_k \rangle. \tag{5}$$

Here, Γ is the set of system links, *i.e.* the pairs (i,j) with $i < j$ and $i,j \in \Lambda$ such that $\eta_{ij} \neq 0$. The stochastic functional $\mathcal{M}^t_{\boldsymbol{n}_0}$ which appears in Eq. (3) actually depends on the jump times $s_1, s_2, \ldots, s_{N_t}$ and on the corresponding sequences $A_0, A_1, \ldots, A_{N_t-1}$ and $V_0, V_1 \ldots, V_{N_t}$.

3. Probabilistic Expectation in the Long Time Limit

Evaluating the expectation $\mathsf{E}\left(\mathcal{M}^t_{\boldsymbol{n}_0}\right)$ over the detailed sequences above specified can be done numerically by a Monte Carlo method.[3] In Ref. 5 we have demonstrated that an analytical expression of $\mathsf{E}\left(\mathcal{M}^t_{\boldsymbol{n}_0}\right)$ can be obtained in the limit of long times. This result is reached in four steps described in the next subsections. The crucial point is that, if one integrates over all the possible jumps times, what matter are not the detailed sequences $A_0, A_1, \ldots, A_{N_t-1}$ and $V_0, V_1 \ldots, V_{N_t}$ but the multiplicities N_A and N_V of the possibles values which the variables A and V may assume. We call $\mathscr{A}$ and $\mathscr{V}$ the sets of these values and $m_{\mathscr{A}}$ and $m_{\mathscr{V}}$ their cardinalities. It is clear that the nature of these sets depends only on the structure of the system Hamiltonian, not on the values of the Hamiltonian parameters. The expectation $\mathsf{E}\left(\mathcal{M}^t_{\boldsymbol{n}_0}\right)$ is reduced to an average over N_A and N_V. For $t \to \infty$, this average can be evaluated analytically by using saddle-point techniques and a central limit theorem.

3.1. *Canonical decomposition of the expectation*

Referring to Ref. 5 for the details, we decompose the expectation as a series of conditional expectations with a fixed number of jumps (canonical averages)

$$\mathsf{E}\left(\mathcal{M}^t_{\boldsymbol{n}_0}\right) = \sum_{N=0}^{\infty} \mathsf{E}\left(\mathcal{M}^t_{\boldsymbol{n}_0} | N_t = N\right). \tag{6}$$

Integrating over the N jumps times, each term of the series (6) can be written as

$$\mathsf{E}\left(\mathcal{M}^t_{\boldsymbol{n}_0} | N_t = N\right) = \left\langle \mathcal{W}_N(t) \prod_{A \in \mathscr{A}} A^{N_A} \right\rangle_N, \tag{7}$$

where $\langle \cdot \rangle_N$ means average over the trajectories with N jumps generated by extracting with uniform probability one of the active links available at the configurations $\boldsymbol{n}_0, \boldsymbol{n}_1, \ldots, \boldsymbol{n}_{N-1}$, and $\mathcal{W}_N(t)$, named weight, is defined as

$$\mathcal{W}_N(t) = \epsilon^N \int_0^t ds_1 \int_{s_1}^t ds_2 \ldots \int_{s_{N-1}}^t ds_N e^{-V_0 s_1 - V_1(s_2 - s_1) - \ldots - V_N(t - s_N)}. \tag{8}$$

3.2. *Evaluation of the weights*

According to their definition, the weights satisfy a recursive differential equation which is easily solved in terms of the Laplace transform $\widetilde{\mathcal{W}}_N(z)$,[5] *i.e.*

$$\widetilde{\mathcal{W}}_N(z) = \epsilon^N \prod_{V \in \mathscr{V}} \frac{1}{(z+V)^{N_V}}. \tag{9}$$

While this expression shows that $\mathcal{W}_N(t)$ depends on the multiplicities N_V for any value of N, the explicit inversion of the Laplace transform can be done analytically only for N large. However, this is the limit we are interested in since the weights $\mathcal{W}_N(t)$ have a maximum at some N which increases by increasing t. By using a complex saddle-point method which is asymptotically exact for $N \to \infty$, we get [5]

$$\mathcal{W}_N(t) = \frac{e^{x_0 t - \sum_{V\in\mathscr{V}} N_V \log[(x_0+V)/\epsilon]}}{\sqrt{2\pi \sum_{V\in\mathscr{V}} \frac{\epsilon^2 N_V}{(x_0+V)^2}}}, \tag{10}$$

where x_0 is the solution of the equation

$$\sum_{V\in\mathscr{V}} \frac{N_V}{x_0+V} = t. \tag{11}$$

3.3. *Canonical averages via a central limit theorem*

To evaluate the canonical averages it is useful to introduce the frequencies, $\nu_V = N_V/N$ and $\nu_A = N_A/N$, which for N large become continuously distributed in the range $[0,1]$ with the constraints

$$\sum_{V\in\mathscr{V}} \nu_V = \sum_{A\in\mathscr{A}} \nu_A = 1. \tag{12}$$

Note that for N large we will not distinguish the different normalizations, $N+1$ and N, of N_V and N_A, respectively. Equation (7) can be then rewritten as

$$\left\langle \mathcal{W}_N(t) \prod_{A\in\mathscr{A}} A^{N_A} \right\rangle_N = \int d\boldsymbol{\nu} \mathcal{P}_N(\boldsymbol{\nu}) \frac{e^{x_0 t + N(\boldsymbol{\nu},\boldsymbol{u})}}{\sqrt{2\pi N \sum_{V\in\mathscr{V}} \frac{\epsilon^2 \nu_V}{(x_0+V)^2}}}, \tag{13}$$

where $\boldsymbol{\nu}$ and $\boldsymbol{u}$ are vectors with $m = m_{\mathscr{V}} + m_{\mathscr{A}}$ components defined as $\boldsymbol{\nu}^T = (\ldots \nu_V \ldots; \ldots \nu_A \ldots)$ and $\boldsymbol{u}^T = (\ldots -\log[(x_0+V)/\epsilon] \ldots; \ldots \log A \ldots)$, respectively. For later use, we also define $\boldsymbol{v}^T = (\ldots (x_0+V)^{-1} \ldots; \ldots 0 \ldots)$ and $\boldsymbol{w}^T = (\ldots (x_0+V)^{-2} \ldots; \ldots 0 \ldots)$. Note that $\boldsymbol{u}$, $\boldsymbol{v}$ and $\boldsymbol{w}$ depend on $\boldsymbol{\nu}$ through $x_0 = x_0(\boldsymbol{\nu})$ and $\boldsymbol{v} = -\partial_{x_0}\boldsymbol{u}$, $\boldsymbol{w} = -\partial_{x_0}\boldsymbol{v}$.

The probability density $\mathcal{P}_N(\boldsymbol{\nu})$ is given by the fraction of trajectories branching from the initial configuration $\boldsymbol{n}_0$ and having after N jumps multiplicities $N_V = \nu_V N$ and $N_A = \nu_A N$. For N large, it can be approximated in the following way. We rewrite the multiplicities as $N_V = \sum_{k=0}^{N} \chi_V(\boldsymbol{n}_k)$ and $N_A = \sum_{k=1}^{N} \chi_A(\boldsymbol{n}_{k-1})$, where $\chi_V(\boldsymbol{n}) = 1$ if $V(\boldsymbol{n}) = V$ and $\chi_V(\boldsymbol{n}) = 0$ otherwise, and similarly for χ_A. Since the configurations $\boldsymbol{n}_k$ form a Markov chain with finite state space, a central limit theorem applies to each rescaled sum $N_V/\sqrt{N}$ or $N_A/\sqrt{N}$.[6] Therefore, these rescaled variables are completely described in terms of the mean values $\overline{\boldsymbol{\nu}}\sqrt{N}$ and of the covariance matrix $\boldsymbol{\Sigma}$, which are easily measured by sampling over trajectories with a large number of jumps. Due to the constraints (12), it is easy to see that

$$\sum_{V\in\mathscr{V}} \overline{\nu}_V = \sum_{A\in\mathscr{A}} \overline{\nu}_A = 1, \tag{14}$$

$$\sum_{V\in\mathscr{V}} \Sigma_{\alpha,V} = \sum_{V\in\mathscr{V}} \Sigma_{V,\alpha} = \sum_{A\in\mathscr{A}} \Sigma_{\alpha,A} = \sum_{A\in\mathscr{A}} \Sigma_{A,\alpha} = 0, \qquad \alpha \in \mathscr{A}\cup\mathscr{V}. \tag{15}$$

In the next two subsections we will describe two different choices for the density $\mathcal{P}_N(\boldsymbol{\nu})$. The two densities differ only for the fact that the first one satisfies the constraints (12) in mean, while the other one identically. We will show that, at least in the limit of large N, the results obtained by assuming the two densities coincide.

3.3.1. *Purely Gaussian density*

In this case, we assume a purely Gaussian probability density

$$\mathcal{P}_N(\boldsymbol{\nu}) = \sqrt{\frac{N^m|\det \boldsymbol{\Sigma}^{-1}|}{(2\pi)^m}}\, e^{-\frac{N}{2}(\boldsymbol{\Sigma}^{-1}(\boldsymbol{\nu}-\overline{\boldsymbol{\nu}}),(\boldsymbol{\nu}-\overline{\boldsymbol{\nu}}))}. \tag{16}$$

Due to Eq. (15), we have that $\det \boldsymbol{\Sigma} = 0$, *i.e.* $\boldsymbol{\Sigma}$ is singular and the density (16) ill defined. Nevertheless, this singularity is eliminable and the calculation with this density has the advantage to be rather simple. We rewrite Eq. (13) as

$$\left\langle \mathcal{W}_N(t) \prod_{A\in\mathscr{A}} A^{N_A} \right\rangle_N = \int d\boldsymbol{\nu} e^{N\phi(\boldsymbol{\nu})} R(\boldsymbol{\nu}), \tag{17}$$

where

$$\phi(\boldsymbol{\nu}) = x_0 \frac{t}{N} + (\boldsymbol{\nu},\boldsymbol{u}) - \frac{1}{2}\left(\boldsymbol{\Sigma}^{-1}(\boldsymbol{\nu}-\overline{\boldsymbol{\nu}}),(\boldsymbol{\nu}-\overline{\boldsymbol{\nu}})\right) \tag{18}$$

and $R(\boldsymbol{\nu})$ is a smooth function. The parametric dependence of ϕ and R on t and N is omitted for brevity. We will perform the integral (17) with a saddle-point method by solving the m-dimensional stationary problem for $\phi(\boldsymbol{\nu})$. The derivatives of ϕ with respect to the V and A components of $\boldsymbol{\nu}$ read

$$\begin{cases} \partial_{\nu_V}\phi(\boldsymbol{\nu}) = \frac{t}{N}\partial_{\nu_V} x_0 - (\boldsymbol{\nu},\boldsymbol{v})\,\partial_{\nu_V} x_0 + u_V - \left(\boldsymbol{\Sigma}^{-1}(\boldsymbol{\nu}-\overline{\boldsymbol{\nu}})\right)_V \\ \partial_{\nu_A}\phi(\boldsymbol{\nu}) = u_A - \left(\boldsymbol{\Sigma}^{-1}(\boldsymbol{\nu}-\overline{\boldsymbol{\nu}})\right)_A. \end{cases}$$

On the other hand, by using the definition of $x_0(\boldsymbol{\nu})$, Eq. (11), it is easy to see that $(\boldsymbol{\nu},\boldsymbol{v}) = t/N$, so that $\phi(\boldsymbol{\nu})$ is stationary for $\boldsymbol{\nu} = \boldsymbol{\nu}^{\mathrm{sp}}$ solution of the equation

$$\boldsymbol{\nu}^{\mathrm{sp}} = \overline{\boldsymbol{\nu}} + \boldsymbol{\Sigma}\boldsymbol{u}(\boldsymbol{\nu}^{\mathrm{sp}}). \tag{19}$$

Note that, due to Eqs. (14) and (15), we have $\sum_{V\in\mathscr{V}} \nu_V^{\mathrm{sp}} = \sum_{A\in\mathscr{A}} \nu_A^{\mathrm{sp}} = 1$ and only $m-2$ equations in (19) are independent. The result of the saddle-point integration then is

$$\left\langle \mathcal{W}_N(t) \prod_{A\in\mathscr{A}} A^{N_A} \right\rangle_N = e^{N\phi(\boldsymbol{\nu}^{\mathrm{sp}})} R(\boldsymbol{\nu}^{\mathrm{sp}}) \sqrt{\frac{(2\pi)^m}{N^m|\det \nabla^2\phi(\boldsymbol{\nu}^{\mathrm{sp}})|}}, \tag{20}$$

where

$$\phi(\boldsymbol{\nu}^{\mathrm{sp}}) = \left[x_0 \frac{t}{N} + (\overline{\boldsymbol{\nu}},\boldsymbol{u}) + \frac{1}{2}(\boldsymbol{\Sigma}\boldsymbol{u},\boldsymbol{u})\right]_{\boldsymbol{\nu}=\boldsymbol{\nu}^{\mathrm{sp}}} \tag{21}$$

and $\nabla^2\phi(\boldsymbol{\nu}^{\rm sp}) = -\boldsymbol{\Sigma}^{-1} - \boldsymbol{A}(\boldsymbol{\nu}^{\rm sp})$, $\boldsymbol{A}$ being the $m \times m$ matrix with elements

$$A_{\alpha,\beta} = \frac{v_\alpha v_\beta}{(\boldsymbol{\nu}, \boldsymbol{w})}, \qquad \alpha, \beta \in \mathscr{V} \cup \mathscr{A}. \tag{22}$$

By writing $\det \nabla^2\phi(\boldsymbol{\nu}^{\rm sp}) = -\det \boldsymbol{\Sigma}^{-1} \det(\mathbf{1} + \boldsymbol{\Sigma}\boldsymbol{A})$ and observing that $\det \boldsymbol{\Sigma}^{-1}$ cancels out, Eq. (26) becomes

$$\left\langle \mathcal{W}_N(t) \prod_{A\in\mathscr{A}} A^{N_A} \right\rangle_N = \frac{1}{\sqrt{|\det(\mathbf{1}+\boldsymbol{\Sigma}\boldsymbol{A})|}} \left. \frac{e^{x_0 t + N\left[(\overline{\boldsymbol{\nu}},\boldsymbol{u}) + \frac{1}{2}(\boldsymbol{\Sigma}\boldsymbol{u},\boldsymbol{u})\right]}}{\sqrt{2\pi N \epsilon^2(\boldsymbol{\nu}, \boldsymbol{w})}} \right|_{\boldsymbol{\nu}=\boldsymbol{\nu}^{\rm sp}}. \tag{23}$$

Note that for $\hat{V} \equiv 0$ the matrix $\boldsymbol{A}$ is uniform and $\det(\mathbf{1} + \boldsymbol{\Sigma}\boldsymbol{A}) = 1$. In general, $\det(\boldsymbol{\Sigma}\boldsymbol{A}) = 0$ so that $\det(\mathbf{1} + \boldsymbol{\Sigma}\boldsymbol{A}) \simeq 1$ up to terms of second order in $\boldsymbol{A}$.

3.3.2. *Density with constraints identically satisfied*

In the previous subsection we have evaluated the canonical averages by exploiting a central limit theorem for the rescaled variables $N_V/\sqrt{N}$ and $N_A/\sqrt{N}$. However, due to the constraints (12), the joint probability for these m rescaled sums is not purely Gaussian. Given an arbitrary set of $m_{\mathscr{V}} - 1$ V-like components and $m_{\mathscr{A}} - 1$ A-like components, we can assume a joint probability density as the product of a Gaussian density for this set of $m - 2$ variables and two delta functions which take into account the constraints (12), *i.e.*

$$\mathcal{P}_N(\boldsymbol{\nu}) = \mathcal{F}_N(\hat{\boldsymbol{\nu}})\, \delta\left(\sum_{V\in\mathscr{V}} \nu_V - 1\right) \delta\left(\sum_{A\in\mathscr{A}} \nu_A - 1\right), \tag{24}$$

where $\mathcal{F}_N(\hat{\boldsymbol{\nu}})$ is the normal density defined in terms of the vector $\hat{\boldsymbol{\nu}}$ having the $m-2$ chosen components of $\boldsymbol{\nu}$

$$\mathcal{F}_N(\hat{\boldsymbol{\nu}}) = \sqrt{\frac{N^{m-2}|\det \hat{\boldsymbol{\Sigma}}^{-1}|}{(2\pi)^{m-2}}}\; e^{-\frac{N}{2}\left(\hat{\boldsymbol{\Sigma}}^{-1}(\hat{\boldsymbol{\nu}}-\hat{\overline{\boldsymbol{\nu}}}),(\hat{\boldsymbol{\nu}}-\hat{\overline{\boldsymbol{\nu}}})\right)}. \tag{25}$$

In the following, with the symbol ˆ over a vector or a matrix we will indicate the projection onto the chosen $m - 2$ dimensional index space.

By using the density (24), we rewrite Eq. (13) as

$$\left\langle \mathcal{W}_N(t) \prod_{A\in\mathscr{A}} A^{N_A} \right\rangle_N = \int d\hat{\boldsymbol{\nu}} e^{N\hat{\phi}(\hat{\boldsymbol{\nu}})} \hat{R}(\hat{\boldsymbol{\nu}}), \tag{26}$$

where $\hat{R}$ is a smooth function of $\hat{\boldsymbol{\nu}}$ and, if V_* and A_* are the two components chosen to reduce to $m-2$ the dimension of the index space,

$$\begin{aligned}\hat{\phi}(\hat{\boldsymbol{\nu}}) = {} & \hat{x}_0 \frac{t}{N} + (\hat{\boldsymbol{\nu}}, \hat{\boldsymbol{u}}(\hat{x}_0)) + \left(1 - \sum_{V\in\mathscr{V}\setminus V_*} \hat{\nu}_V\right) u_{V_*}(\hat{x}_0) + \left(1 - \sum_{A\in\mathscr{A}\setminus A_*} \hat{\nu}_A\right) u_{A_*} \\ & -\frac{1}{2}\left(\hat{\boldsymbol{\Sigma}}^{-1}(\hat{\boldsymbol{\nu}} - \hat{\overline{\boldsymbol{\nu}}}), (\hat{\boldsymbol{\nu}} - \hat{\overline{\boldsymbol{\nu}}})\right).\end{aligned} \tag{27}$$

Note that $\hat{x}_0 = \hat{x}_0(\hat{\boldsymbol{\nu}})$ is the solution of Eq. (11) where the constraints (12) have been explicited. The function $\hat{\phi}$ can be rewritten as

$$\hat{\phi}(\hat{\boldsymbol{\nu}}) = \hat{x}_0 \frac{t}{N} + \left(\hat{\boldsymbol{\nu}}, \hat{\delta \boldsymbol{u}}\right) - \frac{1}{2}\left(\hat{\boldsymbol{\Sigma}}^{-1}(\hat{\boldsymbol{\nu}} - \hat{\overline{\boldsymbol{\nu}}}), (\hat{\boldsymbol{\nu}} - \hat{\overline{\boldsymbol{\nu}}})\right) + u_{V_*} + u_{A_*}, \tag{28}$$

where $\delta \boldsymbol{u} = \boldsymbol{u} - \boldsymbol{u}_0$ and $\boldsymbol{u}_0^T = (u_{V_*} \dots u_{V_*}; u_{A_*} \dots u_{A_*})$. By evaluating the derivatives of $\hat{\phi}$ with respect to the components of $\hat{\boldsymbol{\nu}}$, we get

$$\begin{cases} \partial_{\hat{\nu}_V} \hat{\phi}(\hat{\boldsymbol{\nu}}) = \frac{t}{N} \partial_{\hat{\nu}_V} \hat{x}_0 - \left(\hat{\boldsymbol{\nu}}, \hat{\delta \boldsymbol{v}}\right) \partial_{\hat{\nu}_V} \hat{x}_0 + \delta u_V - \left(\hat{\boldsymbol{\Sigma}}^{-1}(\hat{\boldsymbol{\nu}} - \hat{\overline{\boldsymbol{\nu}}})\right)_V - v_{V_*} \partial_{\hat{\nu}_V} \hat{x}_0 \\ \partial_{\hat{\nu}_A} \hat{\phi}(\hat{\boldsymbol{\nu}}) = \delta u_A - \left(\hat{\boldsymbol{\Sigma}}^{-1}(\hat{\boldsymbol{\nu}} - \hat{\overline{\boldsymbol{\nu}}})\right)_A, \end{cases}$$

where $\delta \boldsymbol{v} = -\partial_{\hat{x}_0} \delta \boldsymbol{u} = -\left(\partial_{\hat{x}_0} \boldsymbol{u} - \partial_{\hat{x}_0} \boldsymbol{u}_0\right)$. Observing that $\left(\hat{\boldsymbol{\nu}}, \hat{\delta \boldsymbol{v}}\right) + v_{V_*} = t/N$, we can rewrite these derivatives as

$$\begin{cases} \partial_{\hat{\nu}_V} \hat{\phi}(\hat{\boldsymbol{\nu}}) = \delta u_V - \left(\hat{\boldsymbol{\Sigma}}^{-1}(\hat{\boldsymbol{\nu}} - \hat{\overline{\boldsymbol{\nu}}})\right)_V \\ \partial_{\hat{\nu}_A} \hat{\phi}(\hat{\boldsymbol{\nu}}) = \delta u_A - \left(\hat{\boldsymbol{\Sigma}}^{-1}(\hat{\boldsymbol{\nu}} - \hat{\overline{\boldsymbol{\nu}}})\right)_A. \end{cases}$$

Therefore, the saddle-point equation is

$$\hat{\boldsymbol{\nu}}^{\mathrm{sp}} = \hat{\overline{\boldsymbol{\nu}}} + \hat{\boldsymbol{\Sigma}} \hat{\delta \boldsymbol{u}}(\hat{\boldsymbol{\nu}}^{\mathrm{sp}}). \tag{29}$$

Finally, we calculate $\hat{\phi}(\hat{\boldsymbol{\nu}}^{\mathrm{sp}})$. By using Eqs. (28) and (29) and the identities $\left(\hat{\overline{\boldsymbol{\nu}}}, \hat{\delta \boldsymbol{u}}\right) + u_{V_*} + u_{A_*} = (\overline{\boldsymbol{\nu}}, \boldsymbol{u})$ and $\left(\hat{\boldsymbol{\Sigma}} \hat{\delta \boldsymbol{u}}, \hat{\delta \boldsymbol{u}}\right) = (\boldsymbol{\Sigma} \boldsymbol{u}, \boldsymbol{u})$, we get

$$\begin{aligned} \hat{\phi}(\hat{\boldsymbol{\nu}}^{\mathrm{sp}}) &= \left[\hat{x}_0 \frac{t}{N} + \left(\hat{\overline{\boldsymbol{\nu}}} + \hat{\boldsymbol{\Sigma}} \hat{\delta \boldsymbol{u}}, \delta \hat{\boldsymbol{u}}\right) - \frac{1}{2}\left(\hat{\boldsymbol{\Sigma}} \hat{\delta \boldsymbol{u}}, \hat{\delta \boldsymbol{u}}\right) + u_{V_*} + u_{A_*}\right]_{\hat{\boldsymbol{\nu}} = \hat{\boldsymbol{\nu}}^{\mathrm{sp}}} \\ &= \left[\hat{x}_0 \frac{t}{N} + (\overline{\boldsymbol{\nu}}, \boldsymbol{u}(\hat{x}_0)) + \frac{1}{2}(\boldsymbol{\Sigma} \boldsymbol{u}(\hat{x}_0), \boldsymbol{u}(\hat{x}_0))\right]_{\hat{\boldsymbol{\nu}} = \hat{\boldsymbol{\nu}}^{\mathrm{sp}}}. \end{aligned} \tag{30}$$

Due to the identity $\left(\hat{\boldsymbol{\Sigma}} \hat{\delta \boldsymbol{u}}, \hat{\delta \boldsymbol{v}}\right) = (\boldsymbol{\Sigma} \boldsymbol{u}, \boldsymbol{v})$, we have $\hat{x}_0(\hat{\boldsymbol{\nu}}^{\mathrm{sp}}) = x_0(\boldsymbol{\nu}^{\mathrm{sp}})$, where $x_0(\boldsymbol{\nu}^{\mathrm{sp}})$ is the value obtained with the purely Gaussian density. Therefore, Eq. (30) shows that, at least at the saddle-point level, the chosen probability density provides the same results obtained in the previous subsection.

3.4. *Resumming the canonical series*

In order to evaluate the expectation $\mathsf{E}\left(\mathcal{M}^t_{\boldsymbol{n}_0}\right)$ we need to resum the series (6). For $t \to \infty$, the sum can be substituted with the integral

$$\mathsf{E}\left(\mathcal{M}^t_{\boldsymbol{n}_0}\right) = \frac{1}{\sqrt{|\det(\mathbf{1} + \boldsymbol{\Sigma} \boldsymbol{A})|}} \int dN \frac{e^{\psi(N)}}{\sqrt{2\pi N \epsilon^2(\boldsymbol{\nu}^{\mathrm{sp}}, \boldsymbol{w}(x_0(\boldsymbol{\nu}^{\mathrm{sp}})))}}, \tag{31}$$

where

$$\psi(N) = x_0(\boldsymbol{\nu}^{\mathrm{sp}}) t + N\left[(\overline{\boldsymbol{\nu}}, \boldsymbol{u}(x_0(\boldsymbol{\nu}^{\mathrm{sp}}))) + \frac{1}{2}(\boldsymbol{\Sigma} \boldsymbol{u}(x_0(\boldsymbol{\nu}^{\mathrm{sp}})), \boldsymbol{u}(x_0(\boldsymbol{\nu}^{\mathrm{sp}})))\right]. \tag{32}$$

The integrand in (31) is exponentially peaked at $N = N^{\mathrm{sp}}$, where N^{sp} satisfies $\partial_N \psi(N^{\mathrm{sp}}) = 0$. For a generic N we have

$$\begin{aligned}\partial_N \psi(N) &= \left[\{t - N\left[(\overline{\boldsymbol{\nu}}, \boldsymbol{v}) + (\boldsymbol{\Sigma u}, \boldsymbol{v})\right]\} \partial_N x_0 + (\overline{\boldsymbol{\nu}}, \boldsymbol{u}) + \frac{1}{2}(\boldsymbol{\Sigma u}, \boldsymbol{u}) \right]_{\boldsymbol{\nu}=\boldsymbol{\nu}^{\mathrm{sp}}} \\ &= \left[(\overline{\boldsymbol{\nu}}, \boldsymbol{u}) + \frac{1}{2}(\boldsymbol{\Sigma u}, \boldsymbol{u}) \right]_{\boldsymbol{\nu}=\boldsymbol{\nu}^{\mathrm{sp}}},\end{aligned}$$

where, due to Eqs. (11) and (19), we noticed that the term $\{t - N\left[(\overline{\boldsymbol{\nu}}, \boldsymbol{v}) + (\boldsymbol{\Sigma u}, \boldsymbol{v})\right]\}$ vanishes for $\boldsymbol{\nu} = \boldsymbol{\nu}^{\mathrm{sp}}$. The stationarity condition for $\psi(N)$ is, therefore,

$$\left[(\overline{\boldsymbol{\nu}}, \boldsymbol{u}) + \frac{1}{2}(\boldsymbol{\Sigma u}, \boldsymbol{u}) \right]_{\boldsymbol{\nu}=\boldsymbol{\nu}^{\mathrm{sp}}, N=N^{\mathrm{sp}}} = 0. \tag{33}$$

Equation (33) is a time independent equation which determines $x_0|_{\boldsymbol{\nu}=\boldsymbol{\nu}^{\mathrm{sp}}, N=N^{\mathrm{sp}}}$ as a function of $\overline{\boldsymbol{\nu}}$ and $\boldsymbol{\Sigma}$. According to Eq. (11), this means that the quantity N^{sp} increases linearly with t so that $x_0|_{\boldsymbol{\nu}=\boldsymbol{\nu}^{\mathrm{sp}}, N=N^{\mathrm{sp}}}$ becomes independent of time in the limit $t \to \infty$. By evaluating the second derivative of $\psi(N)$

$$\partial_N^2 \psi(N) = -(\boldsymbol{\nu}^{\mathrm{sp}}, \boldsymbol{v}(x_0(\boldsymbol{\nu}^{\mathrm{sp}})))\partial_N x_0(\boldsymbol{\nu}^{\mathrm{sp}}), \tag{34}$$

where

$$\partial_N x_0(\boldsymbol{\nu}^{\mathrm{sp}}) = \frac{1}{N} \frac{(\boldsymbol{\nu}^{\mathrm{sp}}, \boldsymbol{v}(x_0(\boldsymbol{\nu}^{\mathrm{sp}})))}{(\boldsymbol{\nu}^{\mathrm{sp}}, \boldsymbol{w}(x_0(\boldsymbol{\nu}^{\mathrm{sp}}))) + (\boldsymbol{v}(x_0(\boldsymbol{\nu}^{\mathrm{sp}})), \boldsymbol{\Sigma v}(x_0(\boldsymbol{\nu}^{\mathrm{sp}})))}, \tag{35}$$

and approximating $\psi(N) \simeq \psi(N^{\mathrm{sp}}) + \frac{1}{2}\partial_N^2 \psi(N^{\mathrm{sp}})(N - N^{\mathrm{sp}})^2$ in the exponent and $N \simeq N^{\mathrm{sp}}$ elsewhere, the integral (31) reduces to a Gaussian one which gives

$$\mathsf{E}\left(\mathcal{M}_{\boldsymbol{n}_0}^t\right) = \sqrt{\frac{1 + \mathrm{tr}(\boldsymbol{\Sigma A})}{|\det(\boldsymbol{1} + \boldsymbol{\Sigma A})|}} \left.\frac{e^{x_0 t}}{\epsilon(\boldsymbol{\nu}, \boldsymbol{v})}\right|_{\boldsymbol{\nu}=\boldsymbol{\nu}^{\mathrm{sp}}, N=N^{\mathrm{sp}}}. \tag{36}$$

Note that, since the peak of the Gaussian at $N = N^{\mathrm{sp}}$ moves to infinity linearly with t while its width increases only as $\sqrt{t}$, the result (36) is asymptotically exact for $t \to \infty$.

According to Eq. (2), the result (36) shows that the ground-state energy of the hard-core boson system is

$$E_{0B} = -\left. x_0\right|_{\boldsymbol{\nu}=\boldsymbol{\nu}^{\mathrm{sp}}, N=N^{\mathrm{sp}}}. \tag{37}$$

Equation (33) is, therefore, the equation for the ground-state energy. It defines E_{0B} in terms of $\overline{\boldsymbol{\nu}}$ and $\boldsymbol{\Sigma}$ and explicitly reads

$$\begin{aligned}0 = &- \sum_{V \in \mathscr{V}} \overline{\nu}_V \log\left(\frac{-E_{0B} + V}{\epsilon}\right) + \sum_{A \in \mathscr{A}} \overline{\nu}_A \log(A) \\ &+ \frac{1}{2} \sum_{V \in \mathscr{V}} \sum_{V' \in \mathscr{V}} \Sigma_{V,V'} \log\left(\frac{-E_{0B} + V}{\epsilon}\right) \log\left(\frac{-E_{0B} + V'}{\epsilon}\right) \\ &- \sum_{V \in \mathscr{V}} \sum_{A \in \mathscr{A}} \Sigma_{V,A} \log\left(\frac{-E_{0B} + V}{\epsilon}\right) \log(A)\end{aligned}$$

$$+\frac{1}{2}\sum_{A\in\mathscr{A}}\sum_{A'\in\mathscr{A}}\Sigma_{A,A'}\log(A)\log(A')\,. \tag{38}$$

In the case $\hat{V}\equiv 0$, the ground-state energy $E_{0B}^{(0)}$ can be solved analytically

$$E_{0B}^{(0)} = -\epsilon\exp\left[\sum_{A\in\mathscr{A}}\overline{\nu}_A\log(A) + \frac{1}{2}\sum_{A\in\mathscr{A}}\sum_{A'\in\mathscr{A}}\Sigma_{A,A'}\log(A)\log(A')\right]. \tag{39}$$

4. Conclusions

By using saddle-point techniques and a central limit theorem, we have exploited an exact probabilistic representation of the quantum dynamics in a lattice to derive analytical approximations for the matrix elements of the evolution operator of a system of hard-core bosons in the limit of long times. The approach yields a simple scalar equation for the ground-state energy. This equation depends on the values of the generalized potentials V and of the kinetic quantities A, and on the statistical moments $\overline{\boldsymbol{\nu}}$ and $\boldsymbol{\Sigma}$ of their asymptotic multiplicities N_V and N_A. In turn, these moments depend only on the structure of the system Hamiltonian, not on the values of the Hamiltonian parameters. This implies that the statistical moments must be measured *una tantum* for a given Hamiltonian structure and, once $\overline{\boldsymbol{\nu}}$ and $\boldsymbol{\Sigma}$ are known, our approach provides the ground-state energy analytically as a function of the Hamiltonian parameters.

The ground state energies obtained with the present formula reveal a small systematic error when compared with exact results.[5] In fact, a more accurate analysis which takes into account the large deviations neglected by the central limit theorem shows that the present approach corresponds to a second order truncation of an exact cumulant expansion for the ground-state energy.[7]

Similar results hold in the case of Hamiltonians with arbitrary kinetic operators.

References

1. G. F. De Angelis, G. Jona-Lasinio and M. Sirugue, *J. Phys.* A **16**, 2433 (1983).
2. G. F. De Angelis, G. Jona-Lasinio and V. Sidoravicius, *J. Phys.* A **31**, 289 (1998).
3. M. Beccaria, C. Presilla, G. F. De Angelis and G. Jona Lasinio, *Europhys. Lett.* **48**, 243 (1999).
4. S. Sorella and L. Capriotti, *Phys. Rev.* B **61**, 2599 (2000).
5. M. Ostilli and C. Presilla, *New J. Phys.* **6**, 107 (2004).
6. P. Billingsley, *Probability and Measure* (John Wiley & Sons, New York, 1986).
7. M. Ostilli and C. Presilla, in preparation.

EFFECTS OF A SINGLE QUANTUM SPIN ON JOSEPHSON OSCILLATIONS

MARINA HRUŠKA*,†, LEV BULAEVSKII†,
ALEXANDER SHNIRMAN†,‡ and DARRYL SMITH†

†*Theoretical Division, Group T-11 Los Alamos National Laboratory, MS B262, Los Alamos, New Mexico NM 87544, United States*
‡*Institüt für Theoretische Festkörperphysik Universität Karlsruhe, D-76128 Karlsruhe, Germany*
* *hruska@viking.lanl.gov*

Received 15 October 2004

We consider a single localized spin-1/2 placed between singlet superconducting leads of a Josephson junction inserted in a superconducting ring, subject to a dc magnetic field $\mathbf{B} \parallel z$. Turning on the tunneling or a time-dependent flux in a superconducting ring, induces oscillations of the Josephson current, with an amplitude sensitive to the initial value of the z-component of the spin, $S_z = \pm 1/2$, which allows for a measurement of the initial spin state. At low temperatures when effects of quasiparticles are negligible, this procedure realizes a quantum-non-demolition (QND) measurement of S_z.

Keywords: Spin dynamics; Josephson effect.

Quantum measurements in mesoscopic systems by use of spin-dependent tunneling have recently attracted great interest,[1] due to the challenges of both the single spin detection and quantum computing where the final state of a qubit after computation must be measured. The detector couples to (and gives information on), usually, a single observable of the qubit, e.g. the spin-half projection S_z onto the direction of an applied magnetic field. If this observable commutes with the spin Hamiltonian and is thus a conserved quantity, the quantum nondemolition (QND) regime is realized, i.e., the two possible eigenvalues of S_z can be measured with proper probabilities even by a weakly coupled detector. Otherwise the initial state may relax quickly and one can only observe the steady state properties of the qubit and the meter performing continuous measurements.[2] In spin detecting tunneling schemes all components of the localized spin are coupled to the tunneling electrons via the exchange interaction. Thus QND measurements of a single spin might seem impossible.

We showed that in fact QND measurements of the spin projection on the direction of the applied magnetic field are possible with the use of spin dependent Josephson tunneling at low temperatures, when effects of quasiparticles are negligible. The Josephson oscillation amplitude depends on the initial spin state and can be measured e.g. in a circuit containing a single localized spin 1/2 between conventional singlet superconducting electrodes as in a superconducting STM setup, a dc-voltage source and a resistor. We have considered such a setup in Ref. 3. The dissipative spin-dependent current in the circuit appears when the voltage exceeds a threshold and it measures the squared amplitude of the Josephson-current oscillations. Alternatively, as we will show here, spin read-out can be achieved in a setup consisting of a spin-1/2 inside a Josephson junction inserted in a superconducting ring. A time-dependent magnetic flux through the ring induces a voltage bias across the junction, and an inductively coupled detector measures the alternating flux induced by Josephson oscillations. Due to the spin conservation in a singlet-Cooper-pair tunneling in the absence of quasiparticles the average spin projection $\langle \hat{S}_z \rangle$ onto the direction of the applied magnetic field is preserved to second order in the tunneling amplitude when voltage $V < 2\Delta$ (Δ being the superconducting gap), even though $\hat{S}_z$ does not commute with the Hamiltonian of the system. Thus using the singlet-pair superconducting tunneling as an intermediate nondissipative macroscopic quantum system allows us to obtain information on the initial value of the spin projection without flipping the spin.

We consider a superconducting ring in the x, y plane with an inserted Josephson junction (see Fig. 1). Electrodes of the junction are made of a singlet-pairing superconductor. Inside the junction a single localized spin 1/2 is placed and electrons can tunnel directly between electrodes as well as via the spin as described by tunneling matrix elements T_0 and $T_s \lesssim T_0$, respectively. In addition to the applied DC magnetic field $\mathbf{B}$ a flux $\Phi_v(t) = 2\phi_0 eVt/h - \Phi_B$ is induced across the ring by a coil with a magnetic field confined inside so that it does not directly act on the spin. (Here $\phi_0 = hc/2e$ is the superconducting flux quantum, and the flux Φ_B is equal to the flux of the field $\mathbf{B}$ through the loop.) The flux $\Phi_v(t)$ leads to Josephson current oscillations across the junction.

The tunneling matrix elements T_0, T_s are activated at time $t = 0$ and at $t > 0$ the Hamiltonian of the system is

$$\mathcal{H} = \mathcal{H}_a + \mathcal{H}_b - \mu B_z S_z + \mathcal{H}_T,\ \mathcal{H}_T = \sum_{n,m,\alpha,\alpha'} a^\dagger_{n\alpha}[T_0\delta_{\alpha,\alpha'} + T_s(\boldsymbol{\sigma})_{\alpha,\alpha'} \cdot \mathbf{S}]b_{m\alpha'} + h.c.,$$

where $\mathcal{H}_a$ and $\mathcal{H}_b$ are the Hamiltonians in superconducting leads a and b, and $a^\dagger_{n\alpha}$ ($a_{n\alpha}$) creates (annihilates) an electron in the lead a in the state n with the spin α. Further, $\boldsymbol{\sigma}$ represents the three Pauli matrices, the localized spin is described by the operator $\mathbf{S} = (S_x, S_y, S_z)$ and μ is the magnetic moment of the spin. In the following we use B for μB_z and V for $eV/\hbar$, using units $\hbar = c = 1$.

We denote the flux across the ring by $\Phi(t)$. As we will show in the following, the flux difference $\Phi(t) - Vt$ oscillates with frequency $\omega_J = 2V$, while the oscillation amplitude depends on the state of the spin at $t = 0$, i.e. on the spin wave function $\Psi_s(0) = \alpha|\uparrow\rangle + \beta|\downarrow\rangle$. For measurements the flux $\Phi(t) - Vt$ is coupled inductively to the classical oscillator $Q(t)$. We assume that the flux $Q(t)$ is confined inside the measuring coil so that noise in $Q(t)$ does not affect the spin directly. Hence, in

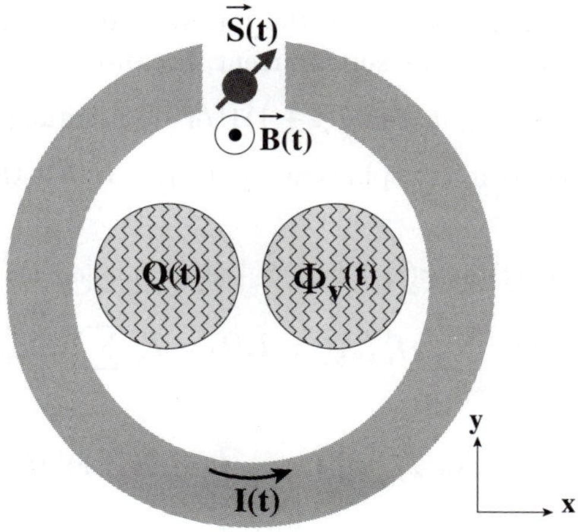

Fig. 1. Superconducting ring with spin inside the Josephson junction. The coil Φ_v induces the voltage V across the junction and causes Josephson oscillations. The coil Q measures flux oscillations across the ring induced by Josephson current oscillations $I(t)$. The amplitude of flux oscillations at time $t > 0$ depends on the S_z-component of localized spin at the moment $t = 0$ at which the spin dependent tunneling matrix element was turned on.

such a setup only the magnetic field $B + \beta\kappa L I(t)$ affects the spin, where $I(t)$ is the current in the ring, L is the ring induction, β is a proportionality coefficient, $\kappa < 1$ characterizes the detector-ring inductive coupling and we assume $\beta\kappa L|I| \ll B$.

First we will derive the expression for the Josephson current across the junction as a functional of $\Phi(t)$ and the spin wave function $\Psi_s(0)$. We neglect the effects of quasiparticles (assuming low temperatures), of detector noise and of self-induction (see below). We neglect quantum fluctuations of the phase difference $\varphi(t)$ across the junction, assuming that charging energy $(2e)^2/C$ is much smaller than the Josephson energy $(T_0^2 + T_s^2)\rho^2\Delta$, where C is the junction capacitance and ρ is the density of states inside leads. We assume that $(T_0^2 + T_s^2)\rho^2 \ll 1$ and $V, B \ll \Delta$ and that all relevant frequencies are much smaller than $\Delta/\hbar$. In a closed loop made of thick superconductor and a tunneling junction the phase difference $\varphi(t) = -2\pi\Phi(t)/\phi_0$ develops at time $t > \Delta^{-1}$. We find the Heisenberg operator for the current using the perturbation theory with respect to the tunneling Hamiltonian $\mathcal{H}_T$ and then average this operator over the electron ground state of the Hamiltonian $\mathcal{H}_a + \mathcal{H}_b$. The current operator $\langle I(t)\rangle_e$ obtained in this way is an operator in spin space and $\langle\Psi_s^*(0)|\langle I(t)\rangle_e|\Psi_s(0)\rangle$ defines the current for any initial state $\Psi_s(0)$.

At time $t > 0$ the current is given to second order in tunneling as

$$\langle\hat{I}(t)\rangle_e = \frac{4ie}{\hbar^2}\rho^2 \int_0^t dt' \sin[\frac{\varphi(t) + \varphi(t')}{2}]\{F^+(t-t')F(t-t')[T_0^2 - T_s^2\tilde{\mathbf{S}}(t)\cdot\tilde{\mathbf{S}}(t')] - h.c.\},$$

where $\tilde{\mathbf{S}}(t) = \exp(-iB\hat{S}_z t)\mathbf{S}\exp(iB\hat{S}_z t)$ and Gor'kov Green functions are related to the Bessel functions as $F^+(t) = F(t) = (\pi\Delta/2)[J_0(\Delta t) - iN_0(\Delta t)]$. We see that the average current is expressed via the spin correlation function $\langle\Psi_s^*(0)|\tilde{\mathbf{S}}(t)\cdot$

$\tilde{\mathbf{S}}(t')|\Psi_s(0)\rangle$. This function depends on the initial state of the spin leading to the corresponding dependence for the supercurrent, because

$$\tilde{\mathbf{S}}(t) \cdot \tilde{\mathbf{S}}(0) = [(2\cos\omega_L t + 1)\hat{\mathbf{1}}]/4 - i\hat{S}_z \sin\omega_L t, \tag{1}$$

where $\hat{\mathbf{1}}$ is the unity matrix in the spin space and $\omega_L = B/\hbar$ is the Larmor frequency.

As a result,

$$\langle \hat{I}(t)\rangle_e = (\pi e/\hbar)\rho^2\Delta\{[a_1(t)\sin\varphi(t) + b_2(t)\hat{S}_z]\sin\varphi(t) + [a_2(t)\hat{\mathbf{1}} - b_1(t)\hat{S}_z]\cos\varphi(t)\},$$

$$a_i(t) = 4T_0^2 f_v^{(i)}(\tau) - T_s^2 \sum_{n=0,\pm1} f_{v+nb}^{(i)}(\tau),\ b_i(t) = T_s^2 \sum_{\pm}(\pm) g_{v\pm b}^{(i)}(\tau),\ i = 1, 2,$$

$$f_k^{(i)}(\tau) = \int_0^\tau dx h_i(kx) J_0(x) N_0(x),\ g_k^{(i)}(\tau) = \int_0^\tau dx h_i(kx)[J_0^2(x) - N_0^2(x)],$$

where $v = V/\Delta$, $b = B/\Delta$, $\tau = \Delta t$, $h_1(x) = -\cos x$ and $h_2(x) = \sin x$. After a transition time of order $100/\Delta$ stationary oscillations are established:

$$\begin{aligned}\langle \hat{I}(t)\rangle_e &= I_0(t)\hat{\mathbf{1}} + 2I_s(t)\hat{S}_z \\ &= -\frac{4\pi e}{\hbar^2}\rho^2\Delta\{[T_0^2 - \frac{3}{4}T_s^2]\hat{\mathbf{1}} + 0.20T_s^2\frac{B}{\Delta}\hat{S}_z\}\sin\omega_J t.\end{aligned} \tag{2}$$

The current amplitude depends on the initial spin state via the operator S_z. In a single measurement at $t > 0$ the current takes one of two values $I_0 \pm I_s$ with probabilities $P_+ = |\alpha|^2$ and $P_- = 1 - |\alpha|^2$, corresponding to $S_z = \pm 1/2$. Since the spin state is conserved in the process of Josephson tunneling on time scales longer than $\hbar/\Delta_0$, consecutive measurements result in the same value until the spin flips due to quasiparticles or other relaxation mechanisms. Indeed, on shorter timescales the spin may be flipped by the first electron of a tunneling Cooper pair, but its state is restored by the second electron since a Cooper pair does not carry spin.

To measure these oscillations, we introduce an inductively coupled detector. The Lagrangian of the system is

$$\begin{aligned}\mathcal{L}\{S_z, \Phi(t), Q(t)\} &= BS_z - E_J\{S_z, \Phi\} + (C/2c^2)\dot{\Phi}^2 \\ &\quad - (1/2c^2L)(\Phi - Vt - \kappa Q)^2 + \mathcal{L}\{Q\},\end{aligned}$$

where $\mathcal{L}\{Q(t)\}$ is the Lagrangian of the classical oscillator used for detection and the Josephson energy E_J is related to the Josephson current as $\langle I\rangle_e = (2e/\hbar)\partial E_J/\partial\varphi$. Minimization over $\Phi(t)$ and $Q(t)$ at given S_z for an overdamped oscillator leads to the equations of motion

$$C\ddot{\Phi} + \Phi/L - cI(\Phi, S_z) = \kappa Q/L + Vt/L,\quad \gamma^{-1}\dot{Q} + Q = \kappa(\Phi - Vt) + f(t), \tag{3}$$

where γ is the dissipation rate of detector, $f(t)$ is the Langevin random force with the correlation function $\langle f(t)f(0)\rangle = \gamma^{-1}c^2LT\delta(t)$, and T is the detector temperature. Solving Eq. (3) we obtain

$$\Phi(t) = Vt + \int_0^t d\tau\omega_0[cLI(\tau) + \kappa Q(\tau)]\sin\omega_0(t - \tau), \tag{4}$$

where $\omega_0^2 = (LC)^{-1}$. At $\omega_0 \gg \omega_J$ the flux in the ring oscillates with the Josephson frequency as described by the term $cLI(t)$. These oscillations are measured by the detector coil as $Q(t) = \kappa cLI(t)$ when $\gamma \gg \omega_J$. The amplitude of oscillation, carrying information on the initial state of the spin, can be extracted by transforming $Q(t)$ into $Q^2(t)$. To distinguish the spin projection $S_z = \pm 1/2$ we need to resolve the difference in corresponding signals accumulated over the measurement time τ_m on the noise background. These accumulated signals are $A(\tau_m) = \int_0^{\tau_m} dt[Q_+^2(t) - Q_-^2(t)] \approx 2\kappa^2 c^2 L^2 I_0 I_s \tau_m$, where I_0 and I_s are the amplitudes of $I_0(t)$ and $I_s(t)$, respectively. The dispersion of $A(t)$ due to thermal fluctuations in the detector is given for $\gamma \gg \omega_0$ as $8(\kappa cL)^2 c^2 LTI_s^2 \tau_m/\gamma$. Taking into account only thermal fluctuations we see that to resolve the signals corresponding to $S_z = \pm 1/2$ the measurement time τ_m should satisfy the condition $\kappa(LI_0^2/2T)^{1/2}(\gamma\tau_m)^{1/2} \gg 1$.

The important point is that the detector introduces noise into the phase of flux oscillations in the ring and then into the spin dynamics causing dephasing of both flux and spin oscillations at the Josephson frequency. From Eq. (4) the random part of the phase difference is $\varphi_r(t) = -(2\pi/\phi_0)\kappa\omega_0 \int_0^t dt' Q(t') \sin\omega_0(t-t')$, Accounting only for thermal fluctuations at $\gamma \gg \omega_0$ we obtain for large t that $\langle \varphi_r^2(t) \rangle \approx \lambda t$, where the dephasing rate $\lambda = 2\pi^2\kappa^2\omega_0^2 c^2 LT/\gamma\phi_0^2$. In the Josephson current after averaging over the Langevin random force contribution to the phase, we get exponential decay of Josephson oscillations, so that $\langle I(t) \rangle \propto \exp(-\lambda t)$. This means that in a single set of measurements the phase of oscillations changes randomly due to thermal noise in the detector with characteristic time $1/\lambda$, though the amplitude remains constant and takes two values corresponding to $S_z = \pm 1/2$. In multiple measurements the average current decays with time.

The demolition rate due to quasiparticles, proportional to their fraction, is exponentially small at low temperature $T \ll \Delta$ and allows for a QND measurement on timescales shorter than $(T_s^2 \rho^2 \Delta)^{-1} \exp(\Delta/T)$. To avoid spin-flipping processes that occur in higher order in tunneling, measurement should be performed on timescales shorter than $(T_s^4 \rho^4 \Delta)^{-1}$.

In conclusion, we have shown that nondemolition tunneling measurements of spin are possible by use of Josephson oscillations in a junction inserted in the superconducting loop. The average amplitude of these oscillations depends on the initial spin state. The use of the singlet-pair tunneling as a mediator between the quantum spin and the dissipative circuit allows one to extract information about the initial value of S_z without flipping the spin.

Acknowledgments

The authors thank I. Martin, D. Averin, G. Ortiz and Yu. Galperin for useful discussions. This work was supported by the US DOE and by the ESF "Vortex" Program and is a part of the CFN of the DFG and the EU IST Project SQUBIT.

References

1. L.M.K. Vandersypen et al, in *Quantum Computing and Quantum Bits in Mesoscopic Systems* (eds. A.J. Legget et al.) Kluwer Academic/Plenum, New York (2003); L.N. Bulaevskii and G. Ortiz, *Phys. Rev. Lett.* **90**, 040401 (2003); L.N. Bulaevskii, M. Hruška and G. Ortiz, *Phys. Rev. B* **68**, 125415 (2003) and references therein.
2. A.N. Korotkov and D.V. Averin, *Phys. Rev. B* **64**, 165310 (2001); A. Shnirman, I. Martin, and D. Mozyrsky, cond-mat/0311325.
3. L. Bulaevskii, M. Hruška, A. Shnirman, D. Smith and Yu. Makhlin, *Phys. Rev. Lett.* **92**, 177001 (2004).

Complex Systems

SPIN TEXTURES AND RANDOM FIELDS IN DIRTY QUANTUM HALL FERROMAGNETS

J. T. CHALKER

Theoretical Physics, Oxford University, 1, Keble Road, Oxford OX1 3NP, United Kingdom
j.chalker1@physics.ox.ac.uk

Received 1 December 2004

Dirty quantum Hall ferromagnets (QHFMs) provide a setting both for new problems in the theory of magnetism with quenched disorder, and for new realisations of old problems. In the first category, the fact that spin textures in Heisenberg QHFMs carry charge leads to a coupling between charged impurities and magnetic order. This coupling drives a zero-temperature transition between a ferromagnet at low disorder and a spin glass at strong disorder, and controls screening and the nature of excitations in the disorder-dominated ground state. In the second category, random fields coupling linearly to the order parameter appear in some Ising QHFMs, and transport measurements appear to indicate field-induced domain states at low temperature.

Keywords: Quantum Hall ferromagnets; disorder; spin textures; spin glass.

1. Introduction

In the twenty-odd years since the discovery of the integer quantum Hall effect, a great deal has emerged about the two-dimensional electron gas in a strong magnetic field that is interesting, beyond quantisation of the Hall conductance. One of the areas that has attracted a lot of attention in the last decade has been various types of itinerant ferromagnetism in these systems. In this paper we set out to provide a brief review of an aspect of that: the interplay between ferromagnetic exchange and disorder in quantum Hall systems.

In one sense, this is obviously a rather specific corner of quantum Hall physics. But from another viewpoint, it is one example from a large class of problems from condensed matter physics, in which one is interested in the interplay between disorder and interactions. Typically, these problems are rather hard, but in the case of quantum Hall ferromagnets (QHFMs) it is possible to produce a basic description of what happens in quite simple terms.

An introduction to ferromagnetism in quantum Hall systems can be found, for example, in Ref. 1, while the work reviewed here is described in more detail in Refs. 2, 3 and 4. Some related work is contained in Refs. 5, 6, 7, 8 and 9.

2. Spin Textures and Spin Glasses

Consider a two-dimensional electron gas in a perpendicular magnetic field: neglecting interactions and in the absence of impurities, the spectrum consists of a sequence of Landau levels, with cyclotron and Zeeman splittings which are different because of band-structure effects. In fact, in GaAs the Zeeman energy is typically about 70 times smaller than the cyclotron energy, and the g-factor can be tuned by varying composition or quantum well width. The physics we are concerned with in this section is most interesting if the g-factor is exactly zero, and so we focus on that case.

Exchange interactions and impurities in the system have competing effects, which can be discussed in the first instance using Stoner theory.[5] In the absence of Zeeman energy, Landau levels for electrons with the same cyclotron energy but opposite spin are split only by exchange. This exchange splitting depends in turn on the magnetisation of the electrons. So if there is an electron density sufficient to fully occupy the Landau level for one spin direction, then when disorder is low that is what happens, and exchange splitting is maximal. But as disorder is increased, the two Landau levels start to overlap and this in turn reduces the polarisation, and therefore also the exchange splitting. At a critical disorder strength, the exchange splitting collapses to zero and there is a zero-temperature transition to a paramagnet. Of course, if the g-factor is non-zero, this transition is smeared, just as for a thermally-driven transition.

Experimentally, one can cross the transition between the disorder-dominated and exchange-dominated regimes by varying magnetic field strength. In the right samples, at low field, when the Fermi energy lies in a high Landau level, disorder dominates and we have a paramagnet, while at high field, when the Fermi energy lies in one of the lower Landau levels, exchange dominates and the system is partially spin polarised whenever an odd integer number of Landau levels is filled. One can see this transition in transport measurements, by looking at Shubnikov-de Haas oscillations in the diagonal part of the resistivity tensor as a function of magnetic field. There are minima whenever the Fermi energy lies in a gap between Landau levels: at high field this happens for every integer filling factor, but below a critical field it happens only at even integer filling factor.[10]

We should next recall the specific feature of QHFMs that makes them of particular interest: the link between spin configurations and charge density.[11] A standard way to introduce it is as follows. Consider a system with filling factor close to $\nu = 1$, and imagine that in a thought experiment we impose a spin texture that is smooth on the scale of the magnetic length, for example by applying a spatially-varying Zeeman field. Now ask what happens to an electron as it moves in this spin background. Exchange interactions keep its spin always parallel to the local polarisation direction. Then it's a basic feature of quantum mechanics that the electron will acquire a Berry phase, and the phase accumulated in following a closed loop can be computed by regarding the spin configuration as a map from real space onto the spin sphere:

the phase is proportional to the area enclosed by the loop on the surface of the spin sphere. This Berry phase adds to the Aharonov-Bohm phase that an electron gets from its orbital motion in the magnetic field: crucially, since the electron density in a filled Landau level is proportional to the flux density, modulations of the Berry phase produce modulations of the charge density, proportional to the topological density of the spin configuration. And so deviations $\delta\rho(\mathbf{r})$ in the charge density from its value in a fully polarised Landau level are proportional to the topological density[12] of the spin configuration. Introducing a classical, three-component unit vector field $\mathbf{S}(\mathbf{r})$ to represent the spin texture, one has[11]

$$\delta\rho(\mathbf{r}) = (8\pi)^{-1}\epsilon_{ij}\epsilon_{\alpha\beta\gamma}S^\alpha\partial_i S^\beta\partial_j S^\gamma\,. \tag{1}$$

A celebrated consequence of this is that there are skyrmion excitations of the system, in which the spin configuration wraps real space once around the spin sphere. They carry the charge of an electron but have a spin much larger the 1/2, and their exchange energy is finite and independent of their size. In contrast, a spin configuration that does not have quantised charge cannot approach uniform ferromagnetic order at spatial infinity and costs divergent exchange energy.

Returning to the competition between exchange and disorder in QHFMs, with this background in mind two alternative scenarios are clear. In Stoner theory, the transition from ferromagnet to paramagnet occurs via a reduction everywhere in the magnitude of the local moment. But there's an alternative, which is to disorder the directions of the local moments, without reducing their magnitude to zero: this is favoured within the lowest Landau level, if disorder is smooth on the scale of the magnetic length.[2] In that regime, the system can be described using a classical Hamiltonian $\mathcal{H}$ with degrees of freedom $\mathbf{S}(\mathbf{r})$ and three contributions to the energy: ferromagnetic exchange of strength J, a coupling between charge density and the impurity potential $V(\mathbf{r})$, and a Hartree interaction of the charge density with itself, taken for simplicity to be local, with coefficient U_0.[2,3]

$$\mathcal{H} = \int \left(\frac{J}{2}|\nabla\vec{S}(\mathbf{r})|^2 + V(\mathbf{r})\delta\rho(\mathbf{r}) + \frac{U_0}{2}[\delta\rho(\mathbf{r})]^2\right) d^2\mathbf{r}\,. \tag{2}$$

Assuming a Gaussian distribution for $V(\mathbf{r})$ with amplitude Δ and correlation length λ, the model is characterised by two energy scales, J and Δ, and two length scales, λ and $L_{\rm H} \equiv (U_0/J)^{1/2}$. The significance of the last of these is illustrated by the fact that for a skyrmion spin configuration of size much smaller than $L_{\rm H}$, Hartree energy is dominant, while in the opposite size limit it is exchange energy that dominates.

A central aim is to understand the phase diagram of this model, as a function of disorder strength and $\langle\delta\rho(\mathbf{r})\rangle$, the deviation of average change density from $\nu = 1$, concentrating on zero temperature since we expect thermal excitations to destroy order. From this perspective, we are concerned with an example of a magnet with quenched randomness. Its symmetries are unusual: the impurity potential breaks time-reversal symmetry (the only relic of the magnetic field), but global spin rotation invariance is intact. In this sense, the model is different from one with random

Zeeman energies, which break both time-reversal and spin-rotation symmetry, and also different from one with random exchange, which breaks neither symmetry.

A complication in treating the impurity potential is that disorder below a certain threshold has no effect at all. The reason for this is that there is a lower bound on the magnitude of the charge density, which is proportional to the exchange energy density: $|\delta\rho(\mathbf{r})| \leq (8\pi)^{-1}|\nabla \mathbf{S}(\mathbf{r})|^2$. As a result, if $|V(r)| < 4\pi J$ everywhere, it doesn't pay to depart from perfect ferromagnetic order. An analysis at small $V(\mathbf{r})$ is therefore not useful.

Instead we present simple arguments giving behaviour in the regime $\lambda \ll L_{\mathrm{H}}$, and supplement these with results from simulations. For $|V(r)| \gg 4\pi J$, a charge distribution is induced that screens the impurity potential. At short distances, by neglecting exchange, one expects $\delta\rho(\mathbf{r}) \simeq -V(\mathbf{r})/U_0$. Corrections to such screening arise at and beyond the scale L_{H}, where exchange is important. We can summarise the effect of exchange by dividing the system into regions of area L_{H}^2, finding for each such area the integral $Q \equiv -\int V(\mathbf{r})/U_0\, d^2\mathbf{r}$, and adjusting the total screening charge within every region to the integer value closest to Q. We argue that these integers are predominantly zero in a ferromagnetic phase, and predominantly non-zero in a phase without ferromagnetic order. To see this, consider a well-ordered ferromagnetic phase, in which $S(\mathbf{r})$ as a function of $\mathbf{r}$ has small fluctuations around a global direction of magnetisation. In this case, the net topological charge in each region has magnitude much less than one. Conversely, in a phase without such order, unit topological charge accumulates in a region of linear size given by the ferromagnetic correlation length. The phase boundary for the ferromagnet is therefore located by setting $\langle Q^2\rangle^{1/2} \sim 1$. Since each area of size λ^2 contributes to Q a charge of magnitude $\lambda^2\Delta/U_0$ with random sign, $\langle Q^2\rangle^{1/2} \sim \lambda L_{\mathrm{H}}\Delta/U_0$ yielding $\Delta_c \sim U_0/(\lambda L_{\mathrm{H}}) = J(L_{\mathrm{H}}/\lambda)$. For $\Delta > \Delta_c$ the ground state has no ferromagnetic order; because within a classical zero-temperature description spins are frozen, we identify this phase as a spin glass.

Consider next the ground state at fixed $\Delta < \Delta_c$, as a function of $\langle\delta\rho\rangle$. Charge is introduced into the system for $\langle\delta\rho\rangle > 0$ as skyrmions of size R. In the presence of Hartree energy alone, R is divergent, but fluctuations of an impurity potential establish an optimal size, by generating random potential wells. For $R \gg \lambda$ these have an average depth $\Delta(\lambda/R)$ (the case $R \lesssim \lambda$ is treated in Ref. 7). The contributions to the skyrmion energy from these two sources are of order U_0/R^2 and $-\Delta\lambda/R$, respectively; minimising their sum, we find $R \sim U_0/(\Delta\lambda) = L_{\mathrm{H}}(\Delta_c/\Delta)$. (Note that, since $R > L_{\mathrm{H}}$ for $\Delta < \Delta_c$, screening as discussed in the previous paragraph does not alter this argument.) We expect ferromagnetic order to persist with increasing $\langle\delta\rho\rangle$ until such skyrmions overlap, so the phase boundary lies at $\langle\delta\rho\rangle_c \sim L_{\mathrm{H}}^{-2}(\Delta/\Delta_c)^2$. In this way we arrive at the schematic phase diagram shown in the inset to Fig. 1.

In order to test these arguments, we have studied ground states of a lattice model numerically.[2,3] In outline, we replace Eq. (2) with a Heisenberg model on a square lattice which has nearest-neighbour ferromagnetic interactions of strength J and a topological charge defined on each elementary plaquette. These charges contribute

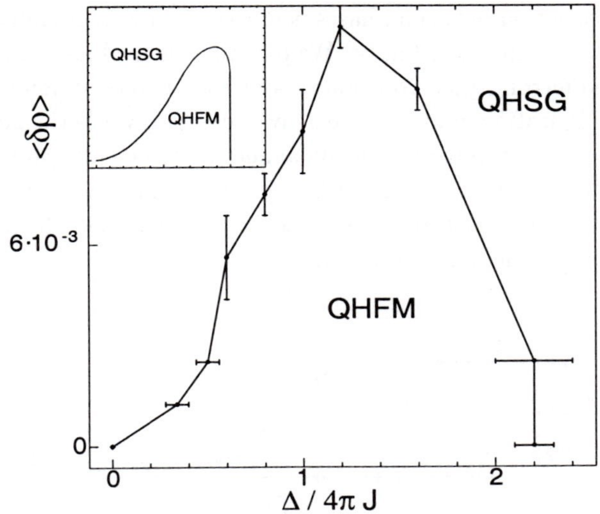

Fig. 1. Phase diagram for $L_{\mathrm{H}} = \sqrt{4\pi}\lambda$, obtained from simulations. Inset: phase diagram for $L_{\mathrm{H}} \gg \lambda$, obtained from analytical arguments described in the text.

to the total energy by a local Hartree interaction with coefficient U_0, and by their interaction with plaquette potentials, independently and uniformly distributed in $[-\Delta, \Delta]$. Thus $\lambda = 1$ in units of the lattice spacing. We find low-energy states using a simulated annealing procedure similar to that of Ref. 13, and can identify states that are equivalent under a global rotation by comparing total energies and also charge distributions. Using system sizes up to 56^2 spins and annealing with 10^6 Monte Carlo steps per spin, for a given disorder realisation in successive runs we repeatedly reach states from a small set: we take lowest of these in energy to be the ground state. In the ground state for each disorder realisation we calculate the site-averaged magnetisation, $M = |\langle \vec{S}(\mathbf{r})\rangle|$. We also determine: the ground-state susceptibility χ, from the response to a Zeeman field applied in the direction of the magnetisation; the ground-state spin stiffness ρ_{S}, from the energy cost of long-wavelength twists imposed on the spin configuration; and the wavevector-dependent compressibility, from the linear response of $\delta\rho(\mathbf{r})$ to a periodic potential added to $V(\mathbf{r})$. Further details of our methods are described elsewhere.[3]

Representative results are shown in Fig. 2. Taking $\langle\delta\rho\rangle = 0$ and $U_0 = 8\pi J$, we present M, χ and ρ_{S} as a function of disorder strength, $\Delta/4\pi J$, using 40^2 spins and an average over three disorder realisations. Three regimes are evident from the behaviour of M. For $\Delta/4\pi J < 1$, the ground state is a fully-polarised ferromagnet, because we use a bounded disorder distribution. For $1 < \Delta/4\pi J \lesssim 2.5$, the ground state is a partially polarised ferromagnet (though with an almost saturated magnetisation for $1 < \Delta/4\pi J \lesssim 1.8$). And for $2.5 \lesssim \Delta/4\pi J$, the ground state is, we argue, a spin glass, with a small non-zero M arising as a finite-size effect. In support of this interpretation, a large peak in χ indicates a phase transition at $\Delta/4\pi J \approx 2.5$. Moreover, the distinction between phases is illustrated by their spin stiffness, specified in full by a 3×3 symmetric tensor. We plot the eigenvalues of this in Fig. 2. For the fully-polarised ferromagnet, rotations about the magnetisation direction do not alter the spin configuration, and so one eigenvalue is zero while the other two are degenerate For the partially-polarised ferromagnet, all three eigenvalues are non-zero, with two remaining degenerate. And at $2.5 \lesssim \Delta/4\pi J$, magnetic isotropy and finite spin stiffness in the spin glass are illustrated by the fact that all three eigenvalues are approximately degenerate and non-zero.

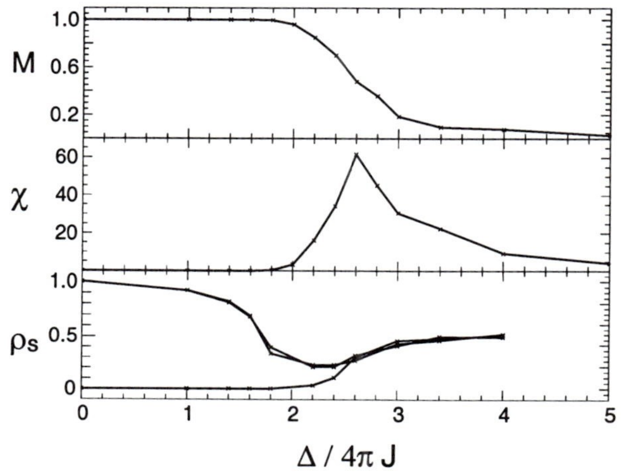

Fig. 2. Ground-state properties as a function of disorder strength: magnetisation M, susceptibility χ, and eigenvalues ρ_S of the spin stiffness tensor.

Repeating such calculations for $\langle\delta\rho\rangle > 0$, we determine the phase diagram shown for $U_0 = 4\pi J$ in the main panel of Fig. 1, which is qualitatively similar to that in the inset.

3. Random Fields

We now turn to a different kind of QHFM, in which another type of disorder is believed to be important. Here we are concerned with two Landau levels having opposite spin and different orbital quantum numbers. In general, of course, the two levels will have different energies, but one can bring them into coincidence by varying the strength of an in-plane component to the magnetic field, because that component couples only through the Zeeman energy, and doesn't change the orbital energy. Naturally, for this to work, one needs a semiconductor with a large g-factor, so we are in the opposite limit to the one considered in the previous section, but this can be achieved with an appropriate choice of sample.

A system with two Landau levels tuned to coincidence in this way, and with an electron density adjusted to fill a one level exactly, would in the absence of interactions have an enormous ground state degeneracy, arising from the different ways of distributing electrons across the two levels. Exchange lifts this degeneracy[14] and energy is minimised if the electrons are either all in one level, or all in the other. Introducing a pseudospin to describe which level is occupied, one has an Ising pseudospin ferromagnet,[15] and expects as a function of sample angle and temperature the phase diagram shown in Fig. 3. Results from low-temperature transport measurements are consistent with this.[16,17] Crucially, the angle difference from the critical value acts as a field conjugate to the order parameter in this Ising system.

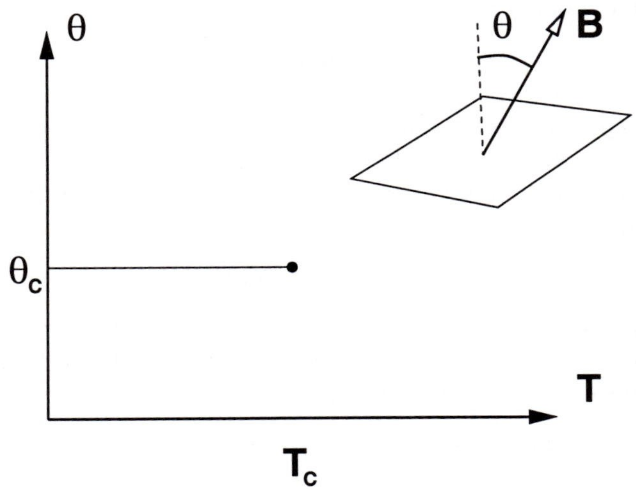

Fig. 3. Schematic phase diagram for an Ising pseudospin QHFM, as a function of temperature T and angle θ between sample normal and magnetic field direction: for $T < T_c$, as θ is varied, there is a first-order transition at θ_c. Inset: view of sample plane and field orientation.

Two striking recent experiments have focussed on transport at the critical angle, finding a large anisotropy in the diagonal components of the resistivity tensor at low temperature, according to whether current flow is parallel or perpendicular to the in-plane magnetic field component.[18,19] Since the anisotropy appears only below a characteristic temperature of about 1 K, it is natural to think that it has a cooperative origin.

An obvious comparison can be made between these Landau level coincidence experiments and the earlier discovery of resistance anisotropy in quantum Hall systems near half-filling of high Landau levels,[20] attributed to the formation of a uniaxial charge density wave with a period set by the cyclotron radius.[21] Some distinctions are, however, clear. Most importantly, the nature of the electron states near the chemical potential and their average occupation is quite different in each case:

two separate orbital Landau levels with opposite spin and a combined filling factor close to unity are involved in the coincidence experiments, as against a single, spin-polarised and roughly half-filled Landau level in the other case. Moreover, within the Hartree-Fock approximation, formation of a modulated phase is preempted by the Ising transition.[14]

An alternative line of explanation is suggested by the fact that in typical samples the plane of the two-dimensional electron gas is not completely flat.[22] Fluctuations in the angle of the sample surface correspond to random field applied to the pseudospins. A twist is that, even if surface roughness is isotropic, the random field it gives rise to will have anisotropic correlations, because the in-plane component of the field defines a preferred direction.

The system can be modelled, again introducing a classical, three-component field $\mathbf{S}(\mathbf{r})$, by the Hamiltonian

$$\mathcal{H} = \int \left(-DS_z^2(\mathbf{r}) + J|\nabla\mathbf{S}(\mathbf{r})|^2 - h(\mathbf{r})S_z(\mathbf{r})\right) d^2\mathbf{r}\,. \tag{3}$$

Here, $D > 0$ represents Ising anisotropy, J is exchange strength and $h(\mathbf{r})$ is the random field. Let $z(\mathbf{r})$ denote height of the sample surface above an average reference plane, and let θ_c be the critical angle at which the Landau level coincidence transition occurs. Then for small-angle roughness

$$h(\mathbf{r}) = \alpha(\theta - \theta_c) + \alpha\partial_n z(\mathbf{r})\,, \tag{4}$$

where α is a proportionality constant and the spatial derivative $\partial_n \equiv \hat{\mathbf{n}}\cdot\nabla$ acts in the direction of the in-plane field component, denoted by $\hat{\mathbf{n}}$. Crucially, by this mechanism surface roughness with a correlation function

$$C(\mathbf{r}) = \langle z(\mathbf{r}')z(\mathbf{r}+\mathbf{r}')\rangle - \langle z(\mathbf{r}')\rangle\langle z(\mathbf{r}+\mathbf{r}')\rangle \tag{5}$$

which is *isotropic* generates a random field with *spatially anisotropic correlations*, since

$$\langle h(\mathbf{r}')h(\mathbf{r}+\mathbf{r}')\rangle - \langle h(\mathbf{r}')\rangle\langle h(\mathbf{r}+\mathbf{r}')\rangle = -\alpha^2\partial_n^2 C(\mathbf{r})\,. \tag{6}$$

For a system at the critical angle $\theta = \theta_c$ and below the critical temperature, a random $h(\mathbf{r})$ will lead to domain formation. While even a weak random field applied to a two-dimensional Ising system is sufficient to induce domains and destroy long-range order,[23] estimates suggest[4] that the systems under discussion are in the strong field limit, in which the domain pattern is simply that of the sign of the random field, so that $S_z(\mathbf{r}) = \operatorname{sgn} h(\mathbf{r})$ almost everywhere.

Transport in such a domain state is expected to take place along domain walls.[24,25,26] To reach a simple estimate of anisotropy, we have studied[4] numerically classical drift along the domains walls generated from a spatially isotropic Gaussian height model, following Eq. 3. As implied by the form of the correlator given in Eq. 6, the domain pattern produced in this way is spatially anisotropic (with domains about three times longer in the direction perpendicular to $\hat{\mathbf{n}}$ than in the direction parallel to $\hat{\mathbf{n}}$). Classical motion along domain walls is diffusive, with a diffusion tensor that is anisotropic, and the ratio $D_{xx}/D_{yy} \sim 8$ is close to the experimentally observed anisotropy in resistivities.[18]

4. Summary

To summarise, the central feature that makes QHFMs interesting is the link that arises in them between spin textures and charge density. One consequence of this is that charged impurities can couple to the spin configuration and drive a zero-temperature transition between a ferromagnet and a spin glass. Apart from this, other types of QHFM and other types of disorder are possible. In particular, transport measurements on certain quantum Hall systems seem to probe aspects of the physics of the random field Ising model.

Acknowledgments

I am grateful to the collaborators with whom the work described in this paper was carried out: D. K. K. Lee and S. Rapsch; and A. F. Evers, A. D. Mirlin, D. G. Polyakov, and P. Wölfle. I am also grateful for support from EPSRC under Grant GR/J78327 and from the A. v. Humboldt Foundation.

References

1. For an introduction to quantum Hall ferromagnets, see: S. M. Girvin, in *Ecole des Houches: Topological Aspects of Low Dimensional Systems*, Eds. A. Comtet, T. Jolicoeur, and S. Ouvry (Les Editions de Physique, Paris, 1999).
2. S. Rapsch, J. T. Chalker and D. K. K. Lee, Phys. Rev. Lett. **88** 036801 (2002).
3. D. K. K. Lee, S. Rapsch, and J. T. Chalker, Phys. Rev. B **67**, 195322 (2003).
4. J. T. Chalker, D. G. Polyakov, F. Evers, A. D. Mirlin, and P. Wölfle, Phys. Rev. B **66**, 161317 (2002).
5. M. M. Fogler and B. I. Shklovskii, Phys. Rev. B **52** 17366 (1995); T. Ando and Y. Uemura, J. Phys. Soc. Jpn. **37**, 1044 (1973).
6. A. G. Green, Phys. Rev. B **57**, R9373 (1998); Phys. Rev. Lett. **82**, 5104 (1999).
7. A. J. Nederveen and Y. V. Nazarov, Phys. Rev. Lett. **82**, 406 (1999).
8. J. Sinova, A. H. MacDonald, and S. M. Girvin, Phys. Rev. B **62** 13579 (2000).
9. E. H. Rezayi, T. Jungwirth, A. H. MacDonald, and F. D. M. Haldane, Phys. Rev. B **67**, 201305 (2003).
10. M. A. Paalanen *et al*, Phys. Rev. B **25**, 5566 (1982); D. R. Leadley *et al*, Phys. Rev. B **58**, 13036 (1998).
11. S. L. Sondhi, A. Karlhede, S. A. Kivelson, and E. H. Rezayi, Phys. Rev. B **47**, 16419 (1993).
12. For an introduction, see: R. Rajaraman, *Solitons and Instantons* (North-Holland, Amsterdam, 1989).
13. M. Hale, O. Schwindt, and T. Weidig, Phys. Rev. E **62**, 4333 (2000).
14. G. F. Giuliani and J. J. Quinn, Phys. Rev. B **31**, 6228 (1985).
15. T. Jungwirth, S. P. Shukla, L. Smrčka, M. Shayegan, and A. H. MacDonald, Phys. Rev. Lett. **81**, 2328 (1998); T. Jungwirth and A. H. MacDonald, Phys. Rev. B **63**, 035305 (2000).
16. S. Koch, R. J. Haug, K. v. Klitzing, and M. Razeghi, Phys. Rev. B **47**, 4048 (1993).

17. A. J. Daneshvar, C. J. B. Ford, M. Y. Simmons, A. V. Khaetskii, M. Pepper, and D. A. Ritchie, Phys. Rev. Lett. **79**, 4449 (1997).
18. U. Zeitler, H. W. Schumacher, A. G. M. Jansen, and R. J. Haug, Phys. Rev. Lett. **86**, 866 (2001).
19. W. Pan, H. L. Stormer, D. C. Tsui, L. N. Pfeiffer, K. W. Baldwin, and K. W. West, Phys. Rev. B **64**, 121305 (2001).
20. M. P. Lilly, K. B. Cooper, J. P Eisenstein, L. N. Pfeiffer, and K. W. West, Phys. Rev. Lett. **82**, 394 (1999); R. R. Du, D. C. Tsui, H. L. Stormer, L. N. Pfeiffer, K. W. Baldwin, and K. W. West, Solid State Commun. **109**, 389 (1999).
21. A. A. Koulakov, M. M. Fogler, and B. I. Shklovskii, Phys. Rev. Lett. **76**, 499 (1996); M. M. Fogler, A. A. Koulakov, and B. I. Shklovskii, Phys. Rev. B **54**, 1853 (1996); R. Moessner and J. T. Chalker, Phys. Rev. B **54**, 5006 (1996).
22. C. Orme, M. D. Johnson, J. L. Sudijono, K. T. Leung, and B. G. Orr, Appl. Phys. Lett. **64**, 860 (1994); R. L. Willett, J. W. P. Hsu. D. Natelson, K. W. West, and L. N. Pfeiffer, Phys. Rev. Lett. **87**, 126803, (2001); K. B. Cooper, M. P. Lilly, J. P. Eisenstein, T. Jungwirth, L. N. Pfeiffer, and K. W. West, Solid State Commun. **119**, 89 (2001).
23. Y. Imry and S.-K. Ma, Phys. Rev. Lett. **35**, 1399 (1975); for a review, see: *Spin Glasses and Random Fields*, Ed. A. P. Young, (World Scientific, Singapore, 1997).
24. V. I. Falko and S. V. Iordanskii, Phys. Rev. Lett. **82**, 402 (1999).
25. A. Mitra and S. M. Girvin, Phys. Rev. B 67, 245311 (2003).
26. L. Brey and C. Tejedor, Phys. Rev. B **66**, 041308 (2002).

DISSIPATIVE QUANTUM DISORDERED MODELS

LETICIA F. CUGLIANDOLO
Laboratoire de Physique Théorique et Hautes Energies, Jussieu,
France and Laboratoire de Physique Théorique,
Ecole Normale Supérieure, Paris, France
leticia@lpt.ens.fr

Received 13 September 2004

This article reviews recent studies of mean-field and one dimensional quantum disordered spin systems coupled to different types of dissipative environments. The main issues discussed are: (i) The real-time dynamics in the glassy phase and how they compare to the behaviour of the same models in their classical limit. (ii) The phase transition separating the ordered – glassy – phase from the disordered phase that, for some long-range interactions, is of second order at high temperatures and of first order close to the quantum critical point (similarly to what has been observed in random dipolar magnets). (iii) The static properties of the Griffiths phase in random Ising chains. (iv) The dependence of all these properties on the environment. The analytic and numeric techniques used to derive these results are briefly mentioned.

Keywords: Quantum spin models; disorder; glassiness.

1. Introduction

Glasses slowly evolve towards equilibrium though never reaching it in observable times scales. Scientific research in this area started more than a century ago. The amount of experimental data gathered is huge. Systems that would reach equilibrium in observable time-scales, such as weakly sheared complex liquids, can be driven out of equilibrium by external perturbations and still evolve slowl.y Powders stay in static metastable states unless externally tapped or sheared: these non-equilibrium perturbations slowly drive them towards more compact configurations. Even if *a priori* very different, these systems share many dynamic properties.

In several cases of practical interest quantum effects play an important role. On the experimental side spin-glass phases have been identified in many condensed matter systems at very low temperature. Among them we can cite the bi-layer Kagome system[1] $S_rCr_sGa_4O_{19}$, the polychlore structure[2] $Li_xZn_{1-x}V_2O_4$, the dipolar magnet[3,4] $LiHo_xY_{1-x}F_4$ and the high T_c compound[5] $La_{1-x}Sr_2Cu_2O_4$. Quantum glassy phases exit also in electronic systems[6] and structural glasses[7] such as Mylar. The driven case is also very important in quantum systems, think of an electronic device driven by an external current.

The non-equilibrium dynamics of classical glasses, weakly driven complex liquids and granular matter have been rationalized within a theoretical approach that is based on the solution of mean-field simple models.[8] How much of the classical glassy phenomenology survives at very low temperatures where quantum effects are important is a question that deserves careful theoretical and experimental analysis. The impossibility of simulating the real-time evolution of quantum systems of moderate size enhances the importance of *solving* simple mean-field or low dimensional models.

On the other hand, peculiar phenomena in quantum phase transitions have been signaled analytically and experimentally in systems with and without quenched disorder.[9,10] For instance, at low temperatures and intermediate dipole concentration, the dipolar-coupled Ising magnet $LiHo_xY_{1-x}F_4$ in a transverse field exhibits a spin-glass-like phase.[3,4] The phase transition is of second order at low transverse field but becomes first order close to the quantum critical point.[4] Morever, the relaxation in the glassy phase is extremely slow and has very strong memory effects.

Another hallmark of finite dimensional disordered quantum spin models are Griffiths-McCoy singularities, that lead to a highly non trivial paramagnetic phase and critical behaviour.

In this contribution we summarize the results of recent studies of the statics, dynamics and critical properties of mean-field and one dimensional quantum disordered spin systems coupled to an environment.[11,16] We also briefly mention related studies on a driven mesoscopic ring,[17] dilute antiferromagnets,[18] and a manifold in an infinite dimensional quenched random potential.[19] In Sect. 2 we review the main questions that we attempted to answer in these papers. In Sect. 3 we recall the definition of the models that we studied. The basic techniques used to study classical glassy models with or without disorder are well documented in the literature (the replica trick, scaling arguments and droplet theories, the dynamic functional method used to derive macroscopic equations from the microscopic Langevin dynamics, functional renormalization, Montecarlo and molecular dynamic numerical methods). On the contrary, the techniques needed to deal with the statics and dynamics of quantum macroscopic systems are much less known in general. We briefly mention the ones that we used to study dissipative disordered quantum models.[20] Finally, in Sect. 4 we list some projects for future research.

2. Questions and Main Results

2.1. *Effect of quantum fluctuations on glassiness*

Since glasses are not expected to reach equilibrium in experimentally accessible times, it is important to device a method to understand the influence of quantum fluctuations on their trully nonequilibrium *real-time* dynamics. Intuitively, one expects quantum fluctuations to only affect the short-time dynamics; however, they are also expected to act as thermal fluctuations. It is then not clear *a priori* whether quantum fluctuations would tend to destroy glassiness or modify it drastically.

The usual methods of equilibrium quantum statistical mechanics are inappropriate to describe this nonstationary situation. We presented a formalism suited to study the real-time dynamics of a general nonlinear, possibly disordered, model in contact with a bath that can also be applied to glassy models.[11] The method is a combination of the Schwinger-Keldysh or closed-time path technique to study real-time phenomena, with the Feynman-Vernon approach to dissipation that consists in modelling the coupling to the environment with an ensemble of quantum harmonic oscillators. As a particular case we studied the relaxation of the spherical version of the quantum p spin fully-connected disordered model. We analyzed the relaxation of 'random initial conditions' in the limit of vanishing coupling strength taken after the long waiting-time (t_w) limit. The same technique was applied to other quantum problems[21−25] and related studies appeared.[26−30] We later studied the effect of a strong coupling to the environment[14] (*i.e.* $t_w \to \infty$ with α finite) as discussed below.

In the disordered phase the dynamics is fast and occurs in equilibrium. The correlation and linear response are stationary, *i.e.* $C(t + t_w, t_w) = C(t)$ and $R(t + t_w, t_w) = R(t)$. They both oscillate with a Γ-dependent frequency that also depends on the characteristics of the bath (α, s) if α is not taken to zero. Correlations and responses are linked by the quantum fluctuation dissipation theorem (FDT). At high temperatures and after a short transient the system decoheres and the dynamics becomes classical (*e.g.* responses and correlations are related by the classical FDT).

In the ordered phase the glassy dynamics persists asymptotically if the thermodynamic limit has been taken at the outset. The dynamics of glassy systems occurs out of equilibrium and the correlations and responses loose time translation invariance. If t_w denotes the time elapsed since a quench into the SG phase, $C(t + t_w, t_w)$ and $R(t + t_w, t_w)$ depend on both t and t_w. The order in which the limits $t_w \to \infty$ and $t \to \infty$ are taken is very important. For sufficiently long t and t_w but in the regime $t \ll t_w$, the dynamics is stationary and the correlation reaches a plateau q_{ea}. A few quantum oscillations exist at very short t (and arbitrarily large value of t_w) and they later disappear. For times $t > t_w$, the system enters an *aging* regime where the correlation function depends on t_w explicitly. In this regime, the correlation vanishes at long times, $\lim_{t\to\infty} C(t + t_w, t_w) = 0$, at a rate that depends on t_w. The behaviour is thus qualitative similar to what observed classically, even if the scaling laws are modified by the quantum fluctuations. One checks that the relaxation of typical (highly energetical) initial conditions approaches a threshold level in phase space with a higher energy-density than the equilibrium one.

The comparison of responses and correlations in the ordered phase is particularly interesting. The quantum FDT is a complicated integral relation between the correlation and the linear response. This relation holds for $t \ll t_w$ when the correlation decays to the plateau. In the second regime the quantum FDT is no longer verified, much as it happens in the classical problem. A comparison of the integrated responses and the symmetrized correlation in a parametric manner[8] shows that the two quantities are related by a *classical* FDT with an effective temperature[31] T_{eff} that depends on the parameters in the problem (T, Γ) [and the characteristics of the bath, α, s]. We proved that $T_{eff} > T$ and it is different from zero even when the environment is at $T = 0$. T_{eff} drives the dynamics at late epochs and it makes the dynamics appear classical in that two-time regime. $T_{eff} > 0$ even when the temperature of the bath is zero. The generation of a non-trivial T_{eff} for the slow part of the decay gives support to the "decoherent" effect observed in the decay of correlations.

The similarity between the second decay in the classical and quantum problem can be argued as follows. The responses decay rather fast to zero when the time difference increases (though integrated over a time-interval of very long length does not vanish.[36]) The explicit dependence on $\hbar$ in the regime of widely separated times comes from factors with higher powers of R that vanish.[11] The effect of quantum fluctuations on the slow time-difference regime is simply to renormalize[22] certain parameters in the equations that otherwise look classical.

Models with $p \geq 3$ interactions have different static and dynamic phase transitions, with the latter surrounding a larger region of the (T, Γ) phase diagram. The static and dynamic ordered-disorder phase transition present a second-to-first order transition.[53,12] Close to the classical critical temperature quantum effects are small and the phase transition is discontinuous but of second order, as in the classical case. There are no discontinuities in the thermodynamic quantities but there is a plateau that develops in the correlation function when the transition is approached from the disordered side. This is the behaviour expected in classical glasses. Spin-glasses instead have continuous transitions (without precursors). Conversely, close to the quantum critical point quantum fluctuations drive the transition first order thermodynamically. Across the first order line the susceptibility is discontinuous and shows hysteresis. This is similar to what has been observed in the dipolar-coupled Ising magnet $LiHo_xY_{1-x}F_4$ in a transverse field.[4]

We also adapted the *Ansatz* of marginal stability[12,50,53] to identify the dynamic critical line that is consistent with the one found using the Schwinger-Keldysh formalism. The analytic continuation of the imaginary-time dependent correlation computed with the AMS in the absence of the bath is identical to the *stationary* part of the non-equilibrium correlation function ($C > q_{ea}$) when one takes the long-time limit first and the limit in which the coupling to the bath goes to zero next.[12,19]

In the classical case, the study of the Thouless-Anderson-Palmer (TAP) free energy landscape has been very useful to understand the behaviour of these systems.[8] A TAP approach can also be developed for quantum problems.[13] It helps understanding the existence of a dynamic and a static critical line as well as the change in nature of the transition close to the quantum critical point.

2.2. *Effect of the bath: Decoherence and localization*

The quantum systems mentioned in Sect. 1 are not totally isolated but in contact with environments of different type.

The low-energy physics of many tunneling systems is well described by the spin-boson model in which the two equivalent degenerate states are represented by the eigenstates $\sigma_z = \pm 1$ of an Ising pseudo-spin. A transverse field coupled to σ_x (say) represents the tunneling matrix element. The coupling to the environment is given in terms of its spectral density $I(\omega) \propto \alpha\, \omega^s$ for $\omega \ll \omega_c$, where α is a dimensionless coupling constant and ω_c a high frequency cutoff. The exponent s characterizes different types of environment. The Ohmic case ($s = 1$) is quite generally encountered but superOhmic ($s > 1$) and subOhmic ($s < 1$) baths also occur in, *e.g.*, the Kondo effect in unconventional hosts.

The coupling of quantum two-level systems (TLS) to a dissipative environment has decisive effects on their dynamical properties. The dilute case, in which interactions between the TLS can be neglected, has been extensively investigated.[32,33] This problem, is related the $1d$ Ising model with inverse squared interactions and the anisotropic Kondo model. In the Ohmic case, at zero temperature, there is a phase transition[34] at $\alpha = 1$. For $\alpha < 1$ there is tunneling and two distinct regimes develop. If $\alpha < 1/2$ the system relaxes with damped coherent oscillations; in the intermediate region $1/2 < \alpha < 1$ the system relaxes incoherently. For $\alpha > 1$ quantum tunneling is suppressed and $\langle \hat{\sigma}_z \rangle \neq 0$ signalling that the system remains localized in the state in which it was prepared. These results also hold for sub-Ohmic baths while weakly damped oscillations persist for super-Ohmic baths. At finite temperatures (but low enough such that thermal activation can be neglected), there is no localization but the probability of finding the system in the state it was prepared decreases slowly with time for $\alpha > \alpha^c$.

The effect of dissipation on the phase transition, critical behaviour, ordered phase, localization and decoherence properties of macrocopic interacting systems is only now starting to be analyzed.[14,15,35] In thermodynamic equilibrium, in the absence of the bath, the interactions between the TLS lead to the appearance of an ordered state at low enough temperature. If the interactions are of random sign, as in the models we considered, the latter will be a glassy state. In this phase the symmetry between the states $\sigma_i^z = \pm 1$ at any particular site is broken but there is no global magnetization, $\sum_i \langle \hat{\sigma}_i^z \rangle = 0$. The coupling to the bath also competes with the tunneling term. We thus expect the presence of noise to increase the stability of the glassy state. The consequences of this fact are particularly interesting when there is localization at some $\alpha = \alpha^c$: a quantum critical point at $J = 0$, $\alpha = \alpha^c$ separates the disordered and the ordered state such that, for $\alpha > \alpha^c$, the glassy phase survives down to $J = 0$.

A system of non-interacting localized TLS and a glassy state *in equilibrium* are in some way similar However, this resemblance is only superficial. The details of the dynamics of the two systems are expected to be quite different, with C saturating at a finite value in the localized state, and C decaying down to zero in the glassy phase.

2.3. *Effect of the bath on interacting mean-field models*

The problem of a single TLS being a difficult one, that of an infinite set of interacting TLS seems hardly solvable. Therefore, as a first step, we focused on the effect of the reservoir on the p-spin *spherical* model,[14] a problem that we studied with the real-time approach and the replica Matsubara technique. The position of the critical lines strongly depends on the strength of the coupling to the bath and the type of bath (Ohmic, subOhmic, superOhmic). For a given type of bath, the ordered glassy phase is favored by a stronger coupling. The classical static and dynamic critical temperatures remain unchanged by the coupling to the environment. The identity between the analytic continuation of the imaginary-time correlation to real-time and the correlation in the Schwinger-Keldysh approach also holds if the strength of the bath is finite.

The spherical model localizes in the absence of interactions when coupled to a subOhmic bath: $C(t + t_w, t_w)$ reaches, for any waiting-time t_w and long enough t, a plateau that it never leaves. When interactions are switched on localization disappears and the system undergoes a phase transition towards a glassy phase.

Similar results were found for the $SU(N)$ random Heisenberg model in the limit $N \to \infty$,[23] the $p = 2$ spherical model[25] and the SU(2) p-spin model.[15]

2.4. *Effect of disorder: Griffiths singularities*

Griffiths singularities[37–39] in classical finite dimensional random systems are so weak that their consequences have not been clearly observed neither experimentally nor numerically. Instead, when quantum fluctuations are introduced they are much stronger and rare regions completely determine the static and dynamic behaviour in the Griffiths phase.

The isolated quantum random Ising chain has been studied in great detail with a decimation technique. [40] It undergoes a quantum phase transition from a PM to a ferromagnet for a special relation between the distribution of exchanges and transverse fields. The quantum phase transition is of second order with the correlation time scaling exponentially with the spatial correlation length (*activated* scaling) Within the renormalization group procedure this is a characteristic of an "infinitely strong disorder" fixed point.[40] Within the PM Griffiths phase the distributions of local linear and non-linear magnetic susceptibilities are large and typical and average values are very different, with the latter being dominated by rare regions. The behaviour is higher spatial dimensions is similar.

The analysis of Montecarlo simulations of the equivalent $d + 1$ classical Ising model is quite tricky. Initially, it was claimed that there was conventional scaling in $d = 1$[41] as well as in $d > 1$[42] but a more careful analysis of the numerical data confirmed the activated scaling in both cases.[43]

On the real-time dynamic side, there have been some studies of the relaxation of special initial conditions of the isolated random Ising chain at constant energy.[44]

2.5. *Disorder and dissipation: Fate of Griffiths phase?*

If the interactions between two-level systems placed in a finite dimensional space are random one may wonder what is the effect of the bath on the Griffiths phase. The answer to this question has been debated over the last years.[45–48] We addressed this problem using Montecarlo simulations of the equivalent $2d$ classical system. [16] Preliminary results for rather small systems ($N_x \leq 32$, $N_\tau \leq 256$) show that an Ohmic bath favors the glassy phase. Our results are compatible with (but we do not prove) activated scaling at criticality, at least for small values of α.

3. The Models

Disordered quantum spin-$\frac{1}{2}$ models with two-body interactions are defined by

$$H_S = -\sum_{ij} J_{ij}\hat{\sigma}_i^z\hat{\sigma}_j^z + \sum_i \Gamma_i\hat{\sigma}_i^x + \sum_i h_i\hat{\sigma}_i^z \ . \qquad (1)$$

$i = 1, \ldots, N$ labels the spins that lie on the vertices of a cubic d dimensional lattice and are represented by Pauli matrices. The interaction strengths J_{ij} couple near-neighbours only and are chosen from a probability distribution, $P(J)$. The average and variance are defined as $[J_{ij}] = J_o$ and $[J_{ij}^2] = J^2/(2c)$, where J_o and J are $O(1)$ and $c = 2d$ is the connectivity of the lattice. The next-to-last term is a coupling to a random quenched local transverse field Γ_i. The last term is the coupling to a longitudinal field that serves to compute local susceptibilities.

Several generalizations that render the model easier to treat analytically are:
– *Fully-connected limit.* One allows each spin to interact with all others, $c \to N-1$.
– *Multi-spin interactions.* In the fully-connected case one can considers

$$H_S = -\sum_{i_1\ldots i_p} J_{i_1\ldots i_p}\hat{\sigma}_{i_1}^z \ldots \hat{\sigma}_{i_p}^z + \sum_i \Gamma_i\hat{\sigma}_i^x + \sum_i h_i\hat{\sigma}_i^z \ . \qquad (2)$$

where the sum runs over all p-uplets,[52,53,12,15] with p an integer parameter, $p \geq 2$. The exchanges are random independent variables with variance $p!J^2/(2N^{p-1})$. This model provides a mean-field description of the structural glass transition and glassy physics that is also intimately related to the mode-coupling approach.[8] In its dilute limit, with no geometry but finite connectivity on each site, this model is related to the K-sat optimization problem. [56]
– *Spherical variables – a particle in a random potential.* One considers the spherical limit, $\sum_{i=1}^N \langle\hat{\sigma}_i^2\rangle = N$, in which the $\hat{\sigma}_i$ may be interpreted as the coordinates of a particle moving on an N-dimensional sphere. A kinetic term, $K = \sum_{i=1}^N \hat{P}_i^2/(2M)$, is then included in the Hamiltonian, with $\hat{P}_i$ the conjugated momentum satisfying the commutation rules $[\hat{P}_i, \hat{P}_j] = 0$, $[\hat{P}_i, \hat{\sigma}_j] = -i\hbar\delta_{ij}$. Other spherical models have been discussed.[49]

The coupling to the environment is modelled by $H = H_S + H_B + H_I + H_{CT}$, where H_B is the Hamiltonian of the bath, H_I represents the interaction between the system and the bath and $H_{CT} = \sum_{l=1}^{\tilde{N}} (2m_l\omega_l^2)^{-1}(\sum_{i=1}^N c_{il}\hat{\sigma}_i^z)^2$ is a counter-term that is usually added to eliminate an undesired mass renormalization induced by the coupling.[32] We assume that each spin is coupled to its own set of $\tilde{N}/N$ independent harmonic oscillators with $\tilde{N}$ the total number of them. For simplicity we consider the bilinear coupling, $H_I = -\sum_{i=1}^{N} \hat{\sigma}_i^z \sum_{l=1}^{\tilde{N}} c_{il}\hat{x}_l$. For $p = 2$ the fully-connected limit with two-body interactions models metallic spin-glasses.[51]

4. Perspectives

In this article we reviewed recent studies of insulating disordered magnets.

The principal merit of the fully-connected models is that they are simple enough to be studied in detail. Yet, many of their properties are generic and expected to hold at least qualitatively for more realistic cases.[8] The analysis of the fully-connected models is by now quite complete, having applied the replica theory, the real-time dynamics approach, and the investigation of the TAP free-energy landscape. A problem that remains not fully developed though is the treatment of the relaxation of initial conditions that are correlated with disorder.[55]

Recently, much progress has been done in the study of the statics of classical dilute spin models, that is to say, models defined on random hyper graphs with finite connectivity. The statics of these models encode problems in combinatorial opimization such as K-sat.[56] An interesting mapping relates (isolated) dilute quantum disordered spin systems to the dynamics of special purpose algorithms used in combinatorial optimization – such as Walk-Sat.[57] It would be interesting to study the latter using tools developed for the former and *vise versa*.

The great challenge remains to understand the behaviour of glassy systems with and without disorder in finite dimensions. In particular, one could try to adapt the decimation technique[40] to study disordered spin$-\frac{1}{2}$ models in contact with an environment. One can also envisage applications of the ideas described in this paper to other quantum systems evolving out of equilibrium. In this respect, we have studied the generation of an effective temperature in a conducting ring threaded by time-dependent magnetic field and coupled to a reservoir;[17] and we plan to analyse quantum dilute antiferromagnets[18] and the low-temperature dynamics of the Bragg glass,[19] as well as other related physical problems.

Acknowledgments

This contribution summarizes work done in collaboration with G. Biroli, D. Grempel, G. Lozano, H. Lozza and C. da Silva Santos. I also wish to thank L. Arrachea, C. Chamon, T. Giamarchi, L. Ioffe, M. Kennett, J. Kurchan, P. Le Doussal, R. Monasson, M. J. Rozenberg and G. Semerjian for collaborations on related problems. I acknowledge financial support from an Ecos-Sud travel grant, an ACI project, the J. S. Guggenheim Foundation and ICTP-Trieste, as well as hospitality from the Universidad de Buenos Aires and Universidad Nacional de La Plata, ICTP, and the KITP.

References

1. Y. J. Uemura *et al*, Phys. Rev. Lett. **73**, 3306 (1994).
2. C. Urano *et al*, Phys. Rev. Lett. **85**, 1032 (2002).
3. D. H. Reich *et al*, Phys. Rev. B **42**, 4631 (1990). W. Wu *et al*, Phys. Rev. Lett. **67** 2076 (1991). T. F. Rosenbaum, J. Phys: Condens. Matter **8**, 9759 (1996). J. Brooke *et al*, Science **284**, 779 (1999).
4. W. Wu *et al*, Phys. Rev. Lett. **71** 1919 (1993).
5. A. Aharony *et al*, Phys. Rev. Lett. **60**, 1330 (1988). F. C. Chou *et al*, Phys. Rev. Lett. **75**, 2204 (1995). M. A. Kastner *et al*, Rev. Mod. Phys. **70**, 897 (1998). E. Dagotto, Rev. Mod. Phys. **66**, 763 (1994). S. Wakimoto *et al*, Phys. Rev. B **62**, 3547 (2000). A. N. Lavrov *et al*, Phys. Rev. Lett. **87**, 017007 (2001).
6. M. Ben-Chorin, Z. Ovadyahu, and M. Pollak, Phys. Rev. B **48**, 15025 (1993). A. Vaknin, Z. Ovadyahu, and M. Pollak, Phys. Rev. Lett. **84**, 3402 (2000). D. R. Grempel, Europhys. Lett. **66**, 854 (2004). A. Kolton, D. R. Grempel, and D. Domínguez, cond-mat/0407734.
7. F. Ladieu *et al*, Phys. Rev. Lett. **90**, 205501 (2003). S. Ludwig and D. D. Osheroff, Phys. Rev. Lett. **91**, 105501 (2003). S. Rogge, D. Natelson, and D. D. Osheroff, Phys. Rev. Lett. **76**, 3136 (1996).
8. L. F. Cugliandolo, in *Slow relaxation and non-equilibrium dynamics in condensed matter*, Les Houches Session 77, J-L Barrat *et al* eds. (Springer-Verlag, 2003), cond-mat/0210312.
9. S. Sachdev, *Quantum phase transitions*, (Cambridge University Press, Cambridge, 1999).
10. T. Senthil *el al*, Science **303**, 1490 (2004). T. Senthil *el al*, cond-mat/0312617. T. Senthil *et al*, cond-mat/0404718. T. Senthil, in this volume.
11. L. F. Cugliandolo and G. Lozano, Phys. Rev. Lett. **80**, 4979 (1998), Phys. Rev. B **59**, 915 (1999).
12. L. F. Cugliandolo, D. R. Grempel, and C. A. da Silva Santos, Phys. Rev. Lett. **85**, 2589 (2000), Phys. Rev. B **64**, 014403 (2001).
13. G. Biroli, and L. F. Cugliandolo, Phys. Rev. B **64**, 014206 (2001).
14. L. F. Cugliandolo, D. R. Grempel, G. Lozano, H. Lozza, C. A. da Silva Santos, Phys. Rev. B **66**, 014444 (2002).
15. L. F. Cugliandolo, D. R. Grempel, G. Lozano and H. Lozza, Phys. Rev. B **70**, 024422 (2004).
16. L. F. Cugliandolo, D. R. Grempel, G. Lozano, and H. Lozza (in preparation).
17. L. Arrachea and L. F. Cugliandolo, cond-mat/0407427.
18. M. P. Kennett, C. Chamon, and L. F. Cugliandolo (unpublished).
19. L. F. Cugliandolo, T. Giamarchi, and P. Le Doussal (unpublisehd).
20. See also L. F. Cugliandolo, Lectures notes in Boulder Summer School, available at www.lpt.ens.fr/leticia/papers.html
21. D. M. Kagan, M. V. Feigel'man, L. B. Ioffe, J. Exp. Theor. Phys. **89**, 781 (1999).
22. M. P. Kennett and C. Chamon, Phys. Rev. Lett **86**, 1622 (2001). M. P. Kennett, C. Chamon and J. W. Ye, Phys. Rev. B **64** 224408 (2001).
23. G. Biroli and O. Parcollet, Phys. Rev. B **65**, 094414 (2002).
24. H. Westfahl, J. Schmalian, and P. G. Wolynes, Phys. Rev. B **68** 134203 (2003).
25. M. Rokhni, and P. Chandra, Phys. Rev. B **69** 094403 (2004).
26. N. Pottier, and A. Mauger, Physica A **282** 77 (2000).
27. N. V. Prokof'ev and P. C. E. Stamp, Rep. Prog. Phys. **63**, 669 (2000).
28. E. Rabani and D. R. Reichman, J. Chem. Phys. **120** 1458 (2004).
29. G. Busiello, R. V. Saburova, and V. G. Sushkova, Eur. Phys. J B **39**, 69 (2004).

30. V. M. Galitski and A. I. Larkin, Phys. Rev. B **66**, 064526 (2002).
31. L. F. Cugliandolo, J. Kurchan, and L. Peliti, Phys. Rev. E **55**, 3898 (1997).
32. U. Weiss in *Series Modern Condensed Matter Physics* Vol. 2, (World Scientific, Sigapore, 1993). R. P. Feynman and Vernon, Jr. Ann. Phys. (NY), **24**, 118 (1963); R. P. Feynman, *Statistical mechanics* (Addison Wesley, 1972). A. Caldeira and A. Leggett, Ann Phys. **149**, 374 (1983).
33. A. J. Leggett *et al*, Rev. Mod. Phys. **59**, 1 (1987), *ibid* **67**, 725 (1995).
34. S. Chakravarty, Phys. Rev. Lett. **49**, 681 (1982). A. J. Bray and M. A. Moore, Phys. Rev. Lett. **49**, 1545 (1982).
35. A. Sandvik, Phys. Rev. E **68**, 56701 (2003). P. Werner, K. Voelker, M. Troyer and S. Chakravarty, cond-mat/0402224.
36. L. F. Cugliandolo and J. Kurchan, Phys. Rev. Lett. **71**, 173 (1993). J. Phys. A **27** 5749 (1994). Phil. Mag. B **71** 501 (1995).
37. R. B. Griffiths, Phys. Rev. Lett. **23**, 17 (1969).
38. A. B. Harris, Phys. Rev. B **12**, 203 (1975).
39. B. M. McCoy, Phys. Rev. Lett. **23**, 383 (1969). Phys. Rev. **188**, 1014 (1969).
40. D. S. Fisher, Phys. Rev. Lett. **69**, 534 (1992), Phys. Rev. **B51**, 6411 (1995).
41. A. Crisanti and H. Rieger, J. Stat. Phys. **77**, 1087 (1994).
42. H. Rieger and A. P. Young, Phys. Rev. Lett. **72**, 4141 (1994), Phys. Rev. B **54** 3328 (1996). M. Guo, R. Bhatt and D. Huse, Phys. Rev. Lett. **72**, 4137 (1994). Phys. Rev. B **54** 3336 (1996).
43. H. Rieger and A. P. Young, Phys. Rev. Lett. **53**, 8486 (1996). C. Pich *et al*, Phys. Rev. Lett. **81**, 5916 (1998).
44. G. M. Schütz and M. Trimper, Europhys. Lett. **47** , 164 (1999). F. Igloi and H. Rieger, Phys. Rev. Lett. **85**, 3233 (2000). S. Abriet and D. Karevski, Eur. Phys. J. B (in press) D. Karevski, Eur. Phys. J. B **27**, 147 (2002). G. O. Berim, S. I. Berim, and G. G. Cabrera, Phys. Rev. B **66**, 094401 (2002).
45. A. H. Castro-Neto and B. A. Jones, Phys. Rev. B **62**, 14975 (2000). E. Novais *et al*, Phys. Rev. Lett. **88**, 217201 (2002), Phys. Rev. B **66**, 174409 (2002).
46. A. J. Millis, D. Morr and J. Schmalian, Phys. Rev. Lett. **87**, 167202 (2001), Phys. Rev. B **66**, 174433 (2002).
47. T. Vojta, Phys. Rev. Lett. **90**, 107202 (2003).
48. V. Dobrosaljevic and E. Miranda, cond-mat/0408336.
49. R. Serral Gracia and Th.M. Nieuwenhuizen, Phys. Rev. E **69** 056119 (2004).
50. T. Giamarchi and P. Le Doussal, Phys. Rev. B **53**, 15206 (1996). O. Parcollet, A. Georges and S. Sachdev, Phys. Rev B **63** 134406 (2001). G. Schehr, T. Giamarchi and P. Le Doussal, J. de Physique IV **12** (PR9) 311 (2002).
51. D. R. Grempel and M. J. Rozenberg, Phys. Rev. B60, 4702 (1999).
52. Y. Y. Goldschmidt and P-Y Lai, Phys. Rev. Lett. **64**, 2467 (1990). P-Y Lai and Y. Y. Goldschmidt, Europhys. Lett. **13**, 289 (1990).
53. T. M. Nieuewenhuizen and F. Ritort, Physica A **250**, 89 (1998).
54. L. F. Cugliandolo and J. Kurchan, J. Phys. Soc. Japan **69**, 247 (2000).
55. L. F. Cugliandolo, D. R. Grempel, and C. A. da Silva Santos (unpublised).
56. R. Zecchina, in this volume.
57. G. Semerjian and R. Monasson, Phys. Rev. E **67**, 066103 (2003) and references therein.

POSSIBLY EXACT SOLUTION FOR THE MULTICRITICAL POINT OF FINITE-DIMENSIONAL SPIN GLASSES

HIDETOSHI NISHIMORI
Department of Physics, Tokyo Institute of Technology, Oh-okayama Meguro-ku, Tokyo 152-8551, Japan
nishimori@phys.titech.ac.jp

KOUJIN TAKEDA
Department of Physics, Tokyo Institute of Technology, Oh-okayama Meguro-ku, Tokyo 152-8551, Japan
takeda@stat.phys.titech.ac.jp

TOMOHIRO SASAMOTO
Department of Physics, Tokyo Institute of Technology, Oh-okayama Meguro-ku, Tokyo 152-8551, Japan
sasamoto@stat.phys.titech.ac.jp

Received 1 November 2004

After briefly describing the present status of the spin glass theory, we present a conjecture on the exact location of the multicritical point in the phase diagram of finite-dimensional spin glasses. The theory enables us to understand in a unified way many numerical results for two-, three- and four-dimensional models including the $\pm J$ Ising model, random Potts model, random lattice gauge theory, and random Z_q model. It is also suggested from the same theoretical framework that models with symmetric distribution of randomness in exchange interaction have no finite-temperature transition on the square lattice.

Keywords: Spin glass; phase diagram; multicritical point; duality; gauge symmetry.

1. Introduction

Spin glasses are magnetic materials in which spins are randomly frozen. Theoretical studies of spin glasses started with the paper by Edwards and Anderson in 1975[1] in which they proposed to model the problem in terms of the following Hamiltonian, now termed the Edwards-Anderson model,

$$H = -\sum_{\langle i,j\rangle} J_{ij} S_i S_j. \quad (1)$$

In the present contribution we will mainly discuss the Ising model ($S_i = \pm 1$) with binary ($J_{ij} = \pm J$) or Gaussian quenched randomness in exchange interactions. The infinite range version of this model was proposed and analyzed by Sherrington and Kirkpatrick soon afterwards using the replica method[2] . These two papers set the landmark of the spin glass theory, and a number of facts are now known, and yet many problems still remain to be solved, particularly in finite dimensions.

Two central issues concerning the static properties of spin glasses are, first, whether or not a spin glass phase exists under a given condition and, second, what characteristics the spin glass phase has if it exists. A successful mean-field theory has been constructed with replica symmetry breaking[3] that solved both of these problems. A spin glass phase, in which spins are randomly frozen, does exist at low temperatures for J_0 (the center of distribution of Gaussian exchange interaction values) close to 0. Ferromagnetic and paramagnetic phases also exist in appropriate parameter regions. In the mean-field model the spin glass phase shows peculiar behavior that the free energy has infinitely many minima, *i.e.* infinitely many stable states, quite differently from the usual ferromagnet which has essentially a unique minimum.

It is important to investigate whether or not these mean-field predictions apply to realistic finite-dimensional systems. Efforts toward this direction have so far clarified that a spin glass phase is very likely to exist in three and higher dimensions for the random Ising system as depicted in Fig. 1. Still under debate are the

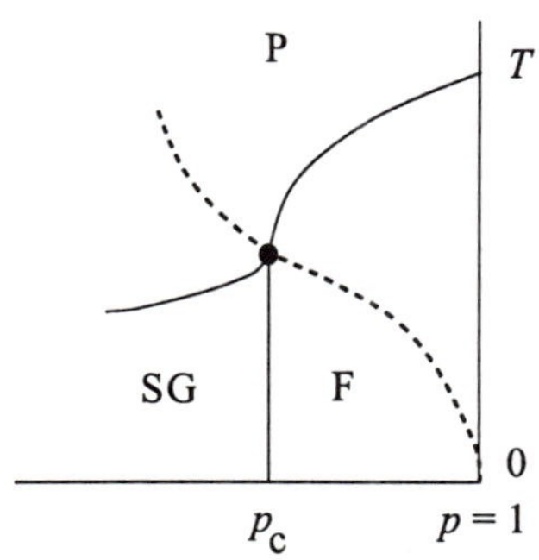

Fig. 1. A typical phase diagram of the finite-dimensional $\pm J$ Ising model, in which the interaction is ferromagnetic with probability p and antiferromagnetic with $1 - p$. The multicritical point is marked as a black circle. The Nishimori line is shown dashed.

nature of the spin glass phase, whether or not it has a very complex mean-field-like structure, and how Heisenberg-like isotropic spin systems show spin glass behavior in experiments.

A serious problem in the research activities on finite-dimensional cases, which is the main topic of the present contribution, is that mathematical analysis is extremely difficult and most of the established (and still debated) facts come from numerical studies. This is in marked contrast with the mean-field theory, where a very sophisticated scheme of replica symmetry breaking and related methods serve as powerful analytical tools.

We were therefore surprised to have found that the well-known technique of duality transformation, in conjunction with arguments using gauge symmetry, enables us to predict the exact (correctly speaking, possibly exact) location of the multicritical point, where the ferromagnetic phase ceases to exist in the phase diagram (see Fig. 1), not only in the two-dimensional $\pm J$ Ising spin glass but also in other systems in two, three, and four dimensions. Our line of reasoning is not completely rigours mathematically, and the results remain to be a conjecture at this moment. We nevertheless believe that a formal proof will come before too long, in which case the present results may serve as an important step toward a comprehensive analytical solution of the spin glass problem in finite dimensions. The following sections describe our results and a summary of the steps to reach them.

2. Phase Diagram in Finite Dimensions: Known Facts

A typical phase diagram is drawn in Fig. 1. The central topic in the present contribution is the identification of the exact location of the multicritical point marked by a black circle. Before presenting the results, let us summarize what has already been known about the phase diagram of finite-dimensional spin glasses, in particular the $\pm J$ Ising model, mainly by symmetry arguments.[4]

An important fact is that the exact value of the energy, averaged over quenched randomness, can be evaluated with the result $E = -N_B J \tanh K$, where N_B is the number of bonds in the system and $K = \beta J$. This formula applies as long as the temperature and probability are related as $e^{-2K} = (1-p)/p$, which defines a line in the T-p phase diagram, called the Nishimori line (NL) shown dashed in Fig. 1. A remarkable aspect is that the exact expression given above is valid for any system with any range of interaction on any lattice, which is shared by all the following results in this section. It should be noted that the energy $-N_B J \tanh K$ has no singularity although the NL clearly crosses the boundary between ferromagnetic and paramagnetic phases; indeed it is believed that the crossing is at the multicritical point. Thus we can focus our attention to the NL to solve the problem of identification of the exact location of the multicritical point. Only the energy is non-singular along the NL, and other physical quantities including the free energy, specific heat and magnetic susceptibility are of course singular at the multicritical point even along the NL.

It can also be shown that the ferromagnetic order parameter m is exactly equal to the spin glass order parameter q on the NL ($m = q$) under quite general conditions. Since the spin glass phase has finite $q > 0$ and vanishing $m = 0$, it immediately follows that the spin glass phase, if it exists, cannot have the NL in it, and is most likely to lie below it as displayed in Fig. 1. This is also remarkable because this rigorous constraint has been shown to hold for the possible region where such a phase is allowed to exist when we do not know for sure by analytical methods if a spin glass phase exists in finite dimensions.

It has been proved that there is no replica symmetry breaking on the NL in the sense that the distribution function of the spin glass order parameter is equal to that of the ferromagnetic counterpart, $P(q) = P(m)$, the latter distribution function being of quite simple trivial structure. Thus the mean-field-like complex phase with non-trivial $P(q)$ can exist, if at all, only away from the NL, again most plausibly below it.

These and many other facts, derived in a mathematically rigorous way without recourse to the replica method but using only gauge symmetry of the system, constitute the almost only set of exact/rigorous results for finite-dimensional spin glass models.[4] The outstanding generality of the theory, *i.e.* its applicability to any lattice, any dimension or any range of interaction, is its strength and, simultaneously, its weakness. The reason for weakness is that the theory is unable to predict lattice-specific properties such as the location of the multicritical point.

Developments in the last few years[5,6,7,8] have completely changed the situation, and we are now able to predict the exact location of the multicritical point for self-dual lattices in finite dimensions. For non-self-dual lattice pairs such as the triangular and hexagonal lattices, our duality argument allows us to relate the locations of multicritical points of two mutually dual systems in a compact formula. We will also develop an argument that the spin glass with symmetric randomness distribution has no finite temperature transition on the square lattice.

3. Results

As has been mentioned, our results have the status of conjecture, not rigorously proved facts. To convince the reader that our conjecture is most likely to be correct, it should be helpful to show the final values that are compared quite favorably with numerical simulation results as in Table 1. A remarkable fact to bear in mind is that

Table 1. Location of multicritical point by numerical studies and our prediction. SQ stands for the square lattice, and TR/HEX for the triangular/hexagonal lattices, respectively. The values are for p_c of the $\pm J$ model except for the Gaussian randomness for which the value is for J_{0c}/J. Spin variables are Ising excepting the three-state Potts model as indicated. Our analysis gives definite values for self-dual systems for which explicit numbers are given in the third column whereas, for mutually dual pairs such as the triangular and hexagonal lattices, the prediction is that the two values are related in a simple formula $H(p_{c1}) + H(p_{c2}) = 1$, where $H(p)$ is the binary entropy $-p\log_2 p - (1-p)\log_2(1-p)$.

Model	Typical numerical result	Our prediction
SQ Ising	0.8894(9)[9]	0.889972
SQ Gaussian	1.00(2)[10]	1.021777
SQ 3-Potts	0.921(1)[11]	0.920269
4*d* lattice gauge	0.890(2)[12]	0.889972
TR	0.835(5)($= p_{c1}$)[8]	—
HEX	0.930(5)($= p_{c2}$)[8]	—
TR + HEX	$H(p_{c1}) + H(p_{c2}) = 1.01(3)$	$H(p_{c1}) + H(p_{c2}) = 1$
3*d* Ising (RBIM)	0.7673(3)($= p_{c1}$)[13]	—
3*d* gauge (RPGM)	0.967(4)($= p_{c2}$)[14]	—
RBIM+RPGM	$H(p_{c1}) + H(p_{c2}) = 0.99(2)$	$H(p_{c1}) + H(p_{c2}) = 1$

the numbers given in the second column of Table 1 have come out of independent numerical investigations whereas the third column has been filled in by a single unified theoretical framework, an essence of which is to be explained shortly. We therefore have strong confidence that the agreement observed in this Table between numerics and analysis is not due to accidental coincidence.

The following are the explicit formulas for the above prediction. For self-dual lattices, the critical concentration p_c for the $\pm J$ Ising model satisfies $2H(p_c) = 1$ irrespective of the lattice structure. This equation is solved to give $p_c = 0.889972$. If the lattice is not self dual, the pair of mutually dual lattices are related by the formula $H(p_{c1}) + H(p_{c2}) = 1$. In the Gaussian case on self-dual lattices, the critical value of J_0 satisfies

$$2\int_{-\infty}^{\infty} du\, P(u) \log_2(1 + e^{-2J_0 u/J^2}) = 1, \tag{2}$$

where $P(u)$ is the Gaussian distribution. The formula for the q-state Potts model is

$$2\{-(1-(q-1)p_c)\log_q(1-(q-1)p_c) - (q-1)p_c \log_q p_c\} = 1. \tag{3}$$

All of these results have been derived using duality tranformation combined with gauge symmetry.

4. Steps to the Results: Duality and Gauge Symmetry

Let us next explain our basic ideas for the Ising system on a self-dual lattice. Details are found in Refs. 6, 7, 8. A preliminary report based on a more naive argument, but leading to the same conclusion, appeared in Ref. 5.

The well-known method to identify the critical point of the ferromagnetic Ising model on a self-dual lattice uses invariance of the partition function under duality transformation of the temperature, $Z(T) = Z(T^*(T))$, where $T^*(T)$ is the dual temperature of the original T and is a monotone decreasing function of T. If the critical point is unique, it should be shared by both sides of $Z(T) = Z(T^*(T))$, and hence the fixed-point condition of duality transformation $T = T^*(T)$ yields the location of the critical point. Since the partition function can also be regarded as a function of the local Boltzmann factors $x_0(= e^K)$ (for parallel spin pair) and $x_1(= e^{-K})$ (for antiparallel pair), the above duality expression is also written as

$$Z(x_0, x_1) = Z(x_0^*, x_1^*). \tag{4}$$

The fixed-point condition $T = T^*(T)$ is equivalent to $x_0 = x_0^*$ as well as to $x_1 = x_1^*$.

A straightforward application of the same idea as above to systems with quenched randomness with positive and negative interactions leads to imaginary couplings in the dual system. To avoid such a difficulty, it is convenient to introduce replicas and average the partition function, replicated n times, over the randomness, which yields the translationally invariant effective partition function Z_n. The goal is to analyze the system properties in the limit $n \to 0$, but we consider integer-n

cases for the moment. The replicated partition function Z_n is completely specified by giving the values of local Boltzmann factors because the system is translationally invariant after configurational average. Since each site has n spins, the number of values of local Boltzmann factors is $n+1$, which we denote as $x_0, x_1, \cdots, x_n$. The leading factor x_0 denotes the Boltzmann weight for all-parallel spin configuration between neighboring sites (see Fig. 2), x_1 is for the case with single antiparallel pair with the remaining being parallel, and so on. The partition function is expressed

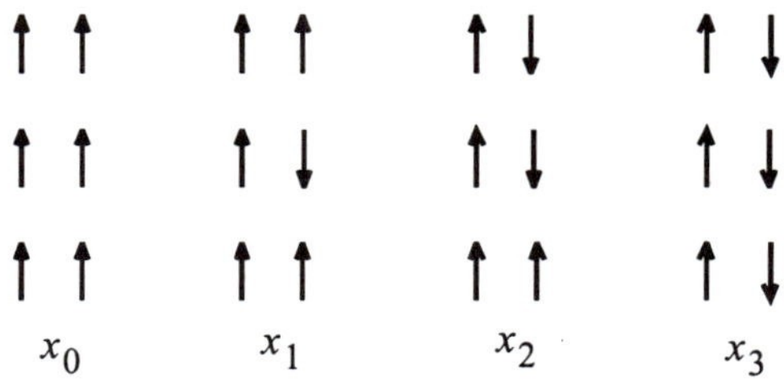

Fig. 2. Possible relative spin configurations between neighboring sites for the $n = 3$ replicated system. x_k stands for the Boltzmann factor for the case with k antiparallel pairs. x_1 and x_2 have three configurations each, and one of them are displayed here for each of them.

in terms of these Boltzmann factors as $Z_n(x_0, x_1, \cdots, x_n)$. It is possible to apply duality transformation to such a case, and the result is the relation

$$Z_n(x_0, x_1, \cdots, x_n) = Z_n(x_0^*, x_1^*, \cdots, x_n^*). \tag{5}$$

The dual Boltzmann factors $x_0^*, x_1^*, \cdots, x_n^*$ are linear functions of the original $x_0, x_1, \cdots, x_n$. Since the original problem has two parameters p and K, all the Boltzmann factors (original and dual) are functions of these two variables.

Unlike the ferromagnetic model, a naive argument using a fixed-point condition as described above does not work in the present case because it is in general impossible to fix all the Boltzmann factors simultaneously, $x_0 = x_0^*, x_1 = x_1^*, \cdots, x_n = x_n^*$. The reason is that these Boltzmann factors are all functions of p and K, and the number of simultaneous equations $x_0(p,K) = x_0^*(p,K), x_1(p,K) = x_1^*(p,K), \cdots, x_n(p,K) = x_n^*(p,K)$ is in general more than sufficient to be solvable for p and K.

We nevertheless proceed further and try the ansatz that the fixed-point condition of the leading Boltzmann factor $x_0 = x_0^*$ may lead to the correct location of transition point if we restrict ourselves to the NL. The results presented in Table 1 were derived this way. We have several reasons to choose the combination of $x_0 = x_0^*$ and NL in our effort to locate the multicritical point, an important one being that the fixed-point conditions of the other variables, $x_k = x_k^*$ $(k \neq 0)$, lead to inconsistency; they violate a rigorous inequality on the location of the multicritical point and, also, the quenched limit of $x_k = x_k^*$ (plus NL) does not have a meaningful solution unless $k = 0$.

Our conjecture that $x_0 = x_0^*$ plus NL gives the exact location of the multicritical point has been confirmed to be rigorously justified explicitly for $n = 1, 2$ and ∞. It has also been shown by numerical simulations that the same is very likely to be true in the case of $n = 3$. As mentioned in the previous section, numerical evidence for the quenched limit $n \to 0$ has also been accumulated extensively.

5. Generalizations

It is possible to generalize the idea described in the previous section for the simple self-dual Ising model to a variety of other systems.

If the model is not self dual, as in the random Ising model on the triangular and hexagonal lattices, the duality relation (5) is replaced with a similar expression with the left hand side being the partition function for the triangular lattice and the right hand side for the hexagonal lattice, for example. As was the case in the non-random systems, it is not possible to exactly identify the multicritical point from such a duality relation, but we may relate the multicritical points of the two systems. Our conjecture for such a problem is that the multicritical points p_{c1} and p_{c2} for mutually-dual systems are related by the symmetrized generalization of $x_0 = x_0^*$,

$$x_0(p_{c1}, K_{c1})x_0(p_{c2}, K_{c2}) = x_0^*(p_{c1}, K_{c1})x_0^*(p_{c2}, K_{c2}), \tag{6}$$

where K_{c1} and K_{c2} are functions of p_{c1} and p_{c2}, respectively, through the NL condition $e^{-2K} = (1-p)/p$. Rigorous validity of this conjecture has been verified for $n = 1, 2$ and ∞, and numerical simulations are in good agreement as in Table 1 for the triangular-hexagonal pair as well as for the three-dimensional problem of the Ising spin glass and random lattice gauge system. Equation (6) can also be applied to a system with anisotropy in the probability variable p on a self-dual lattice. The p_{c1} should then be understood as the critical probability for the vertical bonds and p_{c2} is for horizontal bonds on the square lattice.

Non-Ising spins can also be treated within the present theoretical framework. It is known that the ferromagnetic Potts model is self dual on a self-dual lattice. The random version of the Potts model, with chiral-type randomness, fits very well to our theory using the same condition $x_0 = x_0^*$ plus NL, and the result agrees with analytical (where applicable) and numerical evidence as indicated in Table 1.

The random chiral Z_q model is another example of non-Ising system. In the non-random ferromagnetic case, it is known that the Z_q model has two phase transitions, and therefore three thermodynamic phases, if q is larger than or equal to five.[15] The random version is expected to share this feature on the NL line (and off as well), and the two transition points are related by a condition corresponding to (6).

An interesting consequence can be derived on self-dual lattice, the square lattice for example, that there is no finite-temperature phase transition for the symmetric distribution $p = 1/2$ if we use an identity that relates Z_n and Z_{n-1}. The replicated partition function is a function of p and K through $x_0, x_1, \cdots, x_n$, which will here be indicated in terms of the expression $Z_n(p, K)$. We also write $Z_n(\text{NL}, K)$ when p

is related to K through the NL condition. Then, it is possible to prove an identity $Z_n(p = 1/2, K) = Z_{n-1}(\text{NL}, K)$, up to a trivial factor. The quenched limit of the symmetric system ($n \to 0$) is therefore equivalent to the problem of replica number -1 on the NL. This means that the transition point of the symmetric system in the quenched limit is equal to the location of the multicritical point of the (-1)-replicated system. Thus, if we are allowed to use the replica method to the (-1)-replica case, we combine the fixed-point condition $x_0 = x_0^*$ and the NL relation and apply the resulting formula to $n \to -1$ to obtain the conjectured value of the critical point for the symmetric system. The result is that $K_c \to \infty$ or $T_c \to 0$ as $n \to 0$ for the symmetric case. The same argument applies not only to the Ising model but also to other systems including the random Potts and Z_q models. This is consistent with many numerical investigations and resolves some controversies on the existence/absence problem of a finite-temperature transition in the symmetric system on the square lattice.[16,17]

6. Discussions

After briefly reviewing the present status of the theory of static properties of spin glasses, we have presented our conjecture on the exact location of the multicritical point in the phase diagram. An important point to be emphasized is that all the numerical results of independent calculations by a number of groups can be understood in a single unified framework of our theory that exploits duality and gauge symmetry. If our conjecture is indeed correct, against which we have no evidence so far, it will be the first analytical theory to systematically reveal lattice-specific properties, *i.e.* the value of the critical point, of finite-dimensional systems for which numerical methods have long been the only path to reach reliable conclusions.

A frequently asked question is why we restrict ourselves to the NL. The fixed-point condition of the leading Boltzmann factor $x_0(p, K) = x_0^*(p, K)$ relates p and K and may give the shape of the whole part of phase boundary. This is indeed the case for $n = 1$ as well as for $n = 2$ above the multicritical point. We nevertheless restrict ourselves to the NL at this moment because it is difficult to directly verify numerically this ansatz on the whole part of the phase diagram for many systems with quenched randomness. It should also be kept in mind that we have focused our attention to the point where two completely different types of symmetries, invariance under duality ($x_0 = x_0^*$) and gauge transformation (NL condition), meet. This implies that the multicritical point has clearly distinguished symmetry features, which allows us to discuss this point on a different basis from other points of phase transition. Rigorous proof of our conjecture is awaited for.

Acknowledgments

This work was supported by the Grant-in-Aid for Scientific Research on Priority Area "Statistical-Mechanical Approach to Probabilistic Information Processing".

References

1. S. F. Edwards and P. W. Anderson, *J. Phys.* **F5**, 965 (1975).
2. D. Sherrington and S. Kirkpatrick, *Phys. Rev. Lett.* **35**, 1792 (1975).
3. G. Parisi, *J. Phys.* **A13**, L115, 1101, 1887 (1983).
4. H. Nishimori, *Statistical Physics of Spin Glasses and Information Processing – An Introduction* (Oxford, Oxford, 2001).
5. H. Nishimori and K. Nemoto, *J. Phys. Soc. Jpn.* **71**, 1198 (2002).
6. J.-M. Maillard, K. Nemoto and H. Nishimori *J. Phys.* **A36**, 9799 (2003).
7. K. Takeda and H. Nishimori, *Nucl. Phys.* **B686**, 377 (2004).
8. K. Takeda, T. Sasamoto and H. Nishimori, (in preparation).
9. N. Ito and Y. Ozeki, *Physica* **A321**, 262 (2003).
10. Y. Ozeki, (private communication)
11. A. Honecker, M. Pico and P. Pujol, *Phys. Rev. Lett.* **87**, 047201 (2001).
12. G. Arakawa, I. Ichinose, T. Matsui and K. Takeda, hep-th/0409076.
13. Y. Ozeki and N. Ito, *J. Phys.* **A31**, 5451 (1998).
14. T. Ohno, G. Arakawa, I. Ichinose and T. Matsui, *Nucl. Phys.* **B697**, 462 (2004); C. Wang, J. Harrington and J. Preskill, *Ann. Phys.* **303**, 31 (2003).
15. J. V. José, L. P. Kadanoff, S. Kirkpatrick and D. R. Nelson, *Phys. Rev.* **B16**, 1217 (1977).
16. T. Shirakura and F. Matsubara, *J. Phys. Soc. Jpn.* **65**, 3138 (1996); F. Matsubara, T. Shirakura and M. Shiomi, *Phys. Rev.* **B58**, R11821 (1998).
17. P. Holme, B. J. Kim and P. Minnhagen, *Phys. Rev.* **B67**, 104510 (2003).

FROM STATISTICAL PHYSICS METHODS TO ALGORITHMS

DEMIAN BATTAGLIA
SISSA, Via Beirut 2–3, I-34100 Trieste, Italy
battagli@sissa.it

MICHAL KOLÁŘ
SISSA, Via Beirut 2–3, I-34100 Trieste, Italy
kolarmi@sissa.it

RICCARDO ZECCHINA
ICTP, Strada Costiera 11, I-34100 Trieste, Italy
zecchina@ictp.trieste.it

Received 1 December 2004

After a brief review of the Survey Propagation equations defined over single instances of complex glassy-like Hamiltonians, we discuss their application to the search for minimum energy configurations in difficult combinatorial optimization problems. Furthermore, we show that the application of arbitrary external field is helpful in the investigation of the topology of the configuration space.

Keywords: Combinatorial optimization; complex interacting systems; spin glasses; random MAX-K-SAT; interdisciplinary applications of statistical physics.

1. Introduction

Recent developments in the field of diluted spin-glasses have pointed out a chance to apply the results of the field outside statistical physics, examples being combinatorial optimization[1] and error correcting codes.[9] Similarly to statistical physics models, a generic combinatorial optimization problem is composed of many discrete variables — e.g. Boolean variables, finite sets of colors or Ising spins — which interact through constraints typically involving a small number of variables, that in turn sum up to give the global energy function. When the problem instances are extracted at random from nontrivial ensembles, computer science and statistical physics display common roots: many of the models considered to be of basic interest for Computer Science are nothing but spin glasses defined over finite connectivity random graphs, i.e. diluted spin glasses. Surprisingly enough, there already exist concrete engineering applications: For instance, among the most effective error correcting codes and data compression methods are the so called Low Density Parity Check algorithms which indeed implement an energy minimization of a spin glass energy defined over a sparse random graph. In such problems, the choice of the graph ensemble is a part of the designing techniques, a fact that makes spin glass theory directly applicable.

New algorithmic results as well as many mathematical issues have been put forward by the statistical physics studies, with examples ranging from phase transitions[1,11] and out-of-equilibrium analysis of randomized algorithms[8] to new classes of message-passing algorithms[12,17] named Survey Propagation (SP).

The physical scenario for the diluted spin glasses version of hard combinatorial problems predicts a trapping in metastable states for exponentially long times of local search dynamic process satisfying detailed balance. Depending on the models and on the details of the process — e.g. cooling rate for SA — the long time dynamics is dominated by different types of metastable states at different temperatures. A common feature is that at zero temperature and for simulation times which are sub-exponential in the size of the problem there exists an extensive gap in energy which separates the blocking states from true ground states. Such behavior can be tested on concrete random instances which therefore constitute a computational benchmark for more general algorithms.

As we shall discuss in the following, Survey Propagation algorithms are indeed able to cope with the proliferation of such metastable states and to find optimal ground states efficiently (polynomial time).

2. Random Combinatorial Problems

In this contribution we shall focus our attention on one of the most famous combinatorial problems, the so called K-SAT problem, which has been the object of a large amount of theoretical and experimental investigations since the seminal proof of **NP**-completeness by Cook.[3] However, the SP-based algorithms that we shall discuss are of general validity and could be applied as well to different combinatorial optimization problems like, for instance, the graph-coloring or the spin-glasses.

The K-satisfiability problem consists in deciding whether there is a global truth value assignment for N boolean variables, such that M given logical constraints are satisfied simultaneously. The individual constraints (called clauses) are logical OR function of K variables, that can appear either directed or negated, giving rise to frustration. An optimization form of K-SAT, called MAX-K-SAT can also be defined, if one is interested in finding an assignment which violates as few clauses as possible.

It is possible to establish a precise one to one mapping between K-SAT formulas and glassy Hamiltonians of the type:

$$E = 2\sum_a E_a = 2\sum_a \frac{\prod_{i=1}^{3}(1 + J_{a,i}s_i^a)}{2} \tag{1}$$

where s_i^a are Ising spins corresponding to the i-th variable in the clause a and the couplings $J_{a,i}$ take the value 1 (resp. -1) if the corresponding variable appears directed or negated in clause a. For instance the clause $(x_1 \vee \bar{x}_2 \vee x_3)$ would be associated to a Hamiltonian term $\frac{1}{4}(1+s_1)(1-s_2)(1+s_3)$ (associating Boolean variables $x_i = \{0,1\}$ to Ising spins by the relation $s_i = (-1)^{x_i}$). The energy function (1) is actually proportional to the number of violated clauses. Thanks to this mapping, the techniques of the so called Cavity Method of statistical mechanics of disordered systems can be employed in order to achieve some understanding of the combinatorial problem together with a concrete algorithmic device for studying specific instances.

In the last years the case of random formulas extracted from well-defined statistical ensembles has been considered with special attention, because of the existence of threshold effects in the large N limit. Varying the ratio $\alpha = M/N$ of the number of clauses to the number of variables, a phase transition takes place at a finite value α_c depending on K and on the ensemble definition. The problem is generally satisfiable (SAT) for $\alpha < \alpha_c$, and unsatisfiable (UNSAT) for $\alpha > \alpha_c$.[4,5,6] Even more, it is known experimentally[7] that algorithmically hard-to-optimize instances tend to be concentrated around the critical value α_c. Statistical mechanics studies have been able to provide an explanation of the onset of hardness in terms of the clustered nature of the low-temperature phase. For $\alpha_g < \alpha < \alpha_G$ ($\alpha_g < \alpha_c < \alpha_G$), any two ground states assignments will be either very close to each other, or very far apart, giving rise to the one-step-replica-symmetry-breaking (1-RSB) clustering scenario. For $\alpha_d < \alpha < \alpha_g$ and $\alpha > \alpha_G$ it will become possible to define higher order clusters, "clusters of clusters", and a continuous distribution of the distance scales will be observed, typical of the full RSB (f-RSB) scenario.[10] Accurate analytical

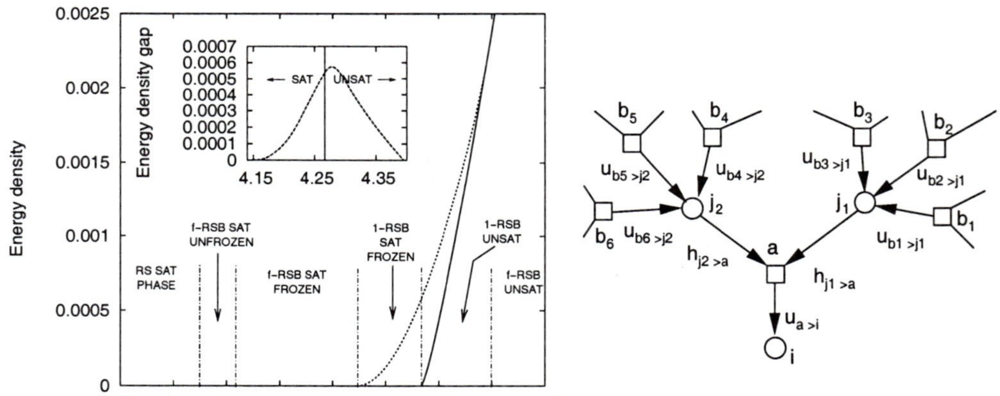

Fig. 1. *Left*: The crossings of the ground state energy (solid line) and of the Gardner energy (dashed line) delimit the 1-RSB clustered region. In this region around the SAT/UNSAT transition point, the difference between the Gardner and the ground state energy is strictly positive and it measures the strength of the metastable states entropic trapping effect for local search heuristics. *Right*: Factor graph representation, Cavity fields and cavity biases. The survey for $u_{a\to i}$ depends on the probability distribution function of the cavity fields $h_{j_1\to a}$ and $h_{j_2\to a}$, functions on the other hand of the surveys incoming to j_1 and j_2.

computations of the thresholds α_d, α_c [11,12], α_g, α_G[13,14] have been performed using the Cavity Method. A more complete cartoon[2] of the many different phases of the 3-SAT problem is presented in Fig. 1.

The normalized logarithm of the number of clusters of states at a given energy E, called complexity and denoted by $\Sigma(E)$, can as well be computed (at a 1-RSB level of approximation). The complexity of the $E = 0$ phase becomes suddenly finite at $\alpha = \alpha_d$ and becomes negative at $\alpha = \alpha_c$, indicating that satisfying assignments clusters disappear and the system enters the UNSAT phase. The complexity $\Sigma(E)$ is an increasing function of energy, and hence many metastable state clusters exist.[2] The energy level represented by the largest number of configurations, e_{th}, is given by $\Sigma(e_{th}) = \max_E (\Sigma(E))$.

The knowledge of the full-RSB form of the complexity would be needed to determine the true position of e_{th}, but it is in any case lower-bounded by the value E_G of the largest energy for which the 1-RSB computation can be considered stable (Gardner energy). One would expect that in the thermodynamical limit traditional local search heuristics, like Simulated Annealing and even highly performing *ad hoc*-devised procedures not properly satisfying detailed balance like WALKSAT undergo an exponential slowdown when approaching E_G, with different layers of states acting as dynamical traps, depending on the details of the algorithm.[15]

As shown in the inset of Fig. 1, the gap between the ground state analytical prediction and the calculated Gardner Energy is rather small and the trapping effect caused by the threshold states will become appreciable only for large enough samples. Hence, really hard random benchmarks for algorithm testing should involve a large number of variables and have $\alpha_g < \alpha < \alpha_G$: in the case of random 3-SAT (for which $\alpha_g = 4.15$ and $\alpha_G = 4.39$) sizes as large as $N = 10^5$ should be taken in order to find satisfactory agreement with theoretical predictions.

3. Cavity Equations for Single Instances

The cavity equations used for the theoretical study of the general properties of the random K-SAT model can be rewritten in a form valid even for single instances,[12] opening the way to the realization of efficient algorithms, which exploit the possibility of retrieving information about the statistical behavior of individual variables in the stable and metastable states at any given energy density.

The factor graph representation of the combinatorial problem can be useful in order to better understand the equation structure (see Fig. 1). The variables $i, j, k, \ldots$ are associated to "variable nodes" depicted as circles, while the clauses $a, b, c, \ldots$ to "function nodes", represented as squares. In the case of random K-SAT, the function nodes have connectivity K, while the variable nodes have a Poisson-distributed connectivity with average $K\alpha$. An important feature of the graphs associated to random formulas is their locally treelike structure, which proves essential in order to justify the uncorrelation hypothesis assumed in the following derivation.

In the SP setting, it is assumed that every function node a sends messages $u_{a\to i} \in \{0,1\}$ to the neighboring variables i: these *cavity biases* are different from 0 if the clause a can be satisfied only in the case in which the variable i assume a specific truth value. Other messages, called *cavity fields*, travel as well from variable nodes j of connectivity Γ_j to neighboring function nodes a, and can assume integer values $-\Gamma_j + 1 < h_{j\to a} < \Gamma_j - 1$. Cavity biases and fields are interdependent. The cavity field $h_{j\to a}$ is the sum of all the cavity biases incoming to the variable j from neighboring function nodes b, except the clause a (hence the name "cavity"): $h_{j\to a} = \sum_{b\in V(j)\setminus a} u_{b\to j}$, where $V(j) \setminus a$ is the set of all the functions nodes b neighboring to j deprived of the element a. If j has no other neighbor than a, then $h_{j\to a} = 0$. On the other hand, a clause a must send a cavity bias to a variable i only if the configuration of the other connected variables is such to not satisfy a yet. In this case, a must try to align i in the right way:

$$u_{a\to i} = \widehat{u}_{a\to i}\left(\{h_{j\to a}\}\right) = -J_i^a \prod_{j\in V(a)\setminus i} \theta\left(h_{j\to a} J_j^a\right) , \tag{2}$$

where $\theta(x)$ is a step function taking values $\theta(x) = 0$ for $x \leq 0$, $\theta(x) = 1$ for $x > 0$, and $V(a)$ is the set of variable nodes neighboring the clause a. If a has no other neighbor than i, then $u_{a\to i} = -J_i^a$.

In a situation in which many disconnected clusters of configurations at the same energy coexist, the probability of picking up a state at random and finding a certain value of $u_{a\to i}$ (referred to in the following as u-survey) can be written as $Q_{a\to i}(u) = \frac{1}{\mathcal{N}_{\text{cl}}} \sum_\ell \delta(u, u^\ell_{a\to i})$ where ℓ runs over all clusters of solutions and $u^\ell_{a\to i}$ is the value assumed by the cavity bias over the edge $a \to i$ in the cluster ℓ; δ denotes the Kronecker delta function. Considering that a cavity bias can assume just three values, this expression can be parameterized in the following way:

$$Q_{a\to i}(u) = \eta^0_{a\to i}\delta(u) + \eta^+_{a\to i}\delta(u-1) + \eta^-_{a\to i}\delta(u+1) \tag{3}$$

The real number $\eta^\pm_{a\to i}$ will give the probability of having a cavity bias traveling along the edge $a \to i$ and forcing i to assume the orientation ± 1, when the existence of many clusters is taken in account. The u-survey is normalized and hence $\eta^0_{a\to i} + \eta^+_{a\to i} + \eta^-_{a\to i} = 1$.

The distribution of the cavity fields trying to drive the system toward states at a predefined energy density can be expressed in terms of h-surveys:

$$\begin{aligned} P_{j\to a}(h_{j\to a}) = C_{j\to a} \int \prod_{b\in V(j)\setminus a} Q_{b\to j}(u_{b\to j})\, du_{b\to j}\, \delta\Big(h - \sum_{b\in V(j)\setminus a} u_{b\to j}\Big) \\ \times \exp\Big(y\big(\big|\sum_{b\in V(j)\setminus a} u_{b\to j}\big| - \sum_{b\in V(j)\setminus a} |u_{b\to j}|\big)\Big), \end{aligned} \tag{4}$$

The Lagrange multiplier y can be considered as the analogue of a temperature.[12] This pseudo-temperature is defined as the inverse of the derivative of the complexity with respect to the energy. Fixing a value of y is equivalent to sample clusters of states at a desired energy density.

Another formal expression for the u-survey of $u_{a\to i}$ is:

$$Q_{a\to i}(u) = \int \prod_{j\in V(a)\setminus i} P_{j\to a}(h_{j\to a})\, dh_{j\to a}\; \delta\left(u - \hat{u}_{a\to i}\left(\{h_{j\to a}\}\right)\right), \tag{5}$$

Inserting the parameterization (3) inside (4) and (5), allows finally to write simple recursions for the evaluation of the convolution integral (4) giving the probability distribution functions (pdf) of the cavity fields. In the case of general y, the full evaluation of $P_{j\to a}(h)$ can be performed in just $(2\Gamma_j - 1)(\Gamma_j - 1)/2$ purely algebraic computation steps.[2] At this point, considering the special form of the function $\widehat{u}(\{h\})$, equation (5) reduces simply to:

$$\eta_{a\to i}^{J_{a,i}} = \prod_{n=1}^{K-1} W_{j_n\to a}^{J_{j_n,a}}, \quad \eta_{a\to i}^{-J_{a,i}} = 0, \quad \eta_{a\to i}^{0} = 1 - \eta_{a\to i}^{J_{a,i}}, \tag{6}$$

where we introduced the integrated pdfs: $W_{j\to a}^{+} = \sum_{h=1}^{\Gamma_j - 1} P_{j\to a}(h)$, $W_{j\to a}^{-} = \sum_{h=-\Gamma_j+1}^{-1} P_{j\to a}(h)$.

A first set of u-surveys is randomly generated and used to evaluate recursively the weights $W_{j\to a}^{\pm}$. By mean of the relations (6) a new set of u-surveys is computed. The whole procedure is repeated until when a fixed-point is reached.

Several quantities can be now evaluated, including free energy and complexity. In particular a slight modification of the recursion for evaluating (4) allows to derive the order parameter of the problem, consisting essentially in the whole collection of the local field distributions:

$$\begin{aligned} P_i(H) = C_i \int \prod_{b\in V(i)} Q_{b\to i}(u_{b\to i})\, du_{b\to i}\; \delta\Big(H - \sum_{b\in V(i)} u_{b\to i}\Big) \\ \times \exp\Big(y\big(|\sum_{b\in V(i)} u_{b\to i}| - \sum_{b\in V(i)} |u_{b\to i}|\big)\Big), \end{aligned} \tag{7}$$

The SP equations allow then to characterize completely the statistical mechanics of a single sample. The next section will show how to exploit this relevant amount of physical information for the realization of a powerful optimization algorithm.

4. The Algorithm and its Performance

Knowledge of the local fields pdf immediately offers us a way how to simplify (decimate) the formula: the variables can indeed be ranked according to their tendency to become oriented in some direction.

Indeed, it is reasonable that a spin tends to align itself with the most probable direction of the local field. A ranking can be realized by finding all the probability weights $W_j^+ = \sum_{H=1}^{\Gamma_j} P_j(H)$, $W_j^- = \sum_{H=-\Gamma_j}^{-1} P_j(H)$, and by sorting the variables according to the values of a bias function:

$$b_{\text{fix}}(j) = |W_j^+ - W_j^-|. \tag{8}$$

One can then decide to fix the most biased variables to their preferred directions. In this way the formula is simplified, the number of variables decreases and, furthermore, we can hope that all variables can be fixed in this way.

The most general algorithm that can be built on the idea of the Eqs. (4, 5) would consider some, correctly chosen, value of the inverse pseudo-temperature. This value can be fixed all over the experiment or, it may be optimized during the simplification. One indeed expects that there is some value y^* at which the probability of convergence is largest. This value corresponds to the maximum of the free-energy function $\Phi(y)$ which can move during the run of algorithm, and that can be efficiently computed form the cavity fields pdfs.[12,2] At each step $\Phi(y^*)$ provides an estimation for the ground state energy: by choosing y^* as value of pseudo-temperature we select only those configurations that are close to the ground state.

When working at finite pseudo-temperature, we have to take in account the possibility that some non optimal fixing is done in presence of "thermal" noise. After several iterations, some biases of fixed spins may become smaller than the values they had at the time when the corresponding spin was fixed, or even invert their orientation.

The occurrence of such badly fixed spins can be tracked by negative or small values of the following index function:

$$b_{\text{backtrack}}(j) = -s_j \left(W_j^+ - W_j^- \right), \tag{9}$$

An "error removal" procedure can then be implemented, alternating fixing or backtracking moves with a fixed ratio[2,16]

In order to test the efficiency of the algorithm we have done several experiments in all the regions of the phase diagram of the random-MAX-K-SAT problem.[2] Among the most interesting results belongs the possibility to access assignments of the SAT formula with energy lower than the Gardner threshold energy in the 1RSB–stable region of the phase diagram. Several 3-SAT sample problems belonging to the 1-RSB stable UNSAT phase have been studied carefully, comparing the best energy values obtained by theWALKSAT algorithm before and after the formula simplification done by Survey Propagation.

As expected, the SP simplification produced always an improvement in the WALKSAT performance, but the combination of runtime y-optimization (i.e. updating of y value to $y*$ after some fixed number of spin fixes) and of error correction was needed in order to overcome the Gardner lower bound. After a rather small number of spin flips, WALKSAT reached a saturation levels strikingly closer to the ground state lower bound than to the Gardner energy, see Fig. 2. A further valuable effect of introduction of the backtracking was the increased decimation efficiency. When backtracking was used, SP was able to determine a truth value for more than 80% of the variables before losing convergence, while without backtracking, the algorithm stopped on average after only 40% of fixings.

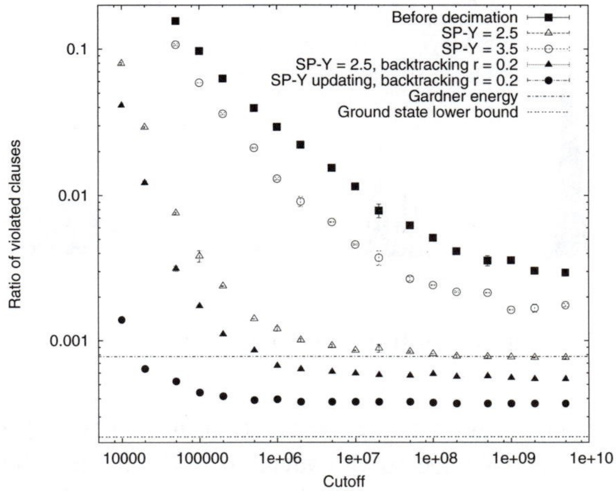

Fig. 2. SP performance in the UNSAT region (single sample with $N = 10^5$ variables and $\alpha = 4.29$). Several simplification strategies are compared: WALKSAT without decimation; with decimation at $y = 2.5$, $y = 3.5$ without backtracking; $y = 2.5$ and runtime optimized y with probability $r = 0.2$ of doing backtrack moves. Need for backtracking is readily visible, and its introduction allows to reach energies closer to the ground state than to the Gardner lower bound.

5. Selection of Glassy Ground States Clusters

The optimization of formulas belonging to the UNSAT region requires the use of a finite pseudo-temperature. Although the recursive evaluation of the integrals (4) is easy to implement in a program, it is a rather slow procedure. On the other hand, in the SAT region true optimal solutions are expected to be fully satisfying assignments, and hence, one has to forbid any configuration of variables that would produce a contradiction. This is done by setting $y \to \infty$, and under this limit, it becomes possible to devise a linear-time algorithm.[17] The new relations for the direct computation of the integrated cavity fields distributions become:

$$W^{\pm}_{j\to a} = C_{j\to a}\left[1 - \prod_{b\in V(j)\backslash a}\left(1-\eta^{\pm}_{b\to j}\right)\right]\prod_{b\in V(j)}\left(1-\eta^{\mp}_{b\to j}\right)$$
$$W^{0}_{j\to a} = C_{j\to a}\prod_{b\in V(j)\backslash a}\left(1-\eta^{+}_{b\to j}-\eta^{-}_{b\to j}\right) \qquad (10)$$

where $C_{j\to a} = W^{+}_{j\to a} + W^{-}_{j\to a} + W^{0}_{j\to a}$ is a normalization constant. As in the case of the finite temperature algorithm, the computations of new cavity surveys with the equations (6) is repeated until a self-consistent fixed point is reached.

It is observed experimentally and there are theoretical evidences that this fixed point is unique if it exists. The usual SP strategy, both at finite and zero temperature, is eventually able to retrieve only one solution of a given formula. It is possible nevertheless to control the fixed point location by mean of an external probability preconditioning.[18] A forcing field $\pi\vec{\xi}$ is introduced, with direction $\vec{\xi} \in \{-1, 0, 1\}^N$

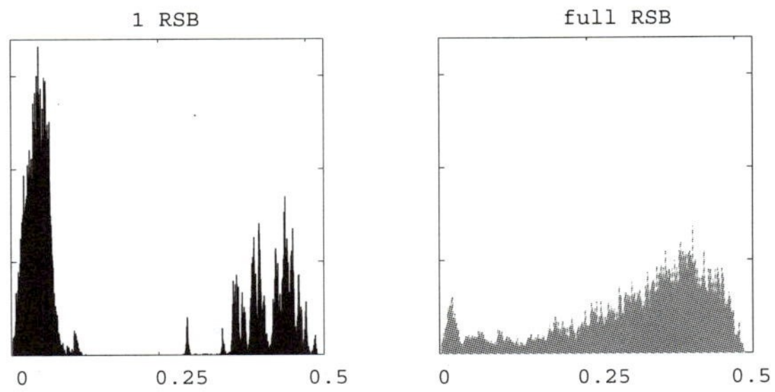

Fig. 3. Distribution of Hamming distances

and intensity π. If the component $\pi\xi_i = \pm\pi$ then an *a priori* probability π of assuming the value ± 1 is assigned to the variable i. If, on the other hand, ξ_i equals 0, the variable i is left completely unconditioned. The introduction of the external field do not modify the structure of the equations (10) and (6). Simply, for each variable an ulterior cavity bias survey is introduced, taking the form

$$Q_i^{(\mathrm{Ext})}(u) = (1 - \pi(1 - \delta_{\xi_i,0}))\,\delta(u) + \pi\,(\delta_{\xi_i,1}\delta(u-1) + \delta_{\xi_i,-1}\delta(u+1)) \tag{11}$$

The forcing drives the evolution of the cavity biases distributions towards the selection of clusters in which the variables are maximally aligned with the externally imposed direction $\vec{\xi}$. A correct choice of the a priori probability π, and the use of different random directions $\vec{\xi}$ allows retrieval of many solutions $\vec{\sigma}_{\pi\xi}$ in many different clusters.

The algorithm can be used successfully for the investigation of the local geometrical structure of the zero energy phase. One first determines a random solution $\vec{\sigma}$. Then a vector $\vec{\xi}_d$ is generated by flipping randomly Nd spins of the solution $\vec{\sigma}$. A new satisfying assignment $\vec{\sigma}_d$ is finally obtained performing a complete decimation using a forcing $\pi\vec{\xi}_d$. These steps are repeated many times, generating several $\vec{\xi}_d$ for each value of d between 0 and 1. To analyze data we computed the reciprocal Hamming distances among all the retrieved solutions and grouped them in a histogram. Looking at Fig. 3, it is evident that, depending on the value of α (and, hence, on the starting solution σ) very different distributions appear. In particular, in the first histogram (black bars) it is possible to observe the existence of the two sharp peaks predicted by the 1-RSB theory, centered around the typical distance among solutions in the same or in different clusters. When the solution space breaks in a more complex hierarchical structure of clusters, the gap is filled (gray bars) and full RSB takes place.

The SP equations with external preconditioning can be used even for the development of a lossy data compressor exploiting the clustered nature of the satisfying assignments space for data quantization purposes. More details can be found elsewhere.[18]

6. Conclusions

In this contribution we have presented an application of the Survey Propagation algorithm as an optimization tool. As an example we have chosen the MAX-K-SAT problem and we have shown that in the hard region of the phase space, the algorithm beats in performance well-established local-search algorithms. Furthermore we were able to reach states well below the Gardner glassy threshold. We show even more that introduction of probability preconditioning helps in finding more than a single satisfying assignment for any sample formula. This allowed us to probe in a detailed way the geometry of the zero-energy phase, and to distinguish among different patterns of replica-symmetry-breaking. Future plans include applications to source coding.

References

1. R. Monasson, R. Zecchina, S. Kirkpatrick, B. Selman, L. Troyansky, Nature **400**, 133 (1999)
2. D. Battaglia, M. Kolář, R. Zecchina, *Phys. Rev. E*, **70**, 036107 (2004)
3. S.A. Cook, Proc. 3rd Annual ACM Symp. on Theory of Computing, 151-158 (ACM Press, CITY, 1971).
4. E. Friedgut, *Journal of the A.M.S.* **12**, 1017 (1999)
5. D. Achlioptas, C. Moore, Random K-SAT: Two Moments Suffice to Cross a Sharp Threshold, preprint (2002)
6. S. Kirkpatrick, B. Selman, *Science* **264**, 1297 (1994)
7. T. Hogg, B.A. Huberman, C. Williams, C. (eds), *Artificial Intelligence* **81** I & II (1996)
8. S. Cocco, R. Monasson, A. Montanari, and G. Semerjian, Approximate analysis of search algorithms with "physical" methods, preprint cs.CC/0302003 (2003)
9. N. Sourlas, Nature **339**, 693 (1989)
10. H. Nishimori, *Statistical Physics of Spin Glasses and Information Processing* (Oxford University Press, 2001)
11. M. Mézard, G. Parisi, R. Zecchina, *Science* **297**, 812 (2002)
12. M. Mézard, R. Zecchina, *Phys. Rev. E* **66**, 056126 (2002)
13. S.Mertens, M. Mézard, R. Zecchina, Threshold values of Random K-SAT from the cavity method, preprint, cs.CC/0309020 (2003)
14. A. Montanari, G. Parisi, and F. Ricci-Tersenghi,*J. Phys. A* **37**, 2073 (2004)
15. A. Montanari, F. Ricci-Tersenghi, On the cooling-schedule dependence of the dynamics of mean-field glasses preprint, cond-mat/0401649 (2004)
16. G. Parisi, A backtracking survey propagation algorithm for K-SATisfiability, preprint cond-mat/0308510 (2003)
17. A. Braunstein, M. Mézard, R. Zecchina, Survey propagation: an algorithm for satisfiability, preprint 2002, to appear in *Random Structures and Algorithms*, cs.CC/0212002
18. D. Battaglia, A. Braunstein, J. Chavas, R. Zecchina, to appear in *Progr. Theor. Phys. Suppl.* (SPDSA04 proceedings).

AF213879

Signals and Communication Technology

Series Editors

Emre Celebi, Department of Computer Science, University of Central Arkansas, Conway, AR, USA

Jingdong Chen, Northwestern Polytechnical University, Xi'an, China

E. S. Gopi, Department of Electronics and Communication Engineering, National Institute of Technology, Tiruchirappalli, Tamil Nadu, India

Amy Neustein, Linguistic Technology Systems, Fort Lee, NJ, USA

Antonio Liotta, University of Bolzano, Bolzano, Italy

Mario Di Mauro, University of Salerno, Salerno, Italy

This series is devoted to fundamentals and applications of modern methods of signal processing and cutting-edge communication technologies. The main topics are information and signal theory, acoustical signal processing, image processing and multimedia systems, mobile and wireless communications, and computer and communication networks. Volumes in the series address researchers in academia and industrial R&D departments. The series is application-oriented. The level of presentation of each individual volume, however, depends on the subject and can range from practical to scientific.

Indexing: All books in "Signals and Communication Technology" are indexed by Scopus and zbMATH

For general information about this book series, comments or suggestions, please contact Mary James at mary.james@springer.com or Ramesh Nath Premnath at ramesh.premnath@springer.com.

Asadullah Shaikh • Abdullah Alghamdi
Qing Tan • Ibrahiem M. M. El Emary
Editors

Advances in Emerging Information and Communication Technology

Springer

Editors
Asadullah Shaikh
Department of Information Systems
College of Computer Science and
Information Systems, Najran University
Najran, Saudi Arabia

Abdullah Alghamdi
Department of Information Systems
College of Computer Science and
Information Systems, Najran University
Najran, Saudi Arabia

Qing Tan
School of Computing and Information
Systems, Faculty of Science and
Technology
Athabasca University
Athabasca, AB, Canada

Ibrahiem M. M. El Emary
Department of Information Science
King Abdulaziz University
Jeddah, Saudi Arabia

ISSN 1860-4862 ISSN 1860-4870 (electronic)
Signals and Communication Technology
ISBN 978-3-031-53236-8 ISBN 978-3-031-53237-5 (eBook)
https://doi.org/10.1007/978-3-031-53237-5

© The Editor(s) (if applicable) and The Author(s), under exclusive license to Springer Nature Switzerland AG 2024, Corrected Publication 2024
This work is subject to copyright. All rights are solely and exclusively licensed by the Publisher, whether the whole or part of the material is concerned, specifically the rights of translation, reprinting, reuse of illustrations, recitation, broadcasting, reproduction on microfilms or in any other physical way, and transmission or information storage and retrieval, electronic adaptation, computer software, or by similar or dissimilar methodology now known or hereafter developed.
The use of general descriptive names, registered names, trademarks, service marks, etc. in this publication does not imply, even in the absence of a specific statement, that such names are exempt from the relevant protective laws and regulations and therefore free for general use.
The publisher, the authors, and the editors are safe to assume that the advice and information in this book are believed to be true and accurate at the date of publication. Neither the publisher nor the authors or the editors give a warranty, expressed or implied, with respect to the material contained herein or for any errors or omissions that may have been made. The publisher remains neutral with regard to jurisdictional claims in published maps and institutional affiliations.

This Springer imprint is published by the registered company Springer Nature Switzerland AG
The registered company address is: Gewerbestrasse 11, 6330 Cham, Switzerland

Paper in this product is recyclable.

Preface

We want to thank all the authors for their participation in the International Conference on Innovation of Emerging Information and Communication Technology, ICIEICT 2023, which was held at Universidad Complutense de Madrid (Complutense University of Madrid) Escuela Profesional de Relaciones Laborales, UCM, on September 11–13, 2023, and organized by CANWEST International Research Congress, Canada. Special support was provided by the University of Almería (Universidad de Almería).

This book constitutes the refereed proceedings of the International Conference on Innovation of Emerging Information and Communication Technology, ICIEICT 2023. ICIEICT aims to provide occasions for international researchers and practitioners to present the most recent advances and future challenges in the fields of Information and Communication Technology. The conference received 128 submissions—confirming a trend of increasing interest in ICIEICT—coming from over 17 different countries including Algeria, Australia, Canada, China, Denmark, France, Japan, Kuwait, Mexico, Morocco, Poland, Romania, Saudi Arabia, Spain, Ukraine, United Kingdom, and USA.

When choosing the technical papers, the Technical Program Committee considered two factors: first, to offer a platform for more global participants by presenting and debating pertinent topics from a wide range of information and communication technologies (ICT) disciplines, reflecting the quickly advancing state of technology we are currently seeing. The second objective is to uphold a fair technical quality through the selection of about 31 technical articles, or about a 16% acceptance rate.

Four parallel-track sessions have been designed into the conference schedule to accommodate paper presentations. Deep learning, cybersecurity, blockchain technologies, health informatics, supervised and unsupervised learning, human digitization, enterprise resource planning, convolutional neural networks, sentiment analysis, endovascular intervention, robotics, and cloud computing were among the multidisciplinary topics covered in the presentation that had reasonable balance between theory and practice. Additionally, five keynote addresses from well-known ICT specialists are included in the program.

This event was only possible with the passion and diligence of several colleagues. We want to thank the members of the steering committee for their encouraging leadership and the general chairs for their help during the entire process. We also express our gratitude to the other Organizing Committee members for their productive collaboration. A special thank you to all of the referees and members of the Technical Program Committee for their essential assistance in reviewing the papers.

Najran, Saudi Arabia — Asadullah Shaikh
Najran, Saudi Arabia — Abdullah Alghamdi
Athabasca, AB, Canada — Qing Tan
Jeddah, Saudi Arabia — Ibrahiem M. M. El Emary
November 2023

Organization

Conference Chairs

Luigi Vladareanu	Institute of Solid Mechanics of Romanian Academy, Romania
Joan Lu	Hudersfield University, UK

Steering Committee

Rajesh Kumar Dhanraj	Symbiosis International, India
Jawad Rasheed	Istanbul Sabahattin Zaim University, Turkey
Sheifali Gupta	Chitkara University, India
Qurat-ul-Ain Mastoi	University of the West of England, UK
Naeem Ahmed Mahoto	MUET, Pakistan
Ahmad Taher Azar	Prince Sultan University, Saudi Arabia
Junzo Watada	Watada University, Japan
Latif Ladid	University of Luxembourg/SnT, Luxembourg

Technical Program Committee Chair

Qing Tan	Athabasca University, Canada
Asadullah Shaikh	Najran University, Saudi Arabia
Abdullah Alghamdi	Najran University, Saudi Arabia

Local Chairs

Juan E. Trinidad Segovia	Universidad de Almería, Spain

Submission and Proceedings Committee

Samar M. Alqhtani	Najran University, Saudi Arabia
Adel Sulaiman	Najran University, Saudi Arabia

Registration and Management Committee

Mohamed El Emary Ecole Technologie Superieure, Montreal, Canada
Ibrahiem El Emary King Abdul Aziz University, Saudi arabia

Web and Publicity Committee

Ibrahiem El Emary King Abdul Aziz University, Saudi Arabia

Contents

Baheta: Balanced and Unbalanced Dataset in Arabic Clickbait Detection Using a Deep Learning Model (LSTM)

Batool Alharbi, Razan Alhanaya, Deem Alqarawi, and Ruwaidah Alnejaidi

1 Introduction

Over the years, the use of social networks has increased until the number of users reached 5.16 billion worldwide in 2023 [1]. As a result, about 40% of fake news [2] is on social media. One method of disseminating fake news is using Clickbait, in which the author manipulates the title head through attractive words or pictures and misleading content to grab the user's attention and persuade him to click the link.

In this paper, we will explain the method of detecting Clickbait using deep learning (DL). DL is the study of artificial neural networks and associated machine learning (ML) algorithms that incorporate more than one hidden layer. It is sometimes referred to as deep structured learning, deep hierarchical learning, or deep machine learning field [3]. There are many DL models used in text classification such as long short-term memory networks (LSTMs), which we will use to detect Arabic clickbait. As Arabs, it is important for us to reduce Clickbait from Arabic content as we see the effect it has on our society and culture, and because there are few papers on detecting Arabic Clickbait using DL models, we will focus on that in this paper.

In Baheta, we will use two datasets, the first is Arabic Clickbait datasets [4] that are unbalanced, as the number of Clickbait is 26% less than the non-Clickbait. The second one is a combined dataset consisting of the Arabic Clickbait datasets and an Arabic fake news dataset [5] to make the Clickbait almost equal to the non-Clickbait.

B. Alharbi (✉) · R. Alhanaya · D. Alqarawi · R. Alnejaidi
Department of Information Technology, College of Computer, Qassim University, Buraydah, Saudi Arabia

© The Author(s), under exclusive license to Springer Nature Switzerland AG 2024

A. Shaikh et al. (eds.), *Advances in Emerging Information and Communication Technology*, Signals and Communication Technology,
https://doi.org/10.1007/978-3-031-53237-5_1

Fig. 1 Logo of Baheta: balanced and unbalanced dataset in Arabic Clickbait detection using LSTM

1.1 What Baheta Means?

Since Clickbait, in short, is "lying" and "misleading" the user, we called this proposal "Baheta," which is an Arabic synonymous with lying, as it denotes Clickbait. We designed a special Baheta identity consisting of two languages because this proposal is in English and Arabic out of pride in our language, the goal of the proposal, and the logic of its work. In Fig. 1, you can see Baheta logo.

1.2 Related Work

Linguists and computer scientists use Clickbait, and it has become a crucial topic for research. In this section, a general review of relevant Clickbait detection techniques has been covered.

According to researchers in [6], LSTM is the most crucial algorithm used to detect Clickbait and rumors. This paper was based on information from Twitter and news organizations. Because of the handling of texts, the text had to be converted into a number matrix. There are two solutions, either TF-IDF or Word2Vec. They must measure the semantic sensitivity correctly, which is why they ignore TF-IDF. With the spread of global Clickbait go to select GloVe over Word2Vec. One of the LSTM benefits is that it can deal with long-range contextual information and with the exploding and vanishing gradient problems in RNNs. They achieved an accuracy of 0.859 ± 0.002 and $\text{MSE} = 0.0296 \pm 0.0002$ when over 10 different runs.

As described in [7], a technique termed the ontology-based LSTM model (OLSTM) is suggested by Vorakit et al. RNNs were employed as a classification model. By actively maintaining self-connecting layers, LSTM expands RNNs. Dataset relies on Twitter and news websites to learn about the model, and they use word embedding information to better Clickbait identification because words are the foundation for deciding whether something is Clickbait or not. According to their observations, the effectiveness of identifying Clickbait has indeed increased by word embedding approaches. The results show that OLSTM significantly outperforms approaches that did not use word embedding characteristics, with an F1-score and precision of 0.8990 and 0.8994.

In the article [8], a two-phase model has been proposed. In the proposed approach, the hybrid CNN-LSTM model is implemented in the first phase to identify Clickbait, which is further fed to the second phase where eight types of Clickbait such as reasoning, number, reaction, revealing, shocking/unbelievable, hypothesis/ guess, questionable, and forward referencing are classified by using the Biterm (BTM). Three datasets were used, which collected data from different sources: Facebook, Twitter, and websites, and dataset 3 was text data and data extracted from images. The highest accuracy achieved by the hybrid CNN-LSTM model is 95.8%.

Zheng et al. [9] proposed a Clickbait convolutional neural network (CBCNN) model that takes into account not only the general features but also the specific features of different article types. The CBCNN model consists of Word2Vec models and a CNN model. CNN understands the input text from different perspectives. Each word has two meanings in the model: its general meaning and its type-related meaning, and it can be used for different natural language processing tasks. CBCNNs are used, for example, for recognizing parts of words, naming entities, extracting relations and events, and computing sentence similarities, and it shows how CNN works in Clickbait detection. One limitation of CBCNNs is that the maximum length of the headline is limited, which can lead to information loss in long headlines. CBCNNs do not include a pre-trained Word2Vec model that could allow the model to better understand the meaning of words.

In addition to the CNN [10], a CNN model is used in this study to provide a method for detecting Clickbait on online social media. The strategy being employed is concentrated on textual features. Word2vec was used to extract features from the data. The results indicate that CNN outperformed other ML methods with a high accuracy of 0.82%. Comparative analysis was carried out using the RF.

There isn't much research focusing on Arabic Clickbait headline detection. In order to better identify Clickbait news on social networks in Arabic, this study [4] built the first Arabic Clickbait headline news dataset and presented a multiple-feature-based strategy. Data collection, data preparation, and ML model training and testing phases are the three key components of the suggested approach. A total of 54,893 Arabic news articles were included in the dataset that was gathered (after pre-processing), and 23,981 of these news stories were Clickbait headlines. The ANOVA F-test was used to choose the most significant features from this pre-processed dataset. To ensure that the best settings were discovered, a number of ML techniques were then used with hyper-parameter tweaking techniques. The evaluation of the ML models was completed, and according to the findings, the support vector machine (SVM) with the top 10% of ANOVA F-test features (user-based features (UFs) and content-based features (CFs)) had the highest accuracy of 92.16%.

Mohammad A. Bsoula et al. [11] also applied ML models to an Arabic dataset of news. The purpose of this dataset is to enable the automatic classification of news headlines as "Clickbait" or "Not Clickbait" using ML models. A total of 3235 records were gathered and 2652 news records were classified as "Not Clickbait" and 583 as "Clickbait." The class distribution within the dataset was unequal with a ratio of "Clickbait" to "Not Clickbait" of 1:4. Therefore, to extract the final dataset needed for this research, the evaluation was performed on the original dataset, oversampled dataset, and under-sampled dataset. The three train datasets were subjected to the application of different seven ML models. Model hyperparameters were fine-tuned by ten-fold cross-validation on the training dataset. The evaluation of these models using the test dataset results produced a Macro F1-Score in the range of 0.18–0.81 for the explored feature combinations, successfully demonstrating the dataset's utility. According to the results, the resampling did not improve the F1-score value.

2 Methodology

We illustrate the general methodology of recognizing the Clickbait detection using DL in Fig. 2, which describes the phases involved in our proposed detection framework. We will explain each phase in detail in this section and illustrate the expected outcomes and our reasoning.

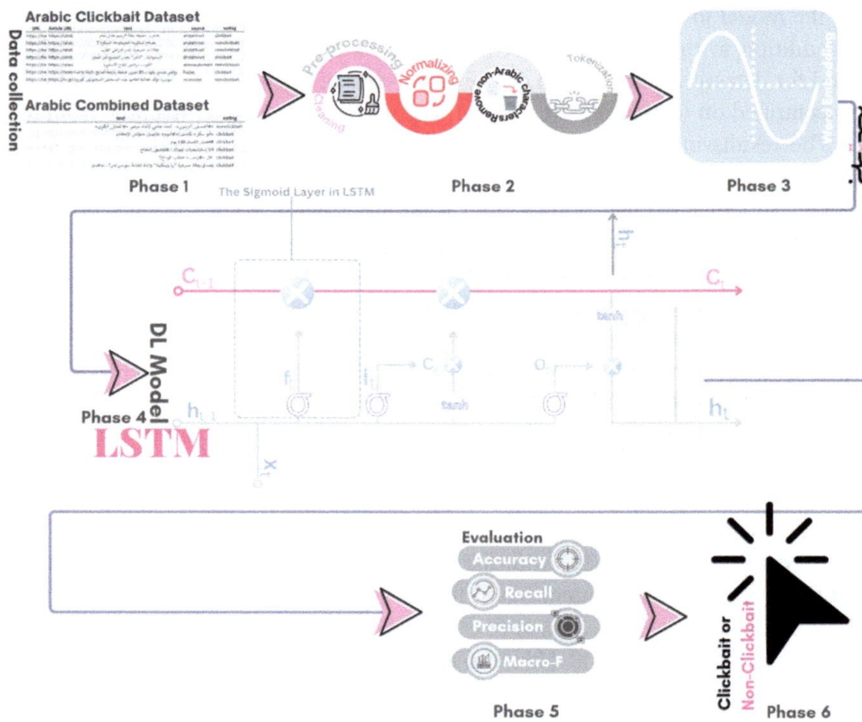

Fig. 2 Overview of basic phases of Baheta methodology approach

2.1 Dataset

The primary stage in the DL model pipeline is gathering data for training the DL model. The accuracy of the predictions provided by DL systems depends on training data. The dataset we will use is [12]; according to our knowledge, there is no other for the Arabic Clickbait. The Arabic Clickbait dataset was collected in this study [4]. This dataset contains 10,564 rows collected in Arabic from Twitter, 7735 were rated as non-Clickbait, and 2829 as Clickbait, which constitutes 26% of all datasets and makes the dataset unbalanced. As shown in Fig. 3, the dataset comprises five columns (features): URL, Article URL, text, source, and voting. Table 1 describes the dataset's features.

We combined the Arabic Clickbait dataset with the Arabic fake news dataset to enhance the Clickbait data ratio until it is equal to the non-Clickbait data because the Arabic Clickbait dataset is unbalanced. Arabic Fake News Dataset [5] contains 606,912 news classified into credible, not credible, and undecided collected from 134 different public Arabic news websites. The combined dataset consists of two columns (features): text and voting as described in Table 2. A sample of it is shown in Fig. 4. This dataset has 15,472 rows, it is an Arabic Clickbait dataset with some Arabic Fake News Datasets that are classified as not credible to become the Clickbait percentage of 50% in the combined dataset, and we published it on GitHub [12].

URL	Article URL	text	source	voting
https://tw	https://ahda	عاجل.. حقيقة وفاة الزعيم عادل إمام	ahdathnet	clickbait
https://tw	https://ahda	نصائح لمقاومة الشيخوخة المبكرة 7	ahdathnet	nonclickbait
https://tw	https://ahda	علامات خارجية تشير أمراض القلب	ahdathnet	nonclickbait
https://tw	https://ahda	السعودية.. "أبشر" تحذر الجميع أمر خطير	ahdathnet	clickbait
https://tw	https://wwv	الكويت ترخص للقاح أكسفورد	almowatennet	nonclickbait
https://tw	https://wwv	رياضي هندي يقوم بـ 85 تمرين ضغط بأربعة أصابع دقيقة واحدة	Ra2ej	clickbait
https://tw	https://is.gd	موديرنا تؤكد فعالية لقاحها ضد النسختين المتحوّرتين كورونا	mzmznet	nonclickbait

Fig. 3 Sample of the Arabic Clickbait dataset [12]

Table 1 Description of the Arabic Clickbait dataset features

Feature	Description
URL	Tweet link
Article URL	Link to the article to which the tweet refers
text	Tweet content
source	The source who posted the tweet
voting	Tweet labeled as "Clickbait or non-Clickbait"

Table 2 Description of the combined dataset features [12]

Feature	Description
text	Tweet content and fake news
voting	Tweet labeled as "Clickbait or non-Clickbait"

text	voting
«#الغسيل_البريتوني».. اتجاه عالمي لإنقاذ مرضى «#الفشل_الكلوي»	nonclickbait
«أبو سكر» تكشف لـ«#اليوم» تفاصيل «مؤتمر الإعاقة»	clickbait
#قضايا_الفساد 100 يوم	clickbait
10 إستراتيجيات تقودك لـ #تحقيق_النجاح	clickbait
قال «#ترامب» خطاب الوداع؟	clickbait
يصدق بطلة مسرحية "ريا وسكينة" والدة الفنانة سوسن بدر؟.. شاهدوا	clickbait

Fig. 4 Sample of the combined Arabic dataset [12]

We now have two datasets: unbalance (Arabic Clickbait dataset) and balance (combined Arabic dataset).

2.2 *Pre-processing*

Raw data from the real world are frequently inconsistent, lacking in specific behaviors or trends, and incomplete. They are also probably full of mistakes. As a result, after being gathered, they undergo pre-processing to create a model-compatible format for the DL model. The data must go through a pre-processing process before classification. The proposed model will undertake data preprocessing tasks, including cleaning, normalization, tokenization, and the removal of non-Arabic characters, prior to its utilization.

- Cleaning includes:
  Convert hashtags to words and remove emails, URLs, mentions, punctuations, repeating chars, stop words, and emojis.
- Normalizing includes:
  Converting the hamza to the near letters, removing diacritics and tatweel characters, and replacing "ة" with "ه".
- Remove non-Arabic characters:
  Remove numbers and non-Arabic characters.
- Tokenization:
  Involves breaking up the sentence into a collection of words known as tokens.

Because we care for the Arabic language, it is important that the pre-processing of Arabic data is using libraries specialized in Arabic natural language processing. So we used Ruqia [13], and pyarabic [14] libraries, as shown in Fig. 5, which we represented in our methodology.

Fig. 5 Text pre-processing phase in Baheta methodology

2.3 *Word Embedding*

The techniques used to map words or phrases to vectors of real numbers are referred to as word embedding. Words are represented as continuous vectors in a low-dimensional space via word embedding techniques. Word2vec technology is arguably the most popular method for using an external feed-forward neural network to learn word embeddings. We used Skip-gram word embedding for the proposed model from AraVec. AraVec [15] is an open-source word2vec project for Arabic. The process of embedding words starts with creating a co-occurrence matrix, and we take into consideration the length of texts in the dataset.

2.4 *Model*

The architecture of the LSTM model we created is composed of several layers. The embedding layer is the first layer followed by the SpatialDropout1D layer with 0.2 units to drop, and SpatialDropout1D will help promote independence between feature maps. The next layer is an LSTM layer with 128 neurons, employing the tanh activation function for cell state and hidden state and the Sigmoid function for LSTM gates, as shown in Fig. 6. GlobalMaxPooling and a Dropout layer with a probability of 0.2 are used. The final dense layer is the output layer, which uses a Sigmoid activation function.

2.5 *Evaluation*

Through the dimension reduction process of feature extraction, a large initial set of raw data is divided into smaller groups for processing, which will be done by the LSTM model. Performance metrics are used to assess the effectiveness or quality of the model. These performance measures aid in our comprehension of how well our

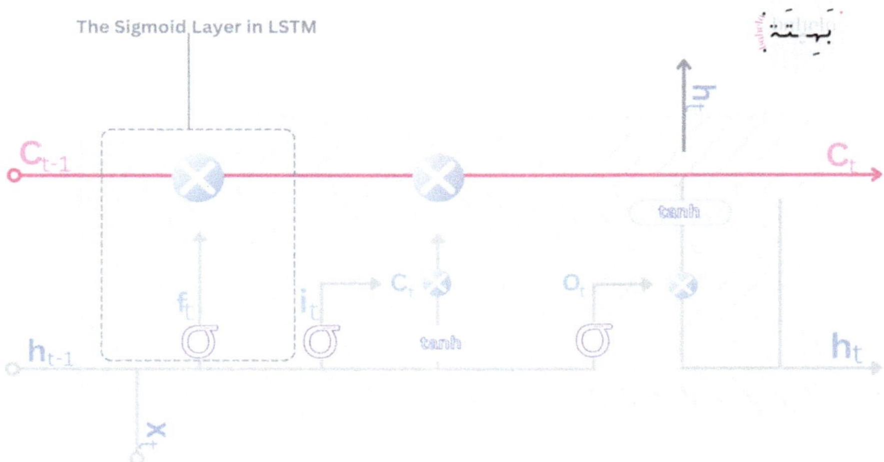

Fig. 6 LSTM layer architecture of the Baheta proposed model

Fig. 7 Confusion matrix for Clickbait detection

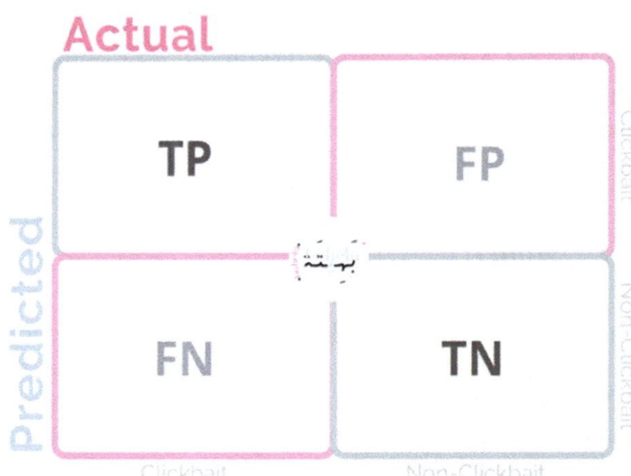

model handled the supplied data. Performance metrics assist measure how well a DL model generalizes on new or previously unexplored data. In our proposed work, we will analyze our chosen set of approaches using performance metrics like accuracy, precision, recall, and Macro-F score. To understand the necessity of using the performance measurements chosen for this study, it is important to take into account the confusion matrix notion. For a prediction task, a confusion matrix summarizes examples of accurate and incorrect predictions. A confusion matrix is graphically represented in Fig. 7, based on [16].

- *True Positives (TP)*: The cases in which the model predicted Clickbait and the actual output were also Clickbait.
- *True Negatives (TN)*: The cases in which the model predicted non-Clickbait and the actual output was non-Clickbait.
- *False Positives (FP)*: The cases in which the model predicted Clickbait and the actual output was non-Clickbait.
- *False Negatives (FN)*: The cases in which the model predicted non-Clickbait and the actual output was Clickbait.

We found several metrics for evaluation during our research, which are explained below:

The accuracy formula is used to determine the percentage of all accurate predictions, and it can be represented by the following equation:

$$\text{Acc} = \frac{TP + TN}{TP + TN + FP + FN} \tag{1}$$

The recall formula is used to determine the percentage of true positive predictions among all possible positive predictions a model may have made, and it can be represented by the following equation:

$$\text{Recall} = \frac{TP}{TP + FN} \tag{2}$$

The precision formula is used to determine the proportion of accurately predicted actual cases to all positively predicted instances, and it can be represented by the following equation:

$$\text{Precision} = \frac{TP}{TP + FP} \tag{3}$$

The macro-F formula is used to calculate the arithmetic mean of the individual class-related f1 score, and it can be represented by the following equation:

$$\text{Macro} - F = \frac{\text{sum}(F1 \quad \text{scores})}{\text{number of classes}} \tag{4}$$

2.6 *Clickbait or Non-clickbait*

After applying the LSTM model to the processed data, which involves feature extraction, the results are evaluated using our model. In this phase, the model determines if the headline is categorized as Clickbait or non-Clickbait based on the final result.

2.7 Hyperparameters and Training

For our model, we chose 500 epochs with a batch size of 512. The dimension of the word embedding is kept at 300. Also, we will use binary-cross entropy as a loss function for binary classification and Adam as an optimizer. The parameters are kept the same for all the experiments for a fair comparison.

3 Results and Discussion

Our model focuses on the text feature to detect Clickbait. Because the dataset is unbalanced, we will adopt an F-macro measurement to compare the results as a fair measure. The model was trained using Word2vec and also without Word2vec using both the balanced and unbalanced datasets. Using random splitting, the datasets will be divided into 20% for test data and 80% for training data. The model obtained the best performance using Word2vec, where the Macro-F value reached 0.79 using the unbalanced (raw) dataset, and the Macro-F value reached 0.76 using the balanced dataset. As we can see the results of the model in Table 3, the results showed that the process of merging to make the dataset balanced did not give better. We can see that the LSTM model performed better with the unbalanced dataset because it achieved a higher Macro-F value of 0.02 than the LSTM did with the balanced dataset. The difference in the performance of LSTM is shown in Fig. 8.

We can see the performance of the LSTM model during training on detecting Clickbait in an unbalanced dataset in Fig. 9, and using the balanced dataset in Fig. 10.

Figure 9 is divided into three subfigures. In (a), the confusion matrix of the model shows the number of correct and false classifications, as we can see the model predicted 1775 of data correctly out of 2113. (b) shows the performance of the model during training, and the model reached a validation accuracy ranging from 0.81 to 0.84. In (c), the amount of loss of the model during training is shown, and we see that the loss value fell to less than 0.4 in the first Epochs and began to rise in Epoch 10.

Figure 10 is divided into three subfigures. As can be seen in (a), the model's confusion matrix displays the amount of true and wrong classifications, and the model correctly predicted 2390 data out of 3095. The model's performance during training is shown in (b), and it achieved a validation accuracy between 0.7 and 0.79. The model's training loss is displayed in (c), and it dropped to less than 0.5 in the early epochs before starting to grow.

Table 3 LSTM model results when trained on unbalanced and balanced datasets

	Without Word2vec				With Word2vec			
	Acc	Recall	Precision	Macro-F	Acc	Recall	Precision	Macro-F
Unbalanced	0.78	0.84	0.85	0.72	0.84	0.92	0.87	**0.79**
Balanced	0.68	0.7	0.67	0.68	0.76	0.71	0.78	**0.76**

Fig. 8 LSTM model results when it's trained on balanced and unbalanced datasets

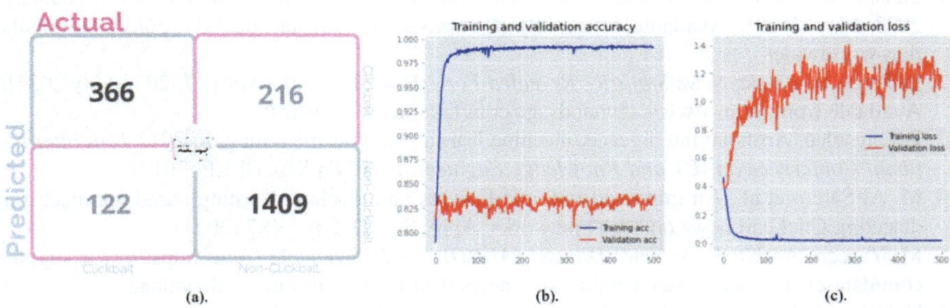

Fig. 9 LSTM performance during training on an unbalanced dataset with 500 Epochs

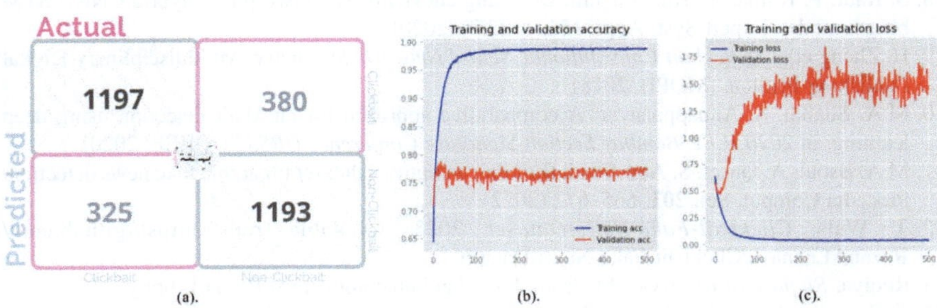

Fig. 10 LSTM performance during training on a balanced dataset with 500 Epochs

4 Conclusion and Future Work

For the first time ever, Arabic Clickbait is detected using DL in this paper, and we offer an LSTM model for its detection. The text is the key clickbait feature, and Word2vec was used to extract the features from the text. The LSTM model was trained on two Arabic datasets: the original, which was unbalanced; and the combined, balanced version of the original and a dataset of Arabic fake news. We conclude that employing Word2vec with LSTM for the Arabic content is preferable for both datasets. The LSTM model showed better performance with the unbalanced dataset, as it obtained a higher Macro-F value of 0.02 than that obtained by the LSTM with the balanced dataset. There are multiple ideas for future development, the most significant of which is to experiment with various deep DL models to find Clickbait in Arabic content.

References

1. Satista. *Number of Internet and Social Media Users Worldwide as of January 2023*. [January 2023 May 2023]; Available from: https://www.statista.com/statistics/617136/digital-population-worldwide/
2. D. Ruby, *69+ Fake News Statistics Revealed For 2023 (Updated)*. [April 17, 2023 May 2023]; Available from: https://www.demandsage.com/fake-news-statistics/
3. P. Ongsulee, Artificial intelligence, machine learning and deep learning, in *2017 15th International Conference on ICT and Knowledge Engineering (ICT&KE)*, (IEEE, 2017)
4. M. Al-Sarem et al., An improved multiple features and machine learning-based approach for detecting Clickbait news on social networks. Appl. Sci. **11**(20), 9487 (2021)
5. M. Yaseen, *Arabic Fake News Dataset (AFND)*. 2022. Available from: https://www.kaggle.com/datasets/murtadhayaseen/arabic-fake-news-dataset-afnd?resource=download
6. C. Oliva et al., Rumor and clickbait detection by combining information divergence measures and deep learning techniques, in *Proceedings of the 17th International Conference on Availability, Reliability and Security*, (2022)
7. V. Vorakitphan, F.-Y. Leu, Y.-C. Fan, Clickbait detection based on word embedding models, in *Innovative Mobile and Internet Services in Ubiquitous Computing: Proceedings of the 12th International Conference on Innovative Mobile and Internet Services in Ubiquitous Computing (IMIS-2018)*, (Springer, 2019)
8. S. Kaur, P. Kumar, P. Kumaraguru, Detecting clickbaits using two-phase hybrid CNN-LSTM biterm model. Expert Syst. Appl. **151**, 113350 (2020)
9. H. Zheng et al., *Clickbait Convolutional Neural Network. Symmetry* (Multidisciplinary Digital Publishing Institute (MDPI), 2018)
10. M.A. Shaikh, S. Annappanavar, A comparative approach for clickbait detection using deep learning, in *2020 IEEE Bombay Section Signature Conference (IBSSC)*, (IEEE, 2020)
11. M.A. Bsoul, A. Qusef, S. Abu-Soud, Building an optimal dataset for arabic fake news detection. Procedia Comput. Sci. **201**, 665–672 (2022)
12. T. Wals, *Clickbait-Fake-News-Dataset*. 2023. Available from: https://github.com/RazanALhanaya/Clickbait-Fake-News-Dataset
13. Ruqiya. *Ruqia Library*. Available from: https://github.com/Ruqyai/Ruqia-Library
14. T. Zerrouki, *Pyarabic Library*. 2023. Available from: https://github.com/linuxscout/pyarabic
15. A.B. Soliman, *AraVec 3.0*. Available from: https://github.com/bakrianoo/aravec
16. A. Luque et al., The impact of class imbalance in classification performance metrics based on the binary confusion matrix. Pattern Recogn. **91**, 216–231 (2019)

Introducing a Vision of Regulation More Complex Than the Traditional One

Olivier Lefebvre

1 Introduction

There is a doxa of regulation, in particular on the crucial topic of the "good" number of competitors in an economic sector. Examples of this doxa are: (1) any entry is good for the consumers; (2) when competition exists, it is easy to maintain it; (3) a monopoly is always bad for the consumers, etc. This doxa is a simplification. There are breaches of this doxa. To show it, a model already presented by the author is used. It is Bertrand competition [1]. In general, when Bertrand competition is concerned, the demand functions are given [2]. One prefers to deduce the demand functions from the utilities of the consumers [3]. Conditions to observe breaches of the doxa are to consider multiproduct firms, demand functions that are not "regular" (twice continuously derivable), and adopt a dynamic view of the equilibrium of the sector (which can evolve through stages).

Examples of these breaches are:

- "Any entry is good for the consumers." Indeed, if the demands are not "regular," an entry can trigger a decrease in the consumers' surplus.
- "When competition exists, it is easy to maintain it." In fact, it is possible that the number of products that are sold decreases, and the consumers' surplus diminishes. We have called this situation a profitable "close down" [4, 5].
- "A monopoly is always bad for consumers." A monopoly is neither good nor bad for the consumers. Sometimes, it is good from the point of view of the diversity of the products sold and bad from the point of view of the prices. In the case of a profitable "close down," the diversity of the products sold is not maintained if there is a competitive equilibrium and is maintained if there is a monopoly. But the prices are high.

O. Lefebvre (✉)
Olivier Lefebvre Consultant, Paris, France

© The Author(s), under exclusive license to Springer Nature Switzerland AG 2024

A. Shaikh et al. (eds.), *Advances in Emerging Information and Communication Technology*, Signals and Communication Technology,
https://doi.org/10.1007/978-3-031-53237-5_2

In this paper, one uses two methods:

- Building up tractable examples, the demand functions being not "regular."
- Making demonstrations, but only supposing that the demands are "regular" functions.

Before setting out the plan of the paper, let us make two remarks:

- One uses the symbol (*i, j*) to describe a firm selling the products *i* and *j* ($1 \leq i \leq 3$, $1 \leq j \leq 3$). And one uses brackets [.] to describe the sector. Thus, [(1, 2), 3] means that there is a multiproduct firm selling products 1 and 2, and a firm selling product 3.
- Several tractable examples are shown in the paper. Thus, the model (Bertrand competition, the demands being deduced from the utilities of the consumers) is set out. The utilities are represented in axes O u_1 u_2 u_3, the demands are deduced, and one calculates the Nash equilibrium.

Now we set out the plan of the paper:

1.1 Methodology

We show an example of entry triggering a decrease of the consumers' surplus. This is because the demand functions are not regular (in a sense that one defines). In the following, the demand functions are supposed "regular."

1.2 Results and Discussion

The following topics are studied:

- *Decrease of the number of the products sold.* Is studied the stability of the equilibrium, in that sense that the number of products sold can decrease (mainly because of profitable "close down"). Also, one demonstrates that when the demand functions are "regular," the profitable close down triggers an increase in the prices (therefore a decrease of the consumers' surplus).
- *One studies this question: in the case of entry [(1, 2)] → [(1, 2), 3], do the prices increase or decrease?* Here the demand functions are supposed "regular." One states that if the demands are linear, the prices decrease. But there is the possibility that the prices increase.
- *One studies the particular case of start-ups, when the question of the diversity of the products sold is concerned.*

1.3 Conclusion

One sets out the hypothesis that the regulators are strategists. There is not a certain doctrine showing what the regulator has to do. No wonder there are two "styles of regulation." The watchdog style of regulation relies on the criteria of welfare of the consumers. The permissive style of regulation relies on the criteria of value. For a regulator of the watchdog style, the stake is competition as intense as possible. For a regulator of the permissive style, some remedy to a saturated market is tolerated. The affair Illumina (a merger approved by some regulators and refused by others) shows that there are two styles of regulation. They are very different and incommensurable. The same phenomenon, the profitable "close down," can be interpreted in two different ways.

2 Methodology

Now one studies the consequences of an entry [1] → [1, 2], or [1, 2] → [1, 2, 3] on the consumers' surplus. If the demand functions are "regular,"[1] the reaction functions exist, and an easy reasoning on these reaction functions allows us to demonstrate that the prices decrease, and the consumers' surplus increases.

But it is easy to build up an example, with demand functions not regular, where the entry triggers a decrease of the consumers' surplus (example 1).

In example 1, the density is on O u_1, linear, homogeneous equal to 2/3, and the weight 1/3 is in A (1/3, 1) (Fig. 1).

If the sophisticated product 2 is not sold, $p_1 = 1/3$. If it is sold, $p_1 = 1/2$ and $p_2 \approx 0, 87$. There is a qualitative explanation: E_2 decreases the price p_2 just enough to discourage E_1 from attracting the consumers in A. E_1 "capitulates" and chooses a price p_1 which is more than 1/3. As one price increases (p_1 from 1/3 to 1/2) and the other decreases (p_2 from more than 1 to 0, 87), one has to calculate the consumers' surplus, in the two situations. If this surplus is S before the entry and S' after, S' is less than S since $S \approx 0, 148$ and $S' \approx 0, 125$.

This example seems artificial because of the concentration of consumers in A. But there is joint purchasing. A batch of products is bought, at some price. Therefore, a unit is bought at the same price as any, in the batch. An example is the online platform Groupon.

Suppose the product is a book, with two variants. One is a standard product (product 1). The other is a luxurious book, sold by a club of readers to its members. This market niche represents 1/3 of the potential consumers. The welfare of these consumers increases when the sophisticated product is sold. But the welfare of the

[1] "Regular" means two times derivable, the reaction functions exist, and are continuous, increasing in the plane O $p_i\, p_j$ (p_k being fixed), and there is a unique stable Nash equilibrium.

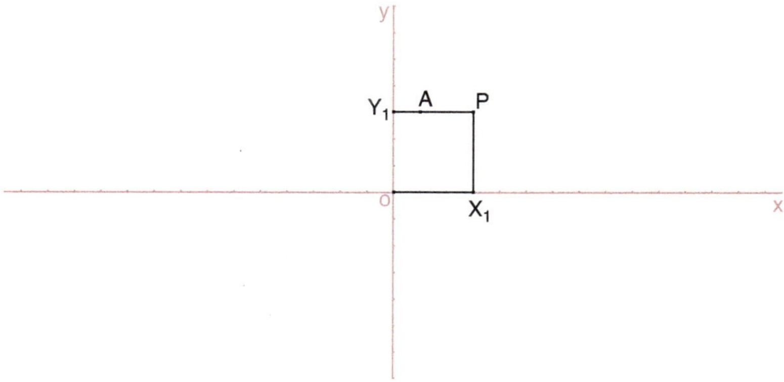

Fig. 1 *Example 1.* The density is on the segment O X_1 (1, 0) and equal to 2/3 and there is the weight 1/3 in A (1/3, 1)

consumers who can buy the standard product only decreases. And the surplus of all the consumers decreases.

To have an example of entry [1, 2] → [1, 2, 3], which triggers a decrease of the consumers' surplus, one can use example 1. The density is: 1/3 on the segment (0, 0), (0, 1) and on the segment (0, 0), (0, 1). And there is a weight of 1/6 in A (1/3, 0, 1) and in B (0, 1/3, 1). The calculations are exactly the same as in example 1.

3 Results and Discussion

One studies the following examples:

3.1 *Study of the Stability of the Equilibrium*

Here we suppose the demand functions "regular." There is an entry [1, 2] → [1, 2, 3]. Is it stable? Can product 3 be withdrawn from the market? Yes, it is possible. But there are degrees of "seriousness." What is possible is this: First, there is a merger [1, 2, 3] → [(1, 3), 2]. It is always profitable: the profit of the merged firm increases and the prices increase. The consumers' surplus decreases. This is easily demonstrated, using the reaction functions. Then, possibly, [(1, 3), 2] → [1, 2]. The asset E_3 is closed down if it is profitable. Finally, product 3 has been withdrawn from the market. The "close down" that is profitable (probably) triggers an increase in the prices (the demonstration is in an Appendix). Therefore, the withdrawal of the product from the market triggers an increase in the prices and should be prevented by the regulator. But it is awkward [5].

Suppose that at the start the profits are P_1, P_2, P_3. After the merger, they are P''_1 and P''_2. And after the close down, they are P'_1 and P'_2. One notes $\Delta P = P_1 + P_2 + P_3$ and $\Delta P' = P'_1 + P'_2$. There are four cases:

Stable equilibrium. It is if the joint profit increases when there is the entry. $\Delta \mathrm{P} > \Delta \mathrm{P}'$. An easy example is when the products are perfectly differentiated. The density is linear, homogeneous, and equal to 1/3, on the axes O u_i between 0 and 1.

Possible withdrawal of the product thanks to lateral payments. If $\Delta P < \Delta P'$, and if the "buy and close down" is not profitable ($P'_1 < P_1 + P_3$), what is possible is a withdrawal of the product thanks to lateral payments. There is a transfer t from E_2 to E_1 such that: $P'_2 - t > P_2$ and $P'_1 + t > P_1 + P_3$, and each firm makes a gain. But it is highly improbable for two reasons:

- A complex "mercato" is needed, with complex exchanges of assets between the firms E_1 and E_2 since the transfer t has to be hidden from the regulator.
- There is a problem of moral hazard. The firm E_1 can accept the transfer, then ... not to close down asset E_3. Its gain is more. Let us imagine an example:

  At the start: $P_1 = P_2 = P_3 = 100$

  E_2 gives 60 to E_1

  E_1 uses 60 to buy E_3 and closes down it.

  After, $P'_1 = P'_2 = 180$. Each firm makes a gain thanks to the lateral payment: 180–60 = 120 > 100.

  But E_1 can betray E_2, not closing down E_3 ... after having cashed out 60.

  The profit of E_1 is 220–60 = 160 > 120 (the profit of E_1 after the merger is 220, which is more than 100 + 100 since the merger is profitable).

  The firm E_2 has lost 60, even if its profit after the merger is more than 100.

It is example 4 (Fig. 2).

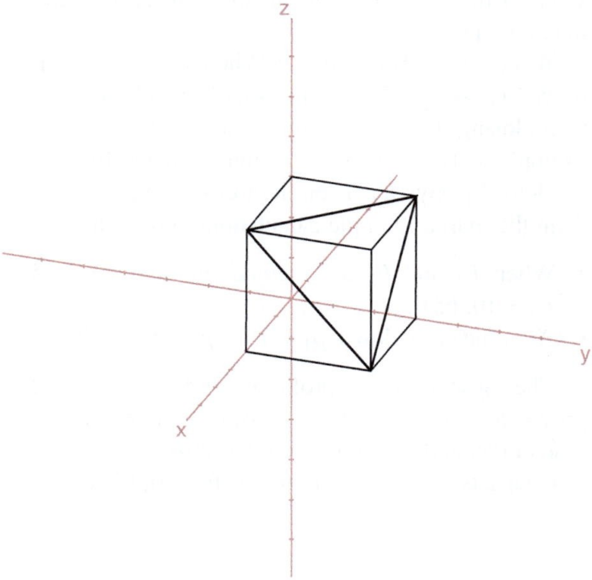

Fig. 2 *Example 4.* The density is linear equal to 1/3√2, on the second bisector of the external faces $u_i = 1$

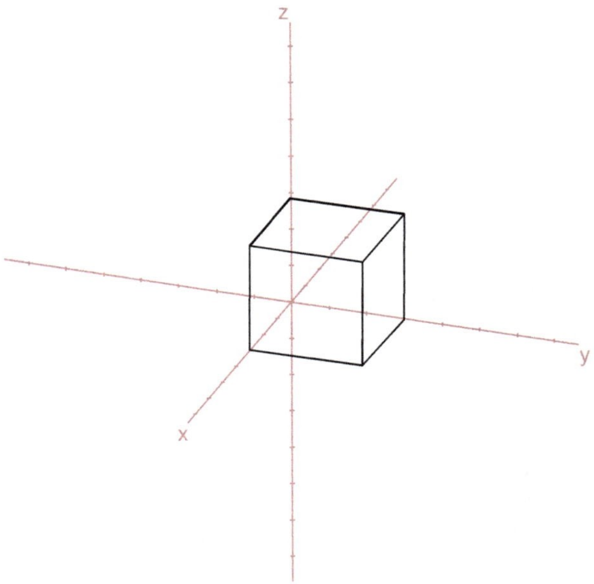

Fig. 3 *Example 5*. In the plane $p_3 = 0.95$, the density is linear, equal to 1/2, on the axes O u_1 u_2

The "buy and close down" is profitable and not the "close down." The buy and close down is profitable ($P'_1 > P_1 + P_3$) and the close down is not ($P'_1 < P''_1$). The merger being always profitable is possible but a second stage, the close down, should not occur. However, if for any reason the managers believe that the merger cannot succeed, the withdrawal of product 3 could occur (since the "buy and close down" is profitable). A merger can fail, for instance, because of a dispute between the two teams of managers. The "buy and close down" profitability is an incentive to merge, since in the case of failure of the merger, one can close down the bought asset, making a profit.

Withdrawal of the product. When the close down is profitable, the "buy and close down" is also profitable, and should be chosen. There are two stages, merger and then closing down the bought asset. Product 3 is withdrawn from the market. It is example 5. The density is distributed like in Fig. 3.

Here, E_1 buys E_2. Then, it closes it, and it is profitable. Product 2 is withdrawn from the market. Simple calculations show that:

- When E_1 and E_2 are merged, $p_1 = p_2 = 0{,}35$, $p_3 = 0, 65$, $P = (0{,}35)^2/2$, $P_3 = (0, 65)^2$.
- After the close down: $p_1 = 1 / 2$, $p_3 = 0, 95$, $P_1 = 1 / 8$, $P_3 = 3 / 4\,(\mathrm{O}, 95) = 0, 71$.

The close down is profitable: $1/8 > (0,\ 35)^2/2$. Notice that the profit of E_3 grows, also: $0, 71 > (0, 65)^2$. And the prices increase (this is demonstrated in the case of demand functions "regular"): p_1 from 0, 35 to 1 / 2, p_2 from 0, 35 to 1 (since product 2 is no more sold) and p_3 from 0, 65 to 0, 95.

Since the equilibrium [1, 2, 3] is not stable, there are two attractors: the monopoly [(1, 2, 3)] and the equilibrium with two products [1, 2]. The regulator has a choice to make. He can allow the monopoly and there is the attractor [(1, 2, 3)]; there is more diversity of the products but the prices are high. Or he can choose the equilibrium with two products, and there is less diversity, but the prices are low.

In this example, obviously the regulator should choose [1, 2]. In the case of monopoly, the diversity is maintained, but the consumers' surplus is very small: $(0, 025)^2/2$. (Notice that if the density is in the plane $p_3 = 1$, the example is simpler, but the consumers' surplus is 0 in the case of monopoly.) The consumers' surplus if product 3 is withdrawn from the market is $1/16 = 0, O62$.

The explanation of the profitable close down is that the product that is withdrawn from the market, 2, creates low utilities compared to 3. It is clear that E_3 is a more efficient seller than E_2. Therefore, when E_1 does E_3 the favor of withdrawing product 2 from the market, E_3 can increase its price. As the prices are strategic complements, E_1 can increase its price, and this allows the "close down," which is profitable (the profit of E_1 increases).

3.2 *Study of the Entry [(1, 2)] → [(1, 2), 3]*

This entry is uneasy to study because a multiproduct firm is involved. If the demands are linear, the prices decrease and the consumers' surplus increases. But it seems that other cases are possible.

Here, we suppose that the demand functions are "regular." One supposes the symmetry of the utilities u_1 and u_2. If p_3 increases from 0 to 1, there are four points moving on the first bisector, like in Fig. 4.

A: point where $\partial P_3/\partial p_3 = 0$.
B: point where $\partial P_1/\partial p_1 = 0$ and $\partial P_2/\partial p_2 = 0$.
M: point where $\partial P_1 + P_2/\partial p_1 = 0$ and $\partial P_1 + P_2/\partial p_2 = 0$.
P: point where $P_1 + P_2 + P_3$ is maximal (for p_3 fixed).

One demonstrates:

- A moves to the NE.
- B moves to the NE.
- M is at the NE of B.
- P is at the NE of M.

0 A B M P 1

Fig. 4 The four points, A, B, M, and P, on the first bisector between 0 and 1, at the start ($p_3 = 0$)

If M moves toward the NE, it is clear. It is the case if the demand functions are linear:

- A catches up B: it is the equilibrium [1, 2, 3].
- Then, A catches up M: it is the equilibrium [(1, 2), 3].
- Then. A catches up P and after (p_3 higher) the maximum of $P_1 + P_2 + P_3$ is reached (in this point, $\partial P_3 / \partial p_3 < 0$).
- Finally, for $p_3 = 1$, it is the monopoly selling 1 and 2, [(1, 2)].

Therefore:

- When the merger occurs the prices increase, the joint profit of the merging firms increases, and the profit of the competitor increases. (Notice that there is a general demonstration of this, without a hypothesis of symmetry. It requires the demand functions "regular.")
- The price p_3 is higher, when the monopoly selling 1, 2, and 3 optimizes its profit.
- Then $p_3 = 1$, M moving toward the NE, the prices increase. The prices decrease when there is the entry [(1, 2)] → [(1, 2), 3].

But if M moves to the SW, p_1 and p_2 decrease, when one moves from [(1, 2), 3] to [(1, 2)] (that is to say, product 3 is withdrawn from the market, there is a monopoly selling 1 and 2). The entry triggers an increase in the prices p_1 and p_2, and the consumers' surplus could decrease.

The direction of the move of M depends on the sign of $\partial^2 P_1 + P_2/\partial p_1 \partial p_3$, equal to $\partial^2 P_1 + P_2/\partial p_2 \partial p_3$. If it is positive, M moves toward the NE. It is the case when the demand functions are linear. If it is negative, M moves toward the SW.

Also, it is not sure that the quantities $\partial^2 P_i/\partial p_i \partial p_j$, exist.

Finally, an entry [(1, 2)] → [(1, 2), 3] triggers a decrease in the prices and an increase in the consumers' surplus, if a condition is fulfilled. It is the case when the demand functions are linear. But if another condition is fulfilled, the entry can trigger an increase in the prices p_1 and p_2 and the consumers' surplus could decrease.

In a particular case, the entry triggers a decrease in the prices: if the utilities of the three products are symmetrical and the close down is profitable. Consider the chain of three operations:

[(1, 2), 3] → [1, 3] → [(1, 3)] → [(1, 3), 2]. The first operation, a close down which is profitable, is supposed to trigger an increase of the prices. Then, the appearance of a monopoly selling two products triggers an increase in the prices. And the entry triggers a decrease in the prices, since they should be at the same level as at the start.

3.3 *Withdrawal of a Product from the Market: The Case of Start-Ups*

Suppose a start-up that will succeed with the probability p or fail with the probability $1-p$. Therefore, we suppose a moderate uncertainty. If the start-up succeeds and enters the market, its profit will be P_3. There is an equilibrium with three firms selling products 1, 2, and 3 and making the profits P_1, P_2, and P_3. The price of the start-up is $p\,P_3$ (to buy it, it is necessary to pay for $p\,P_3$ plus a premium). If E_1 buys E_3, with the probability p it makes the profit P'_1 (when there is [1, 2]), and with the probability $1-p$ it makes the profit P'_1. Therefore, the "buy and close down" is profitable if: $P'_1 > p\,P_1 + (1-p)\,P'_1 + pP_3$. The second member of the inequality is the expectation of a profit of E_1 if it does not buy then close down E_3. It added the price to pay to buy E_3. The condition is: $P'_1 > P_1 + P_3$. It is the same formula as when E_3 is not a start-up. Also, the close down is profitable if: $P'_1 > P''_1$. Again, it is the same formula as when E_3 is not a start-up.

For instance, take example 5. If E_2 is a start-up, it will be bought by E_1, then either E_2 fails (probability $1-p$) or succeeds (probability p) and is closed down. In any way, product 2 is withdrawn from the market. This result does not depend on the probability p.

4 Conclusion

Since the model used shows breaches in the doxa of regulators, it is not surprising if a regulator hesitates between choices he has to make, or that there are different "styles of regulation." Indeed, there has been a discussion on the nature of regulation for a long time [6]. We shall insist on (1) hesitation between two attractors and (2) the existence of two different and incommensurable styles of regulation, the watchdog one and the permissive one. The model used will be useful to explain why there are these two different styles of regulation.

4.1 Hesitation Between Two Attractors

The two attractors are like in example 5, the monopoly selling three products, [(1, 2, 3)], and the equilibrium with two products [1, 2] (this, when the close down is profitable). The monopoly is favorable to the diversity of the products sold but chooses high prices. A monopoly never withdraws a product from the market (its profit will decrease). And it always creates a new product, if it is possible (its profit increases). A firm with competitors will buy another firm and close down the asset if the close down is profitable. A product is withdrawn from the market. And a firm with a competitor will never create a new product (if the close down is profitable), since its profit would decrease. In general, the regulator cannot model the economic sector, calculate the consumers' surplus, and make a comparison. Sometimes, he can oppose the merger, if he suspects that it will be followed by a close down. Of course, he prefers [1, 2, 3] to [1, 2], since the consumers' surplus is more (a product more sold, lower prices). In the examples quoted by Schapiro et al., three are concerned: the "contrafactual" (the prohibited alternative) is buy and close down. Start-ups are concerned. It is the same reasoning as when the firm that is bought sells a product, and the probability p of success of the start-up does not matter.

4.2 Existence of Styles of Regulation

There are two styles of regulation, the watchdog one (for instance, to prohibit a close down) and the permissive one (to tolerate a close down). One reason for the existence of these two styles of regulation is that when a regulator makes a decision, he has to prove that it is justified, and it is often uneasy [7]. He has to describe a contrafactual (an alternative that is virtual) and this is of course uneasy [7]. One can also explain the two styles by examining what is a close down. When a close down that is profitable occurs, the prices increase and the consumers' surplus decreases (this is demonstrated when the demand functions are regular). Therefore, the regulator of the watchdog style attempts to prohibit the close down, since he is sensible to the criterium of consumers' surplus. But there is another interpretation: the product that is withdrawn is not "interesting." There are two possibilities: (1) it is not enough differentiated from the product of the competitor (product 2 in our model). This triggers a lower price of 2 (Bertrand paradox). When it is withdrawn, the price p_2 increases and also the price p_1 (since the prices are strategic complements): (2) it triggers low utilities for the consumers buying it. This obliges E_2 to decrease its price p_2, to not have market shares stolen (by E_3). And E_1? Since products 1 and 3 are inside E_1, there is no stealing of market share [6]. But a unit not sold at a high price p_1 is sold at a low price p_3. It is cannibalization. When product 3 is withdrawn, it ceases, E_1 can increase the price p_1, and E_2 increases p_2. The product that is withdrawn is deemed uninteresting either due to insufficient differentiation compared to the remaining product of a monoproduct firm or because it potentially cannibalizes other products offered by the multiproduct firm. If it is withdrawn from the market, it does not matter. The profits of the firms in the sector will increase. The damage for the consumers is small.

The features of the two styles of regulation are shown in Table 1.

The conflicts between regulatory bodies have often been noticed, for instance, in the field of water. Let us quote an example, the merger Illumina/Grail. Illumina, a leader in the biological sector, wanted to buy a start-up, Grail. The merger was authorized by a US court but refused by the FTC (Federal Trade Commission) and

Table 1 The features of the two styles of regulation

Watchdog style	Permissive style
Criterium of consumers' surplus	Criterium of value
Preserving the diversity of products	Indifference to pseudo-diversity
Consumers' interests	Dynamism of innovative firms, disruption
Governmental agencies, FTC (Federal Trade Commission), European Commission	DOJ (Department of Justice), US courts, firms, disruptive entrepreneurs

the European Commission. It did not occur. It was an attempt of vertical merger, so it is not exactly the topic of the paper. What was feared was too much "market power." But it shows a conflict between regulatory bodies.

The examples quoted in Schapiro et al. can be classified, using the words of this paper. The contrafactual is described.

Buy and manage. Theratec/Heartware, Medtronic/Covidien, Pfizer/Hospira, J and J/Actelion, Applied Materials/Tokyo Electron, Bayer/Monsanto, Halliburton/Baker Hugues, Dow/Dupont.

Buy and close down (case of a start-up). Questor/Synacthen, Novartis/GSK, General Electric/Alstom (turbines).

Appearance of a monopoly. Nielsen/Arbitron, Deutsche Börse/Euronext.

The French economist Philippon has clearly chosen the watchdog style [8, 9].

An example of arguments in favor of the permissive style is in Christensen [10, 11]. The value of the innovative product (the conditions are: novelty, utility, and affordability) is high and can increase, while the value of the old product decreases, notwithstanding the investments made by the providers of this product [12]. In the words of this paper: P_3 is high, P'_1 and P'_2 decrease so $P'_1 > P_1 + P_3$ is impossible. The value of the innovative product 3 is high, and the buy and close down is impossible.

Notice that when a dominant firm accumulates patents (which are neutralized) and start-ups (which are closed down), to stifle the appearance of a rival technology that is not polluting, it is another story. The strategy is coherent. The goal is to avoid new rules, prohibiting pollution. Examples are quoted in chemistry. Exemptions in favor of a product require three conditions: (1) it is proved that it is toxic, (2) there is an essential need, and (3) there is no alternative. Stifling the appearance of a new technology means that the condition (3) will apply. And the exemption can continue. The value of the new product would be high, but a condition for that is the appearance of new rules.

Appendix

Here, one demonstrates that in the case of profitable close down, the prices increase.

The demonstration is a little subtle and relies on a reasoning on the reaction functions.

Let us define:

- $p_i = p_{i0}$ is "adapted" to p_{j0} and p_{k0} if: $\partial Ri/\partial pi\ (p_{i0}, p_{j0}, p_{k0}) = 0$. That is to say, p_{i0} is the best response of E_i to p_{j0} and p_{k0}.
- Let us call O $p_i\, p_j\, [p_k = p_{k0}]$ the axes O $p_i\, p_j$ if $p_k = p_{k0}$, with the reaction functions R_i and R_j (for $p_k = p_{k0}$). If p_{i0} is "adapted" to p_{j0} and p_{k0}, in O $p_i\, p_j\, [p_k = p_{k0}]$ the point (p_{i0}, p_{j0}) is on R_i. Or, in 0 $p_i\, p_k\, [p_j = p_{j0}]$ the point (p_{i0}, p_{k0}) is on R_i.

Suppose E_1 sells product 1, and E sells products 2 and 3.

There is a Nash equilibrium (p_{10}, p_{20}, p_{30}).

In O p_2 p_3 [$p_1 = p_{10}$], p_3 increases as far as 1. The representative point moves vertically as far as $p_3 = 1$.

One supposes a unique, stable Nash equilibrium when product 3 is no more sold.

In O p_2 p_3 [$p_1 = p_{10}$], the representative point moves on the line $p_3 = 1$ as far as the reaction function R_2 ($p_2 = p'_2$, $p_3 = 1$). When $p_2 = p'_2$, E is "adapted" to $p_1 = p_{10}$ and $p_3 = 1$. In O p_1 p_2 [$p_3 = 1$], the representative point (p_{10}, p'_2) is on R_2.

In O p_2 p_3 [$p_1 = p_{10}$], the representative point went to R_2, sliding toward the left or the right:

- If p_2 increased (from p_{20} to p'_2), the best response of E_1 is an increase (because p_2 increased and p_3 also). Therefore, in O p_1 p_2 [$p_3 = 1$], the representative point goes to the Nash equilibrium thanks to increases of p_1 and p_2. At the start, the point is on the lower strand of R_2. The prices p_1 and p_2 increase.
- If p_2 decreased (from p_{20} to p'_2), one does not know the best response of E_1, because p_2 decreased and p_3 increased. If the best response of E_1 is to increase p_1, the reasoning is the same as in the case dealt with. At least the price p_1 increases. The hypothesis of p_1 decreasing is impossible. The representative point would be on the higher strand of R_2. At the start, the profit P_2 is less than Max $P_2 + P_3$ (if $p_1 = p_{10}$). When the point slides alongside R_2, the profit P_2 decreases. Therefore, it cannot be more than Max $P_2 + P_3$ (if $p_1 = p_{10}$).

The three prices increase, or at least two (the price of the product which is no longer sold and the price of the competitor of the multiproduct firm). If the three prices increase, the consumers' surplus decreases. If two only increase, it is not sure. In the symmetrical case (symmetry of products 1 and 2, that of the competitor of the multiproduct firm, and that of the multiproduct firm which is still sold), the three prices increase. In this case, the consumers' surplus decreases.

References

1. M. R. Baye, D. Kovenock, Bertrand competition, in *The New Palgrave Dictionary of Economics,* (2016)
2. B. Guerrien, G. Ozgur, *Dictionnaire d'analyse économique (Dictionary of Economic Analysis)* (La Découverte, Paris, 2012)
3. O. Lefebvre, *Game Theory and the Stakes in the Telecommunications Industry* (LAP Lambert Academic Publishing, 2014)
4. O. Lefebvre, About competition between firms: Equilibrium of disruption? J. Game Theory **10**(1) Scientific and Academic Publishing (2021)
5. O. Lefebvre, Modeling innovation and regulation thanks to game theory: Bertrand competition. Heritage and Sustainable Development **4**(2), 165 (2022) <hsd.ardascience.com>
6. S. Breyer, P. W. MacAvoy, Regulation and deregulation, in *The New Palgrave Dictionary of Economics*, (2016)
7. G. Federico, F. Scott Morton, C. Schapiro, *Antitrust and Innovation: Welcoming and Protecting Disruption* (The University of Chicago Press, 2020)
8. T. Philippon, *The Great Reversal: How America Has Given up on the Free Markets* (Harvard University Press, 2019)

9. T. Philippon, The great reversal: How America has given up on free markets. Conference on line on <parisschoolofeconomics.eu>
10. J. L. Bower, C. M. Christensen, Disruptive technologies: Catching the waves. Harv. Bus. Rev. (January—February 1995)
11. C.M. Christensen et al., *The Clayton M Christensen Reader* (Harvard Business Review Press, 2016)
12. C. Christensen, *The Innovator's Dilemma* (Harvard University Press, 2016)

New Technology for Lean 4.0: Prospects For the Industry of the Future

Dbaich Douae, Aziza Mahil, and Mohamed Tabaa

1 Introduction

Currently, Industry 4.0 is provoking an industrial revolution, based on the digitalization of companies as well as organizational, technological, and cultural changes. This digitization and organization aims to produce quickly, efficiently, and at a lower cost while guaranteeing the security of operations on the one hand, and adapt to new consumption patterns by producing customized services and products in limited and/or reduced quantities. Thus, the industrial unit becomes agile and reactive thanks to the simplification of workers' tasks and the automation of manufacturing operations, which relieves their workload. For a long time, this appeasement during the manufacturing exercise has been neglected in favor of reducing business costs.

This can only be achieved by making full use of new information, intelligent communication, and robotics technologies such as IoT, virtual reality, artificial

D. Douae (✉)
Laboratory Interdisciplinary Business Engineering, Soft Skills, Management and Law (LIASMAD), Faculty of Law, Economics and Social Sciences of Ain Sbaa (FSJESAS), Casablanca, Morocco

Multidisciplinary Research and Innovation Laboratory (LPRI), Moroccan School of Engineering Sciences (EMSI), Casablanca, Morocco
e-mail: d.dbaich@emsi.ma

A. Mahil
Laboratory Interdisciplinary Business Engineering, Soft Skills, Management and Law (LIASMAD), Faculty of Law, Economics and Social Sciences of Ain Sbaa (FSJESAS), Casablanca, Morocco

M. Tabaa
Multidisciplinary Research and Innovation Laboratory (LPRI), Moroccan School of Engineering Sciences (EMSI), Casablanca, Morocco
e-mail: m.tabaa@emsi.ma

© The Author(s), under exclusive license to Springer Nature Switzerland AG 2024

A. Shaikh et al. (eds.), *Advances in Emerging Information and Communication Technology*, Signals and Communication Technology,
https://doi.org/10.1007/978-3-031-53237-5_3

intelligence, big data, and cybersecurity, which are powerful and global management tools, surpassing mechanical production. Mass production is being replaced by automated and intelligent production.

It is important to note that Industry 4.0 was recently conceived as a technological need whose main objective is the production of series while taking into account the golden triangle: cost, quality, and time.

In this article, we first focus on the views of some authors and researchers who have contributed to the richness of this article by providing evidence. These elements have allowed us to understand the essence of the concept of "Industry 4.0" with all the revolution it has caused in the manufacturing sector. We then address the new industrial reality characterized by the change of paradigms at the enterprise level, through the integration of the new technological and digital tools and means mentioned above. These tools and means consist of installing, monitoring, and programming machines and manufacturing equipment in real time throughout the stages of the production process.

Next, it is important to highlight the impact of the new technology on the evolution of Lean 4.0, in order to highlight the digitalization of procedures allowing operators within companies to focus on tasks with high added value. This increases productivity on the basis of manufacturing at the customer's request and offers the customer an ever more qualitative service.

Finally, we focused on the objective expected by Industry 4.0, which is defined by the change in this industry. This mutation will move from a logic of mass production (quantity) toward a logic that we call personalization.

2 State of Art

The challenge of manufacturing methodology involves various mature changes. These changes are defined in the ability to manufacture personalized wealth, moving from human labor to the machine with numerical control (to remote control of production), which must make use of the design of the services enabling it to differentiate itself in relation to manual work in production.

Indeed, this challenge requires a very good control of all levels of interference and reaction of operators, machines, products, and customers in real time. These levels of interference and reaction will make the machine a tool at the service of man to evolve and progress, knowing that man has always needed tools to increase his strength and shape the world.

Now, according to this vision, the tool whether it is a machine or another, there is indeed a reciprocity in the relationship between man and machine. This reciprocity seems to us increasingly identified over the years and thus becomes omnipresent in the personal and professional life of the human being.

The challenge in question comes to overcome some very decisive differences in terms of competitiveness. Indeed, the penetration of the new technology aims to change the context of the so-called quantity production toward that of mass customization in order to highlight the centrifugal place where the customer must henceforth be located. In this vein, the European Commission's Green Paper on the Environment, Public Health, and Consumer Protection will be extended to cover corporate governance and industrial competition policy ideologies. This competition, therefore, will position us at a higher level: customer-oriented relational competition by accompanying it throughout its purchase journey through maximum targeting and customization of shares.

This surpassing, which we describe as privileged, will lead us to an approach that focuses on all the mobilizing capacities of different actors: resources and strategic skills. In a way, this last approach proposes the adoption of an integrative and strategic dynamic based on the development of all kinds of productive projects. One dynamic emphasizes the relationship between the producer and the distributor. Indeed, far from unilateral interests, the interdependence of their activities remains an absolute truth. That said, both actors are forced to better manage each other's objectives and conflicts of interest [1].

This situation, between the two actors in question, even prompted some authors to talk about what can be called conflicts and resistance strategy. This so-called cooperative aspect of the relationship between producers and distributors is meant to be a partial, not a complete, negotiation, by ignoring the importance of their tasks within economic activity as a whole by being closer to the consumer [2].

The cooperation relationship between the two actors will evolve toward a fair and reciprocal partnership. The negotiation, once described as unilateral, became creative and conciliatory between the two actors.

Beyond what is technological, this last perspective calls for a culture defined as inventive adapting to this new sphere of industry called Industry 4.0. Within this sphere three main changes are required:

- The transition from individual and specialized work to another integrative and interdisciplinary work
- The transition from a control and management process, and therefore decision-making, to another process based on decisions of all actors at the operational level
- Moving from a strong, directive mindset to a flexible mindset that can adapt to new technologies

The three elements mentioned above call for a field approach, which remains one of the main conditions for the success of a future industry.

Indeed, the role of managers of industrial companies will lie in the involvement and mobilization of teams on the future progress imposed by the notion of competitiveness of companies in terms of profitability and sustainability.

The implementation of this new management philosophy will allow managers to overcome waste in production, ensuring the retention of tasks that contribute added value as opposed to those that do not. This management philosophy aims to identify unnecessary activities in order to reduce the resources associated with this problem and replace them with industrial resources capable of creating more values: the robot, on the other hand.

In view of the above, it should be stressed that Lean Management presents itself as a tool that tries to build new solutions of organizational order well adapted to the current context. This will allow this tool to anchor a new culture based on operational excellence as a systematic and methodical approach capable of leading any organization to maximize their productivity performance while ensuring the concept of reduction of production costs [3].

If the processes implemented by Lean Management 4.0 highlight a topic that seems to be striking and that lies in the competitive pressure on the one hand and in the development of technology serving the industry on the other hand, they will lead us to an industrial context oriented toward a new sphere: Industry 5.0. In this sphere, the role of the human will once again emerge as an indispensable catalyst in the production process.

This role of the human has been the subject of an appearance in a context characterized by the COVID-19 pandemic. This change, which is described as a paradigm, therefore calls for a sort of juxtaposition between two contexts: the digital processes (Industry 4.0) on the one hand, the human, the work organization, and the technology (Industry 5.0) on the other hand [4].

Indeed, the transition from Industry 4.0 to that of 5.0 will undoubtedly impact the main objectives of firms that will find themselves in an obligation to introduce new functionalities. However, our reflection is at the heart of the strategic orientations of the European Commission in 2018. These guidelines call for several projects aimed at innovation actions located on the strategy of future industrial units. Industrial units that try new types of innovation processes and development of agile production on the one hand and redevelopment of spaces by equipping it with digital manufacturing tools and technologies on the other. According to Laurent Dupont, "Do it together and transition to Industry 5.0," it should be noted that the use of innovative technologies in the "Factories of the Future" project has been claimed in order to carry out the revolutions of industrial orders. This claim will no doubt highlight certain perimeters of actions and experiments oriented toward the materialization of plants 4.0 in the first stage to put it into application as complementarity in the second stage [5].

In fact, Industry 4.0 remains a concept that calls for the integration of certain commercial and organizational technological aspects; this is in order to deploy so-called efficient processes in a way that makes them more accessible and more transferable between different sectors. The integration of the above-mentioned aspects must not under any circumstances neglect human activities because it seems appropriate to combine the two aspects, that is, the technological and human aspects. Moreover, the human–machine interface formalizes this so-called alternative process in order to promote the human values that actively contribute to production within firms and thus to their profitability by interacting between two as a participatory strategy that aims to involve individuals in the process production while integrating other societal and environmental economic issues. Hence, there is a crucial need to conduct an awareness campaign with companies in order to make them adhere to the actions of experiences mentioned on the screen [6].

In the same context, we note that the White Paper published by the European Commission attempts to highlight the expected future of industrial production. It aims to integrate new technologies with a view to bringing about profound change in terms of the gain, productivity, number, and quality required, without forgetting the human role that remains very significant in the overall production process; processes based on digital technologies: Internet of Things, systems, cloud computing [7]. This leads us to the idea that production technology is done around production workers. The latter maintain relations of industrial activities both ordered and coordinated between them. For this reason, we note that, in industrial units around the world, there is an emphasis on the role of people involved in production. These individuals must demonstrate their ability to adapt to this new paradigm, which can be described as fundamental and different. This paradigm necessitates the integration of artificial intelligence and augmented reality to complement human work, thereby enhancing the relevance and achievement of objectives [8]. This report on complementarity leads us to the idea that the lack of digital integration in several production activities is not due to technical problems but to the interaction between man and machine (electronic system). On the other hand, this can lead us to another new conception that we call a socio-technical system, a concept that tries to reconcile the Human–Technology–Organization (MTO) [9].

In effect, socio-technical activities link the social (command and coordination between teams) and the technical system on the one hand and on the other hand bring together people within organizational structures, which leads us to the conclusion that the creative process of liaison already mentioned represents in some way the human talent par excellence. This therefore offers great potential for developing sustainable manufacturing models. This model necessitates a delicate equilibrium among three fundamental elements: product, process, and system. This balance is geared toward optimizing production—a sequence of activities involving measurement, analysis, prioritization, and implementation. The overarching goal is to enhance productivity [10].

Through the optimization of production, the majority of the manufacturing companies allow themselves to better understand the management of these production processes and their productivity systems, which allows them to increase the sales operations and subsequently to consider putting expansion action plans for the future [11]. This cannot be done without the establishment of a rigorous system of performance control by each member of the team in order to be part of a policy aimed at eliminating so-called waste practices in the image of the approach of the scientific organization of work. To do so, the manufacturing company must use an organization respecting a number of steps to follow:

- Identify the problems that put a brake on production and determine the resources needed to achieve the objectives.
- Provide accurate, objective, and quantitative data to improve processes and increase productivity.
- Define objectives that are both achievable and achievable.

- Identify frequency measurement and calibration methods in the production process.
- Manage the obstacles to the optimization process: lack of information about teams, communication, or specific software.

As a result, the organization allows the company to adapt to market demand regardless of the variants they have faced on the one hand and to update optimal parameters especially as products and raw materials change continuously while knowing that the optimal production remains determined by the price, quality, delivery time, and environmental compliance [12].

3 New Technology (Digitalization, Big Data Analytics, Artificial Intelligence, and Virtual Reality)

The development of the future industry has accelerated considerably over the last 10 years. In fact, it is not a question of applying certain innovative tools aimed at modernizing them, but above all of shifting industrial activity in companies toward digitization. Thus, the industry becomes a catalyst for the revolution of an agile model within industrial units highlighting a better use of technologies and data to integrate the manufacturing company into the industrial ecosystem [13].

But cobots, artificial intelligence, the Internet of Things, Big Data, virtual reality, Cloud computing, Metaverse, and other advanced technological tools must contribute to the improvement of the attractiveness of various industrial applications and therefore to the modernization and improvement of the economy in a significant and sustainable way. The Chinese and Japanese model illustrates in this context the properties of this future industry and in particular their economic strategy [14].

It is important to emphasize that this industrial digital transformation of factories was initiated by the Germans through Industry 4.0 or otherwise called Smart Manufacturing (Smart Manufacturing). This term denotes a novel industrial approach utilizing advanced technologies for interconnected means of production. Transformation is seen as a source of inspiration for several global initiatives.

In view of the above, it can be said that this German revolution has brought about radical changes in systems on the one hand and processes on the other, as well as in management strategies, business models, and manpower.

In addition, it is therefore necessary to merge the information and communication techniques with the industrial technologies mentioned above, allowing the virtual space to be linked to the physical world. This merger is intended to give birth to a new generation of connected factories known as "smart factories" capable of improving the performance of companies in terms of means of production, product customization, and on-time delivery on the market [15]. This performance improvement cannot be achieved without the adoption of three types of integration:

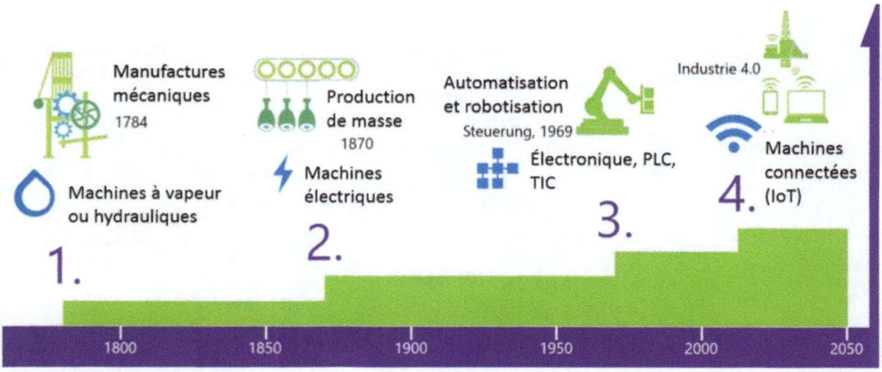

Fig. 1 Industrial revolution

- Horizontal: create or link other occupations in the same occupation. The objective is to penetrate the market and thus benefit from potentially significant economies of scale.
- Vertical: A strategy that allows companies to control their value chain by choosing to diversify or integrate new activities into their production activity.

Point-to-point: It is essential to carry out modifications of various kinds with prudence, method, and vigilance whatever the goal pursued. Indeed, we must always start with the beginning (Strategic Analysis), which explicitly contains strategic objectives in order to simplify things afterward and translate them into action plans [16].

Referring to the above, several countries have adopted the same concept for the overall improvement of their industry. We cite as an example the Chinese industry that has integrated digital and industrial technologies, "the French industry" otherwise called La nouvelle France Industrielle and "Industrual Internet and advanced manufacturing partnership in the USA" (Fig. 1).

The diagram highlights the four different stations of the industrial revolution that gave rise to a series of engagements within manufacturing companies. These commitments have required a new generation of skills able to adapt to this new industrial sphere for a better way of producing (the one best way). In other words, the challenge of companies remains mainly to find the right employee for the right job and the right post, hence setting up a collaborative system between the machine on the one hand and the human being on the other. It is in this same context and in this logic that Industry 5.0 has introduced itself to respond to this problem: human–machine.

Industry 5.0 has been replaced in the Fifth Basic Plan for Science and Technology by the Science, Technology, and Innovation Council. Indeed, the following are the consecutive stages of societies in the history of humanity:

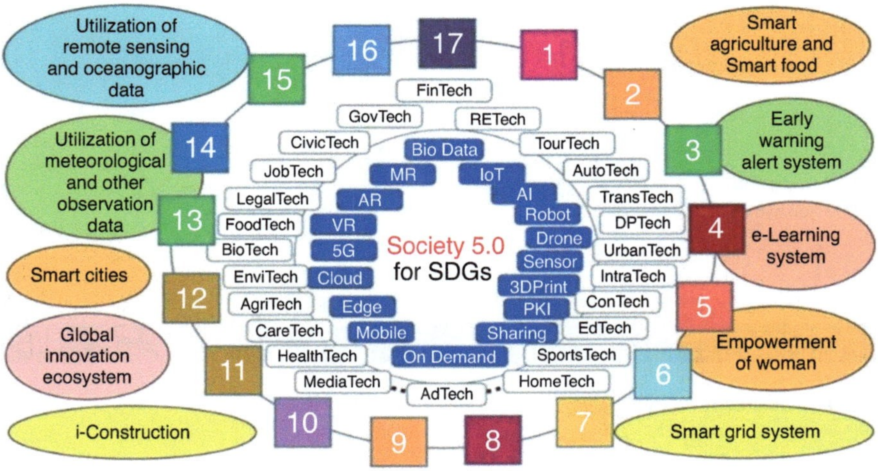

Fig. 2 Japanese industrial model

- Company 1.0: Group of people who hunt and gather in coexistence.
- Company 2.0: Group of people based on an agricultural culture marked by a growing organization and the notion of nation.
- Company 3.0: A company that promotes industrialization with the objective of mass production.
- Company 4.0: Information society that produces significant added value through information networks.
- Company 5.0: Human-centered enterprise 4.0, which has undoubtedly created significant industrial capacity (Fig. 2).

The objective of this development model is to create a model focused on man by offering him another quality of life and thus providing him with the necessary means and services. This innovative Japanese growth strategy has enabled this country to face several strategic problems such as the aging of the population, the decline of creation, and the sharing of solutions with the world through this "Made in China" model as a 10-year project focused on smart manufacturing in diverse sectors, as shown in Fig. 3.

The main objective of the MIC 2025 model is to cope with foreign technological imports, this being said creating competition by focusing on the domestic manufacturing process. This strategy is based on strengthening quality and productivity, increasing investment, creating value, and smart manufacturing.

In view of the above, the plant is becoming intelligent thanks to automation and digitization technologies. These latest technological advances actively contribute to the success of the industry of the future [17]. A few key multidisciplinary elements have contributed to the implementation of global industrial revolutions:

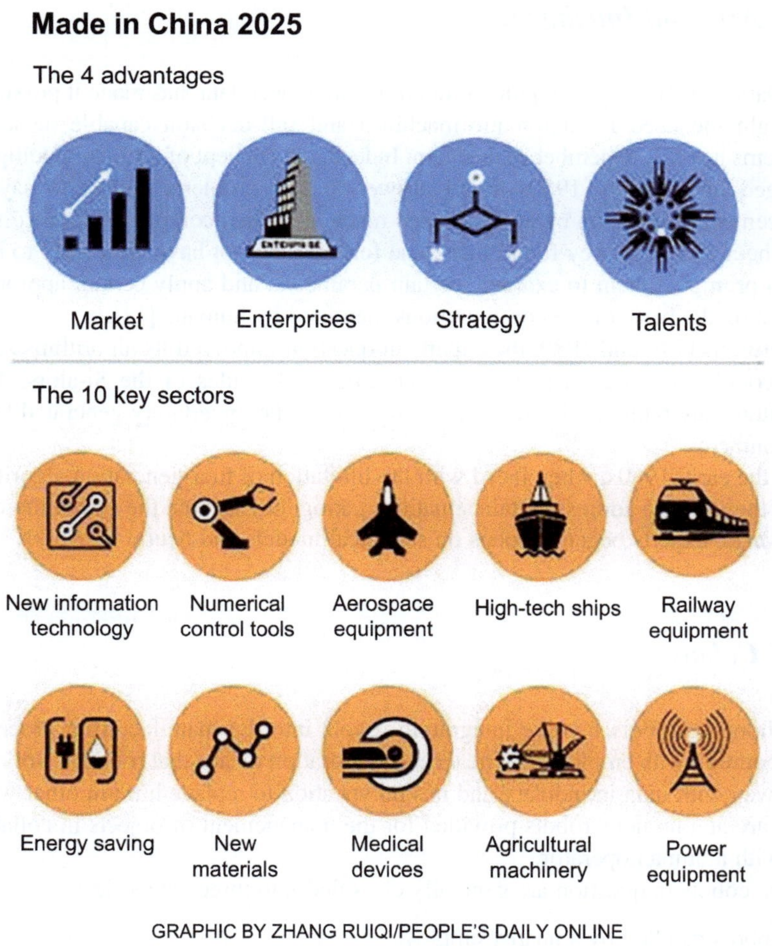

Fig. 3 Made in China 2025 Strategy

3.1 Industrial Internet of Things

The IoT as a connection tool made it possible to communicate all objects to each other. These objects exchange considerable amounts of data conveyed by wireless transmission: objects connected via the Internet. It can be said that IoT remains the indispensable tool that is at the base of the smart factory concept on the one hand and on the other hand an accessibility of useful information for making decisions on the other.

3.2 Artificial Intelligence

Big Data, the ability to compute production and collect data, has made it possible to highlight the need for automatic machines and self-decision capable of solving problems just like a member of the team. Indeed, the concept of artificial intelligence emerged in the early 1960s as an answer to this problem, that is to say, the replacement of a person by an automated machine. When computer experts discovered the algorithms, they found that these formulas do not have the ability to learn, which prompted them to examine certain parameters and apply certain appropriate actions on the basis of the configurations identified by humans [18].

Between 1970 and 1980, the experts in question improved its algorithms so that they could make logical reasoning according to the rules of the Si-alors. These algorithms are built "by hand" and do not contain the uncertainty generated by the environment.

In the early 1980s, when faced with the limitation of first-generation algorithms, when they tried to address certain situations taking into account the variability of the real world, experts began to focus on statistical models and neural networks.

3.3 Cobots

Collaborative robots involve integrating robots into the manufacturing process in collaboration with employees. The term collaboration means that robotics works in a joint way with man in industry and has no vocation to replace him; in other words, these are autonomous robots provided for the management of objects in collaboration with a human operator.

The cobots in question are generally classified into three categories:

(a) Robots handled by a nearby employee
(b) Remotely operated robots
(c) Systems assisting the human being in the form of electromechanical structures

These three categories of cobots assist the human operator in his work by assuring him safety, strength, precision, and comfort without influencing him in the decision-making, which remains a human role par excellence [19].

3.4 Advanced Simulation Tools

In 2019, Gartner the US analytics firm ranked digital twins (Digital Twin) as a strategic technology trend. The concept of digital twins is based on the development of multiphase modeling of a complex system. This complex system takes into account the integration of real objects for real-time monitoring. Among the applications, we refer to production infrastructures such as factories, railway networks, machine tools, and robotic systems.

Digital twins provide information on machine behavior and identify moments or equipment that may fail through preventive maintenance strategies to increase productivity gains and to optimize processes [20].

3.5 Clean Energy

In 2014, according to the UN, more than 54% of the population lived in urban areas; this rate is likely to reach 72% by the year 2050. Consequently, nations facing constraints in energy resources or relying heavily on raw material imports must address this situation intelligently. This has given rise to the emerging concept of the smart city. Indeed, compared to the electricity sector, which is the main source of energy consumed by urban areas, urban planners are forced to move toward distributed generation with the integration of renewable energies as an alternative to the traditional power plant.

4 Impact of New Technology on the Revolution of Lean 4.0: Discussion

In conclusion, Industry 4.0, driven by technologies such as digitalization, automation, and Big Data, as well as interconnected objects, has initiated a shift prompting manufacturing companies to establish innovative production tools and contemplate new methods concerning productivity and market analysis [21].

Indeed, working conditions in such an environment are no longer the same and require new core job profiles in terms of skills characterized by autonomy, adaptation, and agility to cope with the arduousness of work in numbers and qualities required.

Also, it must be said that one of the major challenges of this industrial transformation is to put the human being at the center of this evolution by highlighting future professions surpassing so-called traditional activities [22]. Issues mobilized professionals to improve productivity on the basis of equipment adapted to new industrial production activities. As a result, the penetration of robotics, digital technologies, and the interaction of information between man and machine offers new possibilities for the production system to manufacture better while controlling the cost.

To accompany change in this context that can be described as intelligent, the training and supervision of employees within companies are essential in order to change the employees' workstations, which are an integral part of the company's overall mission, on the one hand, and to raise the awareness of stakeholders in their reports with the new industrial unit called intelligent on the other hand [23]. This awareness-raising operation will act on some reluctance on the part of some employees who are referred to as both normal and human. Hence, it is crucial to instill in the minds of the relevant stakeholders the positive aspects of the new situation, which can work to their advantage, whether in terms of cost savings, time efficiency, or reduced effort compared to outdated methods in the now obsolete context.

5 Conclusion and Perspectives

Industry 4.0 has created an industrial revolution with excellence. Indeed, the entire paradigm of industrial strategy is undergoing a transformation that has had an impact on economic logic and also on all mechanisms aimed at creating value. Also, it must be said that the future industry has been the subject of a mutation based on a logic of mass production (quantity) toward a logic that is called mass customization. In other words, the future industry will no longer opt for the effect of scale and volume but for a flexible production close to demand. This entails manufacturing based on customer requests, placing the utilization of the end product at the core of its rationale.

In conclusion, it can be asserted that the industry of the future will transform organizations from rigid structures, inheritors of traditional scientific work organization, into flexible entities characterized by the appeal and attractiveness of work.

References

1. M. Debabi, Dimensions intégratives et distributives de la négociation entre producteurs et distributeurs. Rev. Sci. Gest. **237-238**(3), 105 (2009). https://doi.org/10.3917/rsg.237.0105
2. F. Ripoll, Espaces et stratégies de résistance: Repertoires d'action collective dans la France contemporaine. Espac. Soc. **134**(3), 83–97 (2008). https://doi.org/10.3917/esp.134.0083
3. R. Saxby, M. Cano-Kourouklis, E. Viza, An initial assessment of Lean Management methods for Industry 4.0. TQM J. **32**(4), 587–601 (2020). https://doi.org/10.1108/TQM-12-2019-0298
4. N. Günther, B. Prell, J. Reiff-Stephan, Industrie 5.0—Von der Vision des menschzentrierten Ansatzes zu sozio- cyberphysischen Produktionssystemen für die Praxis. Open Access (2022)
5. L. Dupont, F. Kasmi, J.M. Pearce, R.J. Ortt, "Do-It-Together" and Innovation: Transforming European industry. J. Innov. Econ. Manag. **40**(1), 1–11 (2023). https://doi.org/10.3917/jie.040.0001
6. R. Souza, H. Ferenhof, F. Forcellini, Industry 4.0 and Industry 5.0 from the Lean perspective. Int. J. Manag. Knowl. Learn. **11** (2022). https://doi.org/10.53615/2232-5697.11.145-155
7. European Commission, Directorate General for Research and Innovation, in *Industry 5.0: Towards a Sustainable, Human Centric and Resilient European Industry*, (Publications Office, LU, 2021) Consulté le: 18 mai 2023. [En ligne]. Disponible sur: https://data.europa.eu/doi/10.2777/308407
8. X. Xu, Y. Lu, B. Vogel-Heuser, L. Wang, Industry 4.0 and Industry 5.0—Inception, conception and perception. J. Manuf. Syst. **61**, 530–535 (2021). https://doi.org/10.1016/j.jmsy.2021.10.006

9. S. Tissir, S.E. Fezazi, A. Cherrafi, Industry 4.0 impact on Lean manufacturing: Literature review, in *2020 IEEE 13th International Colloquium of Logistics and Supply Chain Management (LOGISTIQUA)*, (IEEE, Fez, Morocco, December 2020), pp. 1–5. https://doi.org/10.1109/LOGISTIQUA49782.2020.9353889
10. T. Wagner, C. Herrmann, S. Thiede, Industry 4.0 impacts on lean production systems. Procedia CIRP **63**, 125–131 (2017). https://doi.org/10.1016/j.procir.2017.02.041
11. C.G. Machado, M. Winroth, D. Carlsson, P. Almström, V. Centerholt, M. Hallin, Industry 4.0 readiness in manufacturing companies: Challenges and enablers towards increased digitalization. Procedia CIRP **81**, 1113–1118 (2019). https://doi.org/10.1016/j.procir.2019.03.262
12. A. Sanders, K.R.K. Subramanian, T. Redlich, J.P. Wulfsberg, Industry 4.0 and Lean Management—synergy or contradiction: A systematic interaction approach to determine the compatibility of Industry 4.0 and Lean Management in manufacturing environment, in *Advances in Production Management Systems. The Path to Intelligent, Collaborative and Sustainable Manufacturing*, IFIP Advances in Information and Communication Technology, ed. by H. Lödding, R. Riedel, K.-D. Thoben, G. von Cieminski, D. Kiritsis, vol. 514, (Springer International Publishing, Cham, 2017), pp. 341–349. https://doi.org/10.1007/978-3-319-66926-7_39
13. W.W.-U. Münster et al., Industrie 5.0 – Die Europäische Kommission auf den Spuren der nächsten industriellen Revolution? Ind. 40 Manag. **2021**(6), 20–22 (2021). https://doi.org/10.30844/I40M_21-6_S20-22
14. N. Nikolakis, N. Kousi, G. Michalos, S. Makris, Dynamic scheduling of shared human-robot manufacturing operations. Procedia CIRP **72**, 9–14 (2018). https://doi.org/10.1016/j.procir.2018.04.007
15. M.C. Zizic, M. Mladineo, N. Gjeldum, L. Celent, From Industry 4.0 towards Industry 5.0: A review and analysis of paradigm shift for the people, organization and technology. Energies **15**(14), 5221 (2022). https://doi.org/10.3390/en15145221
16. J. Barcenilla, J.M.C. Bastien, L'acceptabilité des nouvelles technologies: Quelles relations avec l'ergonomie, l'utilisabilité et l'expérience utilisateur. Trav. Hum. **72**(4), 311–331 (2010). https://doi.org/10.3917/th.724.0311
17. A. Mayr et al., Lean 4.0 - A conceptual conjunction of lean management and Industry 4.0. Procedia CIRP **72**, 622–628 (2018). https://doi.org/10.1016/j.procir.2018.03.292
18. L.S. Valamede, A.C.S. Akkari, Lean 4.0: A new holistic approach for the integration of lean manufacturing tools and digital technologies. Int. J. Math. Eng. Manag. Sci. **5**(5), 851–868 (2020). https://doi.org/10.33889/IJMEMS.2020.5.5.066
19. S.D. Wu, M. Erkoc, S. Karabuk, Managing capacity in the high-tech industry: A review of literature. Eng. Econ. **50**(2), 125–158 (2005). https://doi.org/10.1080/00137910590949887
20. E. Garnsey, H.L. Smith, Proximity and complexity in the emergence of high technology industry: The Oxbridge comparison. Geoforum **29**(4), 433–450 (1998). https://doi.org/10.1016/S0016-7185(98)00022-0
21. A.G. Uriarte, A.H.C. Ng, M.U. Moris, Supporting the lean journey with simulation and optimization in the context of Industry 4.0. Procedia Manuf. **25**, 586–593 (2018). https://doi.org/10.1016/j.promfg.2018.06.097
22. H. Gholami, F. Abu, J.K.Y. Lee, S.S. Karganroudi, S. Sharif, Sustainable manufacturing 4.0—Pathways and practices. Sustain. For. **13**(24), 13956 (2021). https://doi.org/10.3390/su132413956
23. F.-Y. Wang, J. Yang, X. Wang, J. Li, Q.-L. Han, Chat with ChatGPT on Industry 5.0: Learning and decision-making for intelligent industries. IEEECAA J. Autom. Sin. **10**(4), 831–834 (2023). https://doi.org/10.1109/JAS.2023.123552

Design and Analysis of a Lower Limb Rehabilitation Robot with Movement in Three-Dimensional Space

Long Yu, Yongfei Feng, Fangyan Dong, Hongbo Wang, Haoyu Li, Tao Shen, Dan Liang, and Victor Vladareanu

1 Introduction

Due to advancements in modern medical technology and the improvement of medical care, the average life expectancy of human beings continues to increase [1]. However, this has given rise to a pressing issue of an aging population. An expanding number of patients are facing impaired motor abilities and hemiplegia due to conditions like stroke, brain injuries, and brain tumors [2, 3]. In addition, there is a noticeable trend of these conditions occurring at younger ages, which poses a significant challenge for post-operative rehabilitation treatment of individuals with lower limb disabilities. Precise and timely rehabilitation can stimulate brain plasticity, allowing patients to reorganize their neural pathways, generate new neuromotor cells, and restore the connection between the injured limb and the central nervous system [4]. Therefore, this greatly enhances the possibility of not only survival but also the recovery of motor function. Traditional physical rehabilitation treatment methods face challenges such as low efficiency, imprecise movement control,

L. Yu · F. Dong · H. Li · T. Shen · D. Liang
Faculty of Mechanical Engineering & Mechanics, Ningbo University, Ningbo, China
e-mail: dongfangyan@nbu.edu.cn; liangdan@nbu.edu.cn

Y. Feng (✉)
Faculty of Mechanical Engineering & Mechanics, Ningbo University, Ningbo, China

Academy for Engineering & Technology, Fudan University, Shanghai, China
e-mail: yongfei_feng@fudan.edu.cn

H. Wang
Academy for Engineering & Technology, Fudan University, Shanghai, China

V. Vladareanu
Robotics and Mechatronics Department, Institute of Solid Mechanics of the Romanian Academy, Bucharest, Romania

© The Author(s), under exclusive license to Springer Nature Switzerland AG 2024
A. Shaikh et al. (eds.), *Advances in Emerging Information and Communication Technology*, Signals and Communication Technology,
https://doi.org/10.1007/978-3-031-53237-5_4

difficulty in assessing rehabilitation outcomes, and individual patient differences. Consequently, the utilization of lower limb rehabilitation robots has emerged as a new hope for hemiplegic patients aiming to regain motor function through rehabilitation training.

In recent years, there have been significant technological advancements in lower limb rehabilitation robotics, leading to the successful commercialization of research findings [5, 6]. ReWalk, an Israeli-designed lower limb rehabilitation robot, consists of two symmetrical mechanical legs, a lumbar support, and a balance crutch [7]. It allows patients with lower limb paralysis to perform activities such as standing, walking, turning, and stair-climbing [8]. Other wearable exoskeleton robots also exist, such as Indego [9], Hal [10], Exo-H2 [11], and Robin [12]. Those robots are often limited to patients with mild-to-moderate lower limb impairment or those in the later stages of rehabilitation.

The use of lower extremity rehabilitation robots for humans' sitting positions has emerged as a means to facilitate rehabilitation training for early bedridden patients by reducing the weight-bearing during exercises. One popular example in this category is the U.S. NuStep four-limb linkage whole-body functional rehabilitation trainer [13], which includes a base, rotating seat, foot pedals, and other components. It enables patients to engage in limb joint training exercises while seated, utilizing both their hands and feet. Moreover, it utilizes an active training mode that offers assistance functions and impedance effects, empowering patients to actively participate in their own recovery process. MotionMaker consists of two mechanical legs, each with three degrees of freedom, a tilt-adjustable seat, a control unit, and an electrical stimulation module [14, 15]. The mechanical legs are symmetrically positioned on both sides of the seat and employ a serial mechanism for driving the linkage movement. This mechanism allows for flexion and extension movements of the hip, knee, and ankle joints in the sagittal plane.

This paper introduces a novel seated lower limb rehabilitation robot that incorporates the features of a 2-PRR parallel mechanism. The robot enables rehabilitation training in both humans' sagittal and coronal planes, facilitating flexion/extension movements of the hip, knee, and ankle joints, as well as hip abduction/adduction movements. The design aims to significantly enhance the efficiency of lower limb rehabilitation while meeting the daily rehabilitation needs of individuals with physical disabilities.

2 Analysis of Range of Joint Motion in Humans' Sitting Positions

The human lower limb comprises three major joints: the hip joint, the knee joint, and the ankle joint. These joints connect the thigh, calf, and foot, forming a complete lower limb. Through the analysis of human lower limb movement, the hip joint allows flexion in the sagittal plane and abduction/adduction in the coronal plane. The knee joint enables flexion/extension in the sagittal plane. For the ease of mechanism design, this study considers only the ankle joint that allows dorsiflexion/plantarflexion movement in the sagittal plane. While the human body's lower limb joints possess the capability for intricate and coordinated movements, the primary objective of rehabilitation is to restore patients' normal walking ability. Table 1 provides information regarding the normal range of joint motion in a healthy human body.

For the purpose of this study, the axis of the ankle joint is chosen as the theoretical endpoint of the lower limb. Consequently, a two-link model was used to simplify each lower limb of the human body, as depicted in Fig. 1.

Let the length of the thigh be l_1 and of the calf be l_2. Set $O_0 - x_0y_oz_0$ as the fixed coordinate system, $O_1 - x_1y_1z_1$ as the hip coordinate system, $O_2 - x_2y_2z_2$ as the knee coordinate system, and $O_3 - x_3y_3z_3$ as the ankle coordinate system. $\theta_i(i = 1, 3, 4)$ is the rotation angle of each joint relative to $z_i(i = 1, 3, 4)$ axis, θ_2 is the rotation angle of lower limb around y_0 axis.. According to Table 1, $0 \leq \theta_1 \leq 45°$, $0 \leq \theta_2 \leq 45°$, and $-150 \leq \theta_3 \leq 0°$ are selected, considering the human body is in a sitting position.

The D-H method was used to calculate the spatial coordinate $O_3(x, y, z)$ at the terminal of the ankle joint. The transformation matrix A_1,A_2, and A_3 can be calculated:

Table 1 Range of motion angle of lower limb joint

Joint	Datum plane	Movement	Angle range (°)
Hip	Sagittal plane	Flexion/extension	0–45
Hip	Coronal plane	Abduction/adduction	0–45
Knee	Sagittal plane	Flexion/extension	−150–0
Ankle	Sagittal plane	Dorsiflexion/plantarflexion	−35–20

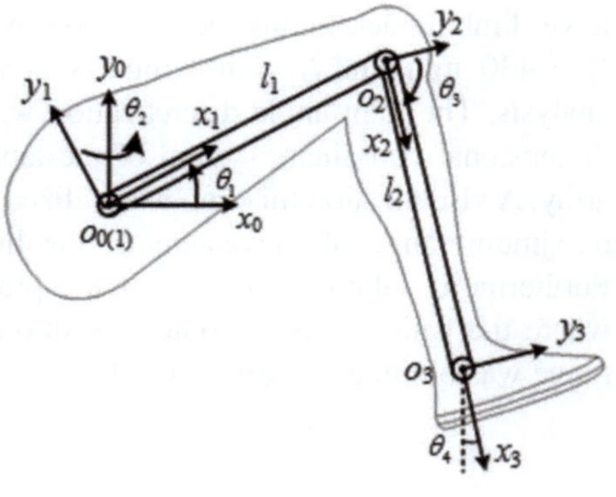

Fig. 1 Lower limb model and linkage coordinate system

$$A_1 = \begin{bmatrix} c\theta_1 & -s\theta_1 & 0 & 0 \\ s\theta_1 & c\theta_1 & 0 & 0 \\ 0 & 0 & 1 & 0 \\ 0 & 0 & 0 & 1 \end{bmatrix} \tag{1}$$

$$A_2 = \begin{bmatrix} c\theta_2 & 0 & s\theta_2 & 0 \\ 0 & 1 & 0 & 0 \\ -s\theta_2 & 0 & c\theta_2 & 0 \\ 0 & 0 & 0 & 1 \end{bmatrix} \times \begin{bmatrix} 1 & 0 & 0 & l_1 \\ 0 & 1 & 0 & 0 \\ 0 & 0 & 1 & 0 \\ 0 & 0 & 0 & 1 \end{bmatrix} \tag{2}$$

$$A_3 = \begin{bmatrix} c\theta_3 & -s\theta_3 & 0 & 0 \\ s\theta_3 & c\theta_3 & 0 & 0 \\ 0 & 0 & 1 & 0 \\ 0 & 0 & 0 & 1 \end{bmatrix} \times \begin{bmatrix} 1 & 0 & 0 & l_2 \\ 0 & 1 & 0 & 0 \\ 0 & 0 & 1 & 0 \\ 0 & 0 & 0 & 1 \end{bmatrix} \tag{3}$$

The pose transformation matrix $\boldsymbol{T}$ of the ankle joint in the fixed coordinate system O_0 is:

$$T = A_1 \cdot A_2 \cdot A_3 \tag{4}$$

Thus, the coordinate expression of the ankle O_3 in the fixed coordinate system O_0 is:

$$\begin{cases} x = l_1 c\theta_1 c\theta_2 + l_2 c\theta_1 c\theta_2 c\theta_3 - l_2 c\theta_2 s\theta_1 s\theta_3 \\ y = l_1 s\theta_1 + l_2 c\theta_0 c\theta_1 s\theta_3 - l_2 c\theta_3 s\theta_1 \\ z = l_2 s\theta_1 s\theta_2 s\theta_3 - l_2 c\theta_1 c\theta_3 s\theta_2 - l_2 c\theta_1 s\theta_2 \end{cases} \tag{5}$$

where $s\theta_i = \sin\theta_i$, $c\theta_i = \cos\theta_i$, $i = (1, 2, 3)$.

Then, the Monte Carlo method was used to analyze the workspace of the human lower limb model. In this section, two groups of extreme value data, $l_1 = 550$ mm, $l_2 = 430$ mm, and $l_1 = 402$ mm, $l_2 = 313$ mm, were selected, respectively, for analysis. The joint angle discretization was applied to Eq. (5) to obtain the three-dimensional coordinate system of the ankle joint, which was stored in an empty array. A visualization function was utilized in MATLAB software to demonstrate the maximum range of movement of the human lower limb in a sitting position. Furthermore, taking into account the practical aspects of rehabilitation training, where the patient's ankle joint is positioned below the hip joint, a portion of the range was omitted, as depicted in Fig. 2.

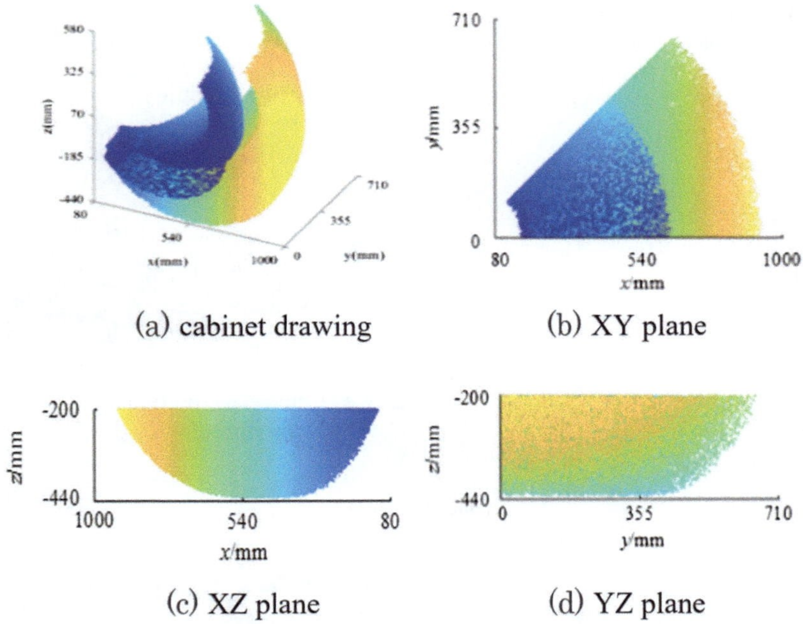

Fig. 2 Lower limb movement space in sitting position (mm). (**a**) cabinet drawing. (**b**) XY plane. (**c**) XZ plane. (**d**) YZ plane

3 Mechanical Design of Lower Limb Rehabilitation Robot

As shown in Fig. 3, the lower limb rehabilitation robot is mainly composed of three modules, including the lower limb rehabilitation training module with 2-PRR parallel mechanism, the seat lifting module based on the electric pushrod, and the underframe built by aluminum.

As shown in Fig. 4, the lower limb rehabilitation training module is mainly composed of pedal assembly, foot rotation pair, square tube I, square tube II, sliding block part rotation pair I, sliding block part rotation pair II, sliding block, linear module I, linear module II, motor I, and motor II. When the robot is running, motor I and motor II pass the torque to the input shaft of linear module I and linear module II through the elastic couplings, respectively, and drive the lead screw of linear module I and linear module II to rotate, so as to realize the forward or backward of the rotating pair of sliding block part I and sliding block part II. When the speed of motor I is consistent with that of motor II, the pedal assembly can move forward/backward in a straight line; when the speed of motor I is inconsistent with that of motor II, the pedal assembly will drive the foot of the patient to achieve abduction/adduction of the hip joint of the lower limb, so as to carry out humans' coronal rehabilitation training.

Fig. 3 Overall structure of the proposed mechanism

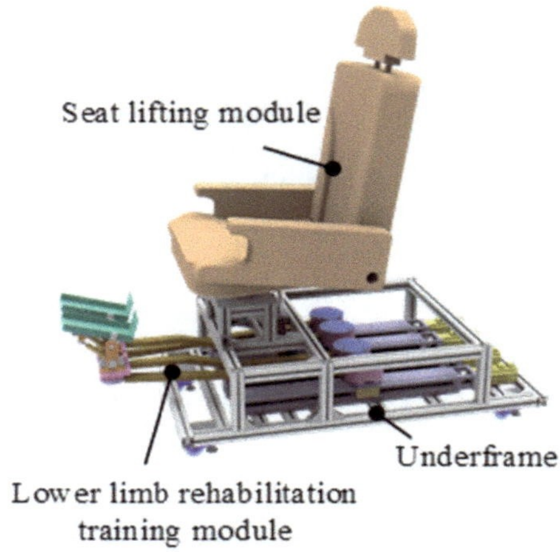

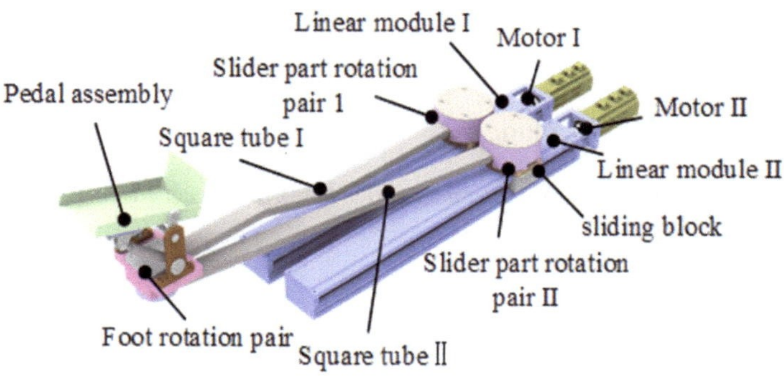

Fig. 4 Lower limb rehabilitation training module

4 Kinematics Analysis of Lower Limb Rehabilitation Training Module

4.1 Forward Kinematics and Jacobian Matrix

The lower limb rehabilitation training module is a 2-PRR parallel mechanism, as shown in Fig. 5, which contains three rotating pairs and two moving pairs. In particular, the axes of rotating pairs among the pairs of rotators are parallel to each other, as are the axes of the pairs of movement.

A new coordinate system is set up. It is denoted as $O - XY$. The relevant points are defined as follows. The coordinate origin O is defined as the lead screw zero of linear module I. H is defined as the linear module II, the distance between linear module I and linear module II is m. C_1 and C_2 respectively represent the slider on linear module I\II. x_1 represents the distance between slider C_1 and the lead screw zero of linear module I. x_2 represents the distance between slide C_2 and linear module II. Link AP and BP, respectively, represent square tube I and square tube

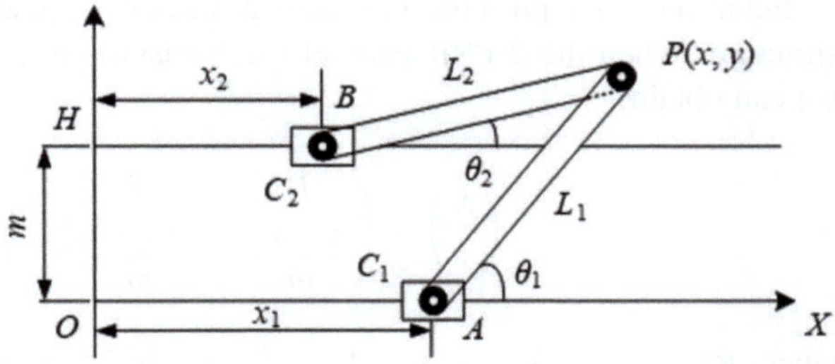

Fig. 5 2-PRR parallel structure

II. θ_1 represents the angle between AP and X axis. θ_2 represents the angle between BP and X axis. L_1 and L_2, respectively, represent the length of the link AP and BP. The end of the connecting rod P represents the foot rotation pair, and its coordinate is (x, y).

The geometric parameters of the mechanism are as follows: $AP = L_1 = 800$ mm, $BP = L_2 = 800$ mm, and the distance between the two parallel guide rails is $m = 600$ mm.

In the coordinate system $O - XY$, it is easy to obtain the inverse kinematics solution according to the geometric relations:

$$\begin{cases} x_1 = x - \sqrt{{L_1}^2 - y^2} \\ x_2 = x - \sqrt{{L_2}^2 - y - m)^2} \end{cases} \tag{6}$$

To meet the spatial requirements and prevent interference between the left and right legs, the kinematic equation must be subject to specific constraints.

$$\begin{cases} x > x_1 \\ x > x_2 \\ x_1 > x_2 \end{cases} \tag{7}$$

After derivation of the time of Eq. (6), we can get:

$$\begin{cases} (x - x_1)\dot{x}_1 = (x - x_1)\dot{x} + y\dot{y} \\ (x - x_2)\dot{x}_2 = (x - x_2)\dot{x} + (y - m)\dot{y} \end{cases} \tag{8}$$

where, $\dot{x}_1$ and $\dot{x}_2$ denotes the motion velocity of the slider C_1 and C_2, respectively; $\dot{x}$ denotes the motion velocity of the terminal point P in the X-axis direction; $\dot{y}$ denotes the motion velocity of the terminal point P in the Y-axis direction. Then:

$$\begin{pmatrix} x - x_1 & 0 \\ 0 & x - x_2 \end{pmatrix} \begin{pmatrix} \dot{x}_1 \\ \dot{x}_2 \end{pmatrix} = \begin{pmatrix} x - x_1 & y \\ x - x_2 & y - m \end{pmatrix} \begin{pmatrix} \dot{x} \\ \dot{y} \end{pmatrix} \tag{9}$$

Equation (9) represents the unified velocity equation of the 2-PRR parallel structure. When the 2-PRR parallel mechanism is in a non-singular configuration, we can obtain:

$$\begin{pmatrix} \dot{x} \\ \dot{y} \end{pmatrix} = \begin{pmatrix} R & S \\ \dfrac{x-x_1}{m} & \dfrac{-(x-x_2)}{m} \end{pmatrix} \begin{pmatrix} \dot{x}_1 \\ \dot{x}_2 \end{pmatrix} \tag{10}$$

where $R = \frac{(y-m)(x-x_1)}{(y-m)(x-x_1)-y(x-x_2)}$, $S = -\frac{y(x-x_2)}{(y-m)(x-x_1)-y(x-x_2)}$.

Finally, the Jacobian matrix of the parallel mechanism can be written as:

$$J = \begin{pmatrix} R & S \\ \dfrac{x-x_1}{m} & \dfrac{-(x-x_2)}{m} \end{pmatrix} \tag{11}$$

Equation (11) reveals two important characteristics of J: (1) it depends on both the end-effector's position and the connecting rod's length; (2) J is unaffected by x, meaning that changes in x do not affect y or J. When evaluating structural performance using the Jacobian matrix, it is essential to consider the characteristics of the workspace in the Y direction.

4.2 Condition Number

In order to quantitatively study the dexterity and motion characteristics of the mechanism, we choose the condition number of the Jacobian matrix as the evaluation index, which is represented by cond(J). The closer cond(J) is to 1, the better the dexterity of the mechanism during operation. However, the closer cond(J) is to positive infinity, the more difficult the mechanism is to move flexibly, and the closer it is to the singular state. When cond(J) = 1, all directions of the mechanism have the same motion ability and the flexibility is the best.

To facilitate observation, the LCI performance index is defined as the reciprocal of cond(J) and denoted as $r$cond(J). This index can be used to measure the dexterity of the mechanism at a given position. Equation (13) defines dexterity as a quantifiable representation of the distortion in the kinematic relationship resulting from the ill-conditioned characteristics of the parallel mechanism's Jacobian matrix.

$\sigma_{max}(J)$ represents the maximum singularity of the Jacobian matrix. In robot kinematics, the maximum singularity reflects the maximum velocity that the robot can generate in a certain direction under some attitude. If a robot has a very small $\sigma_{max}(J)$ for certain attitudes, then the robot's motion will be limited and it may not be able to reach certain positions or attitudes. Therefore, the maximum singularity of the Jacobian matrix can be used to evaluate the motion performance of the robot, and better control algorithms can be designed to improve the motion flexibility of the robot.

$\sigma_{\min}(J)$ represents the minimum singularity of the Jacobian matrix and the minimum execution speed of the robot at a specific position and direction. In robotics, the minimum singularity is an important index, because it reflects the sensitivity and controllability of the actuator under the specific position and direction of the robot. When the minimum singularity is very small or approaches zero, the robot's actuator speed will become very sensitive, resulting in the robot losing control or being unable to perform the task accurately.

According to the structural dimension parameters of the 2-PRR parallel mechanism, $\sigma_{\max}(J)$ and $\sigma_{\min}(J)$ can be obtained as shown in Figs. 6 and 7, respectively.

Then, the expression of the condition number is:

$$\mathrm{cond}(J) = \frac{\sigma_{\max}(J)}{\sigma_{\min}(J)} \tag{12}$$

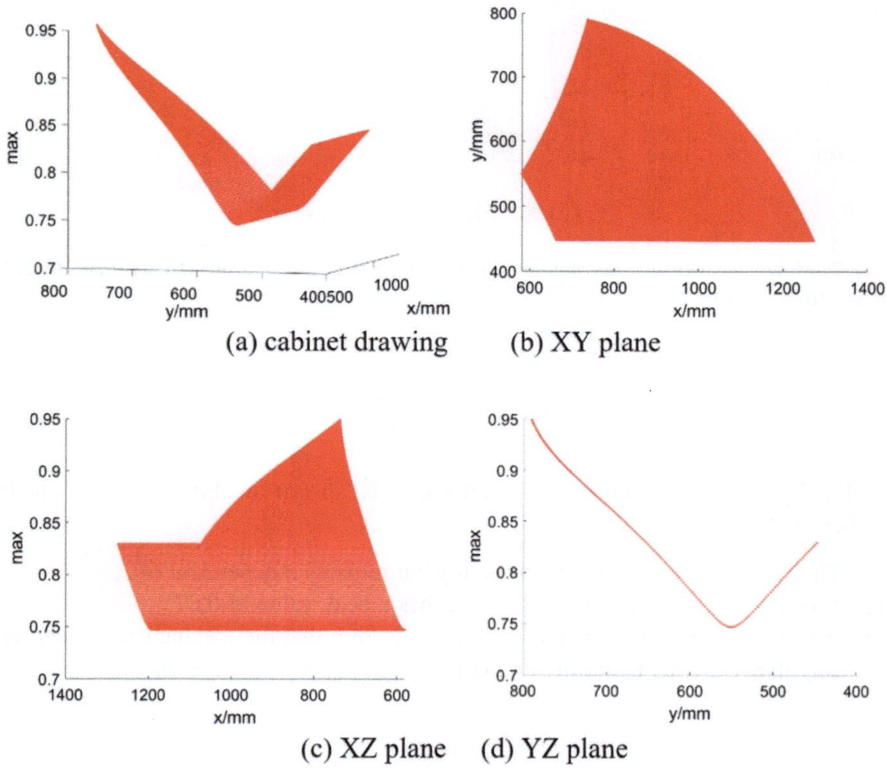

Fig. 6 Maximum singular value of the Jacobian matrix. (**a**) cabinet drawing. (**b**) XY plane. (**c**) XZ plane. (**d**) YZ plane

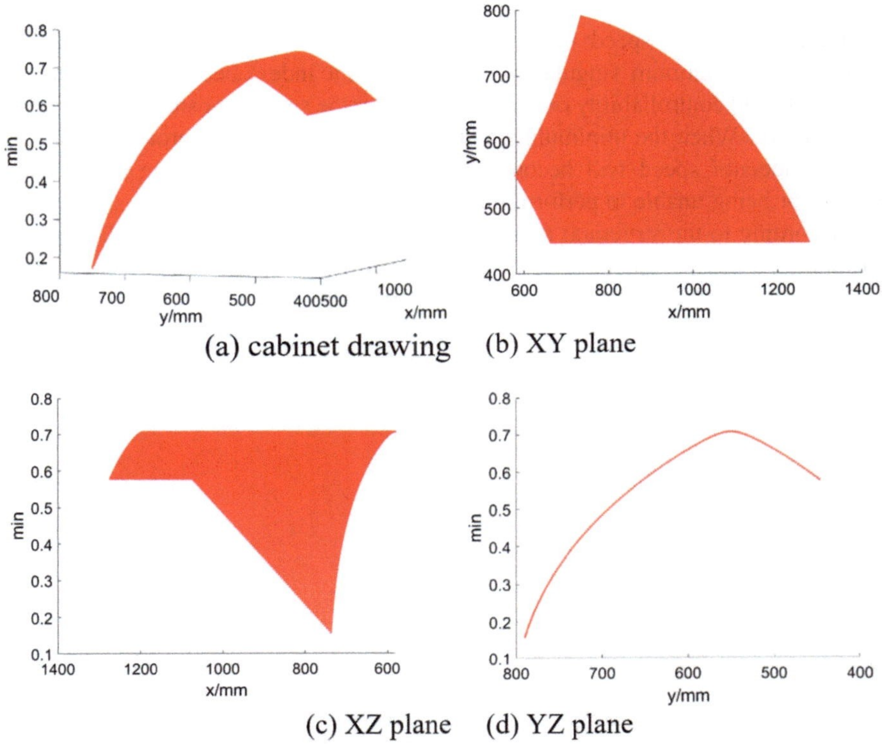

Fig. 7 Minimum singular value of the Jacobian matrix. (**a**) cabinet drawing. (**b**) XY plane. (**c**) XZ plane. (**d**) YZ plane

$$\mathrm{rcond}(J) = \frac{1}{cond(J)} \tag{13}$$

Finally, the *r*cond(*J*) in the workspace is shown in Fig. 8. This graph indicates that:

(1) The *r*cond(*J*) varies with changes in y but remains independent of x.
(2) When $y = 550$ mm, the *r*cond(*J*) reaches a peak value of 0.87.
(3) When $y > 800$ mm, rcond(*J*) < 0.2, indicating that the mechanism is nearing singularity and its flexibility and control stability are diminished.

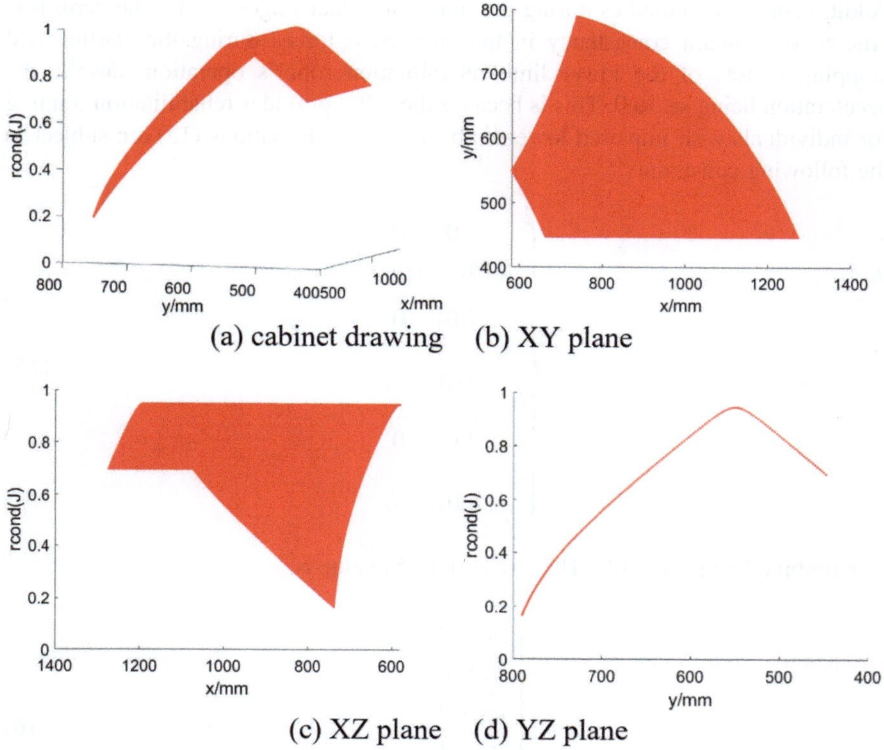

Fig. 8 Condition number of the Jacobian matrix. (**a**) cabinet drawing. (**b**) XY plane. (**c**) XZ plane. (**d**) YZ plane

5 Trajectory Planning

The purpose of trajectory planning is to design the rehabilitation trajectory required by lower limb rehabilitation training. The proposed lower limb rehabilitation robot proposed in this paper belongs to the end traction type, which selects the Cartesian space trajectory planning method.

Take the trajectory planning of a straight path as an example. The specific parameters are as follows: the start point is (990 mm, 0 mm), the end point is (1440 mm, 150 mm), $\beta = 18.43^\circ$, the track length $L = 473.34$ mm, and the movement time $t = 10s$.

Define its displacement in the linear direction as a fifth-degree polynomial:

$$l(t) = a_0 + a_1 t + a_2 t^2 + a_3 t^3 + a_4 t^4 + a_5 t^5 \tag{14}$$

The rehabilitation robot should adhere to the planning trajectory by ensuring that the starting and ending points are in accordance with the specified requirements.

Additionally, the velocities during the initial and final stages need to be zero. It is crucial to maintain consistency in the interactive forces during the starting and stopping phases of the lower limb rehabilitation robot's operation, despite the acceleration being set to 0. This is because the robot provides rehabilitation training for individuals with impaired lower limb movement. Equations (15) are subject to the following constraints.

$$\begin{cases} l(0)=0 \\ l(10)=474.34 \\ \dot{l}(0)=0 \\ \dot{l}(10)=0 \\ \ddot{l}(0)=0 \\ \ddot{l}(10)=0 \end{cases} \tag{15}$$

Combined with Eq. (14), The $a_i(i = 0, 1...5)$ value is:

$$\begin{cases} a_0=0 \\ a_1=0 \\ a_2=0 \\ a_3=4.743 \\ a_4=-0.712 \\ a_5=0.028 \end{cases} \tag{16}$$

The displacement polynomial of the linear trajectory is:

$$l(t)=4.743t^3-0.712t^4+0.028t^5 \tag{17}$$

By differentiating Eq. (17), the velocity expression of the linear trajectory can be written as:

$$\dot{l}(t)=14.229t^2-2.848t^3+0.14t^4 \tag{18}$$

Then, the acceleration expression is:

$$\ddot{l}(t)=28.458t-8.544t^2+0.56t^3 \tag{19}$$

The expressions after differentiation along the X and Y directions are as follows:

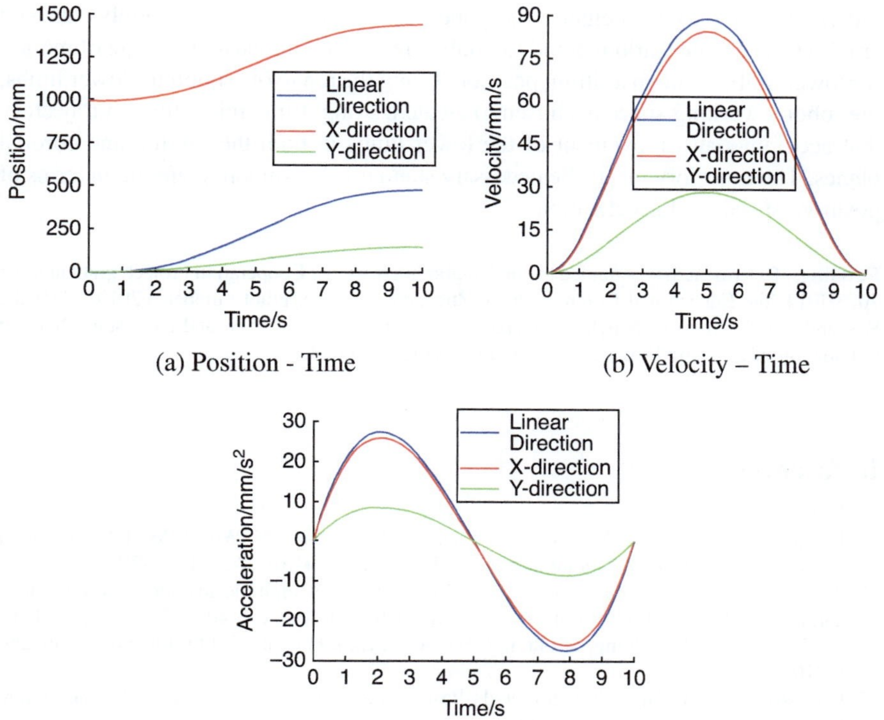

(a) Position - Time

(b) Velocity – Time

(c) Acceleration - Time

Fig. 9 Curve of displacement, velocity, and acceleration. (**a**) Position–Time. (**b**) Velocity–Time. (**c**) Acceleration–Time

$$\begin{cases} x(t) = l(t)\cos(\beta) + 990 \\ y(t) = l(t)\sin(\beta) \end{cases} \tag{20}$$

To provide a more intuitive assessment of the smoothness of the planned line, the x-direction, y-direction, and the line's squared values are analyzed. Fig. 9 depicts the graphical representations of displacement, velocity, and acceleration over time. Upon observation, it is evident that the position–time, speed–time, and acceleration–time curves all adhere to the requirements of smooth planning.

6 Conclusion

This paper proposes a new parallel-structured rehabilitation robot for lower limb training. It analyzes the kinematic model of the 2-PRR parallel mechanism and evaluates its condition number based on the design parameters. To accommodate individuals of varying heights, the paper uses two sets of lower limb data and employs the Monte Carlo method to establish the three-dimensional range of motion for lower limbs while in a sitting position. Using the range of motion for lower limbs, the robot's working space is chosen to create a straight-line rehabilitation trajectory that accommodates movement of the lower limbs in both the sagittal and coronal planes. The trajectory meets the necessary standards for smooth planning in terms of position, speed, and acceleration.

Funding This research was funded by the Ningbo International Cooperation Project, grant number 2023H014; the Educational Commission of Zhejiang Province, grant number Y201941335; the Research Fund Project of Ningbo University, grant number XYL19029; and the Natural Science Foundation of Zhejiang Province, grant number LQ21E050008.

References

1. P. Kowal, S. Chatterji, N. Naidoo, et al., Data resource profile: The World Health Organization study on global ageing and adult health. Int. J. Epidemiol. **41**(6), 1639–1649 (2012)
2. A.J. Cowan, C. Allen, A. Barac, et al., Global burden of multiple myeloma: A systematic analysis for the global burden of disease study 2016. JAMA Oncol. **4**(9), 1221–1227 (2018)
3. Q. Jia, L.P. Liu, Y.J. Wang, Stroke in China. Clin. Exp. Pharmacol. Physiol. **37**(2), 259–264 (2010)
4. P.M. Rossini, C. Calautti, F. Pauri, et al., Post-stroke plastic reorganization in the adult brain. Lancet Neurol. **2**(8), 493–502 (2003)
5. H. Yu, H. Wang, Y. Du, et al., Design and isokinetic training control method of leg press training device. Appl. Sci. **9**(14), 2822 (2019)
6. D. Quy-thinh, Y. Shin-ichiroh, Assist-as-needed control of a robotic Orthosis actuated by pneumatic artificial muscle for gait rehabilitation. Appl. Sci. **8**(4), 499 (2018)
7. G. Zeilig, H. Weingarden, M. Zwecker, et al., Safety and tolerance of the ReWalk? Exoskeleton suit for ambulation by people with complete spinal cord injury: A pilot study. J. Spinal Cord Med. **35**(2), 96–101 (2012)
8. S.H. Hwang, D.I. Sun, J. Han, et al., Gait pattern generation algorithm for lower-extremity rehabilitation–exoskeleton robot considering wearer's condition. Intell. Serv. Robot. **14**, 345–355 (2021)
9. A. Rodríguez-Fernández, J. Lobo-Prat, J.M. Font-Llagunes, Systematic review on wearable lower-limb exoskeletons for gait training in neuromuscular impairments. J. Neuroeng. Rehabil. **18**, 1–21 (2021)
10. B. Chen, C.H. Zhong, X. Zhao, et al., A wearable exoskeleton suit for motion assistance to paralysed patients. J. Orthop. Transl. **11**, 7–18 (2017)
11. J. Vaughan-Graham, D. Brooks, L. Rose, et al., Exoskeleton use in post-stroke gait rehabilitation: A qualitative study of the perspectives of persons post-stroke and physiotherapists. J. Neuroeng. Rehabil. **17**, 1–15 (2020)
12. A. Esquenazi, M. Talaty, A. Jayaraman, Powered exoskeletons for walking assistance in persons with central nervous system injuries: A narrative review. PMR **9**, 46–62 (2017)
13. L.C. Dalleck, E.C. Borresen, A.L. Parker, et al., Development of a metabolic equation for the NuStep recumbent stepper in older adults. Percept. Mot. Skills **112**(1), 183–192 (2011)

14. S. Freivogel, J. Mehrholz, T. Husak-Sotomayor, et al., Gait training with the newly developed"LokoHelp"system is feasible for non-ambulatorypatients after stroke, spinal cord and brain injury: A feasibility study. Brain Inj. **22**(7/8), 625–632 (2008)
15. M.A.M. Dzahir, S. Yamamoto, Recent trends in lower-limb robotic rehabilitation orthosis: Control scheme and strategy for pneumatic muscle actuated gait trainers. Robotics **3**(2), 120–121 (2014)

BC-PUS: A Proposed Blockchain-Based Private User-Driven Mobile Advertising System

Nora Alammari

1 Introduction

Online advertising now accounts for most advertising spending, surpassing broadcast advertising in television, radio, and newspapers. In 2022, according to Statista, online advertising accounted for 62% of all advertising spending globally, while television accounted for 23% [1]. This is largely attributable to the capability of online advertising platforms to customize or personalize advertisements and thus target specific user segments. Big data analytics is the foundation of targeted advertising because it allows for the collection and processing of user data to enable the creation of user segments with varying sizes according to interests, location, or specific characteristics like age, gender, etc. Mobile devices, such as smartphones and tablets, are the most important platforms for collecting personal information, which is subsequently utilized for targeted advertising. This is because of the broad and nearly constant use of the platform by a big audience that may potentially be targeted by advertisements.

The majority of mobile applications have at least one advertising library for targeted advertising [2]. Through the advertising library API calls, information about users' online activities and behavior is gathered [3]. The advertising and analytics companies, such as Google Analytics and Flurry, are attempting to increase their income by offering advertising libraries that application developers use to serve advertisements. As part of monetizing data efforts, advertising and analytics

N. Alammari (✉)
Department of Information Technology, College of Computer, Qassim University, Buraydah, Saudi Arabia

Department of Computers and Control Engineering, Faculty of Engineering, Tanta University, Tanta, Egypt
e-mail: N.alammari@qu.edu.sa; d.hussein@qu.edu.sa

© The Author(s), under exclusive license to Springer Nature Switzerland AG 2024
A. Shaikh et al. (eds.), *Advances in Emerging Information and Communication Technology*, Signals and Communication Technology,
https://doi.org/10.1007/978-3-031-53237-5_5

companies aggressively pursue all possible means of collecting users' personal data, including the purchase of such information from third parties [4]. It is worth noting that personal data collection and use threaten users' privacy.

With regard to privacy, earnest concerns have been raised about the greatly increased practices of intrusion of online advertising. In the past few years, there has been an increase in the number of individuals who are concerned about their privacy over how targeted advertising uses tracking information. According to [4], 2 out of 3 Internet users are concerned that their online activity is scanned without their consent or knowledge. Multiple studies in the same vein reflect the increasing level of ubiquitous and misuse of advertising perceived by users as a serious deterioration of their browsing experience [4]. There are obvious security concerns for users as personal and sensitive information about their behaviors is shared under the umbrella of targeted advertising through an infrastructure that works in the shadows almost without oversight. Consequently, the privacy concerns of users, in turn, inevitably lead to a level of hostility toward advertisers, which could direct to the use of ad-blocking tools, which subsequently affects the business model [5].

Maintaining the privacy of personal data is a civilized concept and an individual requirement. In a step to ensure an adequate level of legal and technical protection for personal data processed electronically, as well as to build trust in the data sector, Saudi Arabia issued on September 16, 2021, the "Personal Data Protection Law—PDPL," the first comprehensive data protection law [6]. The system aims to protect individual privacy by regulating the collection, processing, disclosure, and retention of personal data. It was implemented on the 17th of March 2023 [7]. This system provides a detailed framework that includes standards for data processing, rights of data subjects, obligations of relevant bodies when processing personal data, data sovereignty, and penalties for non-compliance. Many of the features of the Saudi Personal Data Protection Regulation are in line with the standards and principles contained in other international data protection laws, such as the European General Data Protection Regulation (GDPR), the regulation established by European Union law on data protection and privacy in the European Union and the region. This system constitutes an important step forward in the field of data protection, and a starting point for the Kingdom of Saudi Arabia.

Overall, the protection of the personal data of users with effective targeting is crucial for both advertising networks and mobile users. Mobile users would like to see interest-based advertisements when their data are not revealed to the outside world, such as ad networks. Hence, there is a need for such a mobile advertising system that can sustain the current business model of targeted mobile advertising (TMA) and can do so in a way that assists protect the personal information of users instead of just circumventing their privacy concerns.

Our focus in this paper is to design and implement an advertising platform that is secure, reliable, and fair, which complies with applicable privacy protection regulations. This is accomplished through developing a Blockchain-based private user-driven mobile advertising system. The main contributions of the proposed system are as follows:

1. To address the problem of violating users' privacy in order to serve targeted ads, we propose leveraging blockchain technology to decouple all direct interactions between applications, analytics companies that they utilize to serve targeted advertising, and advertising systems (including third parties) for their activities (for both web and applications activity).
2. To address the problem of personal information leakage and disclosure, we propose to encrypt the data using the asymmetric Elliptic Curve Cryptography (ECC) algorithm and store it in off-chain storage and the corresponding address will be returned. The returned address is stored on the blockchain.
3. To address the problem of users' lack of control over their data, we propose the use of smart contracts that allow users to manage and monetize their data to support the delivery of targeted ads in a fair and secure manner.

The remainder of the paper is organized as follows: Section II presents the background of our study and related work. Section III presents the architecture of the proposed system and provides design details. In Sect. IV, the experimental results are discussed. A comprehensive discussion regarding the achievements of the proposed system is given in Sect. V. Finally, the conclusion is drawn in Sect. VI.

2 Background and Related Work

In this section, we review the background of our study as well as related work.

2.1 Background

Blockchain The fundamental idea of the blockchain was suggested by Nakamoto Satoshi [8], it is the underpinning technology behind digital currencies. Since then, blockchain has been widely adopted in a variety of domains including the Internet of Things, Healthcare, Resource Management, and Big Data.

Blockchain is typically described as a distributed data structure that can be duplicated and shared with network nodes and has no central authority or repository [9]. Like a database, information is stored in a blockchain electronically and in digital format. The innovation in a blockchain lies in that it ensures the integrity and security of a data record as well as generates trust without the necessity for a reliable third party.

A blockchain is made up of a series of interconnected blocks, which each store and record all of the transaction data that occurs over a certain time period [10]. In addition, all blocks have a reference, which refers to the block that precedes it except for the first block, which is known as the Genesis block [9]. This serves as a link to the previous block and determines the order over the chain of blocks. The blocks are thus linked in chronological order. The cryptographic hash algorithm guarantees the integrity of the transaction data in each block and prevents any alterations to the linked blocks that comprise the blockchain [11].

The blockchain is maintained by miners. They are referred to as the nodes attempting to produce new valid blocks that are confirmed by the consensus mechanisms (e.g., Proof of Work (POW) and Proof of Stake (POS)), and if successful, they will be rewarded with the cryptocurrency, i.e., Bitcoin. Compared to conventional distributed systems, blockchain provides reward mechanisms to encourage individuals to participate in blockchain maintenance [12]. The blockchain guarantees a reliable and transparent picture of the shared ledger between mutually distrustful entities, thanks to cryptography and consensus mechanisms [12]. As a result, all of the information included in a public blockchain is viewable by each and every node that is a part of the network. However, it is exceedingly difficult to change any block's information in the public blockchain until 51% of nodes are simultaneously compromised [9].

From the above, it can be concluded that a blockchain is an immutable database or ledger that is shared among peers in a network wherein records of transactions or events are added in chronological order. Clearly, blockchain incorporates some interesting features that benefit the privacy issue in targeted advertising applications.

Decentralization is a significant feature of blockchain, as it allows for the implementation of systems that do not depend on central authority. In addition, the fact that blockchain data is shared among all network nodes fosters an environment of transparency, openness, and trust by letting all parties involved—especially the user—see, understand, and control who has access to, and how it is used for, their information [4, 13]. More significantly, since the data in the ledger is shared and replicated throughout several nodes in the network, compromising of any node in the blockchain network has no impact on the ledger's state. Furthermore, because of the nature of blockchain, anonymity can be preserved through Zero Knowledge Proof, which allows the ownership of information to be confirmed without revealing the identity [14]. What is more, using cryptographic techniques to encrypt data kept on the blockchain guarantees that only users with authorized access to the data may decrypt it, enhancing both the security and privacy of the data.

Another feature of the blockchain is smart contracts, which may be used to program the rules that give users control over the sharing and use of their data records [15]. This is in some way related to the Personal Data Protection Law, which makes it illegal to process users' sensitive personal data unless they give their express consent or certain requirements are met [6]. Therefore, blockchain can assist in the building of data management system with Personal Data Protection Law compliance by encoding in the smart contract a group of rules ensuring that sensitive user data cannot be shared or used without the proper authorizations.

Smart Contracts Nick Szabo was first suggested the concept of smart contracts in the early 1990s [16]. He also provided the definition of a smart contract at that time, which was a computerized transaction protocol that carried out the terms of a contract. Szabo proposed in [17] to transform contractual terms into code and embed them in an appropriate running environment where they can operate them automatically, in order to reduce the need for trusted intermediates and the incidence of malicious or inadvertent exceptions.

Ethereum is an open-source decentralized blockchain that incorporates smart contract features that were derived from Bitcoin [18]. In Bitcoin, transaction operations are conducted via a stack-based non-Turing complete programming language that supports only simple logic, restricting its application in many industries [11]. Ethereum is typically described as a programmable system that operates similarly to Bitcoin with Turing complete scripting capabilities [19]. Its distinguishing feature over the Bitcoin network is that it is a programmable blockchain. Users may build, deploy, and execute smart contracts on the blockchain thanks to the platform's support for Turing complete applications [11]. The Ethereum Virtual Machine (EVM), a platform for building smart contracts that supports the execution of smart contracts on blockchain, is where the smart contracts are often stored as scripts and run [18]. A great point about Ethereum is that smart contracts can be programmed using nearly developer-friendly languages. The two best and common languages to build a smart contract are Solidity and Vyper [20, 21].

A smart contract is, in its most basic form, a computer program that is designed to automatically execute its instructions in a secure environment. The smart contract itself is a part of the blockchain network. It responds to incoming information, receives, and stores the value, and then sends the information and value. Whenever the smart contract is implemented, the bytecode constructed by the compiler is placed in the blockchain and will be associated with the storage address [21]. Figure 1 depicts the smart contract model. When a contract condition is met, a transaction is delivered to the contract address, and all nodes in the network implement the instruction set produced by compiling the contract code. Finally, the results of the execution are written to the blockchain.

In contrast to traditional contracts, which must be manually executed by transaction participants and third parties, smart contracts can be implemented autonomously and automatically without the involvement of third parties, and the results of execution are irreversible by each participant on EVM. A further feature of smart contracts is their determinism, meaning that they can provide the same output for each participant who executes them with the same input. Additionally, a smart contract broadcasts all of its output to the whole network; hence, all participants involved in a transaction are able to verify and track the operations on the smart contract. Ideally, there is no downtime, fraud, censorship, interference from a third party, or other problems.

The aforementioned features of smart contracts make them applicable to numerous decentralized contexts, including but not limited to online games, digital assets, online voting systems, and financial services and products. At this end, smart contracts serve as a perfect option to deliver the true value of what the blockchain has to offer.

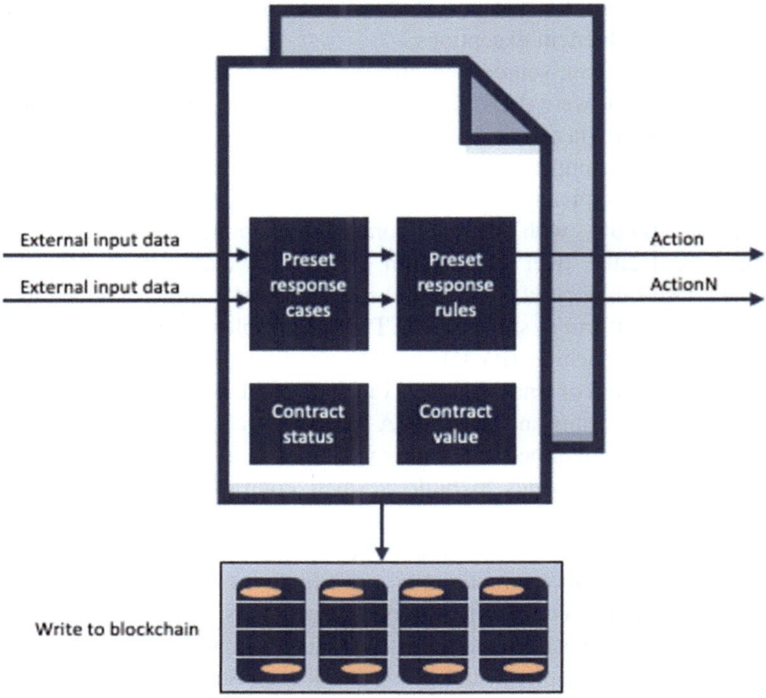

Fig. 1 Smart contact model

2.2 *Related Work*

Blockchain has become a new basis for decentralized business models. In the context of advertising systems, this makes it an ideal choice for regulating contact between mobile applications (possibly a large source of private data leaks) ad-networks, and analytics companies, thereby protecting the privacy of individuals.

To the best of our knowledge, only a few works are available in the literature for Blockchain-based mobile targeted advertising. For example, [22] proposes a decentralized framework for targeted mobile coupon delivery based on Blockchain. In this paper, the authors link the behavioral profiles that meet the vendor-defined criteria for targeting profiles with appropriate advertisements. We should point out that this framework does not cover all advertising system components, such as user profiles and Blockchain-based transactions and operations. In addition, the data are stored on the blockchain, and this is in contravention of the PDPL and GDPR. Where it is stipulated that the user data can be deleted at will, and this cannot be applied in the blockchain, where storage is permanent.

Another proposal, AdBlock [23], introduced a decentralized framework for targeted advertising based on blockchain. This framework serves to deliver advertisements to users whose behavioral profiles exactly fit the given advertisements, as defined in the advertising system. Along with many essential components, the cloud system enables privately computed access policies for mobile users and advertising systems for accessing different resources, private billing, private user profiling, and others. As compared to the processing times of the ad systems already in use, processing delays with various procedures might be evaluated to a level that is reasonably acceptable. Noting that it adopts Bespoke Miner, this unquestionably comes at the expense of a large rise in complexity and deployment as well as high energy consumption. Furthermore, user profiling and tracking are used in the advertising process. Even if this is done locally, the user is not in direct control of their data. Rather, profiling is driving the advertising process.

3 The Proposed BC-PUS for Mobile Advertising

In this section, the proposed blockchain-based private user-driven system (BC-PUS) for mobile advertising is introduced in detail. The following subsections will describe the architecture and roles of the proposed BC-PUS. Also, the design of the proposed system will be explained.

3.1 System Architecture

This subsection presents the proposed BC-PUS for mobile advertising architecture. Figure 2 depicts the proposed system's structure, including the key roles.

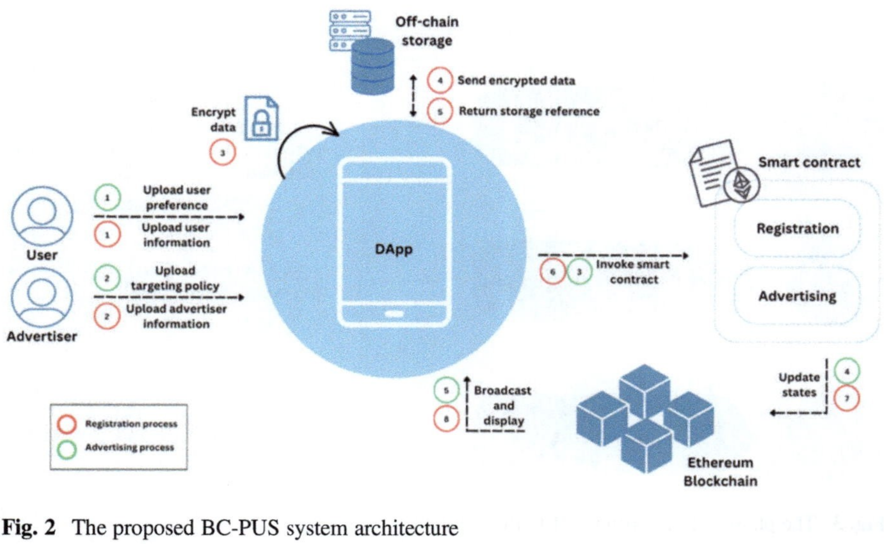

Fig. 2 The proposed BC-PUS system architecture

To facilitate the introduction of the architecture of the proposed BC-PUS, we first list the five roles involved in the proposed BC-PUS as follows.

Users: Users are in our case the data providers. In order to receive highly targeted advertisements, they voluntarily disclose particular information about their interests to advertisers.

Advertisers: Advertisers are the consumers of data in our case, it may be the advertisers themselves or the ad network on their behalf. It is considered a data consumer because it uses users' personal information, interests, and preferences for more effective targeting.

User application: Decentralized applications, or dApps, are similar to user interfaces that communicate directly with the user. It collects user and advertiser data, distributes it on a database or blockchain, and executes the smart contract. The dApps' power comes from smart contracts. As explained earlier, a smart contract is a blockchain-based executable program. The incentive and reward mechanism and transaction actions are both implemented automatically. The transaction actions are then published to the blockchain. Encrypting the uploaded data with the asymmetric Elliptic Curve Cryptography (ECC) algorithm is one of the most crucial application tasks.

Off-chain storage: Off-chain storage is primarily used to store large amounts of encrypted data via trusted third parties, and blockchain for on-chain verification just maintains a small amount of metadata and pointers or indexes (i.e., the location of the raw data in an off-chain database). It stores the user and advertiser information, and the corresponding address is returned to the dApp.

Blockchain: It stores the addresses of user and advertiser information and the transaction actions and gives feedback to dApp.

It consists of three main phases: registration phase, user profile management phase, and advertising phase. These phases are briefly illustrated in Fig. 3.

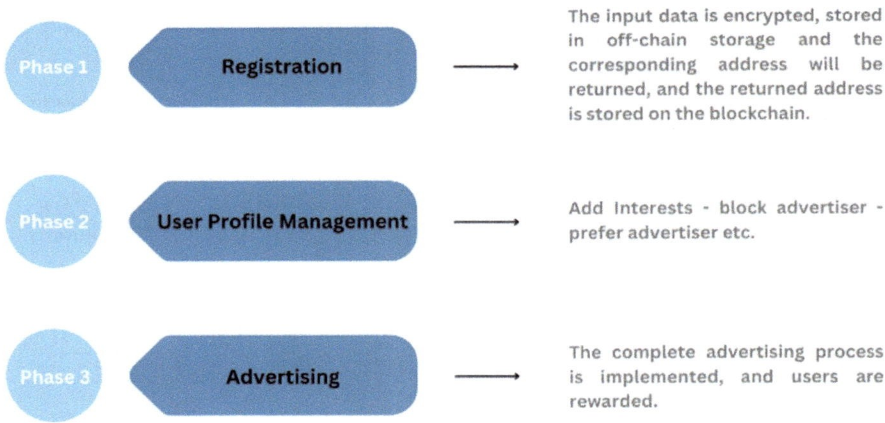

Fig. 3 The phases of our proposed BC-PUS system

3.2 System Design

This subsection details the design of the proposed BC-PUS for mobile advertising. It is composed of three main phases. Each of these phases will be described in more detail below.

Phase 1: Registration Phase. In this phase, both users and advertisers register in the system. Since the blockchain is not suited for storing large amounts of data, less data are stored in the blockchain compared to off-chain storage databases to reduce blockchain overhead. Therefore, the input in this step will be user (or advertiser) data that will be encrypted and stored in off-chain storage. The storage reference will be returned, and the returned reference will be stored on the blockchain. After completing the registration, the system (advertising gateway) gives the user (or advertiser) a unique Ethereum address that will be used as a pseudonym to avoid revealing the identity of the user (or advertiser). Figure 4 shows the workflow of the registration process.

Phase 2: User Profile Management Phase. At this phase, after users have been successfully registered in the system, the user profile management process begins by adding keywords that indicate interests and controlling them. This is done by enabling or disabling access to these keywords. There is also an option to block advertisements from certain advertisers by blocking them and putting them on the blacklist. There is also a list of preferred advertisers called a whitelist. Where the whitelist is considered a fast pass card for the advertiser, that is, the whitelist gives advertisers the advantage of displaying their advertisements regardless of the keyword-matching strategy applied in this system, even if the advertisement is not among the interests.

Phase 3: Advertising phase. In this phase, the complete advertising process is implemented, and users are rewarded. The scenario to accomplish this process is as follows.

1. The advertiser must deposit funds to be able to create an advertisement.
2. The advertiser can now create an advertisement by uploading keywords that describe the targeting policy along with the advertisement content. After creating the advertisement, the status of the advertisement will be pending.
3. The smart contract will apply a keyword-matching strategy between the user profile and the advertiser targeting policy in order to target the appropriate audience. The keyword-matching strategy pseudocode is explained in Algorithm 1.
4. As a result of the previous step, the smart contract fetches the most relevant users and puts them in the list of targeted users.
5. The rewards will be set by the advertiser and the smart contract will check the advertiser's balance funds. Once the amount of rewards is determined, the status of the advertisement changes to rewards.

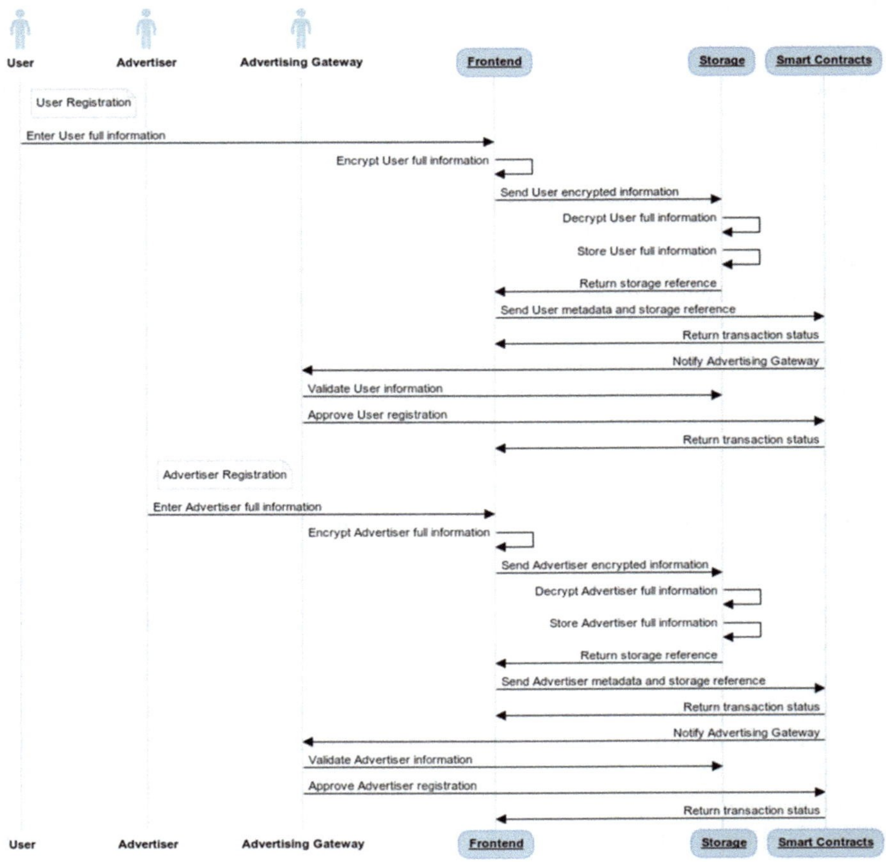

Fig. 4 Sequence diagram of the registration process. The process starts with uploading user (or advertiser) data, encrypting it, sending it to off-chain storage to create a storage reference, and then storing it in the blockchain

6. The system (advertising gateway) will generate claim tokens for each targeted user.
7. The advertisement will be broadcast to all users in the list of targeted users.
8. The advertisement reaches the user containing an advertisement identification code and claim code.
9. Now the next action is the reward payment. To claim the reward, the user will have to enter the claim code along with the identification code, then the smart contract will check the permissions and status of the claim codes and thus pay the reward.

Figure 5 shows the sequence diagram for this phase.

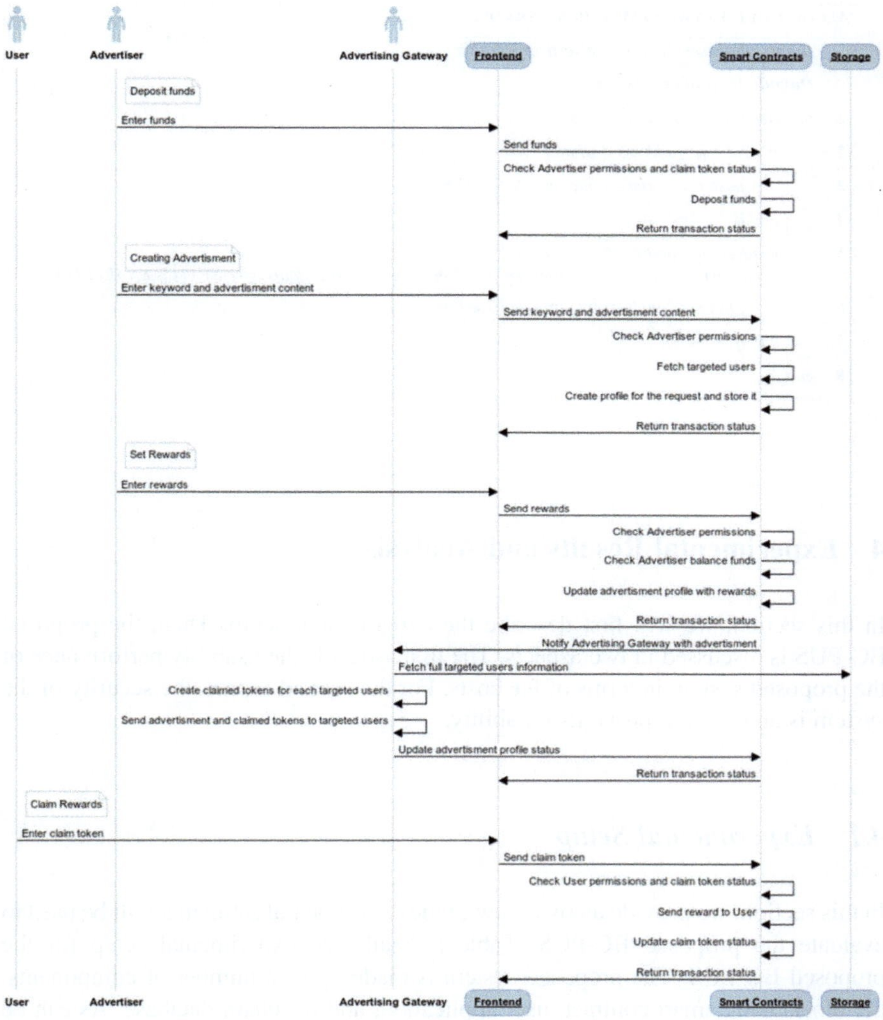

Fig. 5 Sequence diagram of the advertising process

Algorithm 1: Keyword Matching Strategy

```
Input: advertiser_keyword, advertiser_address
Output: Array of targeted user
for each user in the database do
    if advertiser_address ∈ whitelist then
        push user_address into targetedUserList
    end if
    if advertiser_address ∉ blacklist &
      ((advertiser_keyword == user_keyword) & (user_keyword status == STATUS.ENABLED))
        push user_address into targetedUserList
    end if
end for
```

4 Experimental Results and Analysis

In this section, we will first describe the experimental setup. Then, the proposed BC-PUS is discussed in two aspects. The first aspect is the usability performance of the proposed system in terms of fee costs. For the second aspect, the security of the system is analyzed to prove its reliability.

4.1 Experimental Setup

In this section, we provide an overview of the experimental setup that will be used to evaluate the proposed BC-PUS. Table I details the experimental setup for the proposed BC-PUS. The proposed system is made up of a number of components: the blockchain, smart contract, user application, and off-chain database. As can be seen in Table 1, Truffle is a development tool that uses the Solidity programming language. Furthermore, it offers a testing framework and pipeline for the Ethereum blockchain. As for the smart contract, it was written in Solidity using the Visual Studio Code editor. While the user application is written in JavaScript. The user interface is built using a React library. The data were encrypted using the JavaScript

Table 1 The experimental setup for testing the proposed BC-PUS

Name	Tool and environment version
Development Platform	Truffle v5.4.33
UI Interaction	Web3.js and React library
Operating System	Windows 10
Editor Visual Studio Code	Visual Studio Code
Programming Language	Solidity v0.8.11

Fig. 6 Front end of the proposed BC-PUS—user interface

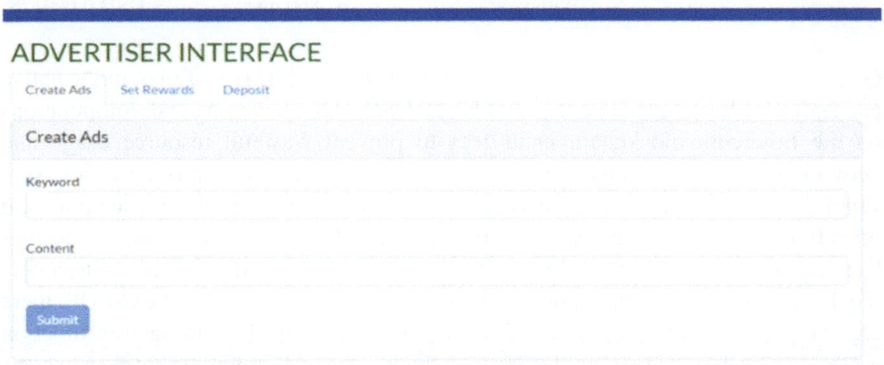

Fig. 7 Front end of the proposed BC-PUS—advertiser interface

Elliptic curve cryptography library. For off-chain storage, the system uses MongoDB to store both users' and advertisers' data.

In the experiment setup, we will evaluate and test the proposed BC-PUS in terms of its usability and reliability. Figures 6 and 7 show the front end of the proposed BC-PUS.

4.2 *Results*

In this subsection, the proposed BC-PUS is discussed in two aspects. The first aspect is the usability performance of the proposed system in terms of fee costs. For the second aspect, the security of the system is analyzed to prove its reliability.

Cost Analysis In this subsection, we test and evaluate the proposed BC-PUS in terms of fee costs in order to determine its usability. The transaction fee in the Ethereum blockchain is referred to as "Gas," where 1 Gas equals 1 Gwei and each

Table 2 Smart contracts deployment gas consumption and cost

Smart contract	Gas used	Transaction fee
Registration	1,119,271 gwei	USD 0.002276
Advertisement	5,013,862 gwei	USD 0.010194

Table 3 System transaction gas consumption and cost

Function	Gas used	Transaction fee
UserRegister	229,426 gwei	USD 0.000466
UserApprove	35,107 gwei	USD 0.000071
AddKeyword	180,905 gwei	USD 0.000368
DisableKeyword	63,589 gwei	USD 0.000129
EnableKeyword	63,524 gwei	USD 0.000129
BlockAdvertiser	58,314 gwei	USD 0.000119
Deposit	77,334 gwei	USD 0.000157
Withdraw	62,860 gwei	USD 0.000128

Gwei equals 0.000000001 ETH (10^{-9} ETH). Ether (ETH) is Ethereum's native currency; 1 ether is equivalent to $1657.21 USD. All operations and computations on the Ethereum blockchain incur fees to prevent wasteful resource usage and network abuse. A user must have enough Gas in their account in order to invoke a smart contract on the Ethereum Blockchain and perform certain transactions and operations. Next, after paying for the operations and transactions, the miner receives Gas in exchange for adding the results of the operations and transactions to a new block and broadcasting that block to the blockchain. In the proposed BC-PUS, there are two types of functions: call and transaction functions. The transaction function writes to the blockchain or database, whereas the call function reads from the database. Depending on the amount of parameters, the function structure, and the types of data used, the cost of each transaction is determined. No cost is associated with the call function.

In order to evaluate the usability performance of the proposed BC-PUS in terms of fee costs, we examine the costs of contract deployment and the functions of smart contracts during the experiments. The costs of each function are listed in Tables 2 and 3. These tables yield the following two observations:

1. The cost fees for the functions of the designed smart contracts, which amount to roughly 0.0003 USD, are obviously quite minimal. The user or advertiser in the proposed BC-PUS is responsible for covering all costs, and the sum of these costs is typically significantly lower than the profit earned by the user or advertiser through the advertising process. Hence, users and advertisers can mostly disregard these costs.
2. Contract deployment costs 0.006 USD, which is slightly higher than the cost of smart contract functions. However, the proposed BC-PUS implements contract deployment just once.

Table 4 Result summary

Test summary	Values
Total managed advertisements	104 advertisements
Total gas used	107884040 gwei
Total transactions fees	USD 0.2193390
Average transaction fee per advertisement	USD 0.0002109

As a summary of these observations, the costs associated with those functions are extremely minimal and acceptable. This proves that the proposed BC-PUS is viable and can be run at a low cost. Table 4 presents a summary of the results.

Security Analysis In this subsection, we will analyze the security of the proposed BC-PUS in order to prove its reliability. The proposed BC-PUS makes it possible to conduct targeted advertising in a secure, private, and fair manner. The solution is secured in many ways, such as using blockchain technology. In addition to the unique characteristics of Blockchain, such as being auditable, irreversible, updated in nearly real-time, and chronological, it does not require a central supervisory authority; for this reason, it is the best option for limiting communications between mobile applications and ad networks in order to protect users' privacy.

In terms of dealing with users' lack of control over their data, smart contracts have been adopted. The security of smart contracts is dependent on the underlying Ethereum blockchain [14]. The security of a smart contract is accomplished if the attacker cannot compromise 51% of the nodes (computing power) simultaneously. This is thanks to the proof-of-work consensus mechanism in the blockchain network. Informally, the Ethereum smart contract offers three security features: (1) Malicious modification of contract data is impossible. (2) In the long term, nodes on the blockchain will concur on a uniform view of the states of smart contracts. (3) A valid contract function call will be checked and executed within a specific time frame (transaction confirmation time).

Moreover, no data should be submitted for storage in plain text in order to guarantee the privacy of personal information and prevent its disclosure and leakage. Therefore, the user's personal information is encrypted before it is stored in the system. Asymmetric cryptography, another security feature of the system, guards against eavesdropping attempts. In particular, the encryption algorithm employed in this solution is ECC, which provides robust security despite using a small key size in comparison to other encryption algorithms.

It is important to emphasize that no personal information is revealed about users because they are referenced through an Ethereum address, which is a completely anonymous pseudo-alphanumeric string. Accordingly, the interests can be regarded as anonymous because they are not associated with any profiles (only an Ethereum address). Hence, it is possible to establish the number of users contacted by a certain advertising campaign without disclosing the identities of those individuals.

In conclusion, the security analysis demonstrates that the proposed system fulfills the required objectives. The system also offers many features that allow it to work with various devices and systems, such as availability and flexibility.

5 Discussion and Evaluation

This study has endorsed several achievements, which can be outlined as follows:

1. *Our proposed system can ensure user control over the sharing of their data.* The proposed system offers several features that allow the user to control his data. Even if the user may be interested in targeted advertisements, it is inappropriate to use their information without their consent. The features offered are enabling or disabling access to a specific interest as well as blocking or favoring advertisers.
2. *Our proposed system can ensure that privacy is preserved.* Using a pseudonym or anonymous address, users can conceal their real identity and prevent it from being tied to their activities on the blockchain. Thus, it aids in preserving the privacy and security of users.
3. *Our proposed system can ensure the protection and security of information.* User data is encrypted to prevent its disclosure or leakage, as well as to protect it from eavesdropping attacks. In particular, the encryption algorithm employed in this solution is ECC, which provides robust security despite using a small key size in comparison to other encryption algorithms.
4. *Our proposed system can ensure compliance with the PDPL.* In the current mobile advertising ecosystem, ad networks rely on user tracking mechanisms to gather information about users' behavior and create user profiles for the delivery of targeted advertisements. The Personal Data Protection Law (PDPL) confers a set of rights on mobile users regarding their personal data. Individuals have the "right to know" about how their data is used by ad networks, as well as the right to have personal data erased when they want to disable personalized advertising services. Our proposed BC-PUS ensures compliance with the PDPL, as the user has the right to fully control his data, whether in determining the desired amount of participation or abstaining completely, as well as deleting the data. BC-PUS is an infrastructure that can efficiently and transparently manage data flow in the mobile advertising ecosystem.
5. *Our proposed system can ensure a fair and reliable advertising mechanism.* Currently, the role of the user in the advertising process is passive, although he is considered the main element in this process. Users are tracked in several ways, and profiles are created and circulated between advertising and data analysis companies without their knowledge. Our proposed system gives the user an active role, as he is the leader of the advertising process and the cornerstone. It allows users in a safe and fair way to share their data and create a vibrant market to monetize this data. All parties involved in the process benefit. The user took a monetary incentive in exchange for sharing the data, and the advertiser got the

Table 5 Comparison between our proposed BC-PUS and related works

Characteristics	[22]	[23]	Our proposed BC-PUS
User-driven	✗	✓	✓
Privacy-preservation	✓	✓	✓
Data-encryption	✓	✓	✓
PDPL-compliance	✗	✗	✓
Low-cost	✗	✗	✓
Monetary incentive	✗	✗	✓
Authentication	✗	✗	✓

information he wanted in order to target advertisements. Also, to effectively resist common attacks such as collusion and identity theft, one-time password (OTP) codes have been used to claim rewards.

As mentioned earlier, only a few works are available in the literature on Blockchain-based mobile targeted advertising. In this section, we will compare the characteristics adopted by our proposed BC-PUS with studies [22, 23]. The results are presented in Table 5.

In [22], the authors proposed a decentralized framework for targeted mobile coupon delivery based on Blockchain. The data is stored in it on the blockchain, and this is in contravention of the PDPL. The PDPL stipulates that user data can be deleted at the user's discretion, a principle that is not easily applicable in blockchain where storage is permanent. This study adopts only two characteristics: user-driven and privacy-preservation.

Furthermore, AdBlock [23], introduced a decentralized framework for targeted advertising based on the blockchain. User profiling and tracking are thus used in the advertising process. Even if this is done locally, the user is not in direct control of their data. Rather, profiling is driving the advertising process. This study contains only three characteristics: user-driven, privacy-preservation, and data-encryption.

The comparison shows that our proposed system adopts all the characteristics: user-driven, privacy-preservation, data-encryption, PDPL-compliance, low-cost, monetary incentive, and authentication that make it a fair and reliable mobile advertising system.

6 Conclusion

In this paper, we have designed, implemented, analyzed, and evaluated a blockchain-based architecture for a mobile advertising system that addresses security and privacy concerns. A substantial awareness campaign is required to support the transition toward a healthier industry for advertising. It may be possible to use PDPL in Saudi Arabia, which is equivalent to GDPR in Europe, to create a sense of urgency about changing the industry and to highlight the solutions at hand. In this paper, we have presented the BC-PUS for mobile advertising. It takes advantage of Ethereum blockchain and smart contracts to control and regulate the advertising process, in addition to preserving the privacy of users. The BC-PUS is a decentralized system that connects users, advertisers, and decentralized applications (dApps) on a blockchain network. Users provide personal information to advertisers in exchange for targeted ads. Advertisers use this information for more effective targeting. The dApps act as the user interface, collecting and distributing user sensitive information on off-chain storage and invoking smart contracts on the blockchain. Smart contracts handle the transactions and monetary incentives and publish the transaction actions to the blockchain. The blockchain stores the addresses of user information and transaction actions and provides feedback to the dApps. Finally, the evaluation results show that the proposed BC-PUS has the required usability and reliability. In comparison with previous studies, our proposed system adopts all the characteristics: user-driven, privacy-preservation, data-encryption, PDPL-compliance, low-cost, monetary incentive, and authentication that make it a fair and reliable mobile advertising system while the previous studies guarantee only the three characteristics: user-driven, privacy-preservation, and data-encryption.

This study developed a blockchain-based private user-driven advertising platform that can accommodate diverse user interests and the business interests of the advertising system. In future work, this implementation could be further improved by either developing an ontology of interests formed by merging interests or utilizing advanced machine learning algorithms, to enhance user targeting.

References

1. Global ad spend distribution by medium 2022 | Statista. https://www.statista.com/statistics/376260/global-ad-spend-distribution-by-medium/. Accessed 30 Apr 2023
2. M.M. Krunz, et al., WiSec'12: proceedings of the fifth ACM Conference on Security and Privacy in Wireless and Mobile Networks: April 16–18, 2012, Tucson, Arizona, USA
3. T. Book, D.S. Wallach, A case of collusion: A study of the interface between ad libraries and their apps, in *Proceedings of the ACM Conference on Computer and Communications Security*, (2013), pp. 79–85. https://doi.org/10.1145/2516760.2516762
4. J. Estrada-Jiménez, J. Parra-Arnau, A. Rodríguez-Hoyos, J. Forné, Online advertising: Analysis of privacy threats and protection approaches. Comput. Commun. **100**, 32–51 (2017). https://doi.org/10.1016/j.comcom.2016.12.016
5. Y. Liu, A. Simpson, *AdSelector: A Privacy-Preserving Advertisement Selection Mechanism for Mobile Devices*
6. Personal Data Protection Law, *Bureau of Experts at the Council of Ministers* (2021, September 24). https://laws.boe.gov.sa/BoeLaws/Laws/LawDetails/b7cfae89-828e-4994-b167-adaa00e37188/1. Accessed 19 Apr 2022
7. "SDAIA," *Twitter* (2022). https://twitter.com/SDAIA_SA/status/1506314435140456452/photo/1. Accessed 19 Apr 2022
8. S. Nakamoto, *Bitcoin: A Peer-to-Peer Electronic Cash System*. [Online]. Available: www.bitcoin.org
9. Z. Zhou, M. Wang, C.N. Yang, Z. Fu, X. Sun, Q.M.J. Wu, Blockchain-based decentralized reputation system in E-commerce environment. Futur. Gener. Comput. Syst. **124**, 155–167 (2021). https://doi.org/10.1016/j.future.2021.05.035

10. M.A. Parssinen, M. Kotila, R. Cuevas Rumin, A. Phansalkar, J. Manner, Is blockchain ready to revolutionize online advertising? IEEE Access **6**, 54884–54899 (2018). https://doi.org/10.1109/ACCESS.2018.2872694
11. S. Wang, Y. Zhang, Y. Zhang, A blockchain-based framework for data sharing with fine-grained access control in decentralized storage systems. IEEE Access **6**, 38437–38450 (2018). https://doi.org/10.1109/ACCESS.2018.2851611
12. D. Liu, C. Huang, J. Ni, X. Lin, X. Shen, Blockchain-based smart advertising network with privacy-preserving accountability. IEEE Trans Netw Sci Eng **8**(3), 2118–2130 (2021). https://doi.org/10.1109/TNSE.2020.3027796
13. D. Liu, *Security and Privacy Preservation in Mobile Advertising* (University of Waterloo, Waterloo, 2020)
14. I. Ullah, R. Boreli, S. S. Kanhere, *Privacy in Targeted Advertising: A Survey* (2020, September) [Online]. Available: http://arxiv.org/abs/2009.06861
15. S. N. Khan, F. Loukil, C. Ghedira-Guegan, · Elhadj Benkhelifa, and Anoud Bani-Hani, "Blockchain smart contracts: Applications, challenges, and future trends," Peer-to-Peer Netw. Appl. 2021, 2901–2925 https://doi.org/10.1007/s12083-021-01127-0.
16. Smart Contracts. https://www.fon.hum.uva.nl/rob/Courses/InformationInSpeech/CDROM/Literature/LOTwinterschool2006/szabo.best.vwh.net/smart.contracts.html Accessed 20 Sept 2022
17. Nick Szabo – The Idea of Smart Contracts. https://www.fon.hum.uva.nl/rob/Courses/InformationInSpeech/CDROM/Literature/LOTwinterschool2006/szabo.best.vwh.net/smart_contracts_idea.html Accessed 20 Sept 2022
18. B. Version, *Ethereum: A Secure Decentralised Generalised Transaction Ledger*
19. Home | ethereum.org. https://ethereum.org/en/. Accessed 20 Sept 2022
20. "Smart contract languages | ethereum.org." https://ethereum.org/en/developers/docs/smart-contracts/languages/. Accessed 20 Sept 2022
21. S. Wang, D. Zhang, Y. Zhang, Blockchain-based personal health records sharing scheme with data integrity verifiable. IEEE Access **7**, 102887–102901 (2019). https://doi.org/10.1109/ACCESS.2019.2931531
22. Y. Gu, X. Gui, P. Xu, R. Gui, Y. Zhao, W. Liu, A secure and targeted mobile coupon delivery scheme using blockchain, in *Lecture Notes in Computer Science (Including Subseries Lecture Notes in Artificial Intelligence and Lecture Notes in Bioinformatics)*, (Springer Verlag, 2018), pp. 538–548. https://doi.org/10.1007/978-3-030-05063-4_40
23. I. Ullah, S. S. Kanhere, R. Boreli, *Privacy-Preserving Targeted Mobile Advertising: A Blockchain-Based Framework for Mobile Ads* (2020, August) [Online]. Available: http://arxiv.org/abs/2008.10479

Contribution of Sustainable and Responsible Finance: Issues and Perspectives

Fatima Bellaali, Abdelhamid El Bouhadi, Mohammed Zaryouhi, and Najah Bouchao

1 Introduction

An investment strategy known as sustainable and responsible finance (SRF) aims to produce financial returns while taking into account environmental, social, and governance (ESG) considerations. As investors grow more conscious of the significance of taking ESG issues into account when making investing decisions, SRF is receiving a lot of attention [1]. By supporting investments in businesses that place a high priority on environmental preservation, social responsibility, and ethical business conduct, SRF aims to advance sustainable development [2].

The original version of the chapter has been revised. A correction to this chapter can be found at https://doi.org/10.1007/978-3-031-53237-5_32

F. Bellaali (✉)
Laboratory for Research and Studies in Management, Entrepreneurship and Finance, National School of Business and Management, Sidi Mohamed Ben Abdellah University, Fez, Morocco

A. El Bouhadi
Laboratory for Research and Studies in Management, Entrepreneurship and Finance, National School of Business and Management, Sidi Mohamed Ben Abdellah University, Fez, Morocco

M. Zaryouhi
Laboratory Management and Formation Système, National School of Business and Management, Abdelmalek Assaadi University, Tanger, Morocco
e-mail: m.zaryouhi@uae.ac.ma

N. Bouchao
Laboratory: Language, Literature, Imagination and Aesthetics, Faculty of Letters and Human Sciences Sais-Fez, Sidi Mohamed Ben Abdellah University, Fez, Morocco
e-mail: najah.bouchao@usmba.ac.ma

© The Author(s), under exclusive license to Springer Nature Switzerland AG 2024, Corrected Publication 2024

A. Shaikh et al. (eds.), *Advances in Emerging Information and Communication Technology*, Signals and Communication Technology,
https://doi.org/10.1007/978-3-031-53237-5_6

SRF is a response to major environmental and social issues such as corporate wrongdoing, social injustice, and climate change. Over time, the idea of SRF has changed from ethical investing, which emphasizes avoiding investments in businesses with adverse social or environmental repercussions, to sustainable investing, which aims to make investments in businesses that advance sustainable development. SRF now includes a wider range of tactics, such as impact investing, which aims to produce financial gains while having a positive influence on society and the environment [3].

The lack of defined ESG measures, the complexity of assessing how investments affect sustainable development, and the trade-offs between financial returns and sustainability goals are just a few of the difficulties that SRF brings. SRF, however, also provides substantial chances for investors to earn money while promoting sustainable development. With more investors looking to match their investments with their values and beliefs, SRF is likely to continue expanding [4].

The Role of Investors in SRF

Investors are essential in advancing SRF. Investors can encourage businesses to improve their ESG performance by taking ESG considerations into account when making investment decisions. Investors can interact with businesses on ESG concerns like climate change, diversity, and corruption by using their ownership rights. A more sustainable and inclusive economy can be transitioned through investors allocating cash to businesses that positively impact sustainable development [5].

The rising demand for investments that are consistent with investors' values and beliefs is what fuels SRF's expansion. According to a survey by the Morgan Stanley Institute for Sustainable Investing, 86% of millennials and just 75% of the general population are interested in sustainable investing. Their preferences for sustainable investments are likely to influence the future of the investing business as millennials take over as the generation with the most wealth [6].

1.1 Regulatory Frameworks and Policies

SRF-related regulatory frameworks and policies are changing quickly, with an increased focus on accountability and openness. With the implementation of the Sustainable Finance Action Plan, which aims to mobilize private resources toward sustainable investments, the European Union has taken the lead in supporting SRF. The strategy calls for developing low-carbon benchmarks, taxonomizing sustainable economic activity, and including sustainability risks in the investment decision-making process [7]. Additionally, the EU has passed legislation forcing businesses to publish their ESG performance, giving investors additional information with which to make educated investment choices.

Other nations are also enacting laws to support SRF, including Canada and Japan. The Sustainable Finance Expert Panel, established by the Canadian federal government, attempts to mobilize private funding for sustainable development. A set of guidelines for the issue of green bonds has been introduced in Japan by the Ministry of the Environment [8].

1.2 ESG Metrics

For SRF, the absence of established ESG measures presents a serious obstacle. ESG measures are essential for investors to assess a company's sustainability performance and to compare investments across various industries and geographical areas. However, there is no agreement on the most important ESG metrics, which has caused an increase in the number of ESG rating agencies and frameworks [9].

1.3 Impact of SRF

Given that the effects of investments may be indirect and long-lasting, measuring the impact of SRF investments on sustainable development is difficult. However, impact measurement is essential for investors to comprehend how well their investments contribute to the achievement of sustainability goals [10].

Investors utilize a variety of techniques, including impact assessment frameworks and impact reports, to assess the social and environmental effects of their investments in order to address this challenge. These tools assist investors in determining the benefits and drawbacks of their investments and in modifying their investment strategy as necessary [11].

1.4 Financial Returns and Sustainability Goals Trade-Offs

In SRF, investors must choose between financial returns and sustainability goals. As organizations that pursue sustainability may incur higher costs in the near future, some sustainable investments may provide poorer financial returns than standard investments. However, long-term advantages like lowered risk exposure and enhanced reputation may also be provided by sustainable investments [12].

Investors can employ a variety of tactics, including diversification, active ownership, and impact investing, to balance financial returns and sustainability goals in order to address this trade-off. Diversification lessens an investor's exposure to unfavorable sustainability outcomes by distributing risk across a variety of industries and geographical areas. While impact investing enables investors to create beneficial social and environmental benefits while earning financial returns, active ownership enables investors to work with businesses to enhance their sustainability performance [13].

SRF is a method of investing that aims to produce profits while taking governance, social, and environmental considerations into account. The lack of defined ESG measures, the complexity of assessing how investments affect sustainable development, and the trade-offs between financial returns and sustainability goals are just a few of the difficulties that SRF brings. SRF, however, also provides enormous chances for investors to profit financially while supporting sustainable growth [14].

By taking ESG problems into account when making investment decisions, interacting with businesses on ESG matters, and providing capital to businesses that advance sustainable development, investors play a critical role in advancing SRF. SRF-related regulatory frameworks and policies are fast changing, with an increased focus on accountability and openness. SRF faces a huge problem due to the lack of defined ESG measures; however, organizations like the SASB and TCFD are attempting to standardize ESG metrics [15].

It can be difficult to gauge how SRF investments affect sustainable development, but impact assessment frameworks and reports can assist investors in assessing the social and environmental effects of their financial decisions. Investors must choose between financial returns and sustainability goals, but diversification, active ownership, and impact investing are some techniques that can help investors strike a balance.

Overall, SRF offers enormous opportunities for investors to match their financial decisions with their values and beliefs and promote sustainable growth. Due to the rising need for investments that take governance, social, and environmental aspects into account, SRF is projected to continue to expand [16].

2 Literature Review

The research piece offers a thorough analysis of SRF, its problems, and its viewpoints. The article starts out by explaining SRF and its primary goals, which are to encourage investments in businesses that place a high priority on environmental preservation, social responsibility, and ethical business conduct. The article emphasizes the growing focus on SRF in recent years as investors become more conscious of the significance of taking ESG issues into account when making investment decisions [17].

The problems and potential related to SRF are then discussed in depth in the essay. The absence of defined ESG indicators is one of the major problems emphasized in the paper. Investors may find it challenging to compare and assess the sustainability performance of various companies as a result. However, efforts to standardize ESG indicators include those of the Sustainability Accounting Standards Board (SASB) and the Task Force on Climate-related Financial Disclosures (TCFD).

Measuring the influence of SRF investments on sustainable development is another issue that the essay mentions. The article points out that impact measurement is essential for investors to comprehend how well their investments contribute to the achievement of sustainability goals. According to the article, investors can assess the social and environmental effects of their investments using a variety of techniques, including impact measurement frameworks and impact reports [19].

The trade-offs between financial returns and sustainability goals in SRF are also covered in the paper. Investments that promote sustainability may incur higher costs in the short term, which could result in sustainable investments providing poorer financial returns than conventional investments. However, long-term advantages like lowered risk exposure and enhanced reputation may also be provided by sustainable investments [20]. The essay makes the case that investors can balance financial returns and sustainability goals through a variety of tactics, including diversification, active ownership, and impact investing.

The importance of investors in the promotion of SRF is emphasized in the article's conclusion. According to the article, investors can support sustainable development by taking ESG considerations into account when making investment decisions, communicating with businesses about ESG matters, and investing money in businesses that support sustainable development [21]. The essay also discusses how SRF regulatory frameworks and policies are changing, with a focus on accountability and openness.

Overall, the study piece offers a thorough analysis of SRF, its difficulties, and its prospects. The essay emphasizes the importance of investors in advancing sustainable development as well as the growing emphasis devoted to SRF in recent years. The article also offers insightful information on methods that investors can employ to strike a balance between financial returns and environmental goals in SRF. Investors, decision-makers, and scholars interested in sustainable finance and responsible investment might find the paper interesting [22].

2.1 The Value of ESG Considerations in Investment Decisions

ESG issues are becoming more widely acknowledged as significant influences on the performance of long-term investments. Investors can recognize and control risks associated with environmental, social, and governance practices by incorporating ESG elements into their decision-making processes. For instance, businesses with poor environmental policies may face increased regulatory and reputational concerns, which over time may have a negative effect on their financial performance. Similar to how individuals are more susceptible to fraud and corruption, organizations with poor governance processes may be less able to maintain their financial stability [23].

Additionally, businesses that place a high priority on ESG variables may be better positioned to take advantage of new opportunities related to sustainable development. For instance, businesses that make investments in clean energy technologies may profit from the rising demand for these resources. Similar to this, businesses that use ethical labor practices may be able to recruit and keep skilled workers, which will ultimately increase their productivity and competitiveness [24].

2.2 The Contribution of Institutional Investors to SRF Promotion

Due to their considerable power over capital allocation in the financial markets, institutional investors are essential in advancing SRF. Large asset pools are managed by institutional investors, such as pension funds, endowments, and insurance companies, on behalf of their beneficiaries. Institutional investors can promote sustainable development while producing long-term financial rewards by including ESG criteria in their investment strategy [1].

Institutional investors can also interact with businesses on ESG problems by using their shareholder power. Shareholder engagement is the process of communication between shareholders and businesses with the goal of enhancing ESG practices. Engagement of shareholders can be a powerful instrument for shaping company conduct and advancing sustainable development. Investors can interact with businesses, for instance, on topics like diversity and inclusion, human rights, and climate change [2].

2.3 Opportunities and Challenges for SRF in Emerging Markets

SRF faces both difficulties and opportunities in emerging markets. On the one hand, poverty, inequality, and climate change are only a few of the environmental and social issues that emerging economies frequently face more than industrialized markets. Additionally, ESG reporting and disclosure requirements might not be as developed in emerging countries, which can make it challenging for investors to assess the sustainability performance of companies operating in these regions.

However, there are also a lot of chances for SRF in emerging markets. For instance, emerging markets frequently give access to new markets and technologies as well as a high potential for growth. Furthermore, because they are frequently more tightly entwined with the neighborhood and environment, businesses in emerging countries might be more eager to implement sustainable business practices [3].

2.4 Corporate Social Responsibility and SRF

Corporate social responsibility (CSR), which refers to a company's moral and social obligations to its shareholders, workers, clients, and community, is closely tied to social responsibility funding (SRF). CSR entails implementing ethical business practices that are advantageous to both the organization and society at large. Investors can encourage businesses to embrace ethical business practices that support sustainable development by including ESG elements in their investment decisions [4].

2.5 Case Studies: Effective SRF Strategy Implementation

Numerous case studies show how institutional investors have successfully implemented SRF methods. For instance, the California Public Employees' Retirement System (CalPERS) has been interacting with businesses on ESG concerns for more than 20 years and incorporating ESG elements into its investment decision-making processes. CalPERS's long-term investment performance has improved as a result of reducing its exposure to businesses with subpar ESG performance.

Similar to this, AP4, the Swedish national pension fund, has adopted an active ownership policy that entails interacting with businesses regarding ESG issues and casting votes on shareholder resolutions [5].

2.6 From Ethical Investing to a Mainstream Investment Strategy: The Evolution of SRF

SRF has developed from a specialized area of socially conscious investing to a widely used investment approach. Historically, values-based investors who aimed to match their investments with their social and environmental ideals have been the main forces behind ethical investing. However, institutional investors that understand the financial significance of ESG considerations are increasingly driving SRF today.

In addition, SRF now covers fixed income and alternative asset classes in addition to equity asset classes. Additionally, SRF has changed from a reactive approach that involved shareholder engagement and ESG integration to a proactive one that now includes a negative screening of businesses with subpar ESG performance [6].

2.7 ESG Factors' Inclusion in Investment Analysis and Decision-Making

A crucial element of SRF is the incorporation of ESG factors into investment research and decision-making. When analyzing and making decisions about investments, ESG integration refers to the systematic consideration of ESG factors along with financial factors. This strategy acknowledges that ESG variables can materially affect a company's financial performance and ought to be taken into account when making investment decisions.

Depending on the investment strategy and asset class, ESG integration can take many different shapes. ESG integration in the stock market, for instance, may entail choosing businesses with high ESG performance or adding ESG aspects to value models. ESG integration in fixed income may entail evaluating issuer creditworthiness in light of their ESG performance. ESG integration may entail assessing an investment's effects on the environment and society in alternative asset classes like real estate and infrastructure [7].

2.8 *The Function of Standards and Ratings in SRF*

By giving investors a framework for assessing the sustainability performance of companies and investments, ratings and standards play a significant role in SRF. ESG ratings are used to assess the sustainability performance of businesses and give investors a comparative evaluation of those businesses. Investors can find a set of ethics and criteria for responsible investing in ESG standards like the UN Global Compact and the Principles for Responsible Investment (PRI).

However, there are issues with ESG criteria and ratings, such as the lack of comparability and uniformity between ratings and the potential for greenwashing. Investors should therefore utilize ESG ratings and standards as one of numerous methods for assessing sustainability performance, keeping in mind their shortcomings and biases [8].

2.9 *SRF in Alternative Asset Classes and Fixed Income*

SRF is not just applicable to equities; it can also be used with alternative asset classes like real estate, infrastructure, and private equity, as well as fixed income and other asset classes. SRF in fixed income may entail examining issuers' ESG performance and determining their creditworthiness based on that performance. SRF might involve assessing an investment's social and environmental impact and encouraging sustainable development through active participation in alternative asset classes [9].

Applying SRF to fixed income and alternative asset classes, however, presents certain difficulties due to the absence of openness and ESG disclosure in these markets. Additionally, it's possible that specialist knowledge and resources are needed for the evaluation of sustainability performance in these markets.

2.10 *The Opportunities and Challenges for SRF's Future*

Opportunities and challenges both exist for SRF in the future. On the one hand, SRF has a lot of chances given the rising relevance of sustainability and ESG factors.

Additionally, the expansion of SRF is being fueled by the mainstream investors' and asset managers' growing adoption of ESG integration, which is related to SRF and needs to be addressed. Making educated investment decisions can be challenging for investors due to the absence of standardization and comparability of ESG data and ratings, which is one of the challenges. Another issue is the potential for "green-washing," in which businesses exaggerate or misrepresent their performance in terms of sustainability [10].

Stakeholders in the SRF community, such as the Task Force on Climate-Related Financial Disclosures (TCFD) and the Sustainability Accounting Standards Board (SASB), are striving to build common frameworks and standards for ESG data and ratings in order to solve these issues. These initiatives seek to increase transparency and accountability in sustainability reporting while also giving investors more standardized and comparable ESG data and ratings [11].

Growing interest in impact investing, which seeks to produce both financial returns and social and environmental benefits, presents another opportunity for SRF. Impact investing is a logical progression of SRF because it aims to alleviate social and environmental issues by funding businesses and initiatives that advance sustainable development [12].

The need to reconcile financial returns with effect objectives as well as the measurement and reporting of impact outcomes are just a few of the difficulties associated with impact investing. Impact investors are creating frameworks for impact measurement and management, such as the Impact Management Project (IMP) and the Global Impact Investing Network (GIIN), to solve these issues.

In order to promote sustainable and ethical finance, investors are essential. Investors are increasingly leveraging their clout to persuade businesses to enhance their sustainability performance as the demand for sustainable investments rises. This can involve speaking up about ESG issues with businesses, casting a vote on shareholder resolutions, and cutting ties with firms that don't live up to their sustainability standards.

The creation of financial products that place a priority on sustainability is another way that investors are fostering change. For instance, there has been a rise in interest in green bonds, which are issued to fund ecologically friendly initiatives. The number of sustainable exchange-traded funds (ETFs), which enable investors to participate in a diverse portfolio of businesses that value sustainability, has also increased [13].

2.11 Approaches and Best Practices for ESG Integration

ESG integration is the process of incorporating ESG considerations into investing analysis and decision-making. Depending on the asset class and investment strategy, there are many ways to include ESG factors. For instance, ESG integration in equity investment may entail identifying companies that place a high priority on sustainability through bottom-up analysis of each one. ESG integration in fixed-income investment may entail evaluating the sustainability records of governments and businesses to determine credit risk [14].

Using trustworthy and standardized ESG data and ratings, identifying significant ESG variables that are relevant to the investment decision, and systematically and consistently incorporating ESG elements into investment research are all examples of best practices for ESG integration.

2.12 *Overview of the Major Players and Frameworks for Ratings and Standards*

By giving investors a framework for assessing the sustainability performance of companies and investments, ratings and standards play a significant role in SRF. The Carbon Disclosure Project (CDP), MSCI, and Sustainalytics are a few of the major participants in the ESG ratings and standards area [15].

Frameworks for reporting and disclosing sustainability data have also been developed to give businesses direction on how to report on their sustainability performance. Two of the most popular frameworks for reporting on sustainability are the Global Reporting Initiative (GRI) and the Sustainability Accounting Standards Board (SASB).

2.13 *Lack of Standardization, Greenwashing, and Other SRF Challenges*

SRF presents a number of difficulties that need to be overcome. Making educated investment decisions can be challenging for investors due to the absence of standardization and comparability of ESG data and ratings, which is one of the challenges. Another issue is the potential for "greenwashing," in which businesses exaggerate or misrepresent their performance in terms of sustainability.

Stakeholders in the SRF community, such as the Task Force on Climate-Related Financial Disclosures (TCFD) and the Sustainability Accounting Standards Board (SASB), are striving to build common frameworks and standards for ESG data and ratings in order to solve these issues. These initiatives seek to increase transparency and accountability in sustainability reporting while also giving investors more standardized and comparable ESG data and ratings [16].

2.14 *Efforts to Promote Sustainable and Responsible Finance and Criticisms*

SRF is becoming more well-known and recognized, but there are also certain concerns and difficulties with it. One significant complaint is that including ESG considerations in investment decisions may lead to worse financial results. ESG criteria are prioritized, according to critics, which might restrict investment possibilities and result in missed chances for higher profits.

The absence of established measures and data for evaluating ESG performance presents another difficulty. This makes it challenging for investors to compare and assess the ESG performance of companies and to base their investment decisions on this data [17].

In addition, there are worries about "greenwashing," which is the practice of businesses exaggerating or misrepresenting their ESG performance in an effort to draw in socially conscious investors. Investors may find it challenging to distinguish between businesses that are actually sustainable and those that are only using sustainability as a marketing tactic as a result.

Finally, some detractors contend that SRF could not be a useful strategy for accomplishing sustainable development objectives. They contend that policies and regulations may be more successful in encouraging businesses to adopt sustainable practices [18].

As investors increasingly understand the significance of taking ESG aspects into account when making investment decisions, sustainable and responsible finance has evolved as a key concept in the financial sector. By supporting investments in businesses that place a high priority on environmental preservation, social responsibility, and ethical business conduct, SRF aims to advance sustainable development.

Implementing SRF can be done in a number of ways, including active ownership, positive screening, and negative screening. Investors should carefully assess which method best fits their financial goals and values since each has advantages and disadvantages.

Despite the fact that SRF has been very popular lately, there are still issues and difficulties with it. These include doubts about the effectiveness of SRF in attaining sustainable development goals, worries about diminished financial returns, the absence of defined measures and data, greenwashing, and issues with greenwashing [19].

Despite these difficulties, SRF is likely to keep gaining significance as investors become more aware of the dangers of unsustainable practices and the potential rewards of backing sustainable businesses. However, more ESG reporting and data standards and openness are required, as well as more robust legislative and regulatory measures to encourage sustainable practices among businesses, in order to fully achieve the promise of SRF [20].

3 Methodology

This study seeks to present a thorough analysis of sustainable and responsible finance (SRF), its problems, and its perspectives. The next strategy is suggested to accomplish this goal:

Conducting a thorough examination of the SRF literature, including scholarly articles, industry reports, and pertinent publications, is the first step. In order to comprehend the current status of research and practice in the field of SRF, it will be helpful to have a solid understanding of the subject and be able to recognize important concepts.

Data collection: To assist the analysis and discussion, the research will require gathering data from a variety of sources. This can entail compiling data on methods for ESG integration, ratings and standards, difficulties, chances, and potential future developments in SRF. Reputable sources of information include academic databases, industry reports, sustainability organizations, and governmental agencies.

Framework Analysis: A framework method will be used to examine the data that has been gathered. The primary themes and concepts found throughout the literature study will serve as the foundation for the framework's development. This analysis will assist in classifying and organizing the data, seeing patterns and trends, and reaching insightful conclusions.

Case Studies: Relevant case studies will be incorporated into the research to offer real-world examples and useful insights. These case studies will demonstrate the SRF techniques used by various businesses, the difficulties they encountered, and the results they obtained. Case studies from a variety of industries, such as finance, business sustainability, and impact investing, may be chosen.

Stakeholder interviews may be undertaken with industry experts, sustainability professionals, investors, and other pertinent stakeholders to acquire a deeper knowledge of their viewpoints and experiences as stakeholders in the SRF field. These interviews will offer insightful information about the difficulties, possibilities, and prospects for SRF's future.

Comparative Analysis: To assess the efficacy and effects of various methods on SRF, a comparative analysis will be performed. The results of various investment strategies will be examined, the success of companies with strong ESG practices will be compared to that of their peers, and the alignment of SRF with sustainable development objectives will be evaluated.

Future Forecast and Suggestions: The research will offer suggestions for investors, policymakers, and other stakeholders based on the findings from the literature review, data analysis, case studies, and stakeholder interviews. It will also present a future forecast for SRF. Addressing issues, improving standards and openness, and promoting the incorporation of SRF principles into investment processes may be the main goals of these recommendations.

4 Results

Sustainable and Responsible Finance (SRF) Overview

The review of the literature gave readers a thorough understanding of SRF and highlighted how important it is becoming to the financial sector [1].

The acknowledgment of the sustainability and long-term worth of investments led to a major rise in knowledge and acceptance of SRF principles among investors and businesses.

ESG Integration Methodologies: The research of several ESG integration methods revealed that negative and positive screening were the two methods that investors used the most [2].

Positive screening meant deliberately choosing businesses with high ESG performance while negative screening involved rejecting businesses that did not match particular ESG criteria [3]. Institutional investors were becoming more involved in active ownership, which focuses on interacting with businesses to enhance their ESG policies, while it was less common [4].

Standards and Ratings: There is a lack of consistency and comparability in measuring and reporting ESG performance, according to the analysis of ESG ratings and standards frameworks [5].

There were differences in the methodology and standards used by various frameworks, which affected how companies' sustainability performance was judged.

This emphasized the requirement for greater uniformity and openness in ESG reporting in order to support better comparisons and more informed decisions [6].

Problems with SRF: Concerns regarding potential trade-offs between financial rewards and sustainability were among the SRF difficulties that were raised.

Recent research, however, has disproved the myth of a trade-off by demonstrating that businesses with high ESG performance can beat their rivals financially [7].

The problem of "greenwashing," in which businesses exaggerate their commitment to sustainability, was also cited as a concern, highlighting the significance of accurate and open reporting [8].

Case Study: The inclusion of case studies illustrated how many businesses and investment funds have successfully implemented SRF techniques.

These examples showed how incorporating sustainability into investment choices can result in increased ESG performance and profitable financial returns [9].

Examples included stakeholder participation, the use of sustainable supplier chains, and investments in renewable energy, all of which improved financial and sustainability outcomes [10].

Discussion: The research's conclusions have several significant ramifications for sustainable and ethical finance: SRF principles are being more widely used, which shows that investor attitudes are changing in favor of sustainable investments [11]. This indicates an increasing understanding of the resilience and long-term worth of businesses that give attention to ESG aspects [12].

The research of various ESG integration methods reveals the wide range of tactics used by investors. The most frequent screening methods were found to be negative and positive, demonstrating a preference for avoiding businesses with weak ESG performance and investing in those with robust sustainability procedures [13].

The necessity for standardized and open ESG reporting was identified through the review of grading and standards frameworks. Investors face difficulties evaluating and contrasting the sustainability performance of different companies due to the absence of uniformity and comparability among frameworks [14].

The SRF's problems, such as the perceived trade-off between financial returns and sustainability, can be addressed by the growing body of research showing that businesses with good ESG performance can make money [15].

The case studies showed how effectively implementing SRF solutions can boost both financial performance and ESG results. This demonstrates the opportunity for businesses to add value by incorporating sustainability into their business plans [16].

The findings of this study highlight the significance of sustainable and responsible finance in the investment sector [17]. The results lend credence to the idea that including ESG considerations in investment decision-making can result in profitable outcomes and enhanced sustainability performance. Insights into areas that need to be addressed for continued development and improvement in SRF practices are provided by the issues that have been discovered, such as the requirement for consistent reporting and tackling greenwashing [18].

5 Discussion

The discussion portion of a study on sustainable and responsible finance (SRF) offers the chance to interpret and analyze the findings, situate them within the body of prior research, and provide insights and field-specific implications. In this section, we will go through the main conclusions and their significance, point out any contradictions or restrictions, and offer suggestions for further study and application of SRF.

Integration of ESG Factors: The analysis of several methods for incorporating environmental, social, and governance (ESG) factors into the selection of investments produced a number of insightful results. Among socially conscious investors, the positive screening strategy, which entails making investments in businesses with high ESG performance, has grown in popularity. However, worries about lower financial returns continue to be difficult. Investors should carefully weigh the long-term advantages and dangers of sustainable investments as well as the possible effects of ESG aspects on financial performance.

Ratings and Standards: The review of the frameworks for ESG ratings and standards has brought to light the need for more consistency and openness in ESG reporting. Investors find it difficult to compare and consistently assess the sustainability performance of companies because there are numerous frameworks and different approaches for evaluating ESG performance. In order to improve comparability between businesses and industries and to promote better-informed decision-making, efforts must be made to create uniform standards and measurements.

Challenges of SRF: The discussion of SRF's difficulties highlighted crucial factors for investors. A major worry continues to be the potential trade-off between financial gains and sustainability. Recent research has refuted the idea of a trade-off, demonstrating that organizations with excellent ESG procedures can outperform their peers over the long term. Maintaining investor confidence in SRF depends on addressing the issues with greenwashing and ensuring a truthful portrayal of companies' sustainability performance.

Case Studies: The inclusion of case studies gave practical illustrations of the effective application of SRF techniques. These case studies showed how businesses and investment funds may make money while also giving sustainability a high priority. They highlighted a variety of organizational actions that enhanced ESG performance and financial outcomes, including investments in renewable energy, supply chain sustainability, and stakeholder involvement.

Viewpoints from Stakeholder Interviews: SRF was better understood from a variety of viewpoints thanks to the insights gleaned through stakeholder interviews. The growing importance of ESG considerations in investment decisions and the necessity for increased stakeholder collaboration to address the difficulties in SRF were underlined by industry leaders, sustainability specialists, and investors. Additionally, they underlined the significance of thorough ESG reporting and data in order to facilitate informed decision-making and precise evaluation of businesses' sustainability performance.

Comparative investigation: The investigation showed that organizations with good ESG practices typically outperform their contemporaries over the long run. This implies that including ESG considerations in investing decision-making can enhance financial success. The linkage of SRF with sustainable development objectives was also clear, highlighting its potential to promote favorable social and environmental consequences.

Future Research: Future SRF research should concentrate on tackling the remaining issues, including creating standardized measurements and procedures for ESG reporting, eliminating greenwashing, and further investigating the connection between financial success and ESG aspects. Further research into the effects of SRF on various industries and geographical areas would shed light on the viability and efficacy of SRF initiatives.

6 Conclusion

As investors increasingly understand the value of taking environmental, social, and governance (ESG) considerations into account when making investment decisions, sustainable and responsible finance (SRF) has emerged as a key concept in the investing sector. By supporting investments in businesses that place a high priority on environmental preservation, social responsibility, and ethical business conduct, SRF aims to advance sustainable development.

Implementing SRF can be done in a number of ways, including active ownership, positive screening, and negative screening. Positive screening entails investing in businesses that perform well in terms of ESG, while negative screening involves rejecting businesses that don't. Engaging in active ownership entails influencing businesses to enhance their ESG performance. Investors should carefully assess which method best fits their financial goals and values since each has advantages and disadvantages.

SRF has grown significantly in popularity in recent years as more investors become aware of the dangers of unsustainable business practices and the possible rewards of backing sustainable businesses. A rising number of institutional investors, including pension funds and sovereign wealth funds, who manage huge sums of money and have the potential to significantly influence firms' ESG practices, have adopted SRF.

SRF, however, is not without its detractors and difficulties. One significant complaint is that including ESG considerations in investment decisions may lead to worse financial results. ESG criteria are prioritized, according to critics, which might restrict investment possibilities and result in missed chances for higher profits. Numerous studies that show that organizations with high ESG performance can outperform their competitors over the long term have addressed this claim.

The absence of established measures and data for evaluating ESG performance presents another difficulty. This makes it challenging for investors to compare and assess the ESG performance of companies and to base their investment decisions on this data. There have been initiatives to create standardized ESG reporting frameworks, such as the Global Reporting Initiative and the Sustainability Accounting Standards Board, to address this issue.

In addition, there are worries about "greenwashing," which is the practice of businesses exaggerating or misrepresenting their ESG performance in an effort to draw in socially conscious investors. Investors may find it challenging to distinguish between businesses that are actually sustainable and those that are only using sustainability as a marketing tactic as a result. Investors must execute extensive due diligence and look beyond marketing materials to analyze companies' actual ESG performance in order to meet this challenge.

Finally, some detractors contend that SRF could not be a useful strategy for accomplishing sustainable development objectives. They contend that policies and regulations may be more successful in encouraging businesses to adopt sustainable practices. While policies and regulations are crucial, SRF may also be a key player in advancing sustainable development by allocating funding to businesses that place a priority on sustainability.

Stronger governmental and regulatory measures are required to encourage sustainable practices among businesses, as well as improved uniformity and openness in ESG reporting and data, in order to fully fulfill the potential of SRF. To achieve this, it will be necessary for investors, businesses, regulators, and other stakeholders to work together to create frameworks and standards that encourage openness and accountability as well as a shared understanding of what constitutes sustainable and responsible finance.

SRF has the potential to be a key player in advancing sustainable development by promoting investments in businesses that place a high value on social responsibility, environmental preservation, and ethical business practices. SRF is not without its problems and detractors, but these can be overcome with more consistency, openness, and regulatory and policy measures in ESG reporting and data. SRF is projected to continue to gain prominence and integrate into the investment business as investors increasingly understand the value of sustainability.

References

1. J. Ali, T. Rasheed, M. Afreen, M.T. Anwar, Z. Nawaz, H. Anwar, K. Rizwan, Modalities for conversion of waste to energy—challenges and perspectives. Sci. Total Environ. **727**, 138610 (2020)
2. A. Amran, M. Nejati, S.K. Ooi, F. Darus, Exploring issues and challenges of green financing in Malaysia: perspectives of financial institutions, in *Sustainability and Social Responsibility of Accountability Reporting Systems: A Global Approach*, (2018), pp. 255–266
3. V. Brescia, A.A. Sa'ad, S.M.B.S.J. Alhabshi, R.B. Hassan, F. Lanzalonga, Exploring sustainability from the islamic finance perspective. Eur. J. Islam. Finance, 45–53 (2021). https://doi.org/10.13135/2421-2172/6107
4. M. Camilleri, Advancing the sustainable tourism agenda through strategic csr perspectives. Tour. Plan. Dev. **11**, 42–56 (2014)
5. Y. Chang, *Green Finance in Singapore: Barriers and Solutions* (Singapore University of Social Sciences, 2019)
6. K. Chatzitheodorou, A. Skouloudis, K. Evangelinos, I. Nikolaou, Exploring socially responsible investment perspectives: a literature mapping and an investor classification. Sustain. Prod. Consumption **19**, 117–129 (2019)
7. T.-L. Chen, H. Kim, S.-Y. Pan, P.-C. Tseng, Y.-P. Lin, P.-C. Chiang, Implementation of green chemistry principles in circular economy system towards sustainable development goals: challenges and perspectives. Sci. Total Environ. **716**, 136998 (2020)
8. D. Deliu, Financial audit vs. Corporate governance practices–a viewpoint on current challenges and perspectives in the time of covid-19. Ovidius Univ. Ann. Ser. Econ. Sci. **20**(1), 909–918 (2020)
9. B. Forte, M. Cerreta, P. De Toro, *The Human Sustainable City: Challenges and Perspectives from the Habitat Agenda* (Routledge, London, 2019)
10. H. Liang, L. Renneboog, *Corporate Social Responsibility and Sustainable Finance: A Review of the Literature*. European corporate governance Institute–finance working paper (2020)
11. C. Malecki, General overview of socially responsible investment: perspectives for sustainable corporate governance, in *Corporate Social Responsibility*, (2018), pp. 395–425
12. A. Maxim, A. Dornean, Trends, challenges and perspectives of sustainable development. An empirical analysis for Romania. Transform. Bus. Econ. **17**, 760 (2018)
13. A. Missbach, The equator principles: drawing the line for socially responsible banks? An interim review from an ngo perspective. Development **47**, 78–84 (2004)
14. R. Prasad, A. Bhattacharyya, Q.D. Nguyen, Nanotechnology in sustainable agriculture: recent developments, challenges, and perspectives. Front. Microbiol. **8**, 1014 (2017)
15. S. Prasad, K. Sheetal, V. Venkatramanan, S. Kumar, S. Kannojia, Sustainable energy: Challenges and perspectives, in *Sustainable Green Technologies for Environmental Management*, (2019), pp. 175–197

16. X.-J. Qiao, A. Kristoffersson, T.B. Randrup, Challenges to implementing urban sustainable stormwater management from a governance perspective: a literature review. J. Clean. Prod. **196**, 943–952 (2018)
17. S. Quatrini, Challenges and opportunities to scale up sustainable finance after the covid-19 crisis: lessons and promising innovations from science and practice. Ecosyst. Serv. **48**, 101240 (2021)
18. S. Rahman, I.H. Moral, M. Hassan, G.S. Hossain, R. Perveen, A systematic review of green finance in the banking industry: Perspectives from a developing country. Green Finance **4**, 347–363 (2022)
19. M. Rodríguez-Fernández, A.I. Gaspar-González, E.M. Sánchez-Teba, Sustainable social responsibility through stakeholders engagement. Corp. Soc. Responsib. Environ. Manag. **27**, 2425–2436 (2020)
20. A.J. Salzmann, The integration of sustainability into the theory and practice of finance: an overview of the state of the art and outline of future developments. J. Bus. Econ. **83**, 555–576 (2013)
21. Z. Shanava, M. Vanishvili, Financial education of the nation: challenges and perspectives. Int. J. Soc. Sc. Econ. Res. **6**, 4646–4672 (2021)
22. M. Situm, O.L. Plastun, I.O. Makarenko, Y.S. Serpeninova, G. Sorrentino, Sdg 3 and financing instruments in Austria and Ukraine: challenges and perspectives. Probl. Perspect. Manag. **19**, 118 (2021)
23. T.T.T. Tu, T.T.H. Yen, Green bank: international experiences and Vietnam perspectives. Asian Soc. Sci. **11**, 188 (2015)
24. J. Tuyon, O.P. Onyia, A. Ahmi, C.-H. Huang, Sustainable financial services: reflection and future perspectives. J. Financ. Serv. Mark. **29**, 1–27 (2022)

End to End Unsupervised Learning-Based Endoscopic View Expansion

Shizun Zhao, Jingjing Luo, Hongbo Wang, Yuan Han, and LiLi Feng

1 Introduction

Image stitching refers to the process of registering and fusing images with overlapping areas to form a panoramic image with a larger field of view. This technology is widely used in medical diagnosis [1, 2]. The focus of this study is the combination of image stitching technology and endoscopic visuals, especially applying image stitching technology to real-time intubation videos of difficult airways. In the clinical application of difficult airway intubation, anesthesiologists obtain visual information through flexible endoscopes, identify key anatomical landmarks in the airway, determine the position and direction of the endoscope, and complete the intubation process. However, when the front end of the flexible endoscope approaches soft tissue, the limited field of view poses a challenge to the identification of anatomical structures. Using video sequences from adjacent time periods, a panoramic image can be formed near the anatomical structure, expanding the doctor's field of view, improving visual perception during the surgical process, and facilitating the progress of the surgery [3].

S. Zhao
Academy for Engineering and Technology, Fudan University, Shanghai, China
e-mail: 21210860103@m.fudan.edu.cn

J. Luo · H. Wang (✉)
Academy for Engineering and Technology, Fudan University, Shanghai, China

Intelligent Robot Engineering Research Center of Ministry of Education, Shanghai, China

Shanghai Intelligent Robot Engineering Technology Research Center, Shanghai, China
e-mail: LuoJingJing@fudan.edu.cn; Wanghongbo@fudan.edu.cn

Y. Han · L. Feng
Eye & ENT Hospital of Fudan University, Shanghai, China
e-mail: hanyuan@eentanesthesia.com

© The Author(s), under exclusive license to Springer Nature Switzerland AG 2024

A. Shaikh et al. (eds.), *Advances in Emerging Information and Communication Technology*, Signals and Communication Technology,
https://doi.org/10.1007/978-3-031-53237-5_7

Image stitching roughly includes two processes: alignment and fusion. In the alignment process, the transformation relationship between the two images often needs to be solved to make the overlapping areas of the two images overlap as much as possible. In the fusion process, a panoramic image needs to be generated from the aligned images, and the artifacts caused by misalignment need to be eliminated to produce high-quality output. Learning-based methods have been applied to image stitching in recent years. Early methods mainly focused on estimating homography transformations. Deep image homography estimation (DH) [4] first proposed a supervised learning method, generating a simulated dataset on a single image and using a VGG network to regress the relative offsets between image vertices and then calculating the homography transformation. Unsupervised deep homography (UDH) [5] proposed an unsupervised learning method, using photometric loss of overlapping regions between transformed images to train the network. Some studies have also used cascaded or hierarchical network structures based on the above two methods to obtain more accurate homography estimates. Recent research has considered more complex scenes. Nie et al. [6] designed a pyramid-structured regression network that estimates homography from coarse to fine, making it applicable to large baseline scenes. Depth-aware multi-grid deep homography [7] designed a contextual correlation layer that explicitly calculates region correlation before inputting the feature map pyramid into the regression network to improve computation speed. This method also uses the idea of grid optimization to estimate local transformations in the last layer of the regression network. Refs. [8–11] focus on the dynamic areas in small baseline scenarios, aiming to reduce or eliminate estimation errors caused by moving objects. What's more, Nie et al. have considered the issue of image fusion in their work [12, 13]. The former is based on supervised learning, and after aligning the images, it adjusts the fusion results finely with the label of the image without artifacts. The latter is based on unsupervised learning and reconstructs panoramic images from feature maps.

However, unlike images captured in natural scenes, the main technical challenges of endoscopic image stitching include the lack of clear texture and contour features in anatomical structures, significant changes in camera perspective and depth of field during surgery, and nonrigid deformation of human tissue. Due to these issues, precise alignment of endoscopic images is often difficult, and there are often artifacts caused by tissue deformation in the alignment area. To address these issues, we propose an unsupervised deep learning network that combines image alignment and fusion. The feature extraction part of the network uses HRNet, which helps to fuse feature maps of different resolutions to obtain better feature representations, making it more robust under low-texture and tissue deformation conditions. The obtained features are used for both the alignment sub-network and the fusion sub-network. The alignment sub-network calculates the local correlation between image regions based on the feature map, regresses the relative offset between image vertices based on the correlation, and solves the homography transformation using the offset. The fusion sub-network reconstructs panoramic images with multi-scale features while preserving the structural information of the images and eliminating artifacts.

The contribution of the unsupervised deep learning network proposed in this study for real-time endoscopic visual field expansion is as follows:

- During the process of image alignment, it has good robustness for low-texture endoscopic images, camera angles and depth of field changes, and tissue deformation and can accurately estimate the transformation relationship between images.
- During the process of image fusion, different resolution output results are reconstructed using multi-scale feature maps, which eliminate artifacts in incompletely aligned areas while retaining high-level semantic information in the image, resulting in high-quality panoramic images.

2 Our Method

2.1 Network Architecture

The network uses HRNet [14] as the feature extraction network, and the input images are denoted as source image I_1 and target image I_2, with a size of 256*256*3. HRNet can obtain multi-scale feature outputs. In this study, only the feature map with a size of 1/4 is selected due to performance and complexity considerations. The outputs are denoted as F_1 and F_2.

Similar to [7], we explicitly match regions in the feature map to accelerate the convergence speed of the network. After normalizing the feature map, we unfold it into small regions of size 3*3 with a stride of 1, using cosine similarity between each small region as a measure of correlation. We then convert the similarity into matching probability through the softmax function to obtain the correspondence between regions. From the correspondence, we can calculate the relative offsets of the small regions in the x and y directions to form a feature flow map and then regress the relative offsets between image vertices based on the feature flow map. Through the vertex offset obtained, we use the tensor direct linear transform (DLT) method to solve the coarse homography transformation H_{coarse}. Based on the obtained transformation H_{coarse}, we use the spatial transformation network (STN) to perform a preliminary transformation on the target image, aligning it roughly with the source image. It should be noted that we believe that region-based matching is sufficient for the network to obtain long-range correlation, and pyramid structures are not necessary. Experiments have also shown that even in large baseline scenarios (overlap areas not exceeding 50%), matching-based methods can still align well.

On the basis of coarse alignment, we further input the source image and the transformed target image into a fine alignment network to obtain a more accurate homography transformation H_{fine}. The reason for using the original image is that we believe it retains more detailed information, which helps to better align and merge the images (Fig. 1).

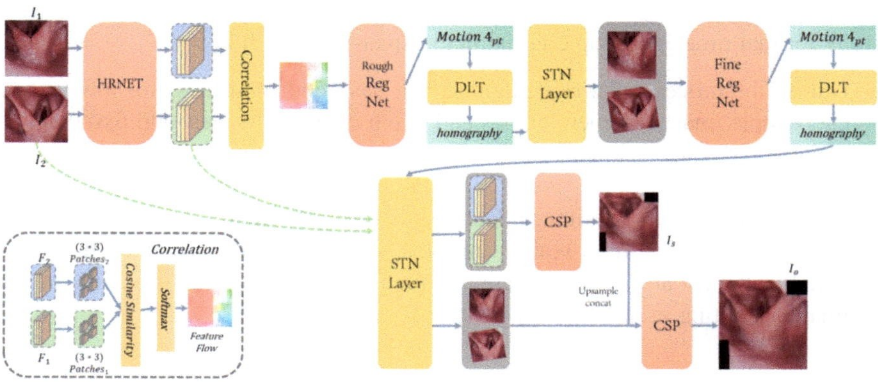

Fig. 1 Network architecture

Afterward, we use the refined transformation matrix H_{fine} obtained to transform both the original image and the feature map through the STN layer. It should be noted that during the transformation sampling process, we project the transformed image onto a 512*512 size area so that the content that exceeds the size of the original image can be retained. The feature map is also transformed in the same way. The transformed original image and feature map are used as inputs to the fusion network. In the fusion part, the feature map is processed by a CSP (cross stage partial network) block and upsampled to the same size as the original image, forming a low-resolution panoramic image I_s. Because it is reconstructed from the feature map, it can reduce the artifacts generated in the incompletely aligned areas while preserving the structural information. After the low-resolution panoramic image is concatenated with the original image in the channel axis, it further undergoes a CSP block to combine the pixel information of the original image with the structural information of the feature map reconstructed image, generating a high-resolution panoramic image I_o without artifacts.

The advantage of the above design is that we can complete the alignment and fusion of images end-to-end, whether in the training or usage process, without using different feature extraction networks separately. By directly applying the obtained transformation H_{fine} to the feature map of the target image, the transformed feature map obtained is aligned with the feature map of the source image itself, which obviously helps the convergence of the fusion network.

2.2 *Unsupervised Loss*

In the alignment stage, we use the L_1 loss function to measure the photometric differences between the source image and the transformed target image in the overlapping area. This loss is used for both the coarse alignment network and the fine alignment network to enable them to converge.

$$\mathrm{Loss}_{\mathrm{coarse}} = L_1(I_1, H_{\mathrm{coarse}}(I_2))$$

$$\mathrm{Loss}_{\mathrm{fine}} = L_1(I_1, H_{\mathrm{fine}}(I_2))$$

In the fusion process, the low-resolution image reconstructed from the feature map is referred to as I_s, and the high-resolution panoramic image reconstructed from the original image is referred to asI_o , M_h is the mask of the overlapping area in high resolution, and M_l is the mask of the overlapping area in low resolution. The fused image should be as similar as possible to the source and target images in the overlapping area, so the reconstruction loss is defined accordingly.

$$\begin{aligned}\mathrm{Loss}_{\mathrm{content}} = {} & L_1(I_o{}^*M_h, H_{\mathrm{fine}}(I_2) * M_h) + L_1(I_o{}^*M_h, I_1{}^*M_h) \\ & + L_1\left(I_s{}^*M_l, H_{\mathrm{fine}}\left(I_{2\_\mathrm{down}}\right) * M_l\right) + L_1\left(I_s{}^*M_l, I_{1\_\mathrm{down}}{}^*M_l\right)\end{aligned}$$

Consistency loss is used to make high-resolution images as similar as possible to low-resolution images, thereby blending structural information with detailed information.

$$Loss_{\mathrm{consist}} = L_1(\mathrm{Upsample}(I_s), I_o)$$

In addition, the reconstructed image should be as smooth as possible to avoid the occurrence of outliers. Therefore, the total variation loss is used as a constraint term., where I and j, respectively, represent the horizontal and vertical coordinates of the pixels in the image.

$$Loss_{\mathrm{tv}}(I_o) = \sum\nolimits_{i,j} \left|x_{i+1,j} - x_{i,j}\right| + \left|x_{i,j+1} - x_{i,j.}\right|$$

The total loss is recorded as below, of which $\alpha_1 \sim \alpha_6$ represents the network parameters.

$$\mathrm{Loss} = \alpha_1 \mathrm{Loss}_{\mathrm{rough}} + \alpha_2 \mathrm{Loss}_{\mathrm{fine}} + \alpha_3 \mathrm{Loss}_{\mathrm{content}} + \alpha_4 \mathrm{Loss}_{\mathrm{consist}} + \alpha_5 \mathrm{Loss}_{\mathrm{tv}}$$

2.3 Datasets and Training

This study used 50 sets of videos from difficult airway intubation procedures as the basic data and sampled 5000 candidate frames with sufficient differences (SSIM less than 0.85). Following a similar dataset production method mentioned in [4], candidate frames were randomly selected and scaled to a size of 400*400. A pair of 256*256 regions were randomly selected and transformed from the selected frames to serve as the source and target images, generating a total of 50,000 pairs of simulation datasets. In addition, we also selected original video frames with intervals of more than 10 frames as the real dataset, totaling 5000 pairs of images, to verify the effect of visual expansion in real situations.

The network was trained for five epochs on the simulation dataset. During the network training, image data augmentation was performed with a probability of 0.3, with the most important methods being motion blur and elastic deformation, which are particularly common in airway images. The Adam optimizer was used with a learning rate of 0.001, a decay step of 10,000, and a gamma of 0.96. For simplicity, the coefficients of the loss function were all set to 1.

3 Results and Discussion

3.1 Evaluation

In this study, the evaluation metrics used for image alignment are NRMSE (normalized root mean square error), PSNR (peak signal-to-noise ratio), and SSIM (structural similarity index) [15].

Root mean square error (RMSE) reflects the degree of variation between variables and is an objective evaluation index of image quality based on pixel error. It is used to measure the pixel difference between the source image and the target image. The smaller the RMSE, the better the image alignment effect. NRMSE is used here to better illustrate the relative error.

$$\mathrm{RMSE}(I_1, I_2) = \sqrt{\mathrm{MSE}(I_1, I_2)} = \sqrt{\frac{1}{n}\sum_{i=1}^{m}\left(y_i - \hat{y}_i\right)^2}$$

In the formula, I_1 and I_2, respectively, represent the source image and the target image, and y_i and $\hat{y}_i$, respectively, represent the pixel values of corresponding points in the overlapping area between I_1 and I_2.

PSNR is used to measure the ratio between effective information and noise in an image, which can reflect whether the image is distorted. The larger the value of PSNR, the better the quality of the fused image.

$$\mathrm{PSNR}(I_1, I_2) = 10 \log_{10}\left(255^2/\mathrm{MSE}(I_1, I_2)\right)$$

SSIM can measure the similarity between two images. Unlike MSE and PSNR, which measure absolute errors, SSIM is a perceptual model that is more in line with human visual perception.

Table 1 Evaluation of airway image stitching results

	Simulation dataset	Real dataset
NRMSE	0.06	0.135
PSNR	39.109	29.281
SSIM	0.986	0.954

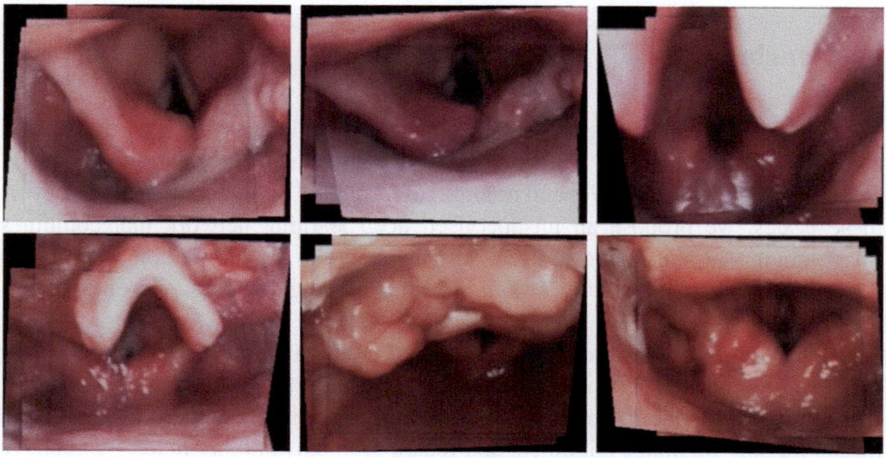

Fig. 2 The visual field extension effect near the glottis and epiglottis

$$\mathrm{SSIM}(I_1, I_2) = \frac{(2\mu_x\mu_y + C_1)(2\sigma_{xy} + C_2)}{\left(\mu_x^2 + \mu_y^2 + C_1\right)\left(\sigma_x^2 + \sigma_y^2 + C_2\right)}$$

In the formula, $\mu_x\mu_y$ represents the mean value of the overlapping area pixels I_1 and I_2, σ_x^2, σ_y^2 represents the variance, and σ_{xy} represents the covariance.

The results on the airway simulation dataset and the real dataset are as follows. The reason why real data has a larger error compared with simulation data is that the images in simulation data come from the same image, so they can be stitched naturally. However, with the movement of the camera, sometimes the two frames in the video may not be aligned (Table 1).

3.2 *Effect of View Expansion*

In the vicinity of some major anatomical structures, where the camera does not have a large depth change, the network can be pieced together in an incremental manner to obtain a large field of view extension (Fig. 2).

By continuously stitching multiple images together, a larger field of view can be obtained compared to a single image. Experiments near some major anatomical structures have shown that continuous stitching of multiple images can expand the effective field of view by over 150%, providing more abundant visual information.

4 Conclusion

Endoscopic image stitching poses many technical challenges, such as the lack of clear texture and contour features in anatomical structures, changes in camera perspective and depth of field, and nonrigid deformation of human tissue. To address these issues, we propose an unsupervised deep learning network that combines image alignment and fusion to accurately estimate the transformation relationship between images, eliminate artifacts in incompletely aligned areas, and obtain high-quality panoramic images.

The results of this research provide more accurate image stitching technology for endoscopic surgery, which is expected to play an important role in clinical practice. At the same time, the methods of this study also provide new ideas and technical means for solving image alignment and fusion problems in other fields. We believe that with the continuous development and application of deep learning technology, there will be more innovative research results emerging.

Acknowledgments This research was funded by the Science and Technology Commission of Shanghai Municipality (21511102000), the Fudan University Medical Engineering Integration Project (YG2021-0008), and the National Natural Science Foundation of China (U1913216).

References

1. A. Pandey, U.C. Pati, Image mosaicing: A deeper insight. Image Vis. Comput. **89**, 236–257 (2019). https://doi.org/10.1016/j.imavis.2019.07.002
2. T. Bergen, T. Wittenberg, Stitching and surface reconstruction from endoscopic image sequences: A review of applications and methods. IEEE J. Biomed. Health Inform. **20**(1), 304–321 (2016). https://doi.org/10.1109/JBHI.2014.2384134
3. Z. Fu et al., The future of endoscopic navigation: A review of advanced endoscopic vision technology. IEEE Access **9**, 41144–41167 (2021)
4. D. DeTone, T. Malisiewicz, A. Rabinovich, Deep image homography estimation. arXiv:1606.03798 (2016)
5. T. Nguyen, S.W. Chen, S.S. Shivakumar, C.J. Taylor, V. Kumar, Unsupervised deep homography: A fast and robust homography estimation model. IEEE Robot. Automat. **3** (2018). https://doi.org/10.1109/LRA.2018.2809549
6. L. Nie et al., Learning edge-preserved image stitching from multi-scale deep homography. Neurocomputing **491**, 533–543 (2022). https://doi.org/10.1016/j.neucom.2021.12.032
7. L. Nie et al., Depth-aware multi-grid deep Homography estimation with contextual correlation. IEEE Trans. Circuits Syst. Video Technol. **32**(7), 4460–4472 (2022). https://doi.org/10.1109/TCSVT.2021.3125736

8. S. Liu et al., Unsupervised global and local Homography estimation with motion basis learning. IEEE Trans. Pattern Anal. Mach. Intell., 1–14 (2022). https://doi.org/10.1109/TPAMI.2022.3223789
9. N. Ye et al., Motion basis learning for unsupervised deep homography estimation with subspace projection, in *2021 IEEE/CVF International Conference on Computer Vision (ICCV)*, (2021). https://doi.org/10.1109/ICCV48922.2021.01287
10. S.H. Lee, H. Chung, N.I. Cho, Exposure-structure blending network for high dynamic range imaging of dynamic scenes. IEEE Access **8**, 117428–117438 (2020). https://doi.org/10.1109/ACCESS.2020.3005022
11. M. Hong et al., Unsupervised homography estimation with coplanarity-Aware GAN, in *2022 IEEE/CVF Conference on Computer Vision and Pattern Recognition (CVPR)*, (2022). https://doi.org/10.1109/CVPR52688.2022.01714
12. L. Nie et al., A view-free image stitching network based on global homography. J. Vis. Commun. Image Represent. **73**, 102950 (2020). https://doi.org/10.1016/j.jvcir.2020.102950
13. L. Nie et al., Unsupervised deep image stitching: Reconstructing stitched features to images. IEEE Trans. Image Process. **30**, 6184–6197 (2021). https://doi.org/10.1109/TIP.2021.3092828
14. J. Wang et al., Deep high-resolution representation learning for visual recognition. IEEE Trans. Pattern Anal. Mach. Intell. **43**(10), 3349–3364 (2021). https://doi.org/10.1109/TPAMI.2020.2983686
15. A. Horé, D. Ziou, Image quality metrics: PSNR vs. SSIM, in *2010 20th International Conference on Pattern Recognition*, (2010). https://doi.org/10.1109/ICPR.2010.579

Research on Man–Machine Contact Force/Position Perception of Wheelchair-Stretcher Robot Based on the Flexible Pressure Sensor

Junjie Tian, Hongbo Wang, Yang Yang, Lianqing Li, Melinte Daniel Octavian, Yu Tian, Lili Zhang, and Jianye Niu

1 Introduction

Patient transfer is one of the most important needs in hospitals and households [1]. With the deepening of population aging and the increasing number of patients, the research and development of transportation equipment that can safely transfer patients has become a research hotspot [2]. Typical transfer equipment includes Resyone [3], AgileLife [4], and wheelchair-stretcher assistive robot [5]. The development of intelligent wheelchair-stretcher robots has brought great convenience to the transportation of patients [6]. However, usually, intelligent wheelchair stretchers lack a way of man–machine interaction perception and cannot obtain the force and position information of man–machine contact [7]. Moreover, there is a lack of research on the intrinsic relationship between the man–machine contact position/

J. Tian (✉) · Y. Yang · L. Li · J. Niu
Parallel Robot and Mechatronic System Laboratory of Hebei Province, Yanshan University, Qinhuangdao, Hebei Province, China
e-mail: jyniu@ysu.edu.cn

H. Wang · Y. Tian
Parallel Robot and Mechatronic System Laboratory of Hebei Province, Yanshan University, Qinhuangdao, Hebei Province, China

Academy for Engineering & Technology, Fudan University, Shanghai, China
e-mail: Wanghongbo@fudan.edu.cn

M. D. Octavian
Institute of Solid Mechanics of the Romanian Academy, Bucharest, Romania

L. Zhang
School of Electrical Engineering, Yanshan University, Qinhuangdao, Hebei Province, China
e-mail: hongbo_w@ysu.edu.cn

© The Author(s), under exclusive license to Springer Nature Switzerland AG 2024
A. Shaikh et al. (eds.), *Advances in Emerging Information and Communication Technology*, Signals and Communication Technology,
https://doi.org/10.1007/978-3-031-53237-5_8

force state and the wheelchair-stretcher robot's posture, which is not conducive to the comfort and safety of patients on the stretcher vehicle. Therefore, it is urgent to carry out research on man–machine interaction force/position perception methods for the intelligent wheelchair-stretcher robot.

Sensor detection technology has developed rapidly in the fields of rehabilitation robotics, aerospace, etc. [8, 9]. The working principle of pressure sensors is to convert pressure changes into electrical signal outputs [10, 11]. In terms of wearable devices, flexible sensors can comfortably attach to the human skin, thus solving the problem of mismatch between traditional rigid sensors and human body curves [12, 13]. Flexible sensors include flexible pressure sensors, flexible strain sensors, flexible humidity sensors, and flexible multifunctional integrated sensors, etc. Therefore, flexible pressure sensors are widely used in fields such as medical treatment [14, 15]. Compared with traditional rigid pressure sensors, flexible pressure sensors have better sensing performance and more suitable applications.

Therefore, this paper conducts research on man–machine interaction force/position perception based on flexible pressure sensors. The rest of this paper is organized as follows. In Sect. 2, the flexible sensor measurement method is introduced in detail. In Sect. 3, the correlation between pressure characteristics and pose angles is studied using gray correlation analysis multivariable correlation theory. In Sect. 4, the pose angle regression prediction is carried out based on the improved neural network method. Section 5 is the summary of the research.

2 Flexible Sensor Measurement Method

The flexible pressure sensor system adopted in this paper consists of an array-type flexible sensor, an acquisition board, and an upper computer. The performance parameters of the flexible pressure sensor are shown in Table 1. The measurement range of the pressure sensors can reach 0–50 kPa. The main components of the experimental measurement system consist of a pair of flexible sensors, a data acquisition board, an upper computer, and a wheelchair-stretcher robot. Two pressure sensors make up a 32 × 32 pressure matrix and 1024 pressure data. In the actual measurement process, the flexible sensors are arranged on both sides of the trunk of the human body to measure the contact force between the trunk of the human body and the robot. The flexible pressure sensors measurement system is shown in Fig. 1.

First, measure the pressure distribution when the trunk of the human body contacts the object at different pose angles. Two flexible sensors are symmetrically pasted on the seat with adjustable angles and treated according to national regulations. The pressure distribution of the pose angle between the trunk and the

Table 1 Performance parameters of the flexible pressure sensor

Parameter	Dimension	Measuring range	Sensitivity	Error range
Value	11 cm × 30 cm × 2	0–50 kPa	14% V/kPa	5%

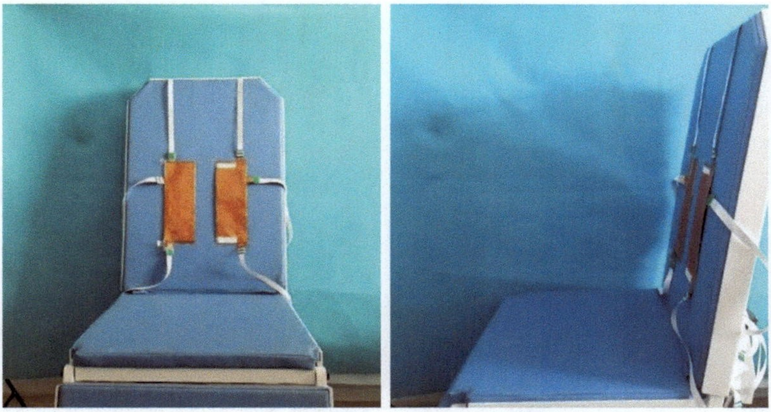

(a) Front view (b) Side view

Fig. 1 Flexible pressure sensor measurement system

horizontal direction of volunteers is 80°, 60°, 45°, and 30°, respectively. The measurement time of each group of experiments is 10 s, and the measurement frequency of the sensors is 20 Hz. To unify the variables and facilitate subsequent analysis, only the change of contact pressure between the trunk and the object caused by the change of the pose angle α between the trunk and the horizontal direction of volunteers is considered, ignoring the possible small rotation movement of the trunk in the measurement of volunteers, and the pressure distribution on the left and right sides can be approximately considered as symmetrical. In the actual measurement process, different volunteers' physical signs, object surface hardness, and surface state can cause the change in pressure distribution. For the same volunteer and the same object contact surface, the change of pressure distribution of the volunteer's trunk and backrest at different angles can still be regarded as the change of pose angle α between the volunteer's trunk and the horizontal direction.

The pressure distribution during the experimental measurement process and at a certain time in the measurement at different angles is shown in Fig. 2. For the wheelchair-stretcher robot, in the process of rehabilitation training or movement, there is no obvious change in the contact characteristics of the trunk and the robot in the process of grasping, touching, different finger movements, and other operations similar to the finger robot. As shown in the heat map of the measurement results in Fig. 2, the pressure distribution between the trunk and the backrest at different angles is similar in image morphology, and it is impossible to identify and distinguish different processes by image morphological characteristics. Therefore, the flexible sensor not only realizes the state perception of man–machine contact information but also analyzes the shape characteristics of the pressure shape, size, center of mass position, and other shape characteristics through the interactive information state. Thus, the motion characteristics of the human body or the trunk of the human body are converted into the corresponding motion state, and the motion state is converted

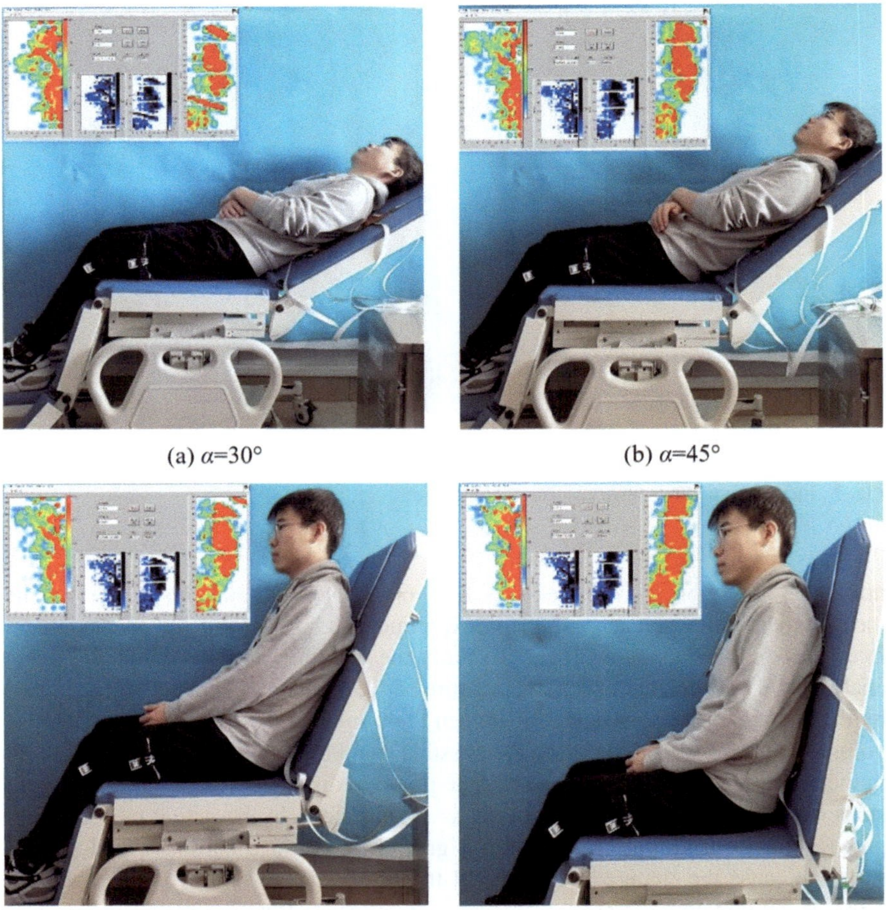

(a) α=30° (b) α=45°

(c) α=60° (d) α=80°

Fig. 2 Pressure distribution between the trunk and the backrest under different angles

into the identifiable information input of the robot control system, to truly realize the trunk rehabilitation robot's real-time state and continuous motion perception of a human body in the process of man–machine contact.

3 Gray Correlation Analysis

To analyze the internal relationship between pressure characteristics and pose angle, it is necessary to analyze the force state of the trunk under different inclinations, obtain evaluation indicators to measure the relationship between pressure

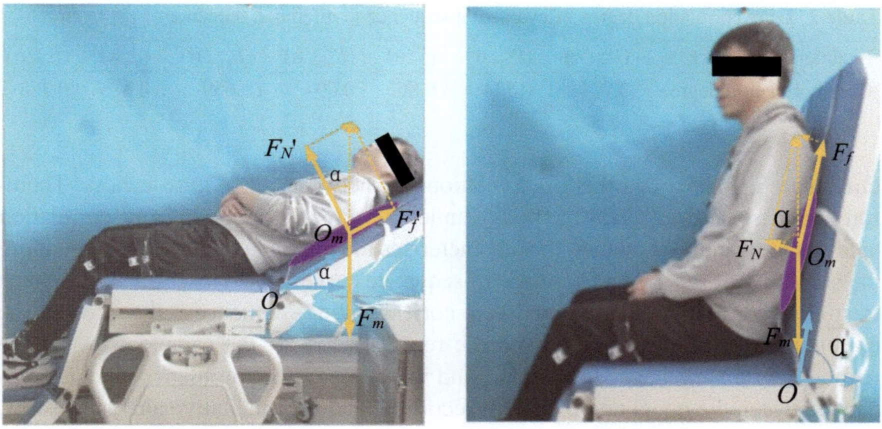

(a) α=30°	(b) α=80°

Fig. 3 Analysis of lumbar forces at two extreme activity positions

characteristics and pose angle, and establish the relationship between pressure characteristics and pose angle according to the evaluation indicator. First, the force on the trunk of the human body is analyzed and the trunk is under two limit states. The force analysis diagram is shown in Fig. 3. From the force analysis, it can be seen that the trunk of the human body is subject to the weight of the trunk F_m, the combined force of the support force of the backrest on the trunk F_N, and the static friction force F_f. When the influence of the coupled motion of the human body is ignored, the types and quantities of the force on the trunk of the human body are the same under different inclinations and are not affected by the change of the pose angle α.

Through force analysis, the resultant force F_N is a monotonic decreasing function of angle α. The resultant force F_N of the support force of the backrest on the trunk can be equivalent to the resultant force of the distributed pressure measured by the flexible sensor array. Due to the flexibility of the human body, when the pose angle α decreases, the contact area S between the trunk and the backrest of the human body and the load q per unit area may change correspondingly under the action of gravity. Based on the analysis of the force and pressure distribution on the trunk of the human body, a set of features $\{F_N, S, q, P_s, P_m\}$ is finally determined as the dependent variable of pose angle α, where F_N is the resultant force of the support force on the trunk of the human body, S is the contact area between the trunk and backrest, q is the equivalent unit area load on the force area of the trunk, P_s and P_m represent the centroid position and pressure center position, respectively.

The system analysis methods commonly used in mathematical statistics include principal component analysis, regression analysis, and variance analysis. When conducting system analysis by using the above methods, the sample itself needs to have typical distribution rules. For the pose angle α and load feature set $\{F_N, S, q, P_s, P_m\}$, the influence relationship between features cannot be determined through

Table 2 Gray correlation degree of evaluation sequence to reference sequence

Evaluation sequence	(P_{sx}, α)	(P_{sy}, α)	(S, α)	(P_{mx}, α)	(P_{my}, α)	(q, α)	(F_N, α)
Grey correlation degree	0.724	0.724	0.863	0.694	0.659	0.678	0.768

qualitative analysis, and there is a relationship between the feature factors. Therefore, such system analysis methods cannot be directly applied. Gray correlation analysis is an analysis method that characterizes the internal relationships by judging the similarity of the geometric shape of sequence curves. The closer the curve shape is to the correlation degree between corresponding sequences, the greater the correlation degree is. It has no specific requirements for the sample size and the regular relationship between samples, and will not have the problem of the discrepancy between the quantitative demerit recording and the qualitative analysis results, which can make up for the shortcomings of principal component analysis and other systematic analysis methods.

According to the definition, the correlation coefficient between features can be expressed as follows:

$$\xi_i(k) = \frac{\min_s \min_t |x_0(t) - x_s(t)| + \rho \max_s \max_t |x_0(t) - x_s(t)|}{|x_0(k) - x_i(k)| + \rho \max_s \max_t |x_0(t) - x_s(t)|} \tag{1}$$

where, x_0 is the reference sequence, x_i is the comparison sequence, ρ is the resolution coefficient and $\rho \in [0,1]$.

The correlation coefficient describes the correlation degree between each value in the comparison sequence and the reference sequence. Compare the sum of the correlation coefficient of each value and make the average processing, and finally get the gray correlation degree of the comparison sequence to the reference sequence, then:

$$r_i = \frac{1}{n} \sum_{k=1}^{n} \xi_i(k) \tag{2}$$

Fifty groups of experimental data were randomly selected from different angles as the dataset. First, the abnormal data are eliminated according to the Pau Ta criterion. Then, forward process the dataset to obtain the forward matrix of the dataset. Finally, the gray correlation degree of the evaluation sequence to the reference sequence is calculated according to Eqs. (1) and (2), and the calculation results are shown in Table 2. The analysis of the correlation matrix shows that the contact area S between the trunk and the backrest has the highest correlation with the pose angle α, the next is centroid position P_s, and centroid position Pm has the lowest impact. In order to clarify the correlation between each variable and the pose angle and verify the gray correlation analysis results, multivariate variable correlation analysis was further carried out to solve the correlation coefficient matrix. The solution results are shown in Fig. 4. The contact area S and pose angle α have the highest correlation degree,

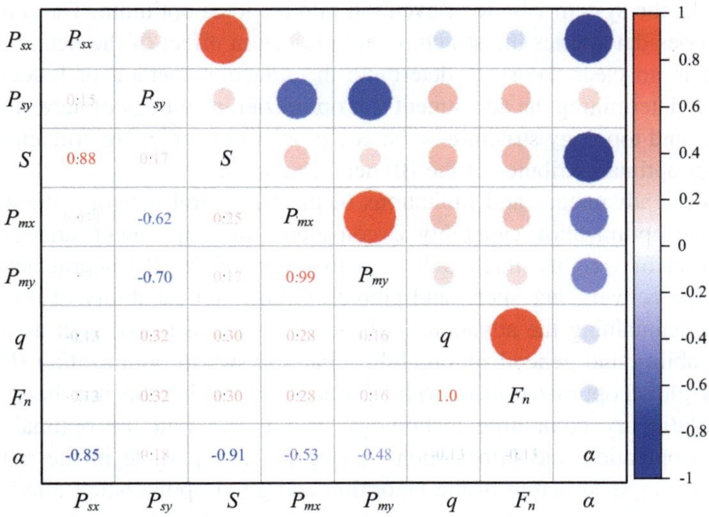

Fig. 4 Multivariate correlation coefficient matrix

and it is a negative correlation, which is consistent with the force analysis results. In addition, in the correlation analysis, the centroid position coordinates (P_{sx} and P_{sy}) have a significant correlation with the attitude angle α. The resultant force $_{FN}$ and equivalent load q show differences between the two methods.

4 Pose Angle Regression Prediction

A back propagation (BP) neural network is a multilayer feedforward network, whose basic principle is to achieve iterative optimization of expected output through gradient descent and achieve optimal approximation and error control about actual output. The multilayer perception structure of the BP neural network makes it have excellent nonlinear mapping ability and generalization ability, and it is widely used in nonlinear regression prediction, gray correlation, and other problems that cannot clarify the system mathematical model or are difficult to solve the model. In the process of building the BP neural network, it is necessary to determine the input/output, hidden nodes, weights, training times, gradient descent direction, and other structural parameters of the network. The input and output nodes are generally determined by the problem demand, but the hidden nodes, weights, training times, gradient descent direction, and other structural parameters need to be adjusted according to the data set to determine the optimal parameter values or network structure. When the training times are insufficient, the prediction effect is poor, and when the training times are too many, the convergence speed will slow down. Improper selection of the gradient descent algorithm and descent direction will easily make the system vibrate or even fall into the local optimum. The selection of hidden nodes determines the structure and prediction effect of the neural network, but there is no clear theory to determine the selection method of hidden nodes. Therefore, determining the adjustment and optimization strategy of network weight, threshold, and topology structure is the key problem to achieving efficient learning and global optimal attributes of the BP neural network.

Given the advantages and limitations of the BP neural network, the algorithm with global optimization capability is considered the adjustment strategy for the optimization of weights, thresholds, and topology of the BP neural network, to improve the convergence speed and global optimization capability of the network based on maintaining the advantages of the BP neural network such as nonlinear mapping ability and generalization ability. Particle swarm optimization (PSO) is a stochastic global optimization algorithm based on swarm intelligence theory derived from evolutionary computing technology, and it can find the optimal solution through cooperation and information sharing of each particle in the population. Due to the simple structure of the algorithm and good optimization effect, PSO is widely used in the optimization process of neural networks, fuzzy systems, and other algorithms. Without losing generality, suppose a dimensional space. Without loss of generality, assume the initial population composed of m particles in a n-dimensional space $\boldsymbol{X} = \{x_1, x_2, \cdots, x_i, \cdots x_m\}$. First, define the initial position x_1 and initial velocity v_i of the i-th particle: after j iterations, the current position and velocity expressions of the particle are shown in Eq. (3). Based on this, the position of the swarm particles is connected to the weights and thresholds of the BP neural network, and the root mean squared error is selected as the fitness function of the PSO algorithm. The weights and thresholds of the BP neural network are optimized through the algorithm, to realize the weight parameter setting of the optimal BP network model in the data set.

$$\begin{cases} v_{mn}^{j+1} = \omega v_{mn}^{j} + c_1 r_1 \left(p_{mn}^{j} - x_{mn}^{j}\right) + c_2 r_2 \left(p_{gn}^{j} - x_{mn}^{j}\right) \\ x_{mn}^{j+1} = x_{mn}^{j} + v_{mn}^{j+1} \end{cases} \tag{3}$$

where m is the particle population size; n is the space dimension; 瑶 is the numb of system iterations; ω is the weight coefficient, c_1 , c_2 are the learning factors, p_{mn}^{i} is the component of the individual extreme value of the m-th particle in the n-dimensional space after the first iteration; p_{gn}^{j} is the component of the global extremum of the particle swarm in the n-dimensional space after the j-generation iteration; r_1, r_2 is a random number on (0,1).Building on the aforementioned concept, the process involves implementing a PSO-BP neural network algorithm with input layer, hidden layer, and output layer networks. The steps are as follows:

1. According to the data set, the training set and test set are divided, and the input and output sample sets of the BP neural network and the number of neurons in the input layer and output layer of the BP neural network are determined.

Table 3 The root mean squared error of different hidden neurons in the training set

Number of hidden neurons	Root mean squared error
3	2.7×10^{-4}
4	7.1×10^{-6}
5	1.5×10^{-5}
6	1.0×10^{-5}
7	1.4×10^{-5}
8	2.3×10^{-6}
9	1.7×10^{-7}
10	1.3×10^{-6}
11	1.1×10^{-5}
12	2.9×10^{-7}

Table 4 Parameter configuration of the BP neural network

Parameter item	Training number	Learning rate	Minimum error	Momentum factor	Minimum gradient
Parameter value	1000	0.01	1×10^{-5}	0.01	1×10^{-6}

Table 5 Initial parameter settings of the PSO algorithm

Parameter item	Population size	Spatial dimension	Maximum number of Iterations	Weight coefficient	Acceleration constant (c_1, c_2)
Parameter value	30	82	100	0.9	2, 2

2. The number of neurons in the hidden layer is determined. According to empirical formulas, for a BP network with m input layer neurons and n output layer neurons, the number of hidden layer neurons p is calculated, as shown in Eq. (4):

$$p = \sqrt{m + n} + a \tag{4}$$

where α is a random integer in the interval [0,10].
3. According to Eq. (4), when $m = 5$, $n = 1$, the number of hidden layer neurons is an integer in [3, 12]. On this basis, calculate and compare the minimum mean squared error of the training set under different numbers of hidden layer neurons and determine the optimal number of hidden layer nodes. The calculation results are shown in Table 3. Finally, determine the number of hidden layer neurons $p = 9$, and the root mean squared error at this time is 1.7×10^{-7}.
4. The BP neural network with the optimal number of hidden layer nodes is constructed, and the network parameters are configured. The parameter configuration of the BP neural network is shown in Table 4.
5. Initialize PSO algorithm parameters. This includes particle population size, spatial dimension, the maximum number of iterations, weight coefficient, acceleration constants, etc. The initial parameter settings are shown in Table 5.

6. Calculate the fitness value of each particle. Here, the root mean squared error function calculated when determining the optimal BP neural network is used. Its function expression is shown in Eq. (5).

$$E = \frac{1}{M}\sum_{j=1}^{M}\sum_{i=1}^{1}\left(Y_{i,j} - t_{i,j}\right)^2 \tag{5}$$

where M represents the number of samples included in the training set; l is the number of neurons in the output layer; $t_{i,\ j}$ represents the j-th expected output of the i-th sample, $Y_{i,\ j}$ is the actual output of sample j for sample i.
7. Compare the fitness value of each particle, record the individual optimal extreme value and global optimal extreme value of each particle, update the speed and position of particles, and calculate the fitness value of new particles.
8. Repeat this step until the expected error is met or the maximum number of iterations is reached.

Referring to the above basic flow and model setting of the PSO-BP neural network, the pose angle α and load feature set obtained $\{F_N, S, q, P_s, P_m\}$ is taken as the data set, including 180 groups of data as the training set and 20 groups of data. As a test set, the fitness function of the PSO algorithm uses the root mean squared error function. When the PSO algorithm reaches the maximum number of iterations or meets the expected error, the calculation process ends. The variation of root mean squared error and optimal fitness value with training times during iteration is shown in Fig. 5. The optimal fitness decreases with the number of iterations and approaches zero indefinitely. Test 20 groups, and the trial data are brought into the trained network model. The expected output and expected output of the model are shown in Fig. 6, and the error distribution is shown in Fig. 7.

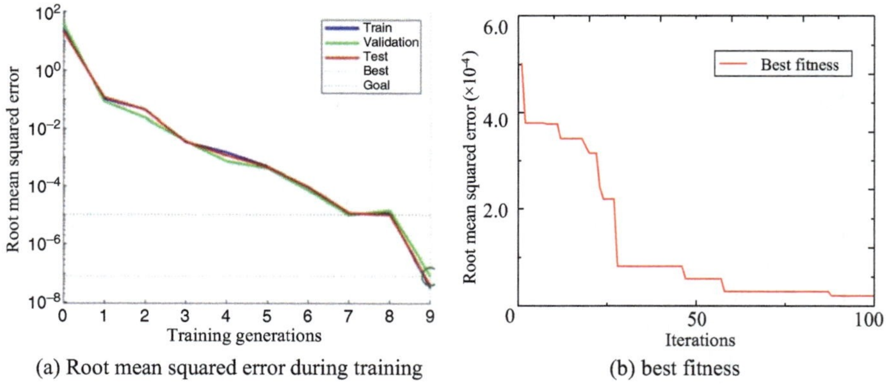

Fig. 5 Change of fitness during iteration (**a**) Root mean squared error during training (**b**) Best fitness

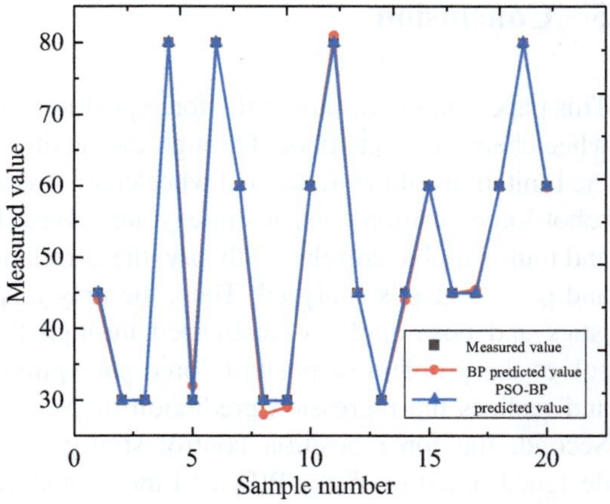

Fig. 6 Measured and predicted values

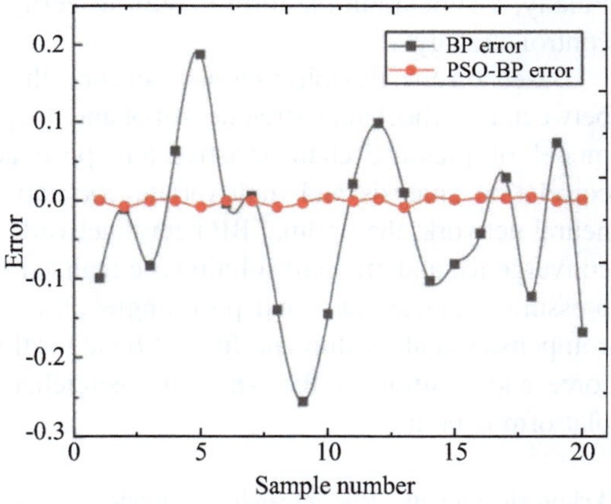

Fig. 7 Predicted value error distribution

The analysis of training results shows that the PSO-BP neural network algorithm can better realize the regression prediction of feature set variables for pose angles. In the test set data, the maximum error of the PSO-BP neural network algorithm is only −0.0084°, which is far higher than the actual requirements of robot control. Compared with the traditional BP neural network algorithm, the improved PSO-BP algorithm has significant advantages in terms of operation speed and optimal fitness, which improves the neural network convergence speed and significantly reduces the prediction error. Since then, based on the pressure detection of the flexible sensor matrix and the regression prediction of pose angles, the key problems such as the sensing detection of the wheelchair-stretcher robot have been solved, and the key breakthrough of the man–machine contact position/force perception of the wheelchair-stretcher robot has finally been achieved.

5 Conclusion

This paper mainly discusses the force/position perception and control strategy of the wheelchair-stretcher robot. Through the flexible pressure sensor, problems such as the limitations of the traditional wheelchair-stretcher robot sensing detection and the robot force position control strategy are solved. First, based on the gray correlation and multivariable correlation theory, the correlation between pressure characteristics and pose angles is analyzed. Then, the gray mapping model of pressure characteristics and pose angle is established through the PSO-BP neural network, which solves the problem of position/force perception of the wheelchair-stretcher robot, and realizes the regression prediction of pressure characteristics and pose angles. Second, the force position control strategy of the wheelchair-stretcher robot is designed based on fuzzy PID, and the simple decoupling of force/position control is realized through the feedforward compensation algorithm of multivariable control. Finally, a simulation platform is built to verify the effectiveness of the model and control strategy.

Based on the flexible pressure sensors, the effective force position perception between the wheelchair-stretcher robot and the patient is realized. The gray mapping model of pressure characteristics and pose angle is established by using gray correlation analysis and multivariate correlation analysis. Based on the PSO-BP neural network, the optimal BP neural network structure is determined, and the fast convergence and efficient solution are realized. Finally, the regression prediction of pressure characteristics and pose angles is completed. Based on the feedforward compensation algorithm and fuzzy PID algorithm, the simple decoupling control of force and position of the wheelchair-stretcher robot is realized and a simulation platform is built.

Acknowledgments This research was funded by the National Natural Science Foundation of China (NSFC) under Grant 52275034 and the Hebei Province Colleges and Universities Science and Technology Research Project under Grant QN2022166.

References

1. L. Andersen, A. Burdorf, N. Fallentin, et al., Patient transfers and assistive devices: Prospective cohort study on the risk for occupational back injury among healthcare workers. Scand. J. Work Environ. Heath **40**, 74–81 (2014)
2. Y. Tian, H. Wang, Y. Zhang, et al., Design and evaluation of a novel person transfer assist system. IEEE Access **9**, 14306–14318 (2021)
3. Y. Kume, S. Tsukada, H. Kawakami, Design and evaluation of rise assisting bed "Resyone®" based on ISO 13482. J. Robot. Soc. Jpn. **33**(10), 781–788 (2015)
4. S. Masroor, H. Bulut, C.Y. Lin, Review on powered mobility and meal preparing assistive devices for physically disabled persons, in *Proceedings of the 6th IFToMM International Symposium on Robotics and Mechatronics (ISRM)*, (2019), pp. 16–28

5. L. Sang, M. Yamamura, F. Dong, et al., Analysis, design, and experimental research of a novel wheelchair-stretcher assistive robot. Appl. Sci. **10**(1), 264 (2019)
6. X.D. Xing, Q.Y. Zhou, P.Y. Fan, *A Practical Wheelchair Structure Used for Patients to Go to the Toilet*. C.N. Patent CN105012087A (2015, November 3) (In Chinese)
7. Y. Tian, H. Wang, Y. Zhang, et al., Structural design and preliminary tests of a novel patient transfer apparatus for medical scenarios. J. Healthc. Eng. **2021**, 2770846 (2021)
8. L. Pan, A. Chortos, G. Yu, et al., An ultra-sensitive resistive pressure sensor based on hollow-sphere microstructure induced elasticity in conducting polymer film. Nat. Commun. **5**(1), 3002 (2014)
9. H. Huang, J. Zhong, Y. Ye, et al., Research progresses in microstructure designs of flexible pressure sensors. Polymers **14**(17), 3670 (2022)
10. B. Zhou, C. Li, Y. Zhou, et al., A flexible dual-mode pressure sensor with ultra- high sensitivity based on BTO@MWCNTs core-shell nanofibers. Compos. Sci. Technol. **224**, 109478 (2022)
11. K. Zhao, H. Zhuang, L. Cui, et al., A packaging technique of pressure sensor for in vivo measurement system. Sens. Actuator A Phys. **340**, 113514 (2022)
12. J. Li, L. Fang, B. Sun, et al., Review—Recent progress in flexible and stretchable piezoresistive sensors and their applications. J. Electrochem. Soc. **167**(3), 037561 (2020)
13. S. Han, H. Peng, Q. Sun, et al., An overview of the development of flexible sensors. Adv. Mater. **29**(33), 1700375 (2017)
14. Z. Xu, D. Wu, Z. Chen, et al., A flexible pressure sensor with highly customizable sensitivity and linearity via positive design of microhierarchical structures with a hyperelastic model. Microsyst. Nanoeng. **9**(1), 5 (2023)
15. X.Y. Hou, C.F. Guo, Sensing mechanisms and applications of flexible pressure sensors. Acta Phys. Sin. **69**(17), 178102 (2020) (In Chinese)

The Digitization of Human Resources Management: What Impact for Public Administrations in Morocco?

Basma El Ouaghlidi, Kamal Yazzif, Fatima El Amrany, and Abderrahmane Ouddasser

1 Introduction

Human resource management (HRM) has long been marked by the difficulty of assessing its contribution to organizational objectives and their success.

Digital technologies are playing an increasingly influential role in employees' working lives than in human resources management (HRM), which must be affected in multiple ways [19] and must be able to anticipate and accompany the transformation of the entity's professions, skills, and culture.

For the Moroccan public administration, human resources management (HRM) represents a major challenge in guaranteeing the quality of its services and fully fulfilling its missions toward the public service user. In Morocco, the last 20 years have been marked by far-reaching reforms of the Human Resources Management system, which are part of a broader framework of reform in public sector management, aimed at improving the efficiency of public administration and enhancing the skills of its workforce.

Today, human resources management is considered a strategic issue in the reform of Moroccan public administration, leading to a paradigm shift. In addition to the decentralization of financial responsibilities and citizen orientation, the digitization of HRM is one of the main elements of the reforms.

As a result, the digitization of HRM in Moroccan public administrations has turned their working methods upside down, forcing them to adapt as a matter of urgency, and putting the spotlight on their HR's ability to reinvent itself, to be flexible and creative, particularly with regard to digital solutions, in other words,

B. El Ouaghlidi (✉) · K. Yazzif · F. El Amrany · A. Ouddasser
Industrial Technologies and Services Laboratory University Sidi Mohamed Ben Abdellah, Fes, Morocco
e-mail: kamal.yazzif@usmba.ac.ma; fatima.elamrany@usmba.ac.ma; Abderrahmane.ouddasser@usmba.ac.ma

© The Author(s), under exclusive license to Springer Nature Switzerland AG 2024

A. Shaikh et al. (eds.), *Advances in Emerging Information and Communication Technology*, Signals and Communication Technology,
https://doi.org/10.1007/978-3-031-53237-5_9

revolutionizing HR management work processes through digitalization. The willingness of public administrations to dematerialize and automate processes and implement new digital solutions enables them to rise to various challenges.

The Kingdom of Morocco, aware of these challenges, has put in place several national programs to support the development of Digital Public Administration, such as the implementation of HRIS within the Human Resources Management departments of several public administrations.

From the above, it is appropriate to ask "**How could the digitalization of human resources management impact Moroccan public administrations?**"

In the first part, we set out the conceptual framework for human resources management in public administrations. We then present a conceptual framework for the digitization of human resources management, its current status in public administrations, and the proposed digital solutions. We conducted an empirical study through an analysis of the functioning of HRIS software, and its effectiveness within the HR function of Moroccan public administrations. We also identified the current situation of HRIS software within Moroccan public administrations and the available digital solutions used by these entities.

2 Human Resources Management in the Public Sector

2.1 The Particularities of Human Resources Management in Public Administrations

There are three main reasons why human resources management is considered a major challenge for public administrations: first, the sheer size of the workforce, which requires the management of a large proportion of the active workforce.

The second reason relates to the challenges posed by the economic situation, which means that public administrations are obliged to adapt quickly to changes in their usual way of operating, triggering tension in terms of staff numbers and the budgets devoted to human resources. As a result, the management of human resource motivation and skills will be more strategic.

The third reason is linked to the quality and efficiency of the services offered by public administrations. In other words, citizens demand services tailored to their own needs, a challenge that public administrations must maintain at the lowest possible cost.

For all these reasons, public administrations need to take a number of actions to increase and improve the functioning of the public service, including human resources management, which is the subject of an exceptional framework.

Human asset management is seen as a particularity of human resources management in that it does not engage in a massive recruitment policy to adapt to changes in the environment, but rather focuses on existing skills and transforms them, making the renewal of human resources skills fundamental and of great strategic importance.

The fact that parts of the government apparatus are made up of central agencies is also a feature that makes human resources management more complex, since these agencies present the link between the bureaucracy and the politicians, while also exerting an influence on government policy-making with regard to human resources management.

2.2 *Key Advantages to Seize*

Even if public administrations have to cope with particular constraints in terms of human resources management, they benefit from relative stability in terms of their missions, so that even if they change or evolve, they always retain the same goals, and the stability of their human resources—something that makes it possible to leverage investments in human resources in terms of simplified planning and skills development.

In addition to the advantage of stability, the motivation aspect is less difficult for public administrations, as they are not obliged to be a real source of motivation.

2.3 *Challenges to Overcome*

Faced with the changes taking place in public administration, human resources management faces the challenge of being innovative and adapting to the new environment, assuming direct roles in the success of public administration, being true agents of change, developing competitive advantages among employees, and positioning itself as a function that creates added value for public administration. The challenges linked to the digitization of the professions of the future are manifold, but the ambitions of human resources management should not be overlooked:

- Improving the employee experience: offering automated and personalized services, facilitating administrative processes, and promoting active employee participation in company decisions.
- Developing equity and diversity: automating recruitment processes and using indicators to evaluate employee performance objectively.
- Encouraging innovation: encouraging active employee participation in innovation projects, by supporting research and development initiatives and fostering the creation of innovation communities.
- Optimizing and measuring performance: the HR function can use digital tools to assess the performance of public administrations and foster a culture of continuous improvement.

In short, to achieve this, public administrations need to take action by updating their HR strategies.

3 The Digitization of the HR Function, a Performance Lever for Public Administrations

3.1 What Do We Mean by the Digitalization of the Human Resources Function?

The digital transformation of human resources management activities is the fruit of a long process marked by successive waves of technology and initially involves digitizing the function itself in its various processes: recruitment, training, compensation, skills management, etc.

It, therefore, aims to use the potential offered by digital technology, with the help of various digital HR solutions already in existence within the entity or yet to be developed, in order to:

- Develop new functions and assets by taking advantage of technological advances: determine and analyze employee performance factors using HR analytics[1]
- Implement new operating modes to meet the expectations of HR managers and staff
- Perfect HR data: managing flows for HR processes (training and payroll plans, etc.), complying with information due to the legislator and data protection, mastering performance measurement
- Optimize HR processes through automation and decentralization, e.g., workflow systems[2]
- Introduce electronic signatures, automatic reporting, and communication tools to answer employees' recurring questions;
- Combine HR content and practices, such as adopting or implementing an HRIS, which requires standardization of practices across several entities where the processes adopted are often not the same.

Implementing digital HR solutions involves two degrees of complexity in terms of managing and supporting change within the HR function:

- In the less complex dimension, the digital transformation of the HR function can be expressed as a set of digital services delivered by a given service provider.
- In the more complex dimension, it involves the integration of an HRIS as a tool for digitalizing the HR function and its implementation in practice.

Noting that several digital HR solutions can be provided in parallel for different HR processes, the HR department can ensure the digitization of various tasks linked to its field of activity.

---

[1]HR analytics consists of conducting analyses based on various internal HR data, to identify performance factors, departure factors, and, more generally, the recurring causes of HR issues.

[2]The Workflow concept encompasses the validation circuit for a request, the tasks to be carried out by the various players in a process, the lead times, and the validation methods.

3.2 Digital Human Resources Management Solutions for Public Administrations

The various digital HR solutions differ from one entity to another but can be summarized as follows [6] adapted by Laudon (translated by Fimbel et al. 2013)):

1. Digital recruitment: refers to the use of various digital solutions to carry out, enrich, facilitate, and/or optimize the actions carried out during the five main stages of the recruitment process: defining needs, sourcing, selecting, deciding, and integrating new employees, not forgetting the administrative management present at every stage. These solutions can range from the simple (use of social networks, and recruitment sites) to the more complex in terms of technological structure and implementation (use of an ATS (Applicant Tracking System)[3]).
2. Digital training: digital training and informal digital learning cover all multimedia teaching methods, enabling employees to learn and develop their skills, formally and informally, during their work activities, and over varying lengths of time. These multimedia teaching methods are designed and distributed using digital design tools (Camtasia, Articulate, adobe Captive, Powtoon, Audacity, Canva, etc.) and distribution tools (LMS, Mooc distribution platform (Fun, SlideShare, YouTube, Ted Talks)).
3. HRIS: a major information system for managing an entity's HR function, it is a computer system used to acquire, store, manipulate, analyze, retrieve, and distribute information related to human resources. An HRIS contributes to information accuracy and greater responsiveness [14].

Added to this, the ability of public administrations to switch rapidly and on a large scale to remote working and delivery of public services—here we're talking about teleworking—enabled them to be a crucial source of resilience in the face of the initial shock of the Covid-19 pandemic, and to guarantee business continuity in the public sector.

These tools were not designed with crisis preparedness in mind, but they gave public administrations a great deal of leeway, enabling them to turn any location into an "administrative counter." The example of teleworking shows just how important it is to retain room for maneuver, while at the same time practicing cutting-edge innovation to ensure that public administrations function smoothly and are able to absorb shocks.

[3]Many public servants provided their own workspace, internet connection, etc., in 2020.

3.3 Why We Should Think to Digitalize the Human Resources Function in Public Administrations?

Whether it's recruitment, administrative processes, or professional development, digital transformation is important for all the missions of the Human Resources function. It has the capacity to support and accompany the actions carried out within the HR function in order to increase the attractiveness of public administrations and their performance.

The main objective of HRM digitization is not just to provide a function for the benefit of the administration, but rather to create a positive effect on the service rendered to citizens. This implies improving the level of skills and attracting caliber profiles; the ambition here is to position the Human Resources Management function of public administrations at the center of public service performance.

There are a number of reasons to digitize your HR function, including:

- Automation: spending more time on high-value-added tasks than on time-consuming ones.
- Dematerialization of processes; less administrative paperwork and more digitized solutions.
- Immediate availability of information; HR officials need information everywhere, all the time. Thanks to remote working or digital solutions offered by HR digitization, HR managers and their staff can access their work files at any time.
- Expose an optimized user experience; thanks to the portal dedicated to HR officials, the latter can request anything (leave requests, training requests, follow distance training courses...). It's worth noting that the more autonomous employees are, the more ambitious they are in their work, and the higher their performance;
- Redefinition of the HR functionaries' role to become the main vectors of organizational change, by supporting and identifying the new skills needed to maintain the administration's development.

Digitizing the Human Resources function involves a number of stages: not everything can be digitized. First, you have to specify and prioritize; i.e., for each public administration, it's essential to determine what's most important to its HR functionaries, and what objectives they aim to achieve. The next step is to specify the appropriate digital solution in line with the HR functionaries' objectives. This solution must be capable of automating a maximum number of processes to meet the various challenges, to ensure the efficiency and positive evolution of public administration.

4 Implementing the Digitization of the HR Function Through the Manipulation of the HRIS in Moroccan Public Administration

4.1 HRIS for Professionals in the Human Resources Function

HRIS has long been at the heart of the digitization of human resources management activities and is the most comprehensive system in terms of HR solutions and processes supported (payroll, personnel administration, training, recruitment, etc.).

"HRIS is a systematic/methodical process used to acquire, store, update, retrieve and validate information needed by a company about its human resources, the activities of its personnel, and the characteristics of its organizational units" ([15], p. 275).

This definition is confined to technical aspects, tools, and practices. "HRIS is not limited to the hardware and software that make up the "technical" part of the system. It also includes the people, policies, practices, and information needed to perform the HR function. As a result, a functional HRIS must create an information system that enables the integration of the policies and practices used to manage the company's human capital, as well as the practices required to use the IT system" ([11], p. 382) (Table 1).

Jean-Marie Peretti ([20], p. 331) distinguishes four different issues:

– "Transparency policy": providing services in the form of information and reports (job offers, news, annual report, sustainable development report...)

Table 1 Human resources management activities supported by HRIS

Process of HR	Activities supported by HRIS
Administrative management	Management of payroll, employment contracts, management of expense accounts, DSN management
GTA	Management of schedules and absences and analytical monitoring of production
Recruitment and mobility	Drafting of job descriptions and profiles Dissemination of job offers on the company intranet and social networks
Formation	Training catalog management, administrative management of presential and e-learning training courses Design, distribution, and evaluation of multimedia content for online training courses
Performance management	Management of individual interviews online and compensation management
Jobs and skills planning	Career planning, succession planning, and employee mobility management

Source: 2022 Human Resources Guidebook

- "Interests sought": processing administrative and payroll tasks with maximum ease (through automation) and security; participating in the introduction of new HR management tools; complementing management tools
- "Performance": productivity gains, optimization and control of information flow, reduction of lead times, cost control, and greater availability for employees
- Refocusing the HR function: free yourself from tedious, repetitive tasks, and concentrate on higher value-added tasks.

4.2 Methodology

A qualitative methodology was selected for this study since the aim of the research is to better understand personal experiences and to explain certain aspects. This study aims to highlight the change in working patterns within human resource management through the adoption of HRIS and its contribution to enhancing the efficiency of Moroccan public administrations. Such in-depth understanding is achieved through semi-structured interviews. The semi-structured face-to-face interviews covered (i) the pre-HRIS work situation, (ii) the measures put in place by HR managers and their roles to drive and support this change, (iii) the constraints and obstacles encountered, and (iv) the digital HRIS solutions while specifying the main advantages and disadvantages of the system to deduce the impact of HRIS on the HR function and its role in making Moroccan public administrations more efficient and value-creating.

The HR managers of the various administrations concerned were sent an e-mail in which we asked them if they would be willing to be interviewed after the purpose of the interview had been explained to them. Anonymity was promised, as no reference was made to the name of the HR manager interviewed or to the administration where he or she worked. Interviewees consented by email and appointments were arranged at their workplaces and through the Microsoft Teams platform. Each face-to-face interview lasted 30 min and was recorded, enabling a more complete transcription. Confidentiality is also guaranteed, as no information contained in the research report allows the participating public administrations to be identified. The method used to analyze the results is thematic analysis, with the aim of understanding the main themes and sub-themes of the interviews, as well as the links between these themes.

4.3 Analysis and Findings

Before describing the impact of HRIS on human resources management in public administrations, and its role in improving performance and ensuring the quality of service for citizens, we need to look at the role of human resources managers in implementing, managing, and supporting change arising from the use of HRIS in human resources management. Textual analysis of the interviewees' speeches has

Table 2 Word frequency query

Word	Number	Weighted percentage (%)
Change	45	0.75
Measures	23	0.38
Support	18	0.30
Reluctance	13	0.22
Impact	10	0.17

Source: Our empirical study, NVivo analysis

enabled us to deduce certain terms that are often used. The following table shows the frequency of use of these terms:

Table 2 clearly shows that the words (change, measures, support, reluctance, and impact) were most frequently adopted by the managers interviewed. It is, therefore, appropriate to carry out a content analysis, which will enable us to specify the context in which these words were cited, and whether these terms reflect the role of HR managers in leading and supporting the change induced by the implementation of the HRIS, which is the focus of the first axis of our interview. To do this, we chose to use a synapsis analysis to frame each term in a precise context. Despite the responsibilities assigned to HR managers to ensure the proper management of human resources management activities, they are responsible for leading and supporting the change in HR practices brought about by the introduction of the HRIS, and for guaranteeing its success. This is why we chose to interview the various human resources managers from different Moroccan public administrations.

The interviewees confirmed that there had been a real change (digitization) in the day-to-day practices and tasks of human resources management. They highlighted that before the introduction of the HRIS, data processing was conducted manually, involving paper-based archiving. Administrative tasks were cumbersome and time-consuming, with a high risk of error, even with the use of IT tools and applications (Excel, Oracle, Arc system, etc.).

The HRIS offers a range of digital solutions to simplify and revolutionize human resources management workflows, including the payroll module (for monthly processing of remuneration and selective processing of several personnel numbers) and the Personnel Administration module (for managing all administrative documents and providing easy access to them via a portal dedicated to all employees with a Login and Password, for all administrative requests and access to information). The training management module enables you to manage training needs and planning, draw up training plans, and monitor the implementation of training actions. In addition to these modules, some of those interviewed have adopted the appraisal management module, which enables them to evaluate pre-set employee objectives using pre-defined forms, not forgetting the hot and cold evaluation of training carried out. They are even considering the use of a mobile HRIS for greater availability of HR managers.

As with any structuring project, measures must be taken by human resources managers. For HRIS implementation, an in-house diagnostic has been drawn up to assess skills and detect needs, for training planning with guides composed of screenshots to help HRIS operators master the system, and for the organization of meetings for all human resources staff in central and regional administrations, to cascade information to managers and support their staff. In terms of support, guidance, and management of digital change within public administrations, the interviewees emphasized the importance of HR managers in guaranteeing the success of HRIS implementation and their ability to delegate the idea, as well as their role in studying and planning the implementation of this system and ensuring its operation.

In terms of the impact of the HRIS, all those interviewed confirmed that, thanks to the HRIS, they had been able to revolutionize their work processes and better meet the expectations of senior management to improve the quality of HR department management. The HRIS is a human resources management software that facilitates the decision-making choices of HR managers and promotes competitive advantage in HR management; in addition, thanks to the digitalization of human resources management, they have been able to withstand the Covid-19 pandemic, provide services for the benefit of citizens at a distance, and ensure the smooth running of their public administrations.

Given its long history, the HRIS is the cornerstone of the digitalization of the HR function, given the very broad scope of HR processes it supports. The results of this study show that HRIS is an essential and indispensable asset, and cannot be considered as an end in itself, but rather as a means for HR departments to meet the performance challenges of public administrations and the expectations of citizens.

From the results, we see that HRIS is perceived by respondents as a technological asset that has made human resources management practices and procedures more effective and efficient. However, HR managers need to play a more active role in this digitization to ensure that all employees are competent and well-trained to meet the digital needs of the present and the future.

The main thrust of the study was that the digitization of human resources management has played a key role in improving the HR activities of Moroccan public administrations, thanks to digital solutions, the sharing of information between different public employees and also between the administration and the citizenry as a whole, who have little need for a physical presence in public administrations. The digitization of human resources management boosts efficiency by increasing the speed and scope of information flows, enabling public administrations to inform their processes with a wealth of data.

It should be noted that the Moroccan government has set up a common repository for human resources management in the Moroccan public administration, consisting mainly of a directory of human resources management concepts; a dictionary of data and associated nomenclatures; and a directory of procedures and associated management rules, providing access to communication protocols and associated print templates (Table 3).

Table 3 Review of the main themes and sub-themes addressed by respondents

Themes	Sub-themes
Work situation before implementation of the Human Resources management system	Manual data processing Use of IT tools and applications (Excel, Oracle, Arc) Archiving in paper form
HRIS implementation planning	The role of Human Resources managers in the implementation and success of the project The various steps planned before and during the introduction of the HRIS (organizing meetings, planning training sessions, drawing up user guides) Internal diagnosis
Constraints and obstacles to adopting HRIS	Employee reluctance The initial setup must be well defined and tied up so as not to disrupt future working methods Data collection (scattered, non-computerized data)
HRIS digital solutions adapted to the HR department's needs	Payroll management Personnel administration Training management Appraisal management
HRIS accessibility	Service accessible only to HR managers HRIS portal accessible to employees Management of parameterization and technical problems by the Information System department
Advantages of HRIS	Savings in terms of time, money, and general resources Dematerialization; savings in paper use Reduced processing time for HR requests and responses Availability and ease of access to information for HR managers and employees Easier preparation of dashboards Data reliability and archiving Simplification and control of HR processes Integrated solution accessible via the web Increased security thanks to a log that traces changes and movements in the system and automatically records them
Disadvantages of HRIS	Failure to master parameterization when starting up the HRIS can lead to blockages and disrupt working methods later on Clear, well-defined requirements are essential for successful parameterization and project start-up Programming difficulties mean dependence on the service provider Unavailability of service provider's time
Impact of HRIS on HR service performance	HRIS offers a major improvement in HR management By simplifying work procedures, it facilitates decision-making by HR managers Promoting competitive advantage in HR management By digitizing HR processes, HR managers are able to focus more effectively on their main objectives, improving the efficiency and productivity of their departments in line with citizens' needs

Source: The authors

5 Conclusion

The implementation of HRIS has brought about profound changes and modifications in the management of human resources and is now seen not just as an opportunity but as a necessity for improving the effectiveness and efficiency of the HR function within Moroccan public administrations and ensuring the optimization of work processes and the ongoing performance of HR activity, as well as meeting citizens' needs as effectively as possible.

In order to answer the question that forms the basis of this article, i.e. how the adoption of HRIS contributes to improving the performance of the HR function, and what impact it has on public administrations in Morocco, it is necessary to summarize this impact by specifying the benefits derived from its implementation, namely, increased accessibility of information, less paperwork, greater speed, easier communication between employees, simplicity of HR tasks, better planning and decision-making support, and finally, improved HR performance in public administrations.

He added that without the crucial role of human resources managers in leading, supporting, and anticipating the introduction and implementation of an HRIS, the success of a digital HRIS project cannot be guaranteed.

Nevertheless, the adoption of an HRIS must be preceded by a precise and well-founded identification of HR needs, as well as technical and financial feasibility, in order to avoid any pitfalls that may arise during the implementation and operation of the system.

With this in mind, the work is an outline of the problem concerning the impact of the digitization of human resources management on Moroccan public administrations through the adoption of the HRIS, which will be studied in greater depth using a larger sample size and will be addressed by a quantitative methodology targeting other public administrations adopting other digital solutions proposed by the HRIS, such as the recruitment component, Job and Competency Management, the Social component, and so on.

Bibliography

1. M. Barabel, J. Lamri, O. Meier, B. Sirbey, *Innovations HR: Moving into Digital and Agile Mode* (Dunod, France, 2018). ISBN:2-10-076198-6
2. E. Baudoin, C. Diard, M. Benabid, K. Chérif, *The Digital Transformation of the HR Function* (Dunod, France, 2019). ISBN:978-2-10-076759-5
3. A. Benraiss et al., Attitude and behavior of HR personnel in the face of change: Case of the implementation of a Human Resources Information System at Marsa Maroc. Int. Rev. Manag. Sci. **2**(2), 609–634 (2019)
4. H. Benziri, The modernization of Moroccan public administration: what place for the human resources function? in *Proceedings of the Second International Conference on Economic Francophonie, Observatory of economic Francophonie*, (University of Montreal and the World, 2020)

5. I. Comtet, S. Chassigneux, *Can HRIS Be a Tool for Proximity Communication? Communication & Management* (ESKA, Paris, 2017)
6. C. Diard, E. Baudoin, S. Berthet, "Human resources checklist", France: Dunod. Commun. Manag. **2**(14), 9–18 (2022)
7. C. Dorkenoo, A. Crespin, L. Lonbardo, F. Klotz, H. Moulin, *Succeeding in your Digital Transformation: HR, Marketing, Data, Logistics* (Eyrolles, Paris, 2016). ISBN: 978-2-212-56520-1
8. A. Dudézert, *The Digital Transformation of Companies* (The Discovery, Paris, 2018). ISBN: 978-2-34-803601-9
9. G. Exbrayat, N. Fisteberg, R. Ronan Fouesnant, *The Human Resources Information System: an asset in the optimization of HRM at the service of the company* (Paris-Dauphine University. MBA Human Resources Management, 2010), pp. 51–60
10. M. Grenier, E. Lallier, *HR Management in the Age of the Pandemic Legal and Practical Aspects* (Norton Rose Fulbright, Canada, 2020)
11. A.R. Hendrickson, Human resource information systems: Backbone technology of contemporary human resources. J. Lab. Res. **24**(3), 381–394 (2003)
12. Z. Hussain, J. Wallace, N.E. Cornelius, The use and impact of human resource information systems on human resource management professionals. Inf. Manag. **44**(1), 74–89 (2007)
13. C. Jacmin, *The Digitization of the Human Resources Department Is a Change to Be Implemented* (Faculty of Economics, Social, Political and Communication Sciences, Catholic University of Louvain, 2019) http://hdl.handle.net/2078.1/thesis:21460
14. N. Kassim, T. Ramayah, S. Kurnia, Antecedents and outcomes of human resource information system (HRIS) use. Int. J. Product. Perform. Manag. **61**(6), 603–623 (2012)
15. K.A. Kovach, C.E. Cathcart Jr., Human resource information systems (HRIS): Providing businesses with rapid data access, information exchange, and strategic advantage. Public Pers. Manag. **28**(2), 275–282 (1999)
16. L. Lemire, Y.-C. Gagnon, *Human Resources Management in Public Administration* (Les Presses de l'Université de Montréal, License Open Edition Books, 2002)
17. A. Nagendra, M. Deshpande, Human resource information systems (HRIS) in HR planning and development in mid to large sized organizations. Procedia. Soc. Behav. Sci. **133**(2014), 61–67 (2014)
18. OECD, *General Government at a Glance 2019* (OECD Publishing, Paris, 2019)
19. E. Parry, S. Strohmeier, HRM in the digital age—Digital changes and challenges of the HR profession. Empl. Relat. Emerald Insight **36**(4) (2014). https://doi.org/10.1108/ER-03-2014-0032
20. J.-M. Peretti, *Human resources* (Vuibert, France, 11th ED, 2007). ISBN: 978-2-7117-7551-4
21. P. Storhaye, *Transformation, HR and Digital* (EMS Management and Society, Paris, 2016)
22. P. Storhaye, *The HRIS-Stakes, Success Factors and Prospects* (Dunod, France, 2013)
23. L. Tamnine, *Contribution of e-HR to the Modernization of the Human Resources Function in Public Administration: The Case of Morocco* (Moroccan Review of Management and Economy, 2016)
24. G. Westerman, D. Bonnet, *Revamping your Business through Digital Transformation* (MIT Sloan Management Review, États-Unis, 2015)

Exploring the Impact of Digital Art Therapy on People with Dementia: A Framework and Research-Based Discussion

Fereshtehossadat Shojaei, Fatemehalsadat Shojaei, Erik Stolterman Bergvist, and Patrick C. Shih

1 Introduction

Roughly 50 million individuals worldwide are afflicted by dementia according to the World Health Organization (WHO). Predictions suggest that this figure is expected to increase to 82 million by 2030 and further escalate to 152 million by 2050 [1]. Therefore, according to Hodge et al. [2] as the population of PWD continues to grow, there is an increasing focus within the field of human–computer interaction (HCI) on developing products with interaction for this demographic. And as a result, designing based on new therapeutic approaches with the help of technologies is one of those subjects. Nonetheless, when it comes to creating novel technologies in either physical or digital forms, it is imperative to gain a comprehensive grasp of the end-users, their behaviors, and their surroundings. This imperative becomes even more pronounced when developing medical services, given that users may be exceptionally susceptible due to their conditions [3].

In this paper, we are trying to find the gap in utilizing digital art therapy for PWD in consideration of prior relevant research. According to de Witte et al. [4], there is a big tendency to employ various artistic techniques to enhance health and well-being. They claimed that these forms of art might be utilized by art therapists as an intervention or in arts programs created for in daily life uses as a side benefit to

F. Shojaei (✉) · E. S. Bergvist · P. C. Shih
Luddy School of Informatics, Computing, and Engineering, Indiana University Bloomington, Bloomington, IN, USA
e-mail: fshojaei@iu.edu; estolter@indiana.edu; patshih@indiana.edu

F. Shojaei
Department of Computer Science, State University of New York, Oswego, NY, USA

© The Author(s), under exclusive license to Springer Nature Switzerland AG 2024
A. Shaikh et al. (eds.), *Advances in Emerging Information and Communication Technology*, Signals and Communication Technology,
https://doi.org/10.1007/978-3-031-53237-5_10

foster health. In addition, Carós et al. [5] state that one of the treatments to prevent the progression of Alzheimer's, as they referred to as a kind of dementia, is the development of therapies. However, the emphasized challenges for PWD to obtain the therapy include cost, availability of the areas where therapy occurs, and commuting to these locations. These factors make it crucial to look into the requirements of, and strategies for, art therapy interventions for PWD.

We will first define dementia and identify concerns of PWD to better understand the difficulties that PWD face. After that, we'll explore the definitions of art therapy and its various forms as interventions. In the last section of this paper, we will give suggestions for adopting art therapy interventions that can address the requirements of persons with dementia.

2 Dementia

2.1 Definition of Dementia

Dementia presents an expanding health concern on a global scale at present. As the population of seniors continues to surge, the likelihood of an increased number of individuals with dementia is also on the horizon. In 2017, the worldwide population aged 60 years or older stood at 962 million, and it is anticipated to reach 2.1 billion by 2050 [6].

According to CDC, "dementia is not a specific disease but is rather a general term for the impaired ability to remember, think, or make decisions that interfere with doing everyday activities. Alzheimer's disease is the most common type of dementia. Though dementia mostly affects older adults, it is not a part of normal aging" [7, 8].

2.2 Challenges of People with Dementia

Dementia affects how people perceive objects, interact with others, and act, as well as how they process information in their brains [8]. Identifying the needs of PWD is more difficult since this population is emotionally frail and sensitive. PWD suffer from a multifaceted ailment that results in the decline of their cognitive capacities, including memory, executive functioning, communication skills, and decision-making capabilities [9]. A significant obstacle that confronts elderly individuals as they navigate the aging process is the heightened prevalence of mobility and sensory impairments. During this stage of life, they often require assistance with a diverse array of responsibilities, ranging from activities of daily living (ADLs) including eating, bathing, and grooming, to instrumental activities of daily living (IADLs) like medication administration, finances, scheduling, and addressing cognitive functions like problem-solving. At present, in nursing facilities or the residences of elderly individuals, caregivers assume responsibility for these duties. Caregivers of PWD

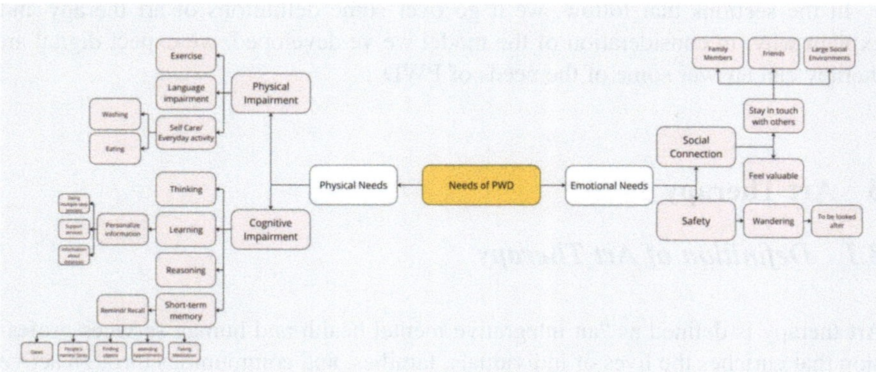

Fig. 1 Model of needs of people with dementia

frequently experience self-blame, as they are compelled to reduce the sense of independence among care recipients due to the diminished cognitive capacities of PWD [10].

We have developed a model [11], to compile all needs of PWD from those that are connected to ADLs to those that are related to IADLs. Based on how other scholars have categorized the demands of PWD using various approaches, we discovered in our earlier study that there are two basic groups into which we may split all of these needs, as seen in Fig. 1: emotional needs and physical needs.

In this model, in accordance with the human-centered design methodology, we place individuals with dementia in the center and attempt to study every aspect of their daily lives, from the simplest to the most critical activities. The aim of our strategy in creating this model was to identify the relationships among the everyday activities, emotions, and difficulties faced by individuals who have dementia. We will elaborate on each component of the model in the paragraphs that follow, outlining how this need was identified and why we believe it needs to be addressed.

This model states that since PWD are roaming and need assistance, one of their "emotional needs" is "safety." Moreover, this group requires "social connection" on an emotional level since they are fearful of being abandoned and need to connect with their relatives, friends, and members of larger social networks.

Additionally, two subcategories of "physical need" for PWD can be determined using this model: "cognitive impairment" and "physical impairment." Some characteristics of the first category, "cognitive impairment," include thinking, learning, reasoning, and short-term memory. Regarding short-term memory impairment, PWD need to be reminded to recall the names, find objects, and take their medications. Furthermore, due to their "learning" intellectual disabilities, they require specific directions on what to do and how to accomplish it. However, exercise, requirements for doing everyday duties, and language disability are all included in the second category, "physical impairment."

In the sections that follow, we'll go over some definitions of art therapy and explain why, in consideration of the model we've developed, we expect digital art therapy can answer some of the needs of PWD.

3 Art Therapy

3.1 Definition of Art Therapy

Art therapy is defined as "an integrative mental health and human services profession that enriches the lives of individuals, families, and communities through active art-making, creative process, applied psychological theory, and human experience within a psychotherapeutic relationship" [12]. The American Art Therapy Association expanded on the notion of Art Therapy by noting that, "Through integrative methods, Art Therapy engages the mind, body, and spirit in ways that are distinct from verbal articulation alone. Kinesthetic, sensory, perceptual, and symbolic opportunities invite alternative modes of receptive and expressive communication, which can circumvent the limitations of language. Visual and symbolic expression gives voice to experience, and empowers individual, communal, and societal transformation" [12].

Patients who participate in art therapy sessions are invited to "express themselves creatively" to improve their psychological well-being [13]. There are six types of creative art therapies (CATs) according to de Witte et al. [4]: art therapy, dance/movement therapy, drama therapy, music therapy, psychodrama, and poetry/bibliotherapy.

Rudolf Arnheim proposed hedonism, pleasure, symbolic communication, trying out, and integrity as factors to address his own query, "What is it that endows the arts with the healing powers you see at work?" [4]. As a praxis theory, Blatner (1992) articulated that individuals mirror their emotions and thoughts onto an artistic medium, which serves as a transitional element. This addresses the fundamental therapeutic principles that form the basis of all creative arts therapies (CATs). In addition, Blatner (1992) claimed that imagery, symbolism, originality, spontaneity, fun, and role expansion are distinctive qualities of CATs. Recently, Jones (2021) claimed that "artistic projection, the triangular relationship, perspective and distance, embodiment, non-verbal experience, the playful space and the informed player, the participating artist-therapist, and the active witness" are fundamental principles that underlie all art therapies [4].

3.2 Traditional Art Therapy

Through the creative process and the final output, the art therapist motivates verbal and nonverbal self-expression as well as reflection. During a psychotherapy session with an art therapist, according to Witte et al., one form of CATs uses a spectrum of 2- and 3-dimensional structured and unstructured visual art instruments such as pencils, paints, chalk, crayons, clay, textiles, etc. [4].

3.3 *Digital Art Therapy*

As Lazar et al. [14] noted, art therapy approaches have evolved along with conceptions of the relationship between art and psychotherapy. In order to employ technologies in art therapy sessions, Peterson noted, participants in their study claimed that assistive technologies from low-tech to high-tech alternatives allowed them to take their hands off of their clients' work. Clients also benefited from this as they felt the full power in their artwork and the creative process [15].

In digital art therapy, online communication tools and the use of digital media for the creation of art are two aspects of the use of technology in art therapy that are equally relevant to in-person practice [13]. However, Zubala et al. [13] stated that there are multiple reasons that digital art therapy is not being used in art therapy sessions currently. The main obstacle for therapists to using digital art mediums in therapy sessions is the absence of comprehensive research on digital art creation. The significance of more specialized technology-oriented ethical codes for practitioners and the little guidance provided by professional groups, a few attempts to create electronic tools specifically for art therapy, which causes the lack of technical input from art therapists, discussion about whether the sensory aspect of materials would be lost if art were created with digital tools, and potential client effects are other concerns.

The use of digital art therapy faces some challenges, according to Zubala et al. [13], some of those limitations include the cost of the equipment, the requirement for extra time to learn and adjust to a new environment, technical difficulties, a lack of training, ethical and privacy issues, and missed sensory quality.

Other obstacles exist for distance art therapy, which uses technology for digital art therapy, in addition to the aforementioned problems. Those include issues with privacy and safety, managing interruptions, technological limitations, image quality, and being unable to watch the creation of art [13].

3.4 *Related Works in Art Therapy*

Technology can have a significant impact on enhancing the well-being of seniors and the lives of their caregivers. Nevertheless, the use of technology, like any novel "object," possesses dual facets. It has the potential to benefit its users greatly, but it also carries the risk of adversely affecting their mental framework and overall health. An underlying component of the design process is frequently participants' capacity to verbally convey ideas, feelings, and desires [14]. Henceforth, as per Bannon and

Bødker [16], "...we have to take the use process, and not the artifact, as the central object for our study" (p. 24). In result, in order to use technology in art therapy in care, it is essential to know the components and process of care. In accordance with Roberts and Mort [17], caregiving involves endeavors in three realms: surveillance, physical caregiving, and providing socio-emotional care.

Making art therapy accessible for everyone, creating social engagement, empowering clients to give them this chance to connect from their own place that makes them more active and lets them feel more autonomous, and reducing opposition to therapy and art creation stand as advantages of employing technology in the field of art therapy [13]. Therefore, art therapy can help to meet the needs of PWD as in line with the guidelines set forth by the American Art Therapy Association, "Art Therapy ... effectively supports personal and relational treatment goals as well as community concerns. Art Therapy is used to improve cognitive and sensory-motor functions, foster self-esteem and self-awareness, cultivate emotional resilience, promote insight, enhance social skills, reduce and resolve conflicts and distress, and advance societal and ecological change" [12]. In addition, some not all of the five clusters of CAT therapeutic factors, according to de Witte et al. [4] "hedonism (pleasure and play), aesthetics, symbolism (non-verbal communication), enactive transitional space, and generativity," could be helpful for PWD. Owing to the obstacles faced by PWD, such as physical impairments of holding objects, digital arts media can create an art-making environment [13] and can be used by them to engage with their surroundings or produce artwork on a tablet device through subtle movements [13]. Digital art creation tools may indeed serve as an optimal platform for individuals who are prone to sensory overload through touch sensations [13], enabling them to maintain a more secure and extended art-making encounter [13].

4 How Art Therapy Can Help Meet Needs of People with Dementia

To summarize the research of other scholars in the fields of dementia, art therapy, and art therapy for PWD, we created a table based on our previous model of PWD's needs (Fig. 1). In order to show how and why we believe that art therapy, and more specifically digital art therapy, could be beneficial for this population, we attempted to connect the needs of PWD and art therapy in Table 1.

Table 1 The connection between digital art therapy and the needs of people with dementia

	Emotional impairments		Physical impairments						
			Cognitive impairments				Physical impairments		
	Safety	Social connection	Short-term memory	Thinking	Learning	Reasoning	Daily activity	Language impairment	Exercise
How "digital art therapy" helps?	1. No wandering as there is no need to commute to a place where art therapy takes place 2. Fostering self-esteem	1. Creating a relationship with an art therapist as a participant in the process 2. Doing group art therapy and interacting with others 3. Enhancing social skills 4. Fostering self-esteem	1. Improving cognitive and sensory-motor function 2. Increasing independence	1. Promoting insights 2. Reducing and resolving conflicts	Reducing and resolving conflicts	–	1. Fostering self-awareness 2. Increasing independence	1. Creating the chance for non-verbal communication 2. Cultivating emotional resilience 3. Increasing independence	1. The ability to make art with tiny gestures 2. Increasing independence

5 Why Digital Art Therapy Can Help People with Dementia in Emotional Impairments

We believe that digital art therapy, which could be done remotely, could become highly beneficial for addressing the demand for "safety" since it eliminates the need for clients—PWD—to commute to other locations where art therapy is practiced. Because of their unique brain health state, PWD are less likely to wander, thus neither clients nor their caregivers need to be concerned. Additionally, employing digital tools in digital art therapy can boost a client's self-esteem because they rarely need to handle tactile materials, which are difficult for them to use on their own and may require aid from others.

Additionally, engaging in digital art therapy alongside a qualified art therapist helps foster a bond between the client and the therapist. PWD may benefit from this interaction in meeting their "social connection" needs. It is important to note that, in our opinion, the involvement of art therapists does not negatively or positively impact PWD's sense of self-worth. This is because art therapists are simply other participants in the process who are creating art, rather than someone who will assist clients with holding objects, etc. Furthermore, group therapy sessions, which are easier to access in the context of digital art therapy, can assist PWD build networks and communicate with others. Furthermore, we discovered that utilizing digital art therapy can help PWD develop their skills and enhance their self-esteem.

6 Why Digital Art Therapy Can Help People with Dementia in Physical Impairments

Our review of the literature led us to conclude that digital art therapy can benefit dementia patients' "short-term memory." They could become more independent, have higher self-esteem as a result, and also have better cognitive and sensory-motor function if digital art therapy is used in art therapy sessions.

Additionally, it can help with "thinking" demands as well. Digital art therapy, as one approach of art therapy, can also provoke insights while at the same time minimizing and resolving conflicts because of the instincts of art.

Furthermore, we consider that digital art therapy can help PWD with their "learning" demands because of the same reason, reducing and resolving conflicts and helping them learn new things easier.

PWD also need to accomplish their "daily activity," as shown in Fig. 1, and doing so while being independent may have a positive impact on their self-esteem. According to the results of our study, digital art therapy can improve self-awareness and therefore may enhance independence in PWD. Digital art therapy might thus be quite helpful for this population, in our interpretation.

According to Fig. 1, the other criteria of PWD are included as "language impairment" under the category of "physical impairment." Like other forms of art therapy, one characteristic of digital art therapy is giving clients a chance to interact nonverbally. As a result, we suspect that using digital art therapy to treat PWD could be quite advantageous. Moreover, since making art digitally doesn't require assistance with speaking, digital art therapy can foster emotional resilience and develop a sense of independence.

The last necessity indicated in "physical impairment" for PWD is "exercise." Clients could create art with the help of digital art therapy, which was previously stated as a sort of art therapy. Since even a small movement can sometimes benefit PWD and motivate them to do more, we believe that digital art therapy can encourage PWD to move and mix their art creation with exercise.

7 Future Study

7.1 Method

Considering PWD's need to be independent and the barriers of holding objects, we propose a study to compare the efficacy of digital art therapy as a modern therapeutic approach in comparison to conventional art therapy. This research will empirically investigate the impact of digital art therapy and traditional art therapy on the emotional stimulation of PWD. In a 2 (Type: Traditional Art Therapy versus Digital Art Therapy) x 2 (Order: Traditional Art Therapy/Digital Art Therapy versus Digital Art Therapy/Traditional Art Therapy) within subjects' experiment, one group will use digital technology in the form of a leap motion application to do mid-air gestures to create a digital painting. The other will use traditional canvas and tactile tools. In order to reduce the effects of being experimented on the results, we are going to divide participants into two groups. In addition, to achieve the goal of this study, we will affix electrodes to the inner side of each participant's foot as they engage in art therapy sessions, with the aim of gauging their arousal and level of serenity while using their hands. To ascertain whether the experiment yields a positive or negative impact on them, we are going to measure valence in addition to arousal. In order to have the result and perform that, electrodes will be placed on the Zygomaticus major on participants' faces.

In conjunction with the aforementioned measurements, we will administer a set of questions to the participants at the conclusion of each art therapy session. These questions will employ a 5-point scale for the subjects to provide self-report responses.

7.2 Hypotheses

For this proposed study, we have two hypotheses:

H1: Implementing digital painting in art therapy sessions is anticipated to elevate the positive emotions experienced by PWD in contrast to using tactile materials for painting during art therapy sessions.

H2: Substituting tactile materials with digital painting applications in art therapy sessions is expected to enhance the feelings of independence among individuals with dementia.

8 Conclusion

In conclusion, this paper sought to emphasize the advantages of utilizing digital art therapy with people with dementia (PWD). The authors underscored the possibilities offered by digital art therapy for this population by evaluating prior research on the needs of PWD and utilizing a developed model that categorized emotional and physical needs. Based on the proposed model, definitions of art therapy, and prior research on various forms of art therapy, a comprehensive table was given that demonstrates the positive aspects of digital art therapy in addition to a psychophysiological experiment.

To explore the impact of digital art therapy on cognitive capabilities, independence, and the development of interactions between PWD and their art therapists, future research projects will concentrate on putting these recommendations into practice in experimental settings.

This study advances knowledge of how digital art therapy can improve the well-being of PWD. It emphasizes how crucial it is to use digital art therapy as a successful intervention to meet the demands of this demographic as digital art therapy has the potential to make a big difference in the lives of people with dementia by enhancing cognitive function and creating meaningful connections.

References

1. A. Webb, *Art Therapy for Older Adults with Dementia: Heuristic Reflection and Strategy of Clinical Practice* ((Publication No. n.d.) [Master's Thesis, Saint Mary-of-the-Woods College]. Woods Scholars, 2020)
2. J. Hodge, K. Montague, S. Hastings, K. Morrissey, Exploring media capture of meaningful experiences to support families living with dementia. Proceedings of the 2019 CHI Conference on Human Factors in Computing Systems (2019). https://doi.org/10.1145/3290605.3300653
3. K. Slegers, A. Wilkinson, N. Hendriks, *Active Collaboration in Healthcare Design* (CHI '13 Extended Abstracts on Human Factors in Computing Systems on – CHI EA '13, 2013). https://doi.org/10.1145/2468356.2468440

4. M. de Witte, H. Orkibi, R. Zarate, V. Karkou, N. Sajnani, B. Malhotra, R.T. Ho, G. Kaimal, F.A. Baker, S.C. Koch, From therapeutic factors to mechanisms of change in the creative arts therapies: A scoping review. Front. Psychol. **12** (2021). https://doi.org/10.3389/fpsyg.2021.678397
5. M. Carós, M. Garolera, P. Radeva, X. Giro-i-Nieto, *Automatic Reminiscence Therapy for Dementia* (Proceedings of the 2020 International Conference on Multimedia Retrieval, 2020). https://doi.org/10.1145/3372278.3391927
6. United Nations, Department of Economic and Social Affairs, Population Division (2017). World Population Ageing 2017 – Highlights (ST/ESA/SER.A/397)
7. Center for Disease Control and Prevention, (2021). About Dementia. https://www.cdc.gov/aging/dementia/index.html
8. F. Shojaei, Design together: Uncovering the impact of co-design and design thinking on designing for people with dementia. In *Advances in Emerging Information and Communication Technology* (ISBN:978-3-031-53236-8). Springer Nature (2024)
9. S. Foley, D. Welsh, N. Pantidi, K. Morrissey, T. Nappey, J. McCarthy, *Printer Pals* (Proceedings of the 2019 CHI Conference on Human Factors in Computing Systems, 2019). https://doi.org/10.1145/3290605.3300634
10. S. Moharana, A.E. Panduro, H.R. Lee, L.D. Riek, *Robots for Joy, Robots for Sorrow: Community Based Robot Design for Dementia Caregivers* (2019 14th ACM/IEEE International Conference on Human-Robot Interaction (HRI), 2019), pp. 458–467
11. F. Shojaei, F. Shojaei, J. Velez. Does game help people with dementia? A guideline to design games for people with dementia's needs. In *HCist 2023 Book of Industry Papers and Poster Papers* (2024)
12. American Art Therapy Association, (2017). https://www.arttherapy.org/upload/2017_DefinitionofProfession.pdf
13. A. Zubala, N. Kennell, S. Hackett, Art therapy in the digital world: An integrative review of current practice and future directions. Front. Psychol. **12** (2021). https://doi.org/10.3389/fpsyg.2021.600070
14. A. Lazar, J.L. Feuston, C. Edasis, A.M. Piper, *Making as Expression* (Proceedings of the 2018 CHI Conference on Human Factors in Computing Systems, 2018). https://doi.org/10.1145/3173574.3173925
15. B. C. Peterson, Art therapists' adoption and diffusion of computer and digital imagery technology [Doctor of Phylisophy's thesis, Florida State University]. (Publisher n.d.), (2006)
16. L.J. Bannon, S. Bødker, *Beyond the Interface: Encountering Artifacts in Use. Designing Interaction: Psychology at the human-computer interface. Carroll, J.M* (Cambridge University Press, England, 1991)
17. C. Roberts, M. Mort, Reshaping what counts as care: Older people, work and new technologies. Alternatives **3**(2), 138–158 (2009). https://doi.org/10.1016/j.alter.2009.01.004

Rethinking the E-HR Function: The Case of ERP at the Hassan II University Hospital Center

Imane Chaaibi, Abderrahmane Ouddasser, and Achraf Baghdad

1 Introduction

The digital revolution refers to the profound changes brought about by the rise of digital technology in our societies and economies. Integrating society in general into the information age has profoundly changed the perception of work and the way companies in all sectors innovate, act, and communicate. Like all industries, health is facing these transitions [1].

In the health sector, we can say that the HR evolution in the digital age is very disparate and depends on several variables, the most important of which are the culture of the organization and its HR Department as well as its financial capacities.

The hospital integrates into its scope a very wide variety of activities that materialize in very different and autonomous services and a very strong diversity of professions and skills.

And when we talk about digital transformation or digital transition we mean a business transformation process via light applications, often based on an Information System and its integrated management packages (ERP) [2].

This computer system, based on new technologies, will guarantee the various departments' efficient management of human capital. It will also help managers make better decisions by adopting a coherent approach to HR policy, which involves modernizing human resources management practices by putting in place effective tools to develop skills, improve human resource indicators, and inform decision-makers to better make decisions.

I. Chaaibi (✉) · A. Ouddasser · A. Baghdad
Industrial Technologies and Services Laboratory, Sidi Mohamed Ben Abdellah University, Fez, Morocco
e-mail: Imane.chaaibi@usmba.ac.ma; abderrahmane.ouddasser@usmba.ac.ma; Achraf.baghdad@usmba.ac.ma

© The Author(s), under exclusive license to Springer Nature Switzerland AG 2024

A. Shaikh et al. (eds.), *Advances in Emerging Information and Communication Technology*, Signals and Communication Technology,
https://doi.org/10.1007/978-3-031-53237-5_11

Through this study, we will try to analyze the E-HR function in the hospital sector precisely in the Hassan II University Hospital Center in Fez and to answer the following question: how does ERP implementation influence the hospital in its digital transformation?

To answer our question, this article proposes to study the E-HR function in the health sector, focusing on the case of the Hassan II University Hospital Center. The first part presents a literature review on the definition and evolution of the E-HR function, possible digital solutions to rethink this function, and the challenges of the digital transformation of organizations. The second part exposes the research methodology based on a qualitative approach precisely a documentary analysis and analyzes the integrated ERP within the Hassan II University Hospital Center. These include the HOSIX Hospital Information System, the MAWARID Human Resources Information System, and the SAGE and COSIWN systems for inventory management and maintenance. The final part discusses the results obtained and the research perspectives.

2 Literature Review

2.1 Definition and Evolution of the E-HR Function

With the digital revolution, the human resource function (HRF) is undergoing profound changes in terms of content, structure, and positioning. We are talking about "E-RH." Several expressions define this phenomenon as "digital transformation," "digitalization," "digitization," or "digital transformation."

This is a real transformation in this area. From a function hitherto recognized as being rather focused on administrative personnel management, HRM progresses in another way, namely a value-generating function, focusing on its strategic contribution and developing value-added services for operational managers [3].

The digital transformation of the HR function requires above all the digital transformation of the HR function itself, in its various processes: recruitment, training, compensation, skills management, etc. First, it can therefore be defined as the use of the potential offered by digital technology, thanks to various digital HR solutions already present on the market or to be developed to: enrich, transform, and automate how HR tasks or processes are performed [4].

In this perspective, the digital transformation of HR is materialized by the implementation of one or more solutions for different HR processes. These solutions are implemented in digital HR projects [4].

In many institutions, human resources management has seen the emergence of major transformation projects centered on redefining the tasks of the function, and the content of the services it provides to its staff, operational managers, and General Management.

Through the digitalization of human resources and the dematerialization of information and data, institutions will significantly improve the management and reliability of their internal processes.

2.2 Rethinking the E-HR Function in the Healthcare Sector: Which Digital Solution?

All sectors, all trades, all functions of the company have been, are, or will be disrupted by the Internet [5]. This declaration is still relevant 20 years later as we have been going through the digital revolution for several years [6].

In the healthcare sector, the digital transformation of the HR function is essentially based on two dimensions: the transformation of the function itself and the accompanying digital transformation of the organization as a whole.

The transformation of the function itself involves exploiting the potential of digital technologies to improve, complement, and modify the functioning of the different HR processes (recruitment, training, compensation, management of jobs and skills, etc.).

Supporting the institution's overall digital transformation means that human resources services support the digital transformation more broadly. In the latter case, we are talking about several topics: developing work organization and management methods, contributing to the regulation of digital uses within the organization, or supporting training and communication actions in the context of managing the change of digital projects [6].

Healthcare institutions are the places where digital technologies are evolving more quickly, and we must support the digital transformation of the organization as a whole. This digital transformation is impossible if health professionals do not develop a real vision and give meaning to all the investments made on this subject [7].

The development of information systems (ISs) represents a major challenge for each health facility, given its general policy of improving the supply of care. The various actors concerned by the IS must therefore possess all the organizational, operational, management, and strategic knowledge and skills contributing through their development to the improvement of administrative procedures, logistics, and care [8].

Right now, the success of organizations depends on good human resource management. When we think of the technological contribution and the digitalization of human resources, we think of the information system in human resources management. It is a "software program that links an organization's human resources to information technology." HRIS was created to facilitate access to information about employees and HR professionals [9].

In this context, there is ERP. An ERP (Enterprise Resource Planning), also known as IMP (Integrated Management Package), is an IT tool that addresses a particular problem: integrating all the data of the establishment into a single database to optimize the resources and the information flows of all its departments, be it the HRM of the maintenance logistics [10].

2.3 *The Challenges of Digital Transformation of Organizations.*

The digital transformation of HR activities is the result of a long process characterized by successive waves of technology. The term digital transformation often gives the mistaken impression that digital development starts from scratch. There was a decade of accelerated development. It should be noted that employees are rather satisfied with these digital transformations of HR processes, and are even interested [11].

While digitalization appears to be an effective response for streamlining human resources management and optimizing workspaces, it is crucial to acknowledge that the challenges associated with digital transformation are both immense and tangible. This means integrating a great deal of flexibility and responsiveness into an organization, with sometimes cumbersome tools and time and people constraints that need to be resolved. This means changing all organizations, sometimes slightly, and profoundly, and we know that changing the way we work is one of the most difficult challenges [12].

Accepting these changes also allows employees to avoid having to change their mindset and habits when entering the company. A world in which employees feel good allows them to focus on their missions and their challenges. Thus, employees are likely to be more creative but also more productive through lighter processes and easier contact with the people they need to communicate with [12].

3 Integration of ERP in the Health Sector: Case of the Hassan II University Hospital Center

3.1 *Research Methodology*

3.1.1 Principles of Qualitative Research

Qualitative research is a way of looking at social reality. Instead of looking for the right answer, qualitative research is also about asking the right questions. Qualitative data collection takes many forms. Qualitative data come from different sources such as observations, document analysis, images or videos and interviews, etc. [13].

The qualitative approach collects data in the natural environment (study of observations or documents, individual or group interviews, etc.).

The methods used to collect information must be strict. They should be specified: structured or open interviews, collection of files or documents, films, minutes of meetings, etc. [14].

The benefits of rigorous qualitative data collection lie in the richness of the data collected and the in-depth understanding of the problem under study. They are not simply intended to describe, but also to help obtain more relevant explanations of a phenomenon [13].

As part of this research, we conducted a documentary analysis strategy, which is a technique used for qualitative research.

3.1.2 Place of Documentary Analysis in Qualitative Research

The choice of qualitative approach should be motivated by the purpose of the research: review procedures, explore, interpret, understand a phenomenon, etc. [14].

Theoretical readings have a significant impact on the formulation of research questions. A documentary analysis allows me to place my work among other studies. In addition, the documentary analysis will facilitate the choice of concepts and concepts during formal categorization, a step that provides access to a theoretical explanation [15].

In addition, the documentary analysis supports the compilation of some of the data, as the research report will have to be based on theoretical work to situate the research issue. Finally, it will identify other empirical research that could be used for triangulation, i.e., confirmation of results with other sources, which partly ensures validation [15].

3.2 Document Analysis

Recently, following organizational modifications, healthcare specialists have encountered human challenges and were capable of depend on their statistics and human sources management (HRM) systems. From the attitude of department managers, and through the years, the implementation of control modules has been discovered. This implementation has ended in significant flexibility in organizing working hours, correctly meeting the necessities of the legal framework.

In the administrative departments and especially the human resources department, this type of technological tool has made it possible to identify and facilitate the management of teams to reduce the administrative burden.

The Hassan II University Hospital Centre consists of 5 hospitals that welcome more than 151,846 people each year in consultation, 111,583 visits to emergency departments, and 111,026 days of hospitalizations.

The Hassan II University Hospital has an Information Systems division responsible for implementing and maintaining operational solutions. It has two main areas: Patient IS, which covers clinical management and medical documentation, and Management IS, which includes HR and logistics systems.

In this section, we will discuss the progress of our intervention at the University Hospital Center, as well as the results of our documentary analysis on information systems. We will present four ERP solutions as case studies. The first one is HOSIX, which is used as a patient information system, followed by MAWARID as a Human Resources Information System (HRIS), and then SAGE and COSIWN as Logistics and Maintenance Information Systems.

3.2.1 Presentation and Operation of the HOSIX Hospital Information System

Introduction to HIS HOSIX

It is an integrated, modular, scalable, interoperable, and standardized system that incorporates administrative, functional, and clinical features, directly connecting key management areas: healthcare, clinical, economic, and services. Screenshots of the application are presented in Fig. 1.

HOSIX is a reliable tool for clinical systems, enabling seamless communication and access to patient history. It serves as a communication tool for organizations and professionals, facilitating activities like appointment scheduling.

Service Designations: The System's Modules

HOSIX is a modern "Full Web" tool with a wide range of functionalities, including comprehensive hospital management, consulting and implementation of hospital solutions, technological developments, telemedicine, and automated billing processes. Service designations within the University Hospital Center:

- Patient care management: Facilitates patient registration and maintains their medical and administrative history.
- Emergency Management: Controls and manages patient entry into the emergency area of the healthcare center.
- Consultation Management: Handles appointment processes, including scheduling, assignment, and relocation.
- Hospitalization Management: Controls administrative processes related to patient hospitalization, such as bed management, transfers, admissions, and billing.
- Waiting List Management: Manages hospitalization requests, consultation, and examination requests based on priority and wait times.
- Billing Management: Captures billing information from various modules, allowing efficient tracking and control of billing status.
- Cash and Collection Management: Handles individual invoices and payment management with insurers.
- Patient Medical Record Archiving: Streamlines patient documentation management with unique clinical record numbers.
- Security/Profiles: Controls user access and manages data security within the integrated administrative and care management system.

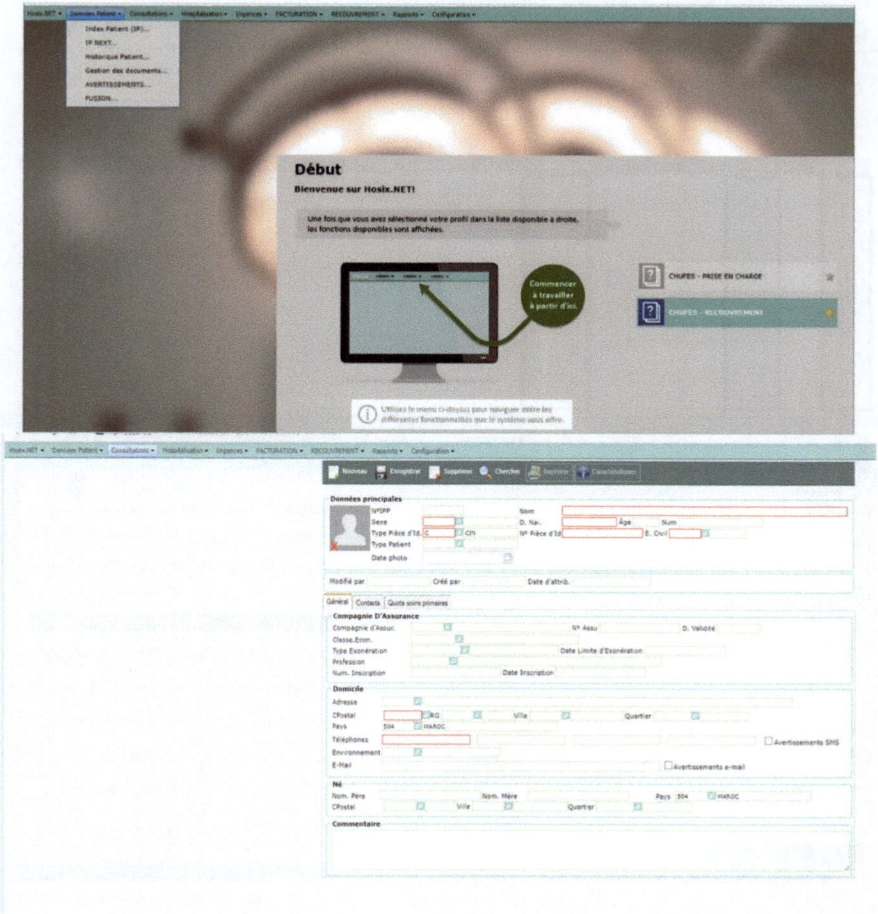

Fig. 1 Screenshots of the patient data and consultations interface of the HIS. (Health Information System HOSIX)

3.2.2 Presentation and Operation of the Human Resources Information System MAWARID

Introduction to HRIS MAWARID

MAWARID is a comprehensive and innovative web-based solution for digitizing and automating HR functions. The goal is to optimize the flow of HR information, reduce delays and lack of coordination in administrative HR processes, and strengthen relationships between HR, top management, and employees. This solution facilitates internal exchange automation, fostering a conducive social climate for employee development. Screenshots of the application are presented in Fig. 2.

Fig. 2 Screenshots of the interface of the homepage and major tasks of the system

The solution is available in Software as a Service (SaaS) mode, accessible through computers, tablets, or mobile phones, providing each HR manager and professional with access to a connected and secure platform.

This solution offers several features that apply to HR management, including absence and leave management, recruitment and interview management, workforce planning, communication, and training.

Operational Aspects of the HRIS MAWARID

The MAWARID Human Resources Information System offers a comprehensive and innovative solution for operational HR management. The different modules, such as absence and leave management, recruitment and interview management, workforce planning and skills management (GPEC), communication, and training, allow for the optimization of HR processes and the strengthening of relationships between HR, top management, and employees. We will subsequently present the various operational aspects of the MAWARID Human Resources Information System. These aspects include:

- Absence and Leave Management: Streamlines the process of managing absences and leaves, saving time and ensuring reliable management.
- Recruitment and Interview Management: Optimizes recruitment and talent management strategies by managing positions, automating application sorting, presenting candidate lists, and handling interviews.
- Onboarding: Provides a personalized welcome brochure, access to the HR knowledge base, training program presentation, and organizational chart access.
- Workforce Planning and Competency Management (WPCM): Tracks employee competency development, including job references, training plans, and competency evaluation.
- Communication: Promotes staff communication through flash information, culture promotion, surveys, and memo distribution.
- Training: Facilitates proposing, evaluating, and organizing training sessions, identifying training needs, and evaluating effectiveness.

In a nutshell, the MAWARID Human Resources Information System presents a dynamic and impactful solution for effective HR management. It enhances operational efficiency, fosters seamless coordination, and empowers employee development, all within a single integrated platform.

3.2.3 Presentation and Operation of the SAGE Information System for Inventory Management

SAGE ERP X3 streamlines inventory management and operational processes for efficient resource planning and order control. Screenshots of the application are presented in Fig. 3.

The ERP system is a platform with modules for different departments. It enables simultaneous use and real-time updates, ensuring accurate information and providing several advantages:

- Information homogeneity: The teams involved have access to the same data in the integrated management software, thereby improving communication and collaboration among the relevant departments daily.
- Time savings and improved productivity: Real-time information provided by the ERP software enables better resource planning, reduces lead times, and speeds up management processes.
- Better organization: Having a comprehensive view of the hospital establishment through the ERP allows for strategic decision-making for the future.
- Cost control: Improved stock planning with the ERP software avoids unnecessary expenses and reduces costs.

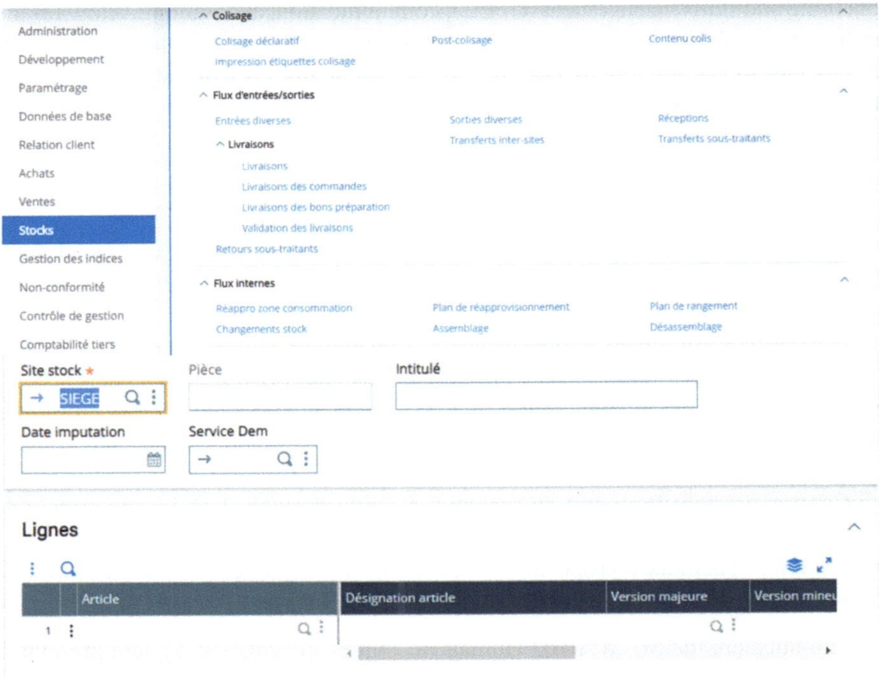

Fig. 3 Screenshots of the interface and components of the SAGE information system for inventory management

3.2.4 Presentation and Operation of the COSIWN Information System for Maintenance Management

COSWIN is a Computerized Maintenance Management System (CMMS) solution that optimizes the management and control of infrastructure and equipment maintenance in healthcare facilities. This solution adapts to the maintenance management needs of hospital technical teams, helping to reduce purchasing costs and stock volumes, thus improving the productivity of maintenance personnel. Screenshots of the application are presented in Fig. 4.

The COSWIN Maintenance Management System is an advanced and comprehensive tool that increases the profitability of the facility and reduces equipment breakdowns, resulting in cost savings and ensuring personnel safety. This solution offers multiple benefits:

- In terms of expenses: reduction in repair-related expenses, decrease in overtime costs (mechanics), and reduction of spare parts inventory.
- In terms of equipment: decrease in mechanical breakdowns, ensures equipment and building performance and reliability, improved maintenance department efficiency, and better organization of work for maintenance teams.
- In terms of personnel safety: decreased risk of work accidents.

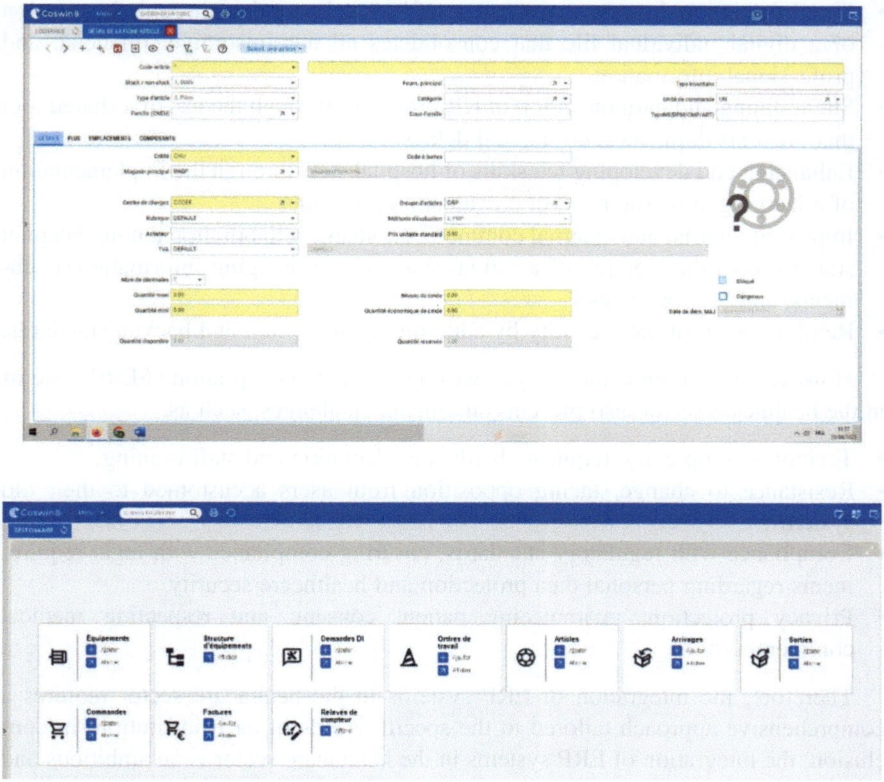

Fig. 4 Screenshots of the COSWIN information system interface and components for maintenance management

3.3 *Analysis of Results and Discussion*

The integration of ERP (Enterprise Resource Planning) systems in the healthcare sector is a strategic challenge for healthcare institutions. ERP systems are information systems that enable the management of all operational processes within an organization. The combination of HRIS (Human Resources Information System) and ERP in a hospital setting offers several advantages, and based on our case study, we have achieved numerous results.

The advantages of integrating ERP systems in the healthcare sector, particularly at Hassan II University Hospital Center in Fes, are manifold:

- Cost optimization by reducing administrative expenses.
- Increased performance and improved quality of care through better allocation of human resources based on the real needs of patients and departments, facilitating access to medical records, and avoiding prescription errors.

- Simplification and secure management of hospital staff data through the creation of a digital individual file that consolidates all administrative, financial, and professional information.
- Streamlining and harmonization of HR processes through the use of a shared tool that prevents duplicates, errors, and delays.
- Enhancing and developing the skills of hospital staff through the implementation of e-learning, e-recruitment, or e-evaluation programs.
- Improved internal and external communication and collaboration among hospital staff through the creation of a virtual space for exchanging information, documents, and best practices.
- Reinforcement of data security by adhering to encryption and backup standards.

However, according to the analysis we conducted, the integration of ERP systems in the healthcare sector also presents significant challenges, such as:

- Technical complexity, requiring hardware adaptation and staff training.
- Resistance to change, facing opposition from users accustomed to their old systems.
- Compliance with regulatory standards, ensuring compliance with legal requirements regarding personal data protection and healthcare security.
- Privacy protection, guaranteeing patient consent, and respecting medical confidentiality.

Therefore, the integration of ERP systems in the healthcare sector requires a comprehensive approach tailored to the specific needs of each institution. In conclusion, the integration of ERP systems in the healthcare sector is an ambitious and beneficial project, but it requires thorough preparation and personalized support.

4 Conclusion

This article has presented a document analysis on the integration of ERP systems in the healthcare sector, using the Hassan II University Hospital Centre as a case study. We have demonstrated that ERP systems offer numerous advantages for improving performance, quality, and safety of care, but they also present challenges and risks related to their complexity, cost, and organizational impact. Thus, we can rethink the E-HR function from strategic, organizational, and technological dimensions, taking into account the specificities of the hospital sector. The goal is to strengthen the role of the E-HR function as a strategic partner, a change agent, and a facilitator of innovation. Furthermore, we have provided examples of four ERP solutions implemented in the institution, considering their operational aspects, context, presented services, and functioning.

According to this study, we can also conclude that accelerated digital transformations, managerial innovations, and new ways of working justify the need to invest in new online platforms. However, based on our analysis, we can sometimes observe shortcomings in these rapid adoptions without clarity, and even in what took healthcare professionals so long to adapt to different services when the technology has been already available and in practice.

Acknowledgments I am deeply grateful to everyone who contributed significantly to this article. Special thanks to my research supervisor, for their invaluable guidance throughout this project.

References

1. L. Seguin, L. Tassy, E-santé, digitalisation ou transformation numérique: impact sur les soins de support en oncologie 2022. Mai **109**(5), 598–611 (2022)
2. M. Zacklad, S.P. Alemanno, M. Ihadjadene, Transition numérique, une approche info-communicationnelle? Approches Théoriques en Information-Communication **1**, 1–7 (2020)
3. D.P. Lepak, S.A. Snell, Virtual HR: Strategic human resource management in the 21st century. Hum. Resour. Manag. Rev. **8**, 215–234 (1998)
4. E. Baudoin, C. Diard, M. Benabid, K. Cherif, Transformation digitale de la fonction RH Année, 2019 (2019)
5. K. Michel, *L'impact de la crise sur le management* (EMS Editions, Business Science Institute, 2020)
6. C. Diard, E. Baudoin, S. Berthet, *Aide-mémoire – Ressources humaines Ed. 3 Editeur* (Dunod, 2022)
7. Y.-M. Herniou, *Informatique de santé: Imagerie, biologie et logistique*, 1st edn. (Eyrolles, 2015)
8. B. Vigneron, *Le management en santé* (2018), pp. 243–266
9. Pieris & Preena, Développements contemporains dans la gestion des ressources humaines L'avenir des SIRH – Tendances émergentes en matière de GRH et d'intégration informatique 2020
10. M. Bidan, C. Godé, DSCG 5 Management des systèmes d'information Manuel et applications, pp. 185–195 (2017)
11. C. Diard, V. Hachard, D. Laroutis, *Bienveillance perçue et télétravail en confinement: une influence sur les risques psychosociaux ? 2022/3*, vol 40 (2022), pp. 107–119
12. J-N. Chaintreuil, RH & Digital: Regards collectifs de RH sur la transformation digitale Editeur: Les Editions Diateino, (2015)
13. L. Kohn, W. Christiaens, Les Méthodes De Recherches Qualitatives Dans La Recherche En Soins De Santé: Apports Et Croyances De Boeck Supérieur. Reflets Et Perspectives De La Vie Economique, 2014/4 Tome LIII, A 82 **Tome LIII**, 67 (2014)
14. G. Borgès Da Silva, La Recherche Qualitative: Un Autre Principe D'action Et De communication Qualitative Research: Another Type of Action and Communication. Rev. Médicale Assur. Mal., 2001 (2001)
15. Y. Comeau, *L'analyse Des Données Qualitatives Cahiers Du Crises Collection Études Théoriques* (N Et9402 Par Département De Counseling Et Orientation Université Laval 1994, 1994)

A Deep Learning CNN Approach Regarding Drone Surveillance in Fire-Fighting Scenarios

Ana-Maria Travediu, Luige Vladareanu, Radu Munteanu, Jianye Niu, Daniel Octavian Melinte, and Ionel Puşcaşu

1 Introduction

In the past, in order to detect a fire, a human presence was needed to sound the alarm and then to intervene in an attempt to extinguish it. Bearing in mind all these details, it is clear that human life was in danger and with the development of technology the amount of people required to handle a fire was reduced, while their awareness regarding extinction was increased.

This study was conducted in order to help firefighters with crucial information regarding victims and also to keep firefighters safe. The unique aspect introduced into the discussion is the consideration of people trapped in fires. The technology advances and keeping the firefighters safe and knowing where to find the victims and their state of health made this paper stand out. Moreover the system developed in this study to help first responders is implemented on the VIPRO (Versatile, Intelligent, and Portable Platform for Emergency Robots) platform and used in the case of blaze and fume occurrence. The VIPRO platform is a research project that incorporates new technologies used by the first responders in order to manage critical situations. It also includes projects such as representing mobile robots, the open architecture and

A.-M. Travediu · L. Vladareanu (✉) · D. O. Melinte · I. Puşcaşu
Romanian Academy, Institute of Solid Mechanics, Bucharest, Romania
e-mail: ana.travediu@imsar.ro; luige.vladareanu@vipro.edu.ro; octavian.melinte@imsar.ro

R. Munteanu
Technical University of Cluj Napoca, Cluj-Napoca, Romania
e-mail: radu.munteanu@mas.utcluj.ro

J. Niu
Yanshan University, Qinhuangdao, China
e-mail: jyniu@ysu.edu.cn

© The Author(s), under exclusive license to Springer Nature Switzerland AG 2024

A. Shaikh et al. (eds.), *Advances in Emerging Information and Communication Technology*, Signals and Communication Technology,
https://doi.org/10.1007/978-3-031-53237-5_12

adaptive networks as an alternative to classic robot control systems or neutrosophic control, adaptive mechatronic humans, etc. [1].

The traditional method involves the manual extraction of characteristics and then using a classifier in order to classify them. A newer approach requires a deep convolutional neural network, which extracts automatically the peculiarities that are classified also automatically and in which regression is used to determine the bounding boxes so the network outputs the original image from the input plus the class of the object that needs to be identified plus a bounding box surrounding the object in the image. This sort of workflow was used to detect and extinguish a fire. There were used aerial vehicles to retrieve information from the environment [2–7]. Thermal images were used to detect people in fire situations [8, 9]. 4K images were used in [3] alongside deep convolutional neural networks. Fire segmentation was used in [10]. Many papers have argued including new firefighting protocols [11–15]. Blazes, heavy smoke, or electricity damage are included in this research. In this paper, the authors use satellite images from Sentinel-1 SAR in order to detect wildfire progression [16]. Deep convolutional neural networks were used to process the images. The authors of [17] used a dataset that includes video recordings and thermal images captured by infrared cameras. An ANN was used to detect whether there are blazes in the image and fire segmentation was also studied with the help of U-Net. This method achieved a 76% classification accuracy. Multispectral images from Himawari-8 satellite were used in [18]. Satellite images were used in [19] also.

In order to attain better performances, we processed the images in different color spectrums (Fig. 1) [20].

We used MATLAB in order to convert the image presented in Fig. 2 in different color spaces to be further processed automatically by the network.

Fig. 1 Firefighting scenario

Fig. 2 The study of color spaces in firefighting scenarios

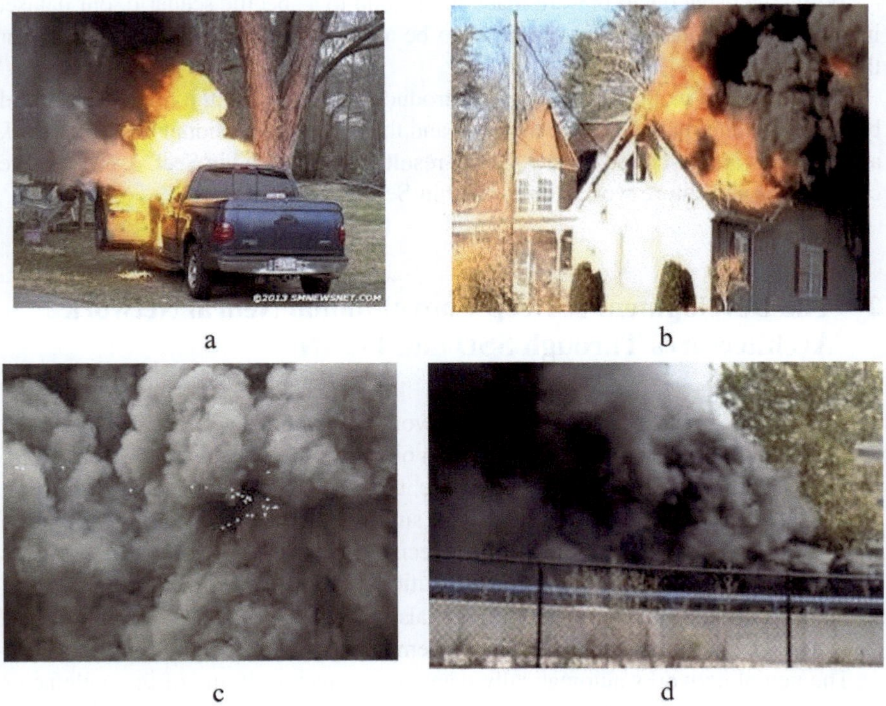

Fig. 3 Examples from our database

Different color spaces, (a) RGB and (b) LAB, in Fig. 3 are presented. We chose to process RGB images due to the unique color scheme found in this color space. In RGB, the fire and the firefighter stand out the most. The dataset, thoroughly analyzed and selected, used in this application is composed of 2393 images for training and 264 images for validation taken from other databases (COCO, ImageNet, FIRE) or from the Internet and manually selected. In order to materialize this experiment, we used three classes: fire, smoke, and person plus background. There are 646 images with people in different joint positions for training and 69 for testing, 900 images of fire without fire classification for training and 100 for testing, and lastly 847 images of smoke to train on and 95 images of smoke for testing. The dataset also contains images where firefighters are labeled as persons.

Examples of images in the dataset are shown in Fig. 3:

In Fig. 3a, b, there are images from the fire class. Two scenarios are presented: a car on fire and a house on fire. These two examples are of a great importance because they are the most common scenarios and have different fuel types and therefore have specific ways to extinguish. In Fig. 3c, d, there are examples of smoke that can occur during a fire. All in all, this shows how diverse is the dataset and how well it responds to fire interventions.

While the usual datasets were made to respond to a specific scenario, our dataset incorporates various scenarios in order to be able to detect human silhouettes near the fire.

Regarding the work, in Sect. 1, we introduced the concept, the related work and background, and materials. The methods and the deep convolutional neural network architectures are described in Sect. 2. The results are discussed in Sect. 3. Finally, the conclusions and future work is presented in Sect. 4.

2 The Development of Deep Convolutional Neural Network Architectures Through SSD and FCNN

Two different architectures of neural networks are trained: an SSD (single shot detector) and a faster RCNN architecture in order to obtain the results we expected. We have chosen these two networks due to the fact that they obtain the best performance in terms of precision in other similar projects. Also, SSD was chosen over faster RCNN, which obtains better precision as a result of a better fps (frames per second) rate. The pre-trained convolutional neural networks on the COCO (Common Objects in Context) dataset were also chosen because of having the person class that we considered crucial in our system.

The neural networks automatically detect in the picture if any of these elements are present and warn the user of the ground control system. The following convolutional neural network architectures were studied and used in this paper: single shot detector (SSD), faster RCNN, architecture obtained through RCNN, and fast RCNN.

Single Shot Detector (SSD) A single shot detector network has two components: the backbone that consists of a convolutional feature extractor network and SSD layers presented in Fig. 4 [21].

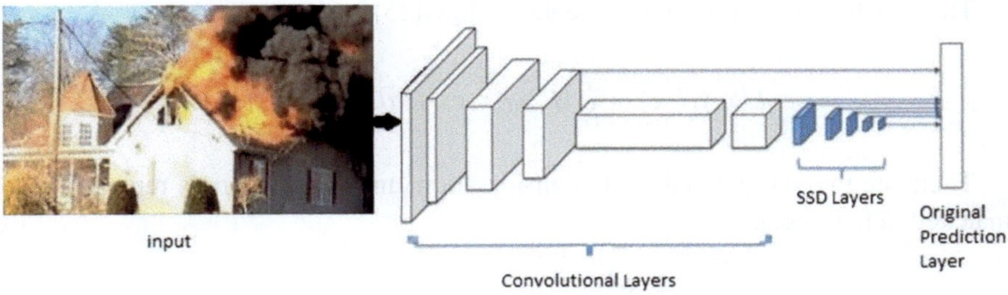

Fig. 4 The architecture of SSD

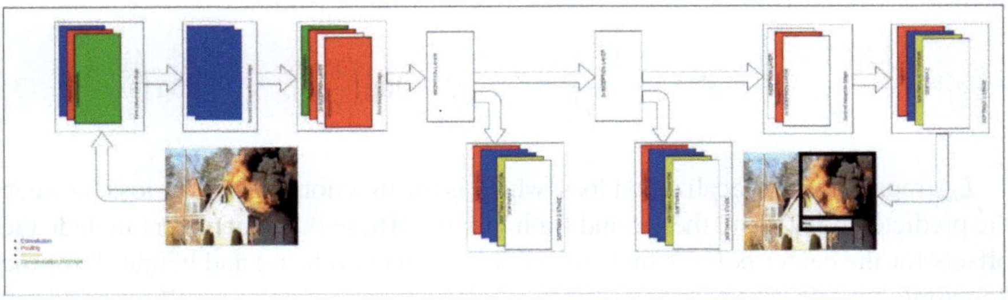

Fig. 5 Inception V1 complete architecture

For this experiment, we chose Inception v2 pre-trained on the COCO dataset. Inception V1 (or GoogLeNet) was the state-of-the-art architecture at ILSRVRC 2014 [22].

The architecture is based on attaining an optimal local sparse structure and covered by readily available dense components [23]. Assuming translation invariance means that the network will be built from convolutional building blocks. While the first layers learn basic contour properties, as we go deeper the layers learn more specific characteristics. Dimension reduction blocks and projections are used wherever the computational requirements would increase too much [23]. The network also includes the use of rectified linear activation, which allows the network to learn. The complete Inception V1 architecture is presented in Fig. 5 [24]. In a later version, the 5×5 convolutions were replaced by 3×3 convolutions and 3×3 convolutions were replaced by 1×3 and 3×1 convolutions in order to be less complex and more computationally efficient.

A model pre-trained on the COCO dataset was chosen due to the fact that it contains a person class, which is the most important class in our dataset. The aim is to automatically detect people in fire and smoke scenarios. The SSD head is just one or more convolutional layers added to this backbone and the outputs are interpreted as the bounding boxes and classes of objects in the spatial location of the final layers activations [21].

The loss function of SSD is presented below [25]:

$$L(x, y, l, g) = \frac{1}{N}\left(L_{\text{conf}}(x, c) + \alpha L_{\text{loc}}(x, l, g)\right) \tag{1}$$

There are two terms to take into consideration: and L_{loc} , where N represents the matched default boxes:

$$L_{\text{loc}}(x, l, g) = \sum_{i \in \text{Pos}}^{N} \sum_{m \in \{cx, cy, w, h\}} x_{ij}^{k} \text{smooth}_{L1}\left(l_i^m - g_j^m\right) \tag{2}$$

$$g_j^{cx} = \frac{\left(g_j^{cx} - d_i^{cx}\right)}{d_i^w}, \quad g_j^{cy} = \frac{\left(g_j^{cy} - d_i^{cy}\right)}{d_i^h}, \quad g_j^w = \log\left(\frac{g_j^w}{d_i^w}\right), \quad g_j^h = \log\left(\frac{g_j^h}{d_i^h}\right) \tag{3}$$

L_{loc} represents the localization loss, which is the function smooth L1 loss between the predicted box (l) and the ground truth box (g). These two parameters include the offsets for the center point, which are cx and cy and width (w) and height (h) of the bounding box [25, 28].

$$L_{\text{conf}}(x, c) = -\sum_{i \in \text{Pos}}^{N} x_{ij}^{p} \log\left(\hat{c}_i^p\right) - \sum_{i \in \text{Neg}} log\left(\hat{c}_i^0\right), \text{where } \hat{c}_i^p = \frac{\exp(c_i^p)}{\sum_p \exp(c_i^p)} \tag{4}$$

L_{conf} represents the confidence loss, which is the softmax loss over multiple class confidences (c). α is set to 1 by the cross-validation. $x_{ij}^p = \{1, 0\}$ which is an indicator for matching i-th default box to the j-th ground truth box of category p [25].

The parameters used for training are presented in Table 1.

Table 1 Parameters used for training

Number of classes:	3
Batch size:	24
Data augmentation:	Random horizontal flip and random crop
Optimizer:	Rms prop optimizer
Learning rate:	Exponential decay learning rate
Initial learning rate:	0.00400000018999
Decay steps:	30000
Decay factor:	0.949999988079
Momentum optimizer value:	0.899999976158
Decay:	0.899999976158
Epsilon:	1.0
Metrics set:	COCO detection metrics

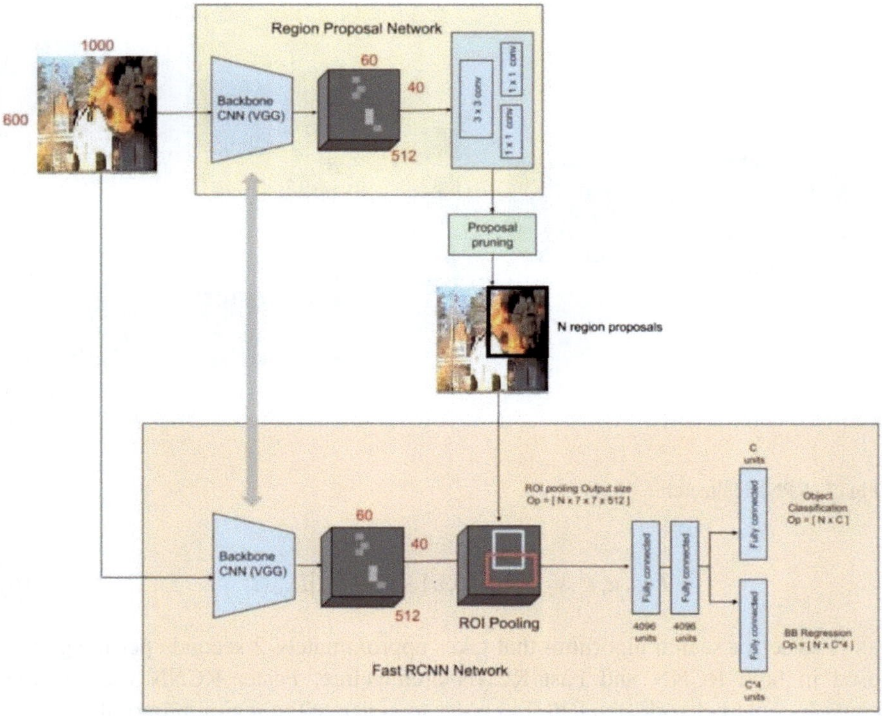

Fig. 6 Faster RCNN architecture

The training parameters presented in the table for SSD were obtained from the literature and through successive experimental attempts applying the previously mentioned SSD, in order to attain a higher precision.

Faster RCNN The classification made by the softmax layer gives, obtained by processing the images of people on fire from the database selected as mentioned in the previous paragraphs, applied to the architecture presented in Fig. 6, probabilities for every ROI over K+1 categories $p = p_0, \ldots p_k$ [27].

The classification loss $Lc_{ls}(p, u)$ is given by $-\log(p_u)$, which is the log loss for the true class u as shown in the relation below. In the context of convolutional neural networks, the Smooth L1 loss is employed for bounding box regression, involving the ranking of parameters. The joint loss is utilized for multitask training [26].

$$L_{\mathrm{loc}}(t^u, v) = \sum_{i \in \{x, y, w, h\}} \mathrm{smooth}_{L_1}\left(t_i^u - v_i\right) \tag{5}$$

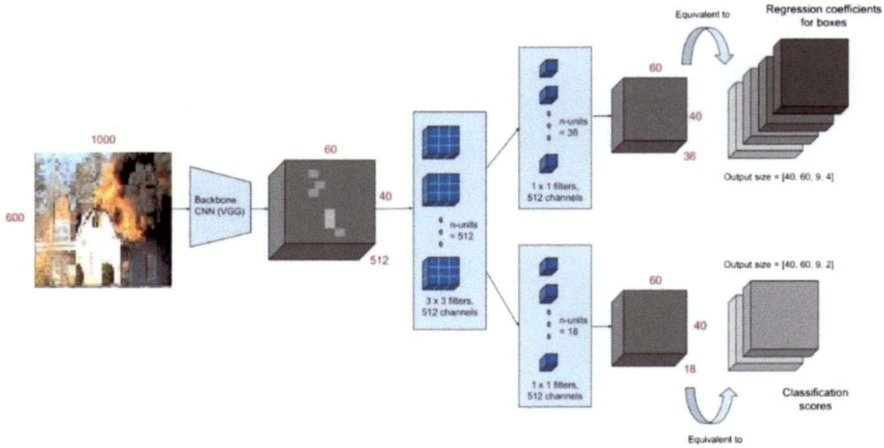

Fig. 7 RPN architecture

$$L(p, u, t^u, v) = L_{cls}(p, u) + \lambda[u \geq 1]L_{\text{loc}}(t^u, v) \tag{6}$$

A selective search algorithm that takes approximately 2 seconds per image was used in both RCNN and Fast RCNN architecture. Faster RCNN uses another convolutional network (the RPN) in order to generate the region proposals.

For every point in the output feature map, the network has to learn whether an object is present in the input image at its corresponding location and estimate its size. The results were obtained by processing the images of people on fire from the selected database, as mentioned in the previous paragraphs, and applied to the architecture presented in Fig. 7 [27]. This is done by placing a set of "Anchors" on the input image for each location on the output feature map from the backbone network. These anchors indicate possible objects in various sizes and aspect ratios at this location [28].

Inception Resnet v2 Faster RCNN is a furthermore improved version of RCNN. It keeps the pipeline, but the specific architecture uses inception resnet v2 as a convolutional network. This network uses residual connections in order to combat the vanishing gradients phenomenon.

The parameters used for training are presented in Table 2.

The training parameters presented in the table for faster RCNN were obtained from the literature and through successive experimental attempts applying the previously mentioned faster RCNN, in order to attain a higher precision. This is the architecture of the faster RCNN we used in this project. Even though this architecture has better results, in terms of precision, 0.75 overall, it was not used because it has a lower time of execution than SSD that we ended up using.

The SSD architecture is less complex than faster RCNN architecture and this leads to a lower precision for SSD, but a higher frames per second rate that makes

Table 2 Parameters used for faster RCNN

Number of classes:	3
Batch size:	1
Data augmentation:	Random horizontal flip
Optimizer:	Momentum optimizer
Learning rate:	Manual step learning rate
Schedule step:	50000
Learning rate:	0.003
Schedule step:	100000
Learning rate:	3.00000010611e-04
Momentum optimizer value:	0.899999976158
Metrics set:	COCO detection metrics

SSD a better network to use for an application that requires automatic detection in real time. For both SSD and faster RCNN, the training was performed on a local deep learning server equipped with a GPU RTX 2080 Ti, 11 GB GDDR6 352-bit RAM, 1.6 GHz processor, and 4352 CUDA cores, while the evaluation took place on the server's CPU equipped with i7 Core 8th gen, 4.6 GHz, 6 cores, and 12 threads.

3 Results and Discussion

We do not only obtain information about fire spreading and victims, but we also receive information regarding the type of fuel that burns making things easier for the firefighters so they can extinct the fire using proper materials in order to eliminate at least one element from the triangle of fire. The information about the location and condition of the people caught in the fire is quite important also. Firefighter's safety is increased due to this type of aircraft device that helps them to detect from a safe distance all the characteristics of the fire.

3.1 SSD

On PascalVoc metrics (mAP of 0.5 IOU (Intersection over Union)), we obtained for each class the following results: fire 0.63, person 0.66, and smoke 0.40. The precision graph is presented in Fig. 8.

The loss measured on COCO metrics is shown in Fig. 9. Examples of detection for SSD are shown in Fig. 10.

It can be observed that the network detects all classes including multiple classes present in an image (Fig. 10a). The system cannot recognize objects smaller than 7×7 pixels because of convolutional filter sizes with which the image is being processed.

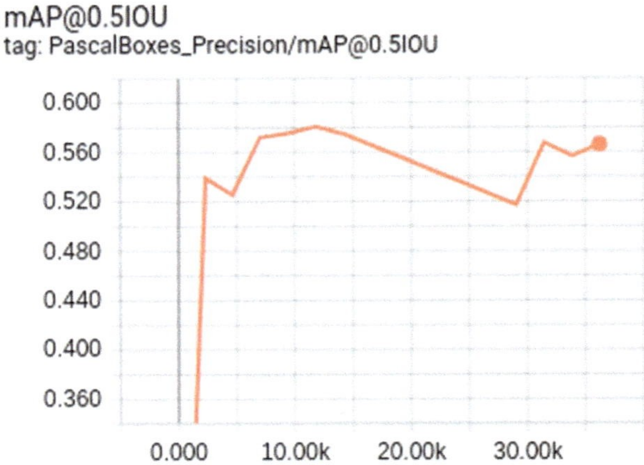

Fig. 8 SSD precision on 50 IoU

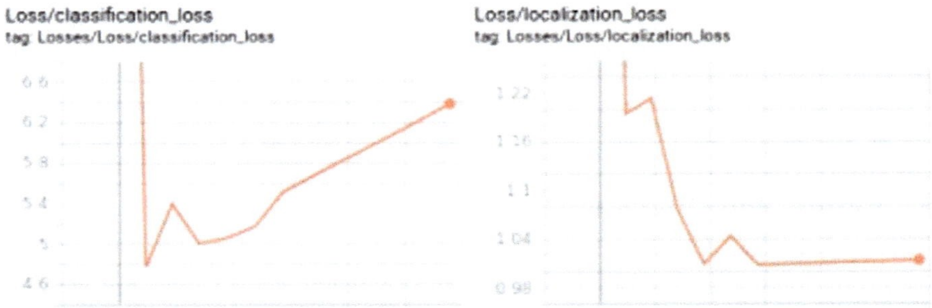

Fig. 9 Loss classification and loss localization graphs

a b

Fig. 10 SSD results

3.2 *Faster RCNN*

The mAP of 50 IoU is shown in Fig. 11.

The total loss decreases to a minimum, as shown in Fig. 12.

Detection examples using faster RCNN are shown in Fig. 13.

It can be observed that even in Faster RCNN there are still false positive detections of fire and that the confidence score is higher.

Faster RCNN has more parameters that conduct to better precision. For instance, the system detects smaller objects than SSD and recognizes objects with a higher confidence score and it has a lower false negative and false positive detection as well; however, it cannot be used in real-time execution programs because it has a very low fps rate.

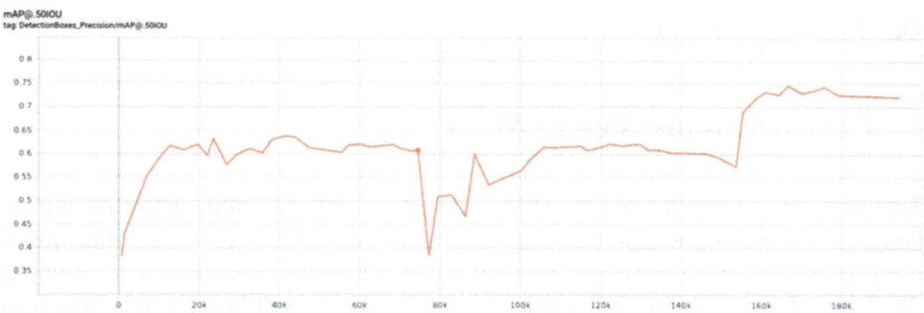

Fig. 11 Faster RCNN mAP 50 IoU

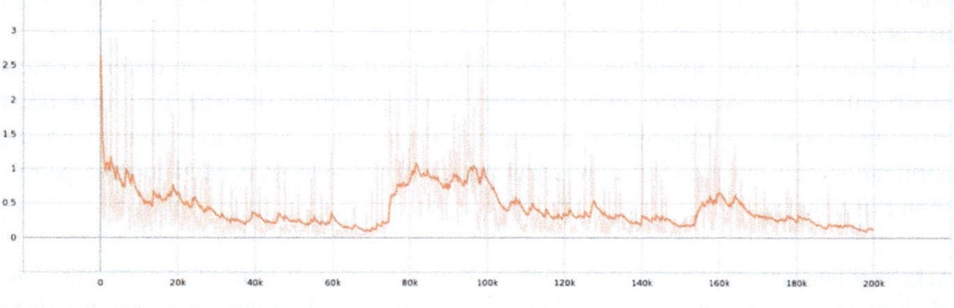

Fig. 12 Faster RCNN total loss

a b

Fig. 13 Faster RCNN results

4 Conclusions and Future Work

The study presented in this paper is focused on automatically detecting human beings in a fire scenario that includes smoke and blazes. Many positions of the joints were taken into consideration because of the variance that gives a human silhouette no matter of posture. Fire and smoke were addressed as well because it is important to detect them as help for the firefighters. Many scenarios were also taken into consideration such as a car on fire, a house on fire, a wildfire, etc. Even though faster RCNN obtained better results in terms of precision, the SSD model was uploaded on the drone due to a lower inference time so that the system could perform well in real time. The operator of the ground control system who sees the images can also observe the fuel of the fire and can better prioritize the mission and extinguish the fire accordingly.

Future work may imply the further development of the dataset in order to detect other calamities. Another approach may be represented by the fact that other institutions such as Emergency Medical Services may come in contact with the application and may lead to new peculiarities of the system needed to save people found in danger.

Acknowledgments The authors gratefully acknowledge the support of the Robotics and Mechatronics Department, Institute of Solid Mechanics of the Romanian Academy and Hebei Provincial Key Laboratory of Parallel Robot and Mechatronic System, Yanshan University, Qinhuangdao, China, and King Abdulaziz University Jeddah, Computer Science and Systems, Saudi Arabia.

References

1. http://vipro.edu.ro/inc/php/menu/aboutVIPRO.php
2. A. Bouguettaya, H. Zarzour, A.M. Taberkit, A. Kechida, A review on early wildfire detection from unmanned aerial vehicles using deep learning-based computer vision algorithms. Sig. Process. (2021). https://doi.org/10.1016/j.sigpro.2021.108309
3. Z. Tang, X. Liu, H. Chen, J. Hupy, B. Yang, Deep learning based wildfire event object detection from 4K aerial images acquired by UAS. AI **1**, 166–179 (2020). https://doi.org/10.3390/ai1020010
4. J.J. Roldán-Gómez, E. González-Gironda, A. Barrientos, A survey on robotic technologies for forest firefighting: Applying drone swarms to improve firefighters' efficiency and safety. Appl. Sci. (2021). https://doi.org/10.3390/app11010363
5. Á. Madridano, A. Al-Kaff, P. Flores, D. Martín, A. de la Escalera, Software architecture for autonomous and coordinated navigation of UAV swarms in forest and urban firefighting. Appl. Sci. **11**, 1258 (2021). https://doi.org/10.3390/app11031258
6. A. Gupta, A. Bhatnagar, A. Mehta, Application of drones in maritime industry (Fire fighting). Bull. Mar. Sci. Technol. **15**, 59–69. ISSN: 0974–8474
7. B. Aydin, E. Selvi, J. Tao, M.J. Starek, Use of fire-extinguishing balls for a conceptual system of drone-assisted wildfire fighting. Drones **3**, 17 (2019). https://doi.org/10.3390/drones3010017
8. J.-H. Kim, S. Jo, B.Y. Lattimer, Feature selection for intelligent firefighting robot classification of fire, smoke, and thermal reflections using thermal infrared images. J. Sens. (Hindawi Publishing Corporation) **2016**, Article ID 8410731, 13 pages (2016). https://doi.org/10.1155/2016/8410731
9. P.-F. Tsai, C.-H. Liao, S.-M. Yuan, Using deep learning with thermal imaging for human detection in heavy smoke scenarios. Sensors **22**, 5351 (2022). https://doi.org/10.3390/s22145351
10. H.-S. Choi, *Automatic Fire Fighting Apparatus using Image Process of Deep Learning* (Department of Computational Science and Technology, Seoul National University, 2020)
11. M.I. Mohd Aliff, N.S. Yusof, A.Z. Sani, Development of fire fighting robot (Qrob). Int. J. Adv. Comput. Sci. Appl. **10**(1) (2019). https://doi.org/10.14569/IJACSA.2019.0100118
12. M. Bhattarai, M. Martínez-Ramón, A deep Q-learning based path planning and navigation system for firefighting environments. arXiv:2011.06450v1 [cs.AI] (2020)
13. M. Bhattarai, Integrating deep learning and augmented reality to enhance situational awareness in firefighting environments. The University of New Mexico. arXiv: 2107.11043v2 [cs.CV] (2021)
14. A. Dhiman, N. Shah, P. Adhikari, S. Kumbhar, I.S. Dhanjal, N. Mehendale, *Fire Fighter Robot with Deep Learning and Machine Vision*. SSRN 15 July 2020
15. V. Nunavath, A. Prinz, T. Comes, Identifying first responders information needs: Sup- porting search and rescue operations for fire emergency response. Int. J. Inf. Syst. Crisis Response Manag. **8**, 25–46 (2016). https://doi.org/10.4018/IJISCRAM.2016010102
16. Y. Ban, P. Zhang, A. Nascetti, A.R. Bevington, M.A. Wulder, Near real-time wildfire progression monitoring with Sentinel-1 SAR time series and deep learning. Sci. Rep. (2020). https://doi.org/10.1038/s41598-019-56967-x
17. A. Shamsoshoara, F. Afghah, A. Razi, L. Zheng, P.Z. Fulé, E. Blasch, Aerial Imagery Pile burn detection using Deep Learning: the FLAME dataset. arXiv:2012.14036v1 (2020)
18. Z. Hong, Z. Tang, H. Pan, Y. Zhang, Z. Zheng, R. Zhou, Z. Ma, Y. Zhang, Y. Han, J. Wang, S. Yang, Active fire detection using a novel convolutional neural network based on Himawari-8 satellite images. Sec. Environ. Inf. Remote Sens. (2022). https://doi.org/10.3389/fenvs.2022.794028
19. A. Larsen, I. Hanigan, B.J. Reich, Y. QinI, M. CopeI, G. Morgan, A.G. Rappold, A deep learning approach to identify smoke plumes in satellite imagery in near real-time for health risk communication. J. Ex-po Sci. Environ. Epidemiol. **31**(1), 170–176 (2021). https://doi.org/10.1038/s41370-020-0246-y

20. https://www.military.com/veteran-jobs/career-advice/transition-to-firefighting-from-military-service.html
21. https://developers.arcgis.com/python/guide/how-ssd-works/
22. https://www.geeksforgeeks.org/inception-v2-and-v3-inception-network-versions/
23. https://towardsdatascience.com/a-simple-guide-to-the-versions-of-the-inception-network-7fc52b863202
24. https://developer.ridgerun.com/wiki/images/e/eb/Googlenet.png
25. https://towardsdatascience.com/review-ssd-single-shot-detector-object-detection-851a9460 7d11
26. https://towardsdatascience.com/fast-r-cnn-for-object-detection-a-technical-summary-a0ff94 faa02
27. https://towardsdatascience.com/faster-r-cnn-for-object-detection-a-technical-summary-474 c5b857b46
28. https://paperswithcode.com/method/inception-resnet-v20

Reinventing Public Health: From LEAN Management to Optimizing Hospital Logistics

Achraf Baghdad, Abderrahmane Ouddasser, and Imane Chaaibi

1 Introduction

Public health is subject to major financial and human pressures, and so evolving in this context is not about working more, but about doing better, and moving together.

Today, public hospitals are confronted with increasing demands in terms of the supply of care. At the same time, the demand for care is increasing due to medical progress, increasing aging rate, or health crises. Second, the instability of the environment as a whole continues to limit the scope of intervention of the states, which cannot finance their healthcare systems, and which must necessarily undertake a stricter rationalization of the expenses of its public hospitals.

On the other hand, the hospital system must evolve its management model and review its adopted strategy to manage the uncertainties and turbulence of the health environment. As a result, it must respond to the new needs of patients, in terms of quality of services.

Lean is a flexible and coherent way of thinking, which will produce profound changes in management models, promoting managerial innovation. That said, lean management contributes to the transition from a bureaucratic system to an open and participatory model. The public sector must also benefit from the adoption of management methods inherited from the industrial world to improve their performance.

The aim of this research is to demonstrate the contribution of lean management to optimizing hospital logistics, and we will attempt to address the following main question:

A. Baghdad (✉) · A. Ouddasser · I. Chaaibi
Management of Organizations, Laboratory of Industrial Technologies and Services - Higher School of Technology - Sidi Mohamed Ben Abdellah University, Fez, Morocco
e-mail: Achraf.baghdad@usmba.ac.ma; abderrahmane.ouddasser@usmba.ac.ma; Imane.chaaibi@usmba.ac.ma

© The Author(s), under exclusive license to Springer Nature Switzerland AG 2024
A. Shaikh et al. (eds.), *Advances in Emerging Information and Communication Technology*, Signals and Communication Technology,
https://doi.org/10.1007/978-3-031-53237-5_13

How can lean management ideas and tools contribute to the smooth running of public hospitals?

We will therefore look at the origins and conceptual framework of lean management, as well as its principles. Subsequently, we will examine the application of lean in public hospitals.

2 The Lean Management Approach

2.1 Origins, Definitions, and Impacts of Lean Management

2.1.1 Appearance of Lean Management?

Before the Second World War, an industrial society founded on mass production whose objective was to minimize costs. Like that of Fordism.

Factories were not aware of overproduction and demand was higher than supply. The USA sent workers to fight in World War II, to support its allies, which led to a shortage of skilled workers, especially in arms manufacturing.

Between 1940 and 1945, the US government authorities decided to launch a program for the development of learning and training practices called "Training Within Industry (TWI)."

After their defeat in the war, the Japanese were looking for a new and reasonable production system that would ensure customer satisfaction. T. Ohno and other researchers collaborated to examine the American approach to innovation and adapted it to create the TPS.

2.1.2 TPS

"Lean manufacturing" was employed by Krafcik (1988), and then he tried to expose the "TPS." The results of the program "IMPV: International Vehicle Program" were then published in the book "The System That Will Change the World," which greatly contributed to its spread [1].

In his book "Toyota Production System" [2], the history of TPS and its implementation was marked. In 1980, researchers at the Massachusetts Institute of Technology, several Lean management researchers, and automotive specialists, in partnership with the government, launched the IMPV to benchmark several factories worldwide.

Lean methods are applied to the global automotive industry and its suppliers. In France, lean started in the automobile production companies [3].

Lean developed in the Japanese automotive industry as an organizational model focused on creating value for customers, and continuous improvement, as well as the fight against waste [4].

Lean is a managerial reference, a way of reviewing and thinking more accurately, which focuses on various concepts, such as the creation of value, the hunting of waste, and continuous improvement...

The tools and principles of lean management allow the elimination of any kind of waste and promote operational excellence of the value chain through a continuous improvement approach [5]. Simply put, lean management aims to fix inefficient processes by ensuring quality and meeting deadlines [6].

Based on these definitions, the concept of "waste" appears in the majority of authors, which reflects its importance in the lean management approach (Table 1).

2.2 *Some Lean Management Concepts and Tools*

There are several concepts of lean that allow the achievement of industrial performance, each based on tools: chasing waste, achieving Just-in-Time, continuous improvement and quality, implementation of visual management, and human resource management [7].

2.2.1 Elimination of Waste

In the definition of lean management, we will observe that the concept of waste is omnipresent. There are three Japanese concepts: "Muda, Mura, Muri," called the "3M." Three forms of "waste" are represented as follows:

Muda: These wastes are invisible to customers and do not represent any value to them. OHNO identifies seven types of waste, in Japanese "Muda": waiting time, transportation, overproduction, inventory, movements, defects, and unnecessary actions. MURI means excess or overload, while MURA means irregularity or variability. In all cases, these anomalies affect process efficiency. In addition to the seven MUDA, an eighth has emerged, characterized by the waste of talents, good ideas, and initiatives.

2.2.2 Just in Time (JIT)

"Just in Time": it contributed to the propagation of lean management thinking. It aims to offer the necessary quantity at the best time, at the right moment by ensuring quality, while optimizing production costs, to satisfy customer expectations and meet their demands in a fair manner.

Table 1 Timeline of the lean management approach

Period preceding 1945	**1911**: Taylor unveils his initial work on his innovative production system, titled "The Scientific Management of Enterprises." **1926:** Ford presents the principles of mass production in an article for the Encyclopedia Britannica.
Between 1945 and 1978: Emergence and development of TPS in Japan	**1937:** Establishment of the Toyoda automobile factory in Koromo, the precursor to Toyota. **1945:** Ohno introduces the foundations of the Toyota Production System (TPS). **1977:** Sugimori and his colleagues publish an academic study on the TPS titled "Toyota Production System and Kanban system: materialization of Just-In-Time and Respect-for-Human System." **1978:** "Toyota Production System," a work by Ohno, released in Japanese. It displays the TPS with two main concepts: just in time…
From 1973 to 1988: The world is growing focus on the TPS	**1973:** Following the first oil shock, Western industrial sectors turn their attention to the TPS, as suggested by Womack et al. in 1990. **Between 1980 and 1986:** Numerous global publications centered on TPS and the Just-In-Time concept, including in France. **1986:** The global IMPV study commences, led by researchers such as Krafcik, Womack, Jones, and Roos, examining 70 factories worldwide. **1988:** Krafcik coins the term "lean" to describe the TPS.
From 1990 to 2000: The initial writings and a plethora of books on Lean emerge	**1990:** Womack et al. release "The System That Will Change the World," a book elaborating on lean without providing a strict definition. **1994:** Womack et al., offering a definition of lean in five key concepts, publish "Lean Thinking." **Late 1990:** Release of "Thinking Beyond Lean" and "After Lean Production" by various authors.
From 2000 onward: The rise of lean institutes on a global scale	**2000:** A surge in publications and specialized books on lean worldwide. **2001:** Establishment of communities and institutes devoted to promoting lean philosophy through publications and training, such as the ean Enterprise Institute in the USA and the Lean Enterprise Academy in the UK. **After 2001:** Creation of lean institutes in numerous countries, including France, Mexico, Brazil, and many more.

Authors: (Derived from Shah and Ward, 2007)

2.2.3 KAIZEN or Continuous Improvement

It is a term that originated in Japan, synonymous with continuous improvement, and is a key operations management tool.

Thus, "Kaizen" is based on the *PDCA (Plan: Plan, Do: Do, Check: Check, Act: Act) or the "Deming wheel."* (Fig. 1).

2.2.4 Perfect Quality

It considers limiting the duration of the production chain, quality control, and risk management.

Therefore, perfect quality can meet the exact requirements of customers [7].

2.2.5 Visual Management

Visual management is a primary method in the lean management approach. The visual management allows for optimizing the workspace and facilitates the flexibility

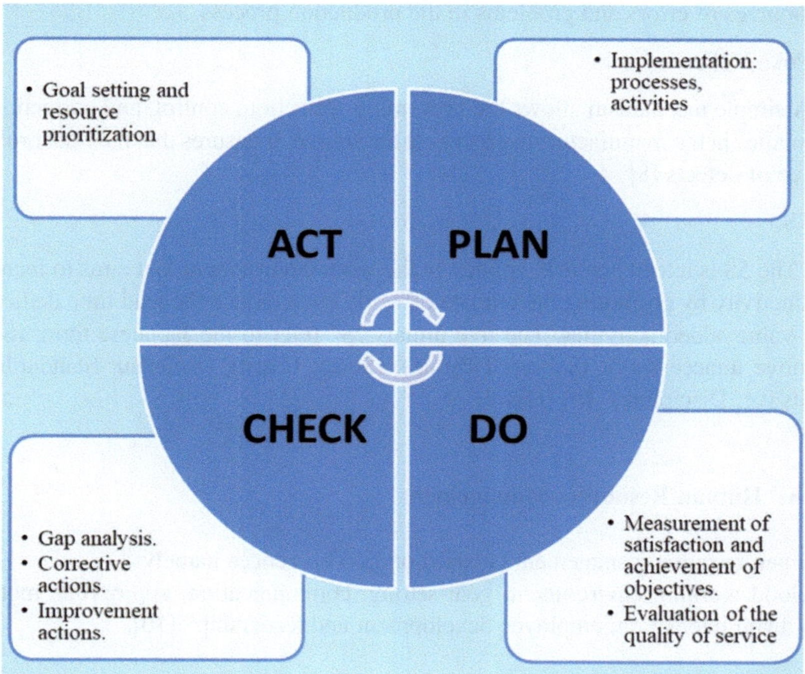

Fig. 1 The Deming wheel or PDCA approach. (Source: Derived from Edward Deming)

Fig. 2 Implementation of 5S. (Source: Guide for implementing 5S-Kaizen in a Healthcare Facility, Ministry of Health and Social Protection, Morocco)

of the work processes. Indeed to succeed in visual management, it is necessary to respect some practices, such as:

- Jidoka:

The main asset of the lean management approach is to ensure quality by detecting the sources of errors and problems in the production process.

- Poka Yoke:

A simple mechanism allows the prevention, detection, control, and correction of anomalies in the manufacturing process at the source. It ensures that the final product is free of defects [8].

- 5S:

"The 5S is a lean practice, applied in the work environment that aims to increase productivity by promoting the climate of work, even reduce the lead time dedicated non-value added activities. The five initials "S" refer to the Japanese term: (Seiri: Remove unnecessary), (Seiton: Tidy up), (Seiso: Clean), (Seiketsu: Standardize), (Shitsuke: Discipline)" [9] (Fig. 2).

2.2.6 Human Resources Management

"Human resource management is based on good practices namely:

Good working environment, goal setting, communication, appropriate motivation, team interaction, employee development and leadership" [10].

2.2.7 VSM

Value Stream Mapping (VSM) is the king tool for analyzing business processes, simplifying and optimizing them. It is used to identify the analyzed processes, their disruptions, and their optimization potential by reducing gaps and improving them.

2.2.8 Lean Six Sigma

Six Sigma is a fact and data-based quality improvement process that prioritizes prevention over error detection. Lean Six Sigma uses existing tools that have been developed based on previous experiences in the world of quality: Visual Management, flow mapping, Pareto and Kano diagrams, the 5 "Why" techniques, Diagram (Ishikawa), Brainstorming and creativity techniques, FMECA (Failure Mode, Effects and Criticality Analysis), and control charts.

2.3 The Fundamental Principles of Lean Management

Like any managerial approach, lean management is based on principles that are articulated as follows:

- Identify value-added activities.

In a warehouse, for example, order preparation goes through several stages, which can be broken down into sub-systems. The VSM allows the analysis of all the components of a specific process.

- Missing the sources of waste

Waste in logistics is generally related to time and space. This phase is mainly based on order, cleanliness, and quality.

- Drawing on optimized flows

This phase can also include investing in new technologies or processes to eliminate time-consuming activities that do not add value to the supply chain. It also aims to implement a more efficient work technique based on timesaving and service quality.

- Adopting the "PULL" strategy to inventory management

Unlike the "push" strategy, which relies on forecast-based production, the main goal of the "pull" method is to meet customers and satisfy demand through supply. In other words, inventory management is a major determinant of value and logistics costs.

- Develop skills and manage talent

To succeed, all employees must be engaged and motivated. This means learning better skills and qualifications, in what we call Talent 4.0.

2.4 Lean Management Returns

Lean leads public organizations to satisfy the customer on its requirements in quality, cost, and time; this means the availability of products or services. In addition, it allows the organization to compete by removing unnecessary costs.

A key success factor of lean is to develop a mindset based on training and involvement of human capital. Lean has many benefits that lead to better results. Operational results have been reported such as reduced cycle time [11], improved inventories, reduced space, and reduced changeover times of up to 70% [12].

These benefits are observed in many sectors of activity such as services. Similarly, in some hospital departments, Dickson states that the rate of patients treated increased, due to the decrease in the average length of stay. The implementation of lean management promotes trust between employees and managers, through several management tools such as stand-up, learning workshops, and continuous training.

The human being is a person who is at the origin of the action, and the main creator of the wealth.

3 Lean Logistics: For a Lean Public Hospital

In the hospital environment, the logistics chain is very particular and interesting, given the complexity of the flows and the diversity of the processes to manage.

3.1 Hospital Logistics: Conceptual Analysis

The etymology of the word "logistics" goes back to the Greek word "logistikos," which, according to Petit Robert, means "a reasoning." In fact, the definition of logistics varies depending on the environment in which it is deployed.

Logistics is the art of directing communications and supplying armies, to ensure their mobility and material condition. It even restricted the survival of several armies that were on the verge of disappearing. The term was introduced into the business world in the 1950s and encompasses functions such as shipping, handling, storage, and even those related to the optimization of business activities.

Years ago, the logistics function was poorly understood in the public health sector and was limited to the activities of hospital technical departments. As a result, managers are recognizing the vital role of the supply chain and its impact on health system improvement and value creation.

Table 2 Hospital logistics

Inventory management	Control and management of medicines, equipment, supplies
Internal transportation	Movement of patients, medical samples, supplies
Waste management	Proper disposal of biomedical waste and others
Catering	Preparation and distribution of meals to patients
Maintenance	Maintenance and repair of equipment and facilities
Procurement	Purchase of medicines, equipment, and other supplies

Source: Authors

In his scientific work [13], he defines hospital logistics as "all the activities of design, planning, supply management, manufacturing (goods and services), delivery and returns management, from the supplier to the beneficiary (patients), while taking into account the trajectories of patients within the hospital without which there is no flow of (pharmaceutical) products. These activities orchestrated by the information flows between the different partners in the chain and lead to financial flows. The aim is to provide an optimal service for the quality and safety of patient care."

On the other hand, hospital logistics can defined as the catalyst that handles the physical flows according to the information and administrative flows of a health establishment.

Then, "the authors have limited the role of hospital logistics in the management of physical flows composed of material flows (classical logistics) and people flows (service logistics). However, each of these physical flows generates information that accompanies the movement of materials or people, either to integrate a change of state in the hospital information system, or to trigger a transfer or production order (for example, the change of service of a patient or the forecast of the number of meal trays for the next day" [14] (Table 2).

"In order to better understand the scope of the supply chain, we therefore remain in a managerial, even hospital vision, whose objective is to consider legitimacy as a key to success in hospital management. Once again, let us note that logistics invites us to have an information system that will allow us to link all the elements of the supply chain, especially when it comes to determining the need until the service is used" [15].

Logistics is an important pivot in the management of the hospital; it has a direct influence on competitiveness. It ensures patient care and synchronizes the hospital supply chain.

3.2 *Lean Management: A Key Vector for Optimizing the Hospital Supply Chain*

Like the new public management, lean management simplifies procedures, drives change, and fosters continuous improvement in public administrations.

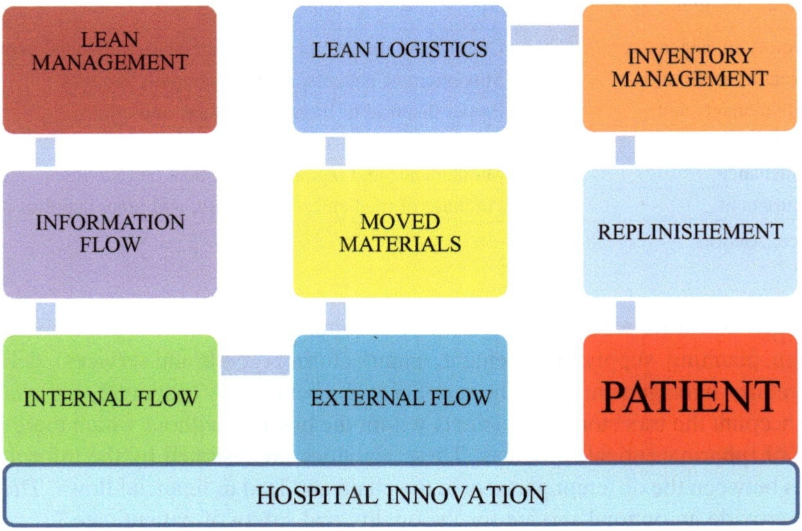

Fig. 3 Combining the different areas of lean logistics. (Source: Authors)

Lean management addresses the different areas of logistics, which can articulated as follows:

- Information flows
- External flows between customers and suppliers upstream and downstream
- The materials moved internally within the organization
- Inventory and warehouse management
- Planning and forecasting of each activity
- Replenishment [16]

After sorting and rereading recent articles on the application of lean management in hospital pharmacy, we note that all the approaches discussed seek to improve the business process.

The VSM [17] has shown the improvement of the sterilization process. Thus, this study is based on the reorganization of manufacturing in terms of schedule and workspace.

The Kaizen approach led to a rationalization of the process and an optimization of the dispensation of ambulatory patients (Fig. 3).

Finally, lean logistics aims to identify actions that no longer add value to the hospital supply chain. It is about creating more and keeping the right tools, at less time, cost, and space.

3.3 *Participatory Management: Toward Continuous Improvement of Public Hospitals*

Motivation, involvement, an active policy of staff development, decentralization, and autonomy in problem-solving and conflict resolution are fundamental principles that make participatory management work.

Continuous improvement is a process of small changes in a production cycle, whose objective is to reduce costs and guarantee quality by ensuring good working conditions. The key characteristics of continuous improvement initiatives are motivation and involvement.

In addition, valuing people is one of the levers of democracy at work and is a means by which public hospitals implement lean management. A bottom-up participative management style therefore allows the project/department manager to communicate objectives and identify priorities, which means that the team is free to participate in solving problems and carrying out a lean project (Fig. 4).

4 Conclusion

Several studies and scientific contributions have shown that the lean management approach has a positive impact on working conditions. Moreover, the implementation of a lean approach leads to a better employee atmosphere and favors the work climate. Contrarily, studies show that lean management negatively influences human capital, leading to increased workload and work intensity. Its sole focus on cost reduction appears to neglect the crucial aspect of ergonomics.

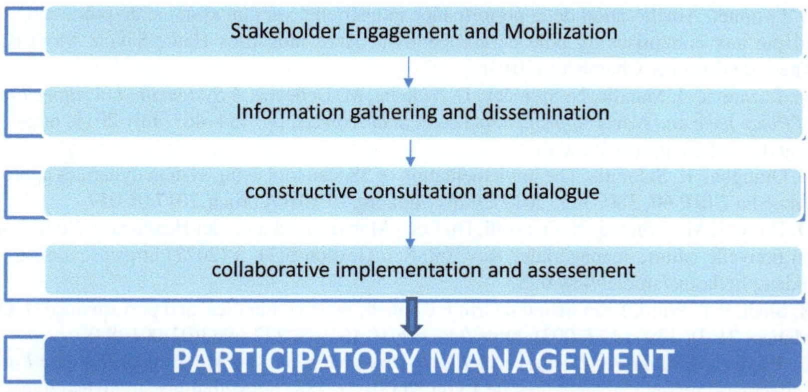

Fig. 4 Structuring of the participatory approach. (Source: Authors)

From this confusion, we can deduce that lean management has never been about cost reduction. The latter is what is achieved with the role of lean logistics.

Cost reduction is a consequence, not an objective. In other words, lean management is a box of ideas, before being a toolbox.

Whereas the hospital is an environment with enormous waste. This is due to the increase in healthcare and hospitalization needs (drugs, equipment, supplies, linen, catering ...). Thus to chase away waste, executives and health managers are called to believe in the lean management approach in order to continuously improve health facilities and meet patients' expectations.

Finally, lean methods cannot be considered as a simple cost reduction methodology, but as an institutionalized, collaborative, and flexible behavior, able to help healthcare facilities, to promote the hospital supply chain.

References

1. J.P. Womack, D.T. Jones, D. Ross, The machine that changed the world, Rawson Associates, New York, 1990, 323 PP. Int. J. Human Factors Manuf. **4**(3), 341–343 (1994). https://doi.org/10.1002/hfm.4530040310
2. T. Ohno, Au-delà de la production à grande échelle. New York. https://doi.org/10.4324/9780429273018
3. J.P. Womack, D.T. Jones, *Système Lean : penser l'entreprise au plus juste* (Village Mondial, 2005)
4. A. Pearce, D. Pons, Advancing lean management: The missing quantitative approach. Oper. Res. Perspect. **6**, 100114 (2019). https://doi.org/10.1016/j.orp.2019.100114
5. G.L. Tortorella, G.A. Marodin, F.S. Fogliatto, R. Miorando, Learning organisation and human resources management practices: An exploratory research in medium-sized enterprises undergoing a lean implementation. Int. J. Prod. Res. **53**(13), 3989–4000 (2015). https://doi.org/10.1080/00207543.2014.980462
6. J. Bicheno, M. Holweg, *The Lean Toolbox, 5th Edition. A Handbook for Lean Transformation* (2016)
7. B. Lyonnet, Amélioration de la performance industrielle: vers un système de production Lean adapté aux entreprises du pôle de compétitivité Arve Industries Haute-Savoie Mont-Blanc, These de doctorat, Chambéry (2010)
8. M. Lazarevic, J. Mandic, N. Sremčev, D. Vukelic, M. Debevec, *A Systematic Literature Review of Poka-Yoke and Novel Approach to Theoretical Aspects*, pp. 454–467, July 2019, https://doi.org/10.5545/sv-jme.2019.6056
9. O. Omogbai, K. Salonitis, The implementation of 5S lean tool using system dynamics approach. Procedia CIRP **60**, 380–385 (2017). https://doi.org/10.1016/j.procir.2017.01.057
10. D. Oulih, E.M. Errachiq, N. Ouarsafi, Du Lean Management au Lean Healthcare: Boîte à outils ou nouvelle culture managériale? Rev. Int. Sci. Gestion **5**(3), 3 (2022) https://revue-isg.com/index.php/home/article/view/995
11. R. Shah, P.T. Ward, Lean manufacturing: Context, practice bundles, and performance. J. Oper. Manag. **21**(2), 129–149 (2003). https://doi.org/10.1016/S0272-6963(02)00108-0
12. G. Baglin, M. Capraro, *L'entreprise LEAN Production ou la PME compétitive par l'action collective* (Presses Universitaires de Lyon, 2000), p. 288. https://hal-hec.archives-ouvertes.fr/hal-00655490

13. T. Hassan, Logistique hospitalière: Organisation de la chaîne logistique pharmaceutique aval et optimisation des flux de consommables et des matériels à usage unique, phdthesis, INSA de Lyon, 2006. https://theses.hal.science/tel-00378591
14. F. Jawab, *L'enjeu logistique dans les établissements hospitaliers* (2007)
15. A. Ouddasser, A. Mellouki, Y. Belyagou, The importance of automated cabinets in the management of drug circuits: The case of Hassan II Fez University Hospital Center, in *2020 IEEE 13th International Colloquium of Logistics and Supply Chain Management (LOGISTIQUA)*, (Dec 2020), pp. 1–6. https://doi.org/10.1109/LOGISTIQUA49782.2020.9353950
16. R. Demetrescoux, La boîte à outils du Lean Ed.2. Dunod, 2019. https://international.scholarvox.com/catalog/book/docid/88870395?searchterm=La%20bo%C3%AEte%20%C3%A0%20outils%20du%20Lean
17. M. Rother, *Learning to See: Value Stream Mapping to Create Value and Eliminate Muda* (1999)

Coffee Leaf Diseases Quadruple Classifier (CLQC) Model Using Deep Learning

Jameela F. AL-Rashidi, Lena A. AL-Enazi, Rawan F. AL-Mutairi, Shahd Y. AL-Dukhayil, Wiaam A. AL-Abas, and Dina M. Ibrahim

1 Introduction

Coffee is a beverage that is consumed by one-third of the world's population, making it the second most traded commodity after petroleum. It is a key commercial commodity that is grown in more than 50 nations worldwide. The estimated daily global coffee consumption is 2.25 billion cups [1]. There are more than 125 species of coffee in the world. Brazil is the largest producer of coffee in the world [2], as it produces about 30% of the world's production of coffee and comes after Vietnam, Indonesia, Colombia, and Africa.

The history of coffee is shrouded in ancient legends that lack concrete evidence about the original discoverers. These myths, while imaginative, shed light on the cultural significance of coffee. One such account involves Kaldi, a ninth-century Ethiopian goat shepherd who observed the stimulating effects of coffee beans on his sheep. After trying the beans, himself, he brought them to a nearby monastery, but the monk declined to use them. The beans were roasted on a fire, producing a delightful aroma that intrigued the other monks. They were then crushed and steeped in hot water, resulting in the world's first cup of coffee. At first, this black water was confined to councils of scholars, but after some time it spread in homes and thus

J. F. AL-Rashidi (✉) · L. A. AL-Enazi · R. F. AL-Mutairi · S. Y. AL-Dukhayil · W. A. AL-Abas
Department of Information Technology, College of Computer, Qassim University, Buraydah, Saudi Arabia

D. M. Ibrahim
Department of Information Technology, College of Computer, Qassim University, Buraydah, Saudi Arabia

Department of Computers and Control Engineering, Faculty of Engineering, Tanta University, Tanta, Egypt
e-mail: dina.mahmoud@f-eng.tanta.edu.eg; d.hussein@qu.edu.sa

© The Author(s), under exclusive license to Springer Nature Switzerland AG 2024
A. Shaikh et al. (eds.), *Advances in Emerging Information and Communication Technology*, Signals and Communication Technology,
https://doi.org/10.1007/978-3-031-53237-5_14

Yemen became the first source of coffee. Thus, coffee was born in Ethiopia and grew up in Yemen! [3]. Coffee has a rich history of how it spread across the world. In Mecca, coffee was introduced by a Yemeni jurist. Yemenis who studied at Al-Azhar brought coffee to Egypt, while Sheikh Ali bin Muhammad bin Iraq introduced it to Al-Sham. Syrian merchants brought coffee to Istanbul in 962 AH and opened the first cafe there. Austrians learned how to make coffee from Turkish soldiers they had captured after defeating the Turks in Vienna in 1094 AH. Coffee then spread across Europe through travel, war, spoils, and trade. The important thing is that it entered the Western world through its widest doors [3].

Coffee trees take a year to develop and 5 years to mature and bear fruit. Their peak producing years are between 7 and 20, during which time they are most productive. Each year, a coffee tree needs 2,000 cherries to produce 1–2 pounds of coffee beans. Peaberries, which have a sweet flavor, are produced when only one seed is found in a cherry [4]. Coffee is highly versatile and used in food processing industries due to its unique flavoring. It takes 18 coffee trees annually to produce one pound of coffee beans, equivalent to two cups of coffee per day. Arabica L. is the most traded coffee species, accounting for 60% of global trade, followed by C. Canephora (Robusta coffee) and C. Barako (Liberica coffee beans). The taste of coffee beans varies depending on where they are grown and environmental factors such as sunlight, soil, and water type. Coffee trees can live for over a century.

The coffee industry has a significant positive impact on society globally. It provides employment opportunities for families and contributes to our social and economic growth. Over 15 million people rely on coffee as their primary source of income, with most being peasant farmers. Furthermore, 15 billion trees have been planted to meet the needs of 25 million farmers across multiple countries.

However, farmers face challenges in getting their coffee products to market due to terrible coffee leaf diseases that afflict their farmlands. Plant diseases and injuries harm 10% of the global plant economy, with Miner, Rust, Cercospora (brown eye spot), and Phoma (brown leaf spot) being the most well-known ailments that cause harm to crops. The leaves, among other parts of the plant, may exhibit symptoms of a disease. Coffee leaf diseases pose a significant challenge for farmers, particularly in underdeveloped countries. Limited resources and expertise make managing and diagnosing these diseases a tedious and time-consuming task, with late diagnosis resulting in the spread of disease and severe damage to coffee leaves or the entire crop. In one country, the loss of all crops due to one of these diseases led to an economic emergency being declared. While pesticides are commonly used in many coffee-producing regions, their indiscriminate use can lead to reduced quality of coffee beans [1, 5–10].

Artificial intelligence has been making strides in the agricultural industry, particularly in helping small farmers identify diseases and pests that affect their yield. With the transition to a smarter future happening faster than ever before, initiatives have been put in place to utilize AI to accurately identify symptoms and severity of these issues. One of the most significant advancements in AI is deep learning, which has surpassed the capabilities of the human brain. After years of study and development, deep learning has greatly enhanced the accuracy and efficiency of identifying and addressing agricultural problems, it has been a game-changer in solving everyday issues, and plant diseases are no exception. By utilizing convolutional neural networks (CNNs), a deep learning technique that has been proven effective in numerous studies, researchers can perform exceptionally well in image recognition tasks and identifying features from an image. Additionally, it has been shown to improve the accuracy of classification problems as more data is analyzed [6, 8].

In line with this, this research is now aiming to detect common coffee diseases by using deep learning convolutional neural network technology. Our research aims to achieve several goals:

- First, we will review the literature to gain a comprehensive understanding of the concepts, principles, and technologies of image processing techniques, deep learning, and convolutional neural networks.
- Second, we will assemble a dataset of images of both healthy and unhealthy coffee leaves.
- Finally, we will construct a system or technique that utilizes deep learning convolutional neural networks to identify the early stages of coffee leaf diseases.

This approach is expected to provide a more accurate and efficient way of identifying and diagnosing coffee diseases. With the help of deep learning, this research can train the algorithm to recognize specific features of the disease, making it easier to detect and diagnose.

The remaining part of this paper is organized as follows. Section 2 presents this research-related work and background on predicting coffee leaf diseases. The details of the materials and methods, the dataset for the study, data preprocessing, and the proposed deep learning models are described in Sect. 3. Section 4 incorporates the results, including the experimental parameters and the performance metrics for our proposed system, with comparisons to current state-of-the-art systems of coffee leaf disease prediction. Comparative analysis and discussion are drawn in Sect. 4. Finally, conclusions with possible ongoing future work are given in Sect. 5.

2 Background: Related Works on the Classification of Coffee Leaf Diseases

This research by [1] was a contribution to farmers in their fight against diseases in coffee leaves; it will help in detecting diseases in time. It involved working on a dataset that had been preprocessed to get rid of unwanted data. The images contain disease-free and diseased leaves. The dataset is divided into two parts: the leaf dataset consists of the original photos of the leaves, and the other contains the cropped images of the symptoms. The entire set was 1747 images divided into 5 images: Healthy leaves, Phoma spots infected leaves, Miner, coffee leaf rust, and Cercospora spots. A convolutional neural network model for this work and several techniques used data augmentation and transfer learning, and it achieved an accuracy of 97.61% and a loss of 0.35. In the end, the research indicated that in future work, more requirements and changes will be necessary for the neural network to be able to accommodate various new diseases in coffee plants and also take into account the changing environmental factors.

Samhitha [11] found that pests cause great harm to crops and coffee leaves, which affect the nation's agricultural output. Usually, farmers notice leaf disease with their eyes and must identify diseases on the coffee leaf as soon as possible to prevent these pests. But, this method may be time-consuming and expensive. For that reason, they advise utilizing automatic detection of different image processing methods because they deliver quick and precise results, such as convolutional neural networks (CNNs) quickly processing a huge number of inputs and providing results related to diseases.

Montalbo and Hernandez [10] proposed a study that classified Barako leaf disease; their work involved a dataset with 3958 images divided into four classes: Healthy, Rust, Infested, and Spots. Their study helps create an optimal deep-learning model that can accurately classify Barako leaf diseases. The study only used a small-scale model and a tiny data class, and it did not cover any additional types of diseases. The outcome suggests that the less complex version of VGG networks can nonetheless offer dependability when it comes to classifying plant diseases. Several deep CNN models were used to gauge the degree of disease severity on coffee leaves. They experimented with a set of three configuration settings each had different data, setting 1 (S1), setting 2 (S2), and setting (S3) determining the number of epochs, FC layer neurons, dropout rate, learning rate, momentum, enhancer, and batch size. This configuration proved effective in the development of a Barako classifier for leaf disease using VGG16. Notably, the model achieved 100% accuracy with Setting 2 (S2).

The model presented by [12] in this paper, "Coffee Leaf Diseases Identification and Severity Classification Using Deep Learning," provided a method for using neural networks to identify and classify leaf diseases and pests in Brazilian arabica coffee leaves in the year 2021. Deep neural networks have proven to be highly effective tools for improving classification accuracy. In order to determine if a particular image of a leaf is healthy or infected with the Miner, Phoma, Cercospora, or Rust pests, a machine-learning model was created and trained using the Common Image Dataset (BRACOL). The neural networks were trained to reduce loss rather than maximize accuracy. The percentage of each leaf that was infected was also determined using a classification technique. After this model achieved an accuracy of almost 90%, it was discovered that these sets were unbalanced after being divided. This issue was handled through the use of data augmentation. Finally, this model was successful in achieving an accuracy of 98.04%.

Montalbo FJP [13] proposed a study that classified seven coffee leaf conditions of Robusta, Arabica, and Liberica coffee images that had different perspectives captured from either controlled or uncontrolled environments. His work aggregated three state-of-the-art pre-trained and fine-tuned DCNNs that included the EfficientNetB0, DenseNet121, and VGG16 formed into an ensemble called the

T-DCNN, performed in a stage-wise fashion, which means narrowing down the classification complexity by simplifying a broad classification task into stages with fewer options and increasing the number of features in a neural network classifier. With a 95.98% accuracy, the SWAT-DCNN model showed that it can outperform commonly used state-of-the-art and classic DCNN models.

This work by [14] presents the application of a convolutional neural network (CNN) to detect and classify coffee diseases present in coffee arabica. The researcher used the VGG19 model, optimized it, and fine-tuned it together with augmentation to detect and classify different coffee diseases to help the farmers with the use of appropriate treatment. Their work involved a dataset with 5600 images divided into seven classes: Leaf Miner, Normal Leaf, Leaf Rust, Predominant stress, CWD, and CBD. The test dataset is used to demonstrate how the proposed model may generalize to new data or how well the model can operate in real-world circumstances. The VGG19 model begins the training process using a training dataset that includes the original image as well as several augmented images. The suggested VGG19 model achieved 97% accuracy with the first hyperparameter configuration. Even with a lower number of neurons, the optimized VGG19 model became much more efficient with less training time required, producing excellent results. Moreover, the assistance of the offered collection of hyperparameters considerably boosted generalization for the classification job and minimized occurrences of overfitting. This work developed a new profound method of diagnosing different coffee plant diseases.

A study was proposed by [8] that classified coffee diseases and assessed severity, and the work included using the largest dataset called PlantVillage, which contains 54 thousand images of several plants, including coffee and 38 categories divided into five categories: Healthy, Leaf miner, Rust, Brown leaf spot, Cercospora leaf spot. In this work, the images were divided into two groups: a group containing the original images and a group containing only symptom images. The study helps create an effective multi-tasking system that can classify coffee diseases and estimate severity at the same time accurately. The study used several techniques of deep learning (AlexNet, GoogLeNet, ResNet50, and VGG19) and the results were 94.05% and 84.76%. They were also taught one task, which provided results of 94.17% and 83.26%. After comparing the results, the results indicated that the symptom group provided better results with an accuracy of 96.63% in both groups. In the end, it was the ResNet50 architecture that provided the best performance and higher accuracy.

Coffee is a significant commercial crop that has yearly productivity losses due to pests and disease. It is essential to identify crop pests and diseases early to implement the proper management measures and minimize the loss of economic production. Research published by [15] on coffee leaf disease, discovered that the most dangerous pest of Arabica coffee in India is the stem borer (Xylotrechus quadriceps), which significantly reduces production each year. In this study, a convolutional neural network (CNN) model is used to detect white stem borer (WSB) infestations early using a set of images from the coffee board. The entire methodology used techniques for unsupervised auto-encoding, supervised transfer learning, image processing, and machine learning. By two points, they were able to get over the restrictions of earlier solutions. The first is that it consistently performs even on uncontrolled, raw photos, and the second is that it incorporates other information, such as environmental data points, to increase prediction accuracy. Consequently, their accuracy reached 85.5%. Citing the article studied [5] in 2021, the primary goal of this thesis study is to create an automated system for identifying diseases on Ethiopian coffee leaves using deep learning features and image processing. Brown eye spot (BES), coffee berry disease (CBD), coffee leaf rust (CLR), and coffee wilt disease (CWD) are the four main types of coffee leaf diseases that reduce coffee production in Ethiopia. A model architecture that includes preprocessing, segmentation, feature extraction, and identification phases has been proposed for this study. Ethiopian coffee infections are identified through conventional inspection and past experiences, which are arbitrary and ineffective; therefore, in this research, a method was suggested for identifying Ethiopian coffee leaf diseases using deep learning features. The photographs of the coffee leaf ailments were taken in Ethiopia, where more coffee is grown. The steps in the proposed model architecture for identifying Ethiopian coffee leaf diseases are picture capture, preprocessing, segmentation, feature extraction, and identification (classification). With 95.75% accuracy, the hybrid of the two filtering procedures has produced better results. K-means clustering was also used to segment the data, and CNN was used to extract features. In the end, they compared the CNN-Softmax classifier to the CNN-SVM classifier. According to the experimental findings, SVM outperforms Softmax in terms of performance and computational speed, whereas the suggested CNN model's Softmax classifier training took a total of about 6 hours or 950 seconds per epoch. When used with the suggested model and SVM classifier, it can attain an overall classification accuracy of 96.5%.

This study was presented by a group of researchers by citing [6]. In this study, a system for visualizing coffee diseases utilizing several visualization techniques was presented. In order to understand what the model "sees" when it learns to distinguish between good and unhealthy images, the goal of this study is to depict several features of a coffee sickness. The ability to automatically identify and categorize plant diseases has improved thanks to deep learning. Disease classification issues, however, are typically handled using black-box methods. Because the model could just use unrelated surroundings for classification rather than accurately identifying the disease's location in the image, it is difficult to trust it. By highlighting the region responsible for the classification, visualization approaches can assist in identifying key areas for the model. Furthermore, visualization enabled this study to identify misclassifications and inspired this study to propose a guided approach for categorizing coffee diseases. Where the Nave Approach is used in this experiment, the network was trained using only the original photos, which can result in overfitting if it learns similarities unrelated to the disease. For instance, by emphasizing background soil structure, the model of this study is unable to generalize well from the training data of this study to unobserved data. Overfitting is the term for this, and it occurs when a model fits a small dataset too closely. This study shows that the guided strategy performs better on unobserved data than the naive approach and generalizes better. The outcomes of the visualization assisted this study in identifying where the naive approach went awry. Changing the background to something more significant and understandable was how this issue was resolved. To instruct the neural network, separate the region of interest (the leaf) from background noise sources like soil and tiny leaves. On the Robusta coffee leaf picture dataset, the guided strategy outperformed the naive approach with a classification accuracy of 98% compared to 77%. Grad-CAM, Grad-CAM++, and Score-CAM were the visualization techniques taken into consideration in this study.

The study was suggested by [16] in 2021, and in this study, multiple techniques for diagnosing plant diseases with scant data were created and assessed. For crop security, automated plant disease identification is crucial. The majority of automated techniques try to create classification models from photos of leaves or fruits. In order to use these methods, a lot of photos must often be collected and annotated, which is a time-consuming and expensive operation, particularly when dealing with novel or uncommon diseases. The goal of this study is to create a method that can learn from little data. Few-shot learning (FSL), a technique used to solve this issue, can make use of the knowledge gained from a sizable source domain dataset. Several methods for categorizing plant diseases that can be learned from small amounts of data were created in this work. These methods include Deep Adversarial Metric Learning (DAML), Triplet Network, and Transfer Learning. The superior capacity of convolutional neural networks (CNNs) to transfer learning is the reason for their use. Two baseline models, a triplet network and a deep adversarial metric learning (DAML) strategy, were built using three CNN architectures (ResNet18, ResNet34, and ResNet50). These techniques were developed to identify novel diseases from a few photos, spanning from 5 to 50 photographs per disease, after being trained on a sizable source domain dataset. The dataset included 1747 images of Arabica coffee leaves, captured using various cell phones. It includes both healthy leaves and leaves that have been infected with one or more diseases (Leaf Miner, Rust, Brown Leaf Spot, Cercospora Leaf Spot). The suggested methods were also assessed in cases where the disease and plant species were recognized simultaneously or simply in cases where the disease was detected without reference to the affected plant. The evaluation outcomes showed that a baseline model that was developed using a large collection of source photos may be modified to categorize new diseases from a limited quantity of images. Additionally, it can benefit from the availability of more photographs. Additionally, they discovered that the baseline model has greater transferability when the source domain images and the target domain images differ noticeably or are taken under various conditions by contrasting them with metric learning techniques. It outperformed all other competing algorithms, with an accuracy of 99% when the shift from the source domain to the target domain was minimal and 81% when it was large. Based on the disease term, a classification for the dataset of coffee leaves is produced. In this instance, the baseline model had the lowest error rate 19.44 for 50 shots and 34.28 for 5, while DAML had the highest 22.2 for 50 shots and 42.6 for 5.

As suggested by [17] in 2019, to create an expert system to help coffee producers identify diseases in their early stages, this project aims to utilize computational approaches to identify the primary diseases in coffee leaves. Cercospora and rust are two diseases that have no defined shapes, which motivates a textural attribute extraction approach for pattern detection. In this work, statistical features and local binary patterns were two texture attributes that were taken into consideration. For a set of photos of coffee leaves, the texture attribute vector was calculated and fed into a feed-forward neural network. The outcomes were contrasted with the rate of recognition achieved by a convolutional neural network using deep learning that was applied directly to the same set of photos without the extraction of texture attributes. Unexpectedly, the second method outperformed the texture extraction technique in terms of outcomes. It might be accounted for by the limited number of diseases they sought to identify and the adequate quantity of training samples they utilized for deep learning. The three output classes of Cercospora, Rust, and Healthy leaves were the subjects of the experiments. Two methods were compared: TBDR, which applied a convolutional neural network with deep learning to the test images, and DLDR, which used a convolutional neural network with deep learning applied directly to the test images. To facilitate comparisons, the Kappa coefficient and sensitivity rates were calculated for each experiment. With kappa = 0.900, which is regarded as a strong agreement, and a sensitivity of 0.933, the local binary pattern texture attribute computation produced the best result utilizing the TBDR approach. The optimal Kappa result using DLDR was 0.970, which is regarded as almost perfect agreement, while the optimal sensitivity result was 0.980. These findings demonstrated that both approaches are workable for creating a system to identify coffee leaf diseases and help coffee farmers. As a result, the best Kappa coefficient and sensitivity were 0.970 and 0.980, respectively.

With a huge decrease in Barako Coffee production in the Philippines because of widespread diseases, Montalbo FJ and Hernandez AA proposed in [7], a study that used deep convolutional models to diagnose Coffee Leaf Rust (CLR), Cercospora Leaf Spots (CLS), and Sooty Molds (SM). The image samples were collected from a local farm in a controlled environment. With a total of 4,667 images, 4023 of them were training samples, and the rest, 644 were used for validation, where this limited quantity caused high hesitation. Because of some restrictions and problems, such as a lack of data and background noise, they applied to preprocess techniques such as data augmentation, hyper-parameters, transfer learning, and fine-tuning. They found overall accuracies of 97% for VGG16, 95% for Xception, and 91% for ResNetV2-152.

After reading the previous research, we gathered the most important data, including the proposed method, performance measures/metrics, and the dataset that was used in the previous findings. Table 1 lists those highlights.

2.1 The Common Coffee Leaf Diseases

- Cercospora Leaf Spot (CLS):

When plants are stressed, a fungus called Cercospora leaf spot (or brown eye spot) develops on the leaves. Both in seedbeds and after plants have been moved into bags, the fungus can grow. It is the most prevalent disease in nurseries and a symptom of

Table 1 Summary of related works in coffee leaf disease classification

Ref	Proposed method	Performance measures/metrics					Dataset name
		Accuracy (%)	F1-score (%)	Precision (%)	Recall (%)	Other	
[1]	transfer learning Inception V3	97.61	–	–	–	Loss values=0.35	BrACoL
[11]	VGGNet	98	–	–	–	–	–
[10]	VGG16	100	–	–	–	–	LiCoLe
[12]	GLCM LBP VGG16	98.04	98.33	98.66	98.00	Standard Deviation: 0.0063	BrACoL
[13]	EfficientNetB0	94.35	94.27	94.75	94.35		RoCoLe
	DenseNet121	93.22	91.52	93.33	93.22	–	BrACoL
	VGG16	92.09	91.93	92.11	92.09		LiCoLe
[14]	VGG19	97	–	–	–	–	–
[15]	Transfer Learning Inception v3 model	85.50	–	–	–	-	Coffee Boardo
[5]	CNN-Softmax CNN-SVM	95.75 96.50	–	–	–	–	–
[8]	AlexNet GoogLeNet VGG16 – Resnet50* Resnet50	Single-task: Biotic stress: 94.17 Multi-task: Biotic stress: 94.05 Single-task: Severity: 83.26 Multi-task: Severity: 84.76 Symptom 96.63	– – – – –	Single-task: Biotic stress: 92.87 Multi-task: Biotic stress: 92.69 Single-task: Severity: 76.40 Multi-task: Severity: 79.26 Symptom 96.28	Single-task: Biotic stress: 91.27 Multi-task: Biotic stress: 90.34 Single-task: Severity: 74.77 Multi-task: Severity: 75.50 Symptom 96.44	–	PlantVillage

(continued)

Table 1 (continued)

Ref	Proposed method	Performance measures/metrics					Dataset name
		Accuracy (%)	F1-score (%)	Precision (%)	Recall (%)	Other	
[8]	AlexNet GoogLeNet VGG16 – Resnet50* Resnet50	Single-task: Biotic stress: 94.17 Multi-task: Biotic stress: 94.05 Single-task: Severity: 83.26 Multi-task: Severity: 84.76 Symptom 96.63	– – – – –	Single-task: Biotic stress: 92.87 Multi-task: Biotic stress: 92.69 Single-task: Severity: 76.40 Multi-task: Severity: 79.26 Symptom 96.28	Single-task: Biotic stress: 91.27 Multi-task: Biotic stress: 90.34 Single-task: Severity: 74.77 Multi-task: Severity: 75.50 Symptom 96.44	–	PlantVillage
[6]	Guided approach Näıve approach	98 77	–	–	–	–	RoCoLe
[16]	Baseline (ResNet50)	5-Shots: 85.58 25-Shots: 96.02 50-Shots: 97.52	–	–	–	–	PlantVillage BrACoL
	Triplet Network (ResNet50)	5-Shots: 81.86 25-Shots: 93.26 50-Shots: 95.20					
	Few-Shot Learning approach	5-Shots: 72 25-Shots: 85 50-Shots: 88					
[17]	Statistical Texture Attribute, Local Binary Pattern, Deep Learning AlexNet	–	–	–	0.980	Kappa index: 0.970	–
[7]	VGG16 Xception ResNetV2-152	97 95 91	–	–	–	–	LiCoLe

subpar management. The symptoms are brown spots on leaves with a reddish-brown edge and spots on both sides of the leaf. With a black center and a yellow halo, brown, circular necrotic lesions are the result of brown eye spots. The afflicted leaves drop quickly, and the affected branches become dry, which lowers plant output and the quality of the fruit. Additionally, this disease results in heavy leaf loss and weak plant development. It develops at temperatures of 17 °C in the UK, but for rapid spread, it needs 25 °C and heavy humidity [18, 19].

- Phoma Leaf Spot (PLS):

One of the major coffee diseases in Brazil, phoma leaf spot (or brown leaf spot), is brought on by the fungus phoma tarda Stewart, Boerema, and Bollen and has significantly reduced crop quality and productivity. The following symptoms have been noted: mummified berries, tip and branch dieback, necrosis of rosettes, and leaf spots. The initial symptoms are white to fawn circular lesions that become dotted with small black fruiting bodies. They occasionally cause partial leaf death before winter but have little effect on growth until spring. The installation of windbreaks, balanced fertilization with nitrogen (N), calcium (Ca), and micronutrients, as well as chemical control when necessary, are among the recommended control approaches. Planting in locations less susceptible to cold winds is also advised. Symptoms of Phoma leaf spot infection can begin in September as crops emerge. Finally, it is a leaf that dies from the tip area toward the other sides, which is a clear sign that it has phoma, according to the pathologist [18, 20].

- The Coffee Leaf Rust (CLR):

The pathologist recognized coffee leaf rust by examining the leaves that had tiny, yellow-orange lesions that were powdery in texture. The pathologist examined the leaves, which had chlorotic spots on the top surface and rust pustules on the bottom. While the margins of the lesions continue to grow and release urediniospores, the centers of the spots eventually dry out and turn brown. The pathologist outlined how wind and rain carry rust to the underside of leaves, which is caused by the fungus Hemileia vastatrix [21–23].

- Sooty Molds (SM):

Black mycelia covered superficially by sooty mold covers fruits and leaves, slowing the host plant's photosynthesis. There haven't been many attempts to quantify their economic significance. Sooty mold colonies include significantly more fungi than anticipated, according to next-generation sequencing. SM is also left behind by other insects such as mealy bugs, scale, and aphids. SM can cover the entire surface of the leaf, preventing it from absorbing enough sunlight, despite not being contagious and being as harmful if treated right away [24].

- The Coffee Leaf Miner (CLM):

The Leucoptera coffeella moth, sometimes known as the "coffee leaf miner," is a danger to coffee production. The feeding habits of the larvae on the leaf are a contributing factor in insect damage. When the insect is in its immature life stages, it feeds on the mesophyll, producing necrosis loss of photosynthetic potential, defoliation, and a considerable loss in production for coffee plantations. Although the coffee production chain is supported by chemical control, market demands are shifting toward conscious consumption and calling for more sustainable approaches [21, 25].

- Red Spider Mites (RSM):

The Tetranychidae family includes the little red spider mite. Both indoor and outdoor leaf and flower plants are attacked by it. Since this is a mite, the label "spider" is a little deceptive. It does, however, have eight legs, make webs, and belong to the same taxonomic order as spiders, the Arachnida. Only a few red spider mites will do much harm. However, if there are many, they could cause the plant to die. When it's hot and dry outside, red spider mites are more likely to proliferate and destroy plants. Furthermore, these spiders aren't insects, and using insecticides improperly promotes their spread. Killing insects carelessly promotes the growth of spider mites since many other insects serve as their predators. Red spider mites respond swiftly to chemical dangers. They develop resistance to numerous insecticides. Red spider mites consume plant sap by sucking it from leaves and absorbing the cells of the leaves. They multiply swiftly, and if an invasion is not stopped in its tracks, it may be disastrous. Furthermore, they can easily move from one plant to the next, where if one becomes infected, they can infect many plants at once [26, 27].

2.2 *The Coffee Leaf Disease Datasets from Previous Related Studies*

The dataset BrACoL [28] contains images of Arabica coffee leaves affected by the main biotic stresses that affect the coffee tree: Leaf Miner, Leaf Rust, Phoma Leaf Spots, and Cercospora Leaf Spot. Figure 1 show some samples of 1747 images available in this dataset. Images in the folder are not divided according to the disease. All diseases are in one file, so the folder needs an expert to classify the pictures.

The datasets JMuBEN [29] and JMuBEN2 [30] form a collection that includes 58,555 Arabica leaf images altogether, distributed among five leaf classifications (Healthy, Phoma, Cercospora, Rust, and Miner) captured on a natural background. Figure 2 displays some samples of it. Three compressed files with photos are included in the JMuBEN dataset. The first file contains 6572 photographs of Phoma Leaf Spot, the second 8337 photographs of Coffee Leaf Rust, and the third 7682 photographs of Cercospora Leaf Spot. JMuBEN2 comes with two compressed files, the first of which has 18,985 photographs of healthy leaves and the second of which has 16,979 images of Coffee Leaf Miner.

The dataset RoCoLe [31] contains two compressed files one for images and the second for annotations. A total of 1560 images of Robusta coffee leaf were captured on the field. They were divided into three classes: Healthy, Coffee Leaf Rust, and

Fig. 1 Samples of BrACoL dataset

Fig. 2 Samples of JMuBEN and JMuBEN2 datasets

Red Spider Mites. Figure 3 shows some samples of those. Coffee Leaf Rust images are classified into four levels based on the severity of their spread.

In study [13], Francis prepared an augmented dataset that combines the three discussed datasets: BrACoL, RoCoLe, and LiCoLe. This dataset contains six coffee leaf diseases, which are Coffee Leaf Rust, Sooty Molds, Cercospora Leaf Spot, Phoma Leaf Spot, Coffee Leaf Miner, and Red Spider Mites of coffee Arabica, Robusta, and Barako. Image samples reached 4675 images captured in both controlled and uncontrolled environments, wherefore their backgrounds vary between white and natural, as shown in Fig. 4.

Fig. 3 Sample of BrACoL dataset

Fig. 4 Sample of Francis prepared dataset

3 Materials and Methods

We proposed the Coffee Leaf Quadruple Classifier (CLQC), a framework that is divided into three distinct stages, as shown in Fig. 5. In the first stage, the leaf will be first checked if it is healthy or not. If it is healthy, the classifier will directly output the result without involvement in other stages. Otherwise, if the image matches any of the unhealthy leaf features, the image will be passed into the second stage, the Disease classifier. This stage predicts if the leaf is infected with either Coffee Leaf Rust (CLR), Sooty Molds (SM), or Brown Spot Lesions (BSLs). BSLs include Cercospora Leaf Spot (CLS), Phoma Leaf Spot (PLS), Coffee Leaf Miner (CLM), and Red Spider Mites (RSM). Likewise, if the leaf is infected with one of the Brown Spot Lesions (BSLs), it will move to its classifier, the third stage to diagnose the leaf with the corresponding lesion. Here, we'll go through each step in more detail and

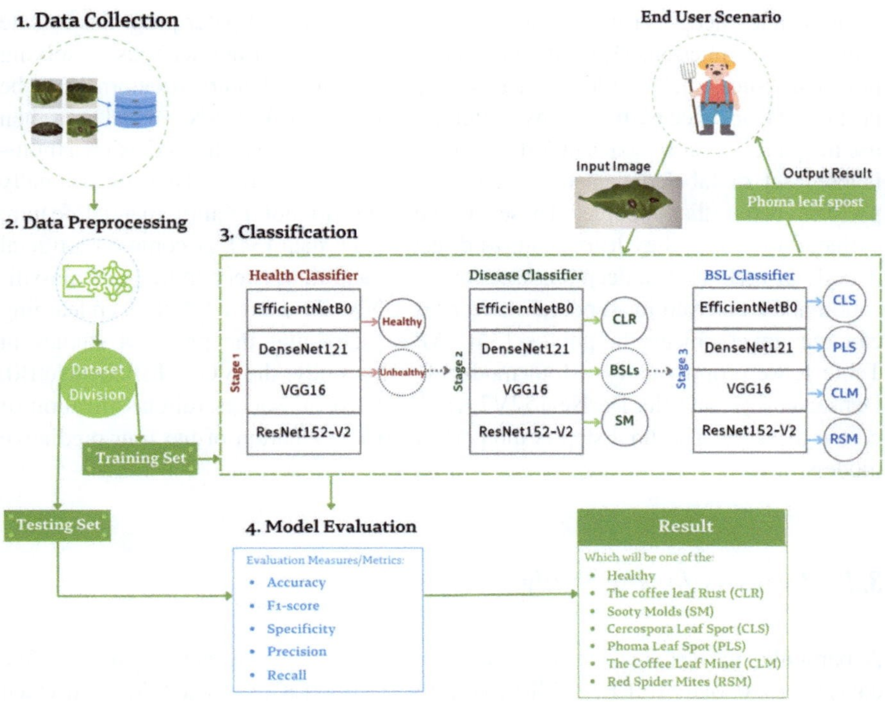

Fig. 5 The proposed Coffee Leaf Quadruple Classifier (CLQC) framework

explain how we made our decisions and the results we expect. As illustrated, data collection is the first step. The dataset was compiled from online resources and categorized based on a number of criteria: diverse diseases, controlled capturing environments or backgrounds (white background), uncontrolled/field environments (natural background), types of coffee beans, and specific issues. Coffee leaf images were gathered throughout multiple nations and during various times of the year. As for before we use images for model training and inference, we format them using a procedure called image preprocessing. Prior to being put into a deep learning or neural network method, image data are also modified. Examples of preprocessing include cropping, scaling, and color conversion. Pre-processing of photos is frequently used to eliminate reflection, normalize the intensity of individual particles on a given image, remove low-frequency background noise, and mask a portion of the image because noises make it more difficult to detect coffee plant diseases [18, 32]. Augmentation: For deep learning, we require a sizable dataset. We use data augmentation as a result in order to prevent the unbalance problem in the dataset. Data augmentation, which is a data-space solution, can solve the issue of limited data. The phrase "data augmentation" refers to a variety of techniques for enhancing the size and quality of training datasets so that more powerful Deep Learning models can be created using them [18]. And for the classification, we used what is called deep learning, which is the study of how computer programs imitate human brain function. Deep learning has a wide range of applications in solving practical issues, such as plant diseases [33]. An image's features (patterns) can be peeked out or detected by a convolutional neural network (CNN), which can then use the pattern to complete a task. Deep learning models can be trained using a multi-layered set of labeled data or a neural network architecture. They occasionally perform better than people. These architectures do not require manual feature extraction because they learn features directly from data [34]. A common artificial neural network used in deep learning for object and image recognition and classification is the convolutional neural network or CNN. By using a CNN, deep learning can recognize objects in photos [35]. After examining the previous studies in Table 1, we propose a set of four models for this study: the VGG, EfficientNetB0, DenseNet 121, and the ResNet152V2, as a result of its successful classification of coffee diseases. The final step is the performance evaluation of the four predictive models.

3.1 Datasets for the Study

A particular dataset that we have created for ourselves is a merging of the five datasets: BrACoL, JMuBEN, JMuBEN2, RoCoL, and LiCoLe. The folder is divided into three files, as shown in Fig. 6. For more information about our dataset, you can browse our dataset and take an overview [47].

Our division of the folders into training, testing, and validation folders is as follows:

- Training set: This set will be utilized during the model's training and fitting processes, during which the model will absorb the labeled data. This folder consists of three files: the first file, called Stage 1, contains two files. The first file is 0 Healthy and contains 4,000 images. The second file is 1 Unhealthy, which contains 4,000 images, for a total of 8000 images in folder Stage 1. The second file, named Stage 2, comprises three files: the first file is 0 Rust and contains 2666 images; the second file is 1 Brown Spots and contains 2667 images; and the third file is 2 Sooty Molds, which contains 2667 images. Thus, the total number of images in file Stage 2 is 8000 images. The third file is known as Stage 3 and contains 4 files: the first file is 0 Cercospora and contains 2000 images; the second file is 1 Phoma and contains 2000 images; the third file is 2 Miner and contains 2000 images; and the fourth file is 3 Red Spider Mite and contains 2000 images. Thus, the total number of images in Stage 3 is 8000.
- Testing or validation set: The test set will be used to test the model during training in order to improve its performance, while this set will be utilized for the training and fitting of the model. This folder consists of three files: the first file, called Stage 1, contains two files. The first file is 0 Healthy and contains 308 images. The second file is 1 Unhealthy and contains 625 images. Thus, the total number of

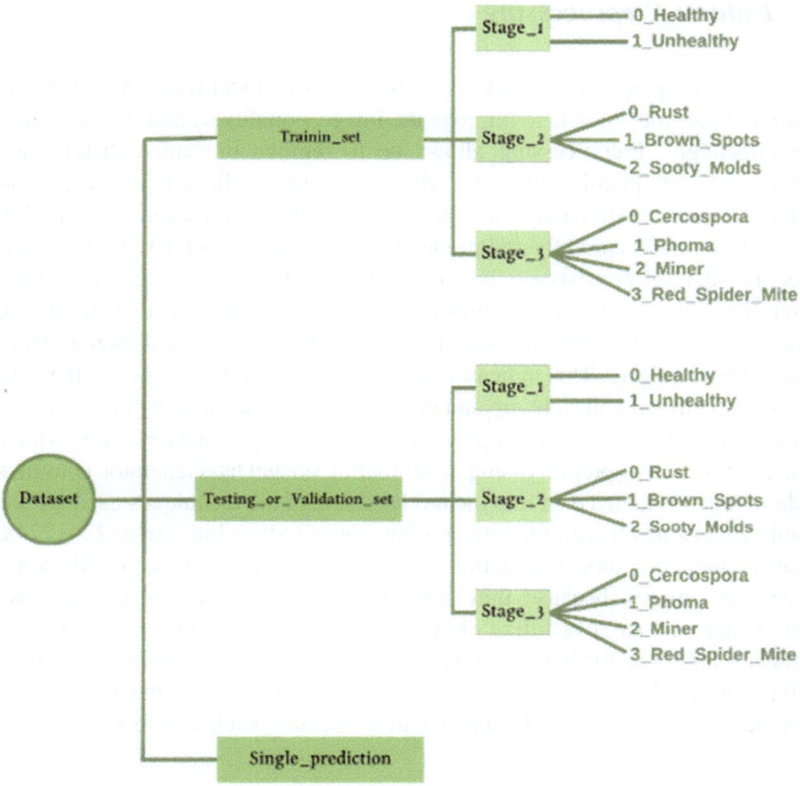

Fig. 6 The architecture of the dataset used for our proposed CLQC model

images in file Stage 1 is 933 images. The second file, named Stage 2, comprises three files: the first file is 0 Rust and contains 266 images; the second file is 1 Brown Spots and contains 177 images; and the third file is 2 Sooty Molds, which contains 183 images. Thus, the total number of images in file Stage 2 is 626 images.

The third file is known as Stage 3 and contains four files: the first file is 0 Cercospora and contains 18 images; the second file is 1 Phoma and contains 68 images; the third file is 2 Miner and contains 64 images; and the fourth file is 3 Red Spider Mite and contains 27 images. Thus, the total number of images in Stage 3 is 177.

- Single prediction: This set will be used to assess, validate, and test our trained model after it has been finished. It has four images in total; all of them are random and come from various stages.

3.2 Dataset Preprocessing

Images must be preprocessed before being used for model training and inference. The goal of preprocessing is to improve the image's quality so that we can analyze it more effectively. Preprocessing allows us to remove unwanted distortions and improve specific qualities that are critical for the application we are working on. These characteristics may vary depending on the application in order for the software to function correctly and produce the desired results [36]. As we discussed earlier, the dataset consists of five separate datasets that were grouped into one dataset. This process requires a lot of concentration to ensure that the dataset is balanced to prevent overfitting issues. We used the first manual augmentation to perform preprocessing. This indicates that we maintained the balance of the datasets by rotating them left and right, up and down, without requiring any code. After that, we used the code to perform preprocessing using ImageDataGenerator, which we imported from Keras preprocessing. The goal of ImageDataGenerator is to make it simple to import labeled data into the model; some of the techniques used are shown in Table 2. As a result, all of the images have been reduced in size to 224x224x3 as the target size. We took this action because there would be a size discrepancy between the images because they were obtained from several datasets, which might result in an overfitting issue. After that, we create an object of ImageDataGenerator for both training and testing data and pass the folder that has training data to the object train datagen and similarly passes the folder that has testing data to the object valid datagen and we set the batch size to 8.

3.3 The Classification of CLQC Framework

Before developing deep convolutional neural network models, there are several preparatory tasks that must be completed. The initial step is to import a dataset already clarified in the previous section, and then thereafter preprocess the data. After that, the model can be trained. This section will include illustrations of the project's conceptual framework and implementation process for VGG16, EfficientNetB0, DenseNet121, and ResNet152-V2 in each stage to provide a better understanding and performance.

Table 2 The augmentation settings for our dataset

Augmentation	Values
Rescale	1./255
Horizontal flip	True
Shear range	0.2
Zoom range	0.2
Rotation range	40
Width shift range	0.2
Height shift range	0.2

3.3.1 Stage 1 of the Proposed CLQC Framework

We will carry out the work using a stage-wise approach. Figure 7 depicts the block diagram of the first stage of the Coffee Leaf Quadruple Classifier (CLQC). That includes each of VGG16, EfficientNetB0, DenseNet121, and ResNet152-V2 trained and tested independently.

The first classification stage called the Healthy Classifier determines the status of the leaf by classifying it into two classes: Healthy or Unhealthy. The Unhealthy leaf samples will be passed into the second classification stage to start diagnosing the disease.

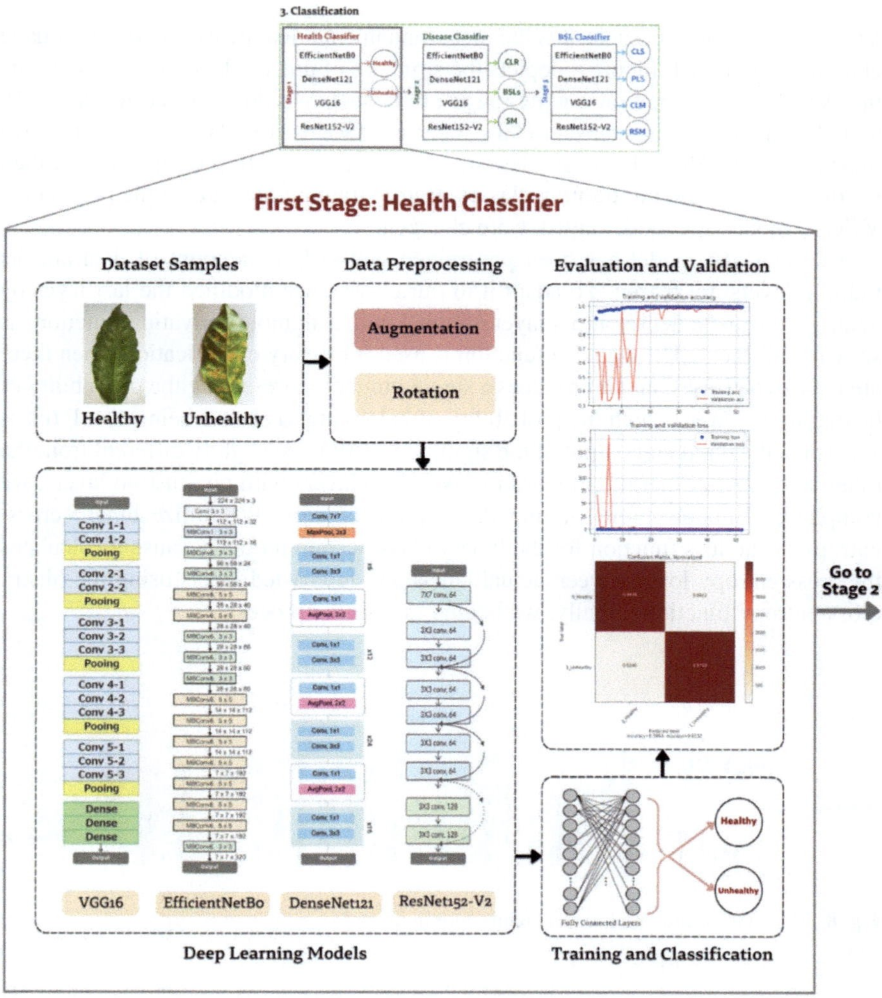

Fig. 7 Block diagram of the first classification stage of CLQC

Before developing deep convolutional neural network models, there are several preparatory tasks that must be completed. The initial step is to port a dataset already clarified in the previous section, and then thereafter should preprocess the data. After that, the model can be trained. This section will include illustrations of the project's conceptual framework and implementation process for VGG16, EfficientNetB0, DenseNet121, and ResNet152-V2 in each stage to provide a system that can diagnose and classify coffee leaf diseases and for a better understanding and performance of the project.

Building VGG16 Model in Stage 1

VGG16 was chosen because it is the most fundamental feature extraction and image classification model. Figure 8 depicts the structure of this model (and we also know that VGG-16 is trained on a wide range of classes). We downloaded the VGG-16 with the top layer parameter set to false, because if we use (top layer = true), then we must retrain the VGG-16 on all classes, whereas if we use (top layer = false), then we only need to add our training classes. Before training the data on the pre-trained VGG16 model, the model layers were changed.

The VGG-16 model has been pre-trained with a large amount of data and its training layers are frozen. To adapt it to our classes, we modified the last layer by adding a flat and a dense output layer of unit 1 with a sigmoid activation function, as shown in Table 3. The sigmoid function is used for binary classification when there are only two classes, and it produces a single number representing the probability of belonging to class 1 (with the probability of belonging to class 2 being 1 − P (class 1)). Thus, the input and output of the sigmoid function are slightly different from the other layers. After merging the original VGG-16 layers with our custom layers, we compile the model before starting the training process. We utilize binary cross-entropy as the loss function for the binary classification model because it calculates the cross-entropy loss between actual labels and forecasted labels using the binary cross-entropy function. Finally, we begin our training process.

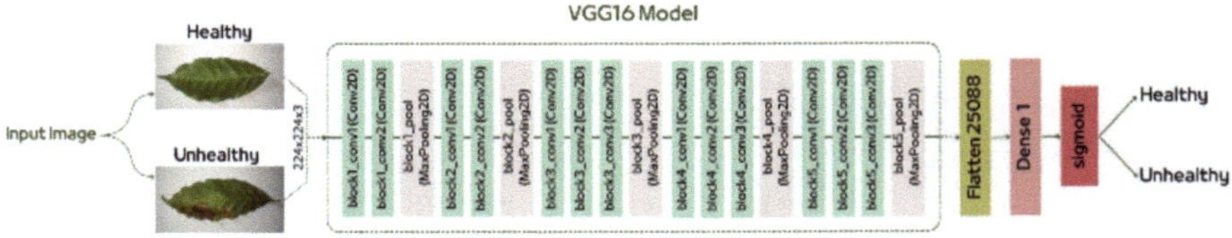

Fig. 8 The VGG16 proposed model architecture in Stage 1

Table 3 The summary of the VGG-16 model in Stage 1 after the modified layers

Layer(type)	Output Shape	Parameters
input_1 (InputLayer)	[(None, 224, 224, 3)]	0
block1_conv1 (Conv2D)	(None, 224, 224, 64)	1792
block1_conv2 (Conv2D)	(None, 224, 224, 64)	36928
block1_pool (MaxPooling2D)	(None, 112, 112, 64)	0
block5_conv1 (Conv2D)	(None, 14, 14, 512)	2359808
block5_conv2 (Conv2D)	(None, 14, 14, 512)	2359808
block5_conv3 (Conv2D)	(None, 14, 14, 512)	2359808
block5_pool (MaxPooling2D)	(None, 7, 7, 512)	0
flatten (Flatten)	(None, 25088)	0
dense (Dense)	(None, 1)	25089
Total params: 14,739,777		
Trainable params: 25,089		
Non-trainable params: 14,714,688		

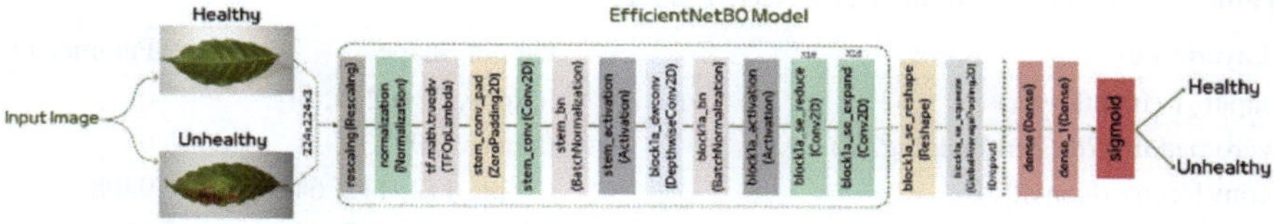

Fig. 9 The EfficientNetB0 proposed model architecture in Stage 1

Building EfficientNetB0 Model in Stage 1

With the objective of achieving accurate classifications and cost efficiency, the proposed model architecture for EfficientNetB0 can be seen in Fig. 9. Following the rescale layer, the subsequent layers include normalization, tfoplambda, zero padding in two dimensions, two dimensions of convolution, Batch Normalization, activation, and two dimensions of depthwise convolution layer. A sigmoid activation function is used for binary classification and Table 4 offers more information about the model's architecture, which illustrates that the total number of parameters of the EfficientNetB0 is 4,213,668 and these parameters are split into two categories. The first one is trainable parameters, which have 4,171,645 parameters and the second one is non-trainable parameters, which have 42,023 parameters.

Building DenseNet121 Model in Stage 1

Each layer in a DenseNet architecture is connected to every other layer, hence the name "Densely Connected Convolutional Network." The feature maps of all preceding layers are used as inputs for each layer, and its own feature maps are used as inputs for each subsequent layer [37]. Two dimensions of global average pooling

Table 4 The summary of the EfficientNetB0 model in Stage 1

Layer(type)	Output Shape	Parameters
input_1 (InputLayer)	[(None, 224, 224, 3)]	0
rescaling (Rescaling)	(None, 224, 224, 3)	0
normalization (Normalization)	(None, 224, 224, 3)	7
tf.math.truediv (TFOpLambda)	(None, 224, 224, 3)	0
stem_conv_pad (ZeroPadding2D)	(None, 225, 225, 3)	0
top_bn (BatchNormalization)	(None, 7, 7, 1280)	5120
top_activation (Activation)	(None, 7, 7, 2048)	0
global_average_pooling2d (GlobalAveragePooling2D)	(None, 1280)	0
dense (Dense)	(None, 128)	163968
dense (Dense)	(None, 1)	129
Total params: 4,213,668 Trainable params: 4,171,645 Non-trainable params: 42,023		

Table 5 The summary of the DenseNet121 model in Stage 1

Layer(type)	Output Shape	Parameters
input_1 (InputLayer)	[(None, 224, 224, 3)]	0
zero padding2d (ZeroPadding2D)	(None, 230, 230, 3)	0
conv1/conv (Conv2D)	(None, 112, 112, 64)	9408
conv1/bn (BatchNormalization)	(None, 112, 112, 64)	256
conv1/relu (Activation)	(None, 112, 112, 64)	0
dense (Dense)	(None, 1024)	1049600
dense_1 (Dense)	(None, 512)	524800
batch_normalization_1 (BatchNormalization)	(None, 512)	2048
dropout_1 (Dropout)	(None, 512)	0
dense 2 (Dense)	(None, 1)	513
Total params: 8,614,465		
Trainable params: 1,744,065		
Non-trainable params: 6,870,400		

with batch normalization and a dense layer, a dropout layer with a sigmoid activation function, are used for binary classification. We have used the output of the first three layers to the last layers of the model for activation, and we have frozen the weights of all the layers except for the last nine, which will be trained for classification. The proposed architecture of the DenseNet121 model can be seen in Table 5. Detailed information can be found in Fig. 10, and the total number of parameters for DenseNet121 in Stage 1 is 8,614,465; these parameters are divided into two categories: trainable parameters, which have a value of 1,744,065; and non-trainable parameters, which have a value of 6,870,400.

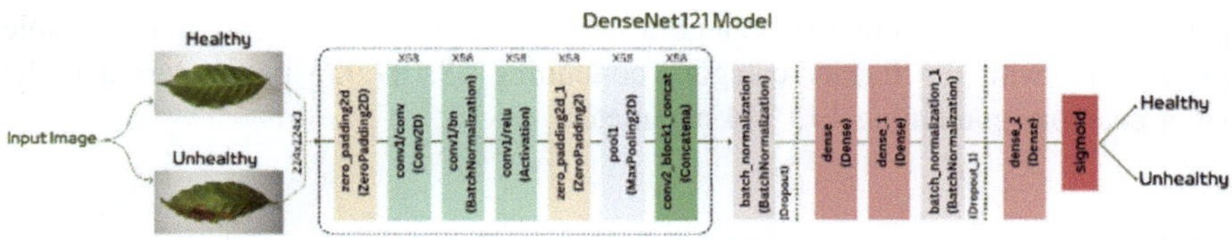

Fig. 10 The DenseNet121 proposed model architecture in Stage 1

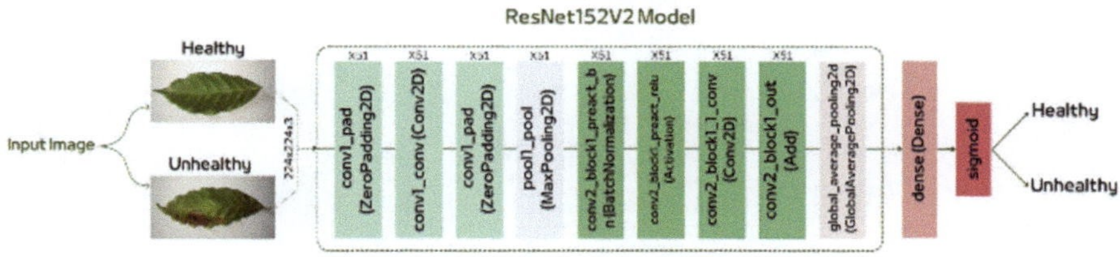

Fig. 11 The ResNet152V2 proposed model architecture in Stage 1

Table 6 The summary of the ResNet152-V2 model in Stage 1

Layer(type)	Output Shape	Parameters
input_1 (InputLayer)	[(None, 224, 224, 3)]	0
conv1_pad (ZeroPadding2D)	(None, 230, 230, 3)	0
conv1_conv (Conv2D)	(None, 112, 112, 64)	9472
pool1_pad (ZeroPadding2D)	(None, 114, 114, 64)	0
pool1_pool (MaxPooling2D)	(None, 56, 56, 64)	0
conv5_block3_out (Add)	(None, 7, 7, 2048)	0
post_bn (BatchNormalization)	(None, 7, 7, 2048)	8192
post_relu (Activation)	(None, 7, 7, 2048)	0
global_average_pooling2d (GlobalAverageagePooling2D)	(None, 2048)	0
dense (Dense)	(None, 1)	2049
Total params: 58,333,697 Trainable params: 58,189,953 Non-trainable params: 143,744		

Building ResNet152-V2 Model in Stage 1

A CNN model called ResNet has thousands or hundreds of strong performance layers. The ResNet152V2 model, which is pre-trained and pre-equipped with initial weights, can outperform a normal CNN in terms of accuracy and speed.

Global average pooling in two dimensions is employed, followed by batch normalization and a dense layer with a sigmoid activation function for binary classification. The proposed architecture of the ResNet152V2 model can be seen in Fig. 11 and Table 6 contains more information regarding the summary of the model. The total number of parameters for the ResNet152V2 is 58,333,697 and these parameters are split into two categories: trainable parameters and non-trainable parameters, with the former category containing 58,189,953 parameters and the latter category containing 143,744 parameters.

3.3.2 Stage 2 of the Proposed CLQC Framework

Figure 12 depicts the block diagram of the second stage of the Coffee Leaf Quadruple Classifier (CLQC). In this classification stage, which is called Disease Classifier,

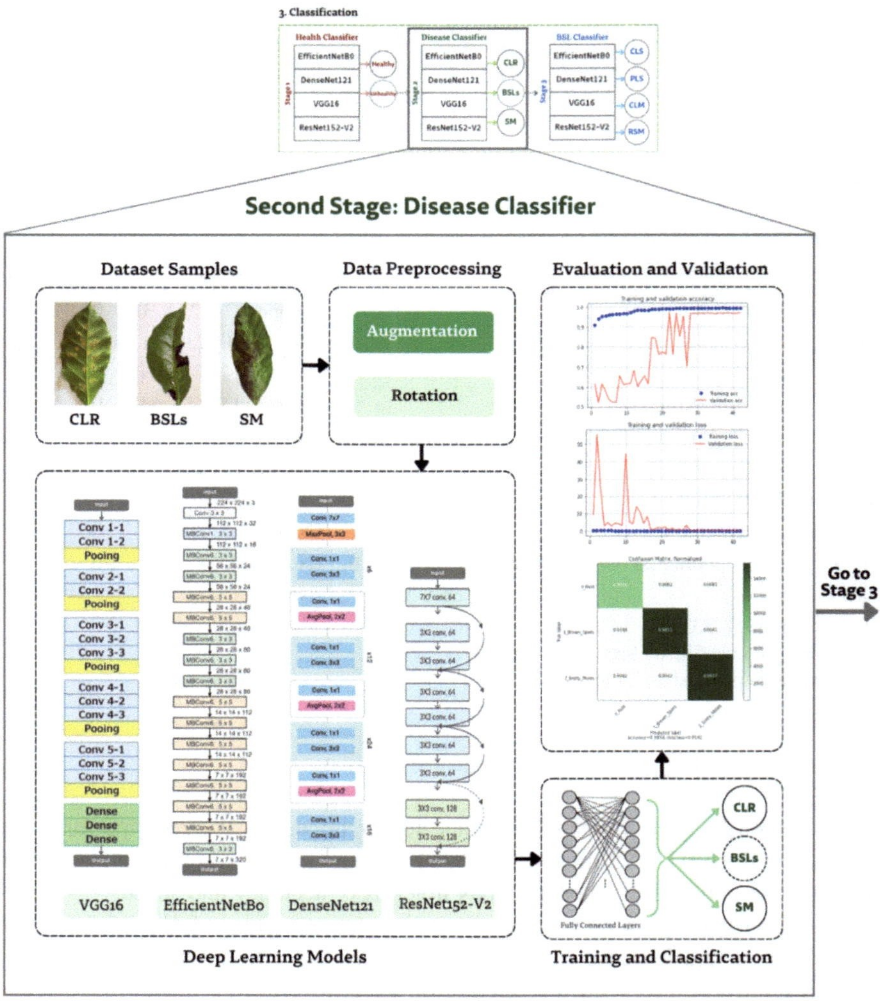

Fig. 12 Block diagram of the second classification stage of CLQC

the unhealthy images will be diagnosed as Coffee Leaf Rust (CLR), Sooty Molds (SM), or Brown Spot

Lesions (BSLs). Where Brown Spot Lesions (BSLs) include multiple diseases that have visual similarities aggregated in third different classification stages and have their own classifier.

For Stage 2, we followed the same procedures as in Stage 1, but with a few modifications. The class mode was changed from binary to multiclass, and the output layer was changed from a sigmoid activation function to a SoftMax activation function, with the number of units increased from 1 to 3. This was done as SoftMax is applicable to multiclass situations and is an extension of the sigmoid function. Additionally, the models' compilation was modified before training, with the loss value changed from two classes to multiple classes. In Stage 2, we trained all four DCNN models but in this section, we will talk about the best model EfficientNetB0 and the lowest model VGG16 in accuracy.

Building VGG16 Model in Stage 2

As previously mentioned, we will be following the procedures from Stage 1, with a few modifications. As seen in Table 7, the unit has changed from 1 to 3 in Stage 2, which is a result of the alterations made to the dense layer. This difference between Stage 1 and Stage 2 of the model has been clearly outlined as simply the number of classes. We will be adhering to the procedures from Stage 1, taking into consideration the changes that were explained in the previous paragraph.

Table 7 The summary of the VGG-16 model in Stage 2 after the modified layers

Layer(type)	Output Shape	Parameters
input_1 (InputLayer)	[(None, 224, 224, 3)]	0
block1_conv1 (Conv2D)	(None, 224, 224, 64)	1792
block1_conv2 (Conv2D)	(None, 224, 224, 64)	36928
block1_pool (MaxPooling2D)	(None, 112, 112, 64)	0
block5_conv1 (Conv2D)	(None, 14, 14, 512)	2359808
block5_conv2 (Conv2D)	(None, 14, 14, 512)	2359808
block5_conv3 (Conv2D)	(None, 14, 14, 512)	2359808
block5_pool (MaxPooling2D)	(None, 7, 7, 512)	0
flatten (Flatten)	(None, 25088)	0
dense (Dense)	(None, 3)	75267
Total params: 14,789,955		
Trainable params: 75,267		
Non-trainable params: 14,714,688		

Table 8 The summary of the EfficientNetB0 model in Stage 2

Layer(type)	Output Shape	Parameters
input_1 (InputLayer)	[(None, 224, 224, 3)]	0
rescaling (Rescaling)	(None, 224, 224, 3)	0
normalization (Normalization)	(None, 224, 224, 3)	7
tf.math.truediv (TFOpLambda)	(None, 224, 224, 3)	0
stem_conv_pad(ZeroPadding2D)	(None, 225, 225, 3)	0
top-bn(BatchNormalization)	(None, 7, 7, 1280)	5120
top-activation (Activation) global_average_pooling2d (GlobalAveragePooling2D)	(None, 7, 7, 1280) (None, 1280)	0 0
dense (Dense)	(None, 128)	163968
dense (Dense)	(None, 3)	387
Total params: 4,213,926 Trainable params: 4,171,903 Non-trainable params: 42,023		

Building EfficientNetB0 Model in Stage 2

As demonstrated in Table 8, compared to Stage 1, we have experienced a few changes. An activation function has been modified from sigmoid to softmax to indicate that we begin dealing with multi-class classification.

3.3.3 Stage 3 of the Proposed CLQC Framework

Figure 13 depicts the block diagram of the third stage of the Coffee Leaf Quadruple Classifier (CLQC). Finally, in the third classification stage or in the Brown Spot Lesions Classifier, the samples diagnosed with BSLs in Stage 2 will be diagnosed now with their specific disease, which will be one of the Cercospora Leaf Spot (CLS), Phoma Leaf Spot (PLS), Coffee Leaf Miner (CLM), or Red Spider Mites (RSM).

In Stage 3, we repeated the same steps as we did in Stage 1 and made the same changes that we made in Stage 2. The only difference between Stages 2 and 3 is the number of classes. In Stage 2, we modified the output layer, which is dense, by changing the unit from 3 to 4. In this stage, we trained all four DCNN models but in this section, we will talk about the best model EfficientNetB0 and the lowest model VGG16 in accuracy.

Building VGG16 Model in Stage 3

Stage 3 of the model VGG16 layers has been modified from the original layers. As seen in Table 9, the layers in Stage 3 are different from the layers in Stages 1 and 2. In summary, VGG16 is a convolutional neural network model composed of 16 layers. In Stage 3, the layers of the VGG16 model can be modified to suit the

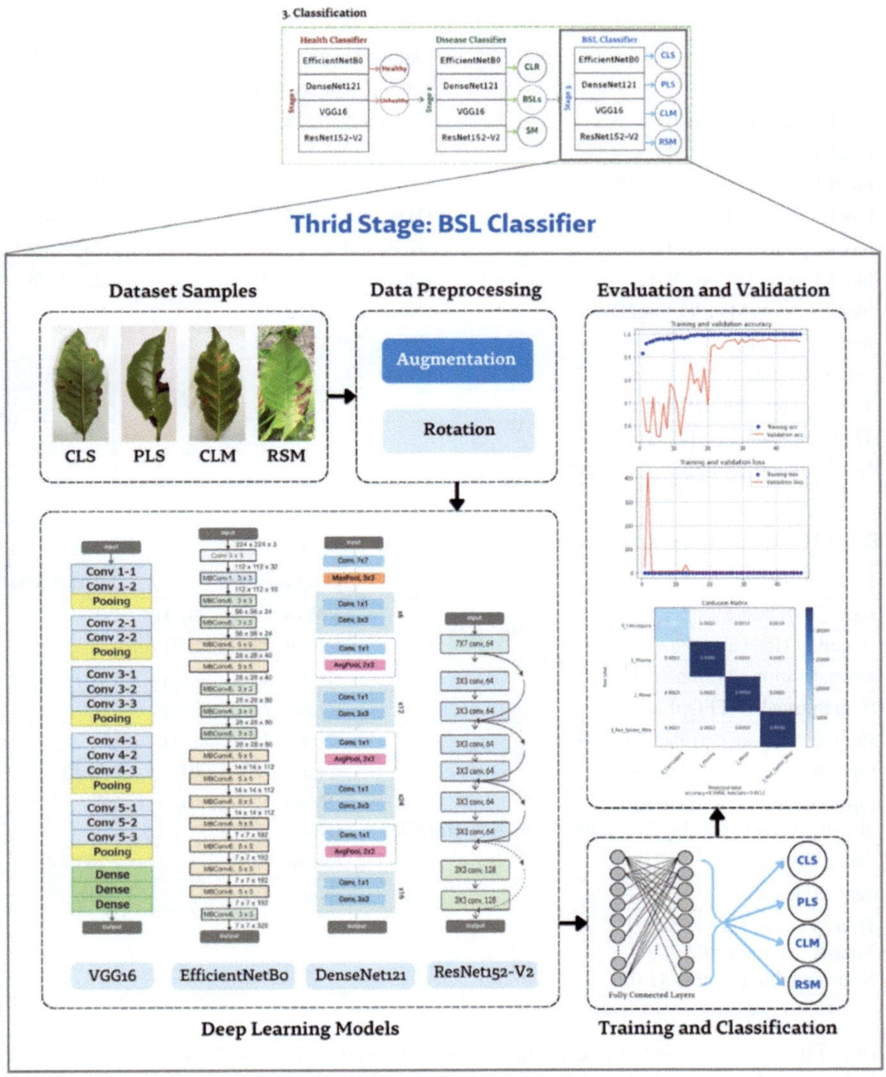

Fig. 13 Block diagram of the third classification stage of CLQC

task at hand. The number of classes in each stage is dependent on how many classes we have in each stage. Table 9 provides an overview of the model VGG16 layers following layer modification in Stage 3. The number of classes in each stage is dependent on the task at hand. For example, if the task is to classify images into two classes, then the number of classes in Stage 3 would be two. However, if the task is to classify images into four classes, then the number of classes in Stage 3 would be

Table 9 The summary of the VGG-16 model in Stage 3 after the modified layers

Layer(type)	Output Shape	Parameters
input_1 (InputLayer)	[(None, 224, 224, 3)]	0
block1_conv1 (Conv2D)	(None, 224, 224, 64)	1792
block1_conv2 (Conv2D)	(None, 224, 224, 64)	36928
block1_pool (MaxPooling2D)	(None, 112, 112, 64)	0
block5_conv1 (Conv2D)	(None, 14, 14, 512)	2359808
block5_conv2 (Conv2D)	(None, 14, 14, 512)	2359808
block5_conv3 (Conv2D)	(None, 14, 14, 512)	2359808
block5_pool (MaxPooling2D)	(None, 7, 7, 512)	0
flatten (Flatten)	(None, 25088)	0
dense (Dense)	(None, 4)	100356
Total params: 14,815,044		
Trainable params: 100,356		
Non-trainable params: 14,714,688		

Table 10 The summary of the EfficientNetB0 model in Stage 3

Layer(type)	Output Shape	Parameters
input_1 (InputLayer)	[(None, 224, 224, 3)]	0
rescaling (Rescaling)	(None, 224, 224, 3)	0
normalization (Normalization)	(None, 224, 224, 3)	7
tf.math.truediv (TFOpLambda)	(None, 224, 224, 3)	0
stem_conv_pad(ZeroPadding2D)	(None, 225, 225, 3)	0
top_bn(BatchNormalization)	(None, 7, 7, 1280)	5120
top_activation (Activation)	(None, 7, 7, 1280)	0
global_average_pooling2d (GlobalAveragePooling2D)	(None, 1280)	0
dense (Dense)	(None, 128)	163968
dense (Dense)	(None, 4)	516
Total params: 4,214,055 Trainable params: 4,172,032 Non-trainable params: 42,023		

four. Thus, the number of classes in each stage is dependent on how many classes we have in each stage.

Building EfficientNetB0 Model in Stage 3

In the last stage, as previously discussed, a dense layer with a softmax activation function has been modified from 3 to 4 to indicate that we have four classes, which are Cercospora Leaf Spot (CLS), Phoma Leaf Spot (PLS), Coffee Leaf Miner (CLM), and Red Spider Mites (RSM). More information can be seen in Table 10.

3.4 Implementation Experiments

We used the Colab platform to train our four models, which are a product of Google Research and well suited to machine learning, data analysis, and education. Colab allows anybody to write and run Python code through the browser. It is a free service, but in this project, we dealt with a significant amount of datasets, which required us to subscribe to the paid version because of the limitations of the free version. We observed after subscribing a rise in training speed and ability to work for more than 12 hours, unlike the free version [48]. To deal with the photos, this study used a variety of open-source Python modules and frameworks [49].

We take benefit of two Keras callbacks:

- Reduce Learning Rate

The learning rate is an important hyper-parameter that determines how much to refresh the model weights with concern to the loss gradient. While a small learning rate may result in stuck training, a large learning rate drifts away from optimal training [38, 39]. Keras has the Reduce Learning Rate callback that observes the model performance and reduces the learning rate when the improvement stops at a specific number of epochs [40, 39].

- Early Stopping

In our first training attempts, we tried different epoch numbers for each of the VGG16, Efficient-NetB0, DenseNet121, and ResNet152V2 models.

Once we raised the number of epochs, the accuracy increased significantly. Later we realized that we had fallen into the Overfitting problem. This means the model is over-trained until it learns the noise or irrelevant patterns in the dataset, so it will fail against unseen data [41]. On the other hand, if the model is not trained enough, the Underfitting problem will occur. To exactly "fit" the appropriate epoch number for the model, we used Keras's callback Early Stopping. It monitors the model's validation set and stops the training before overfitting appears when poor performance or stuck enhancement is seen, as a large number of epochs is specified [42, 43].

3.5 Performance Metrics

In the evaluation phase, we will evaluate the model by using several evaluation measures to understand the performance of our model and to determine how the model is accurate [44]. The evaluation metrics are essential to determining the best classifier. Thus, a selection of suitable evaluation metrics is a critical step in making a distinction and getting the best classifier. Evaluation measures are derived from a confusion matrix, which contains the number of cases classified as positive and negative for each class displayed in the confusion matrix [45, 46].

- True Positive (TP): When the coffee leaf is healthy and is classified from the model as healthy.
- True Negative (TN): When a coffee leaf is unhealthy and the model predicts that it is unhealthy.
- False Positive (FP): The model predicts a coffee leaf is healthy, but it is unhealthy.
- False Negative (FN): The model predicts a coffee leaf is unhealthy, but it is healthy.

The confusion matrix can be used to create a number of metrics to evaluate the performance of the classifier.

4 Results

In this section, we train all previous models at each stage to detect and classify coffee leaf diseases. After you've finished training, it's now time to show and discuss the results of all the models at each stage. By comparing the performance and accuracy of the models in each stage, we can find out which one is better in terms of performance and accuracy in detecting and classifying coffee leaf diseases. With this information, we can make better decisions on which model to use for detecting and classifying coffee leaf diseases.

4.1 Results of CLQC Stage 1

We set up a giant experiment in order to get high outcomes in the classification of coffee leaf diseases. In our experiment, we tested four different models as shown in Table 11, each of which is covered in Sect. 3 in great detail. For the sake of displaying the findings, we created tables for each stage of the experiment. Every table had columns that showed the classifier models' performance in terms of loss, accuracy, validation precision, recall, specificity, and f1-score; we went into great detail about these evaluations in the preceding Sect. 3. Additionally, we added a column for the active function to distinguish between each stage in the type of function and the number of units, and we added a column for the epoch number, which shows the best result for the rest of the trained epoch numbers. This number is determined by looking at the performance of the model over all the epochs that have been trained and selecting the epoch that has the highest accuracy or lowest loss. This number can then be used to identify the best model, which can then be used to make predictions on new data. The gray row was used to select the highest value in each column. This table allowed us to compare the different parameter methods and determine which was the most effective.

Table 11 Performance measures of CLQC Stage 1 using different evaluation metrics

CLQC Stage 1								
			Performance Measures/Metrics					
Models	Activation Function	Epochs	Accuracy(%)	Loss(%)	Precision(%)	Recall(%)	Specificity(%)	F1-score(%)
VGG16	1/Sigmoid	46/300	91.36	21.73	91.62	91.05	91.67	91.33
EfficientNetB0	1/Sigmoid	51/300	99.91	0.35	99.90	99.92	99.9	99.90
DenseNet121	1/Sigmoid	29/300	97.8	8.04	98.04	97.55	98.05	97.79
ResNet152-V2	1/sigmoid	50/300	98.85	3.77	99.12	98.58	99.12	98.98

4.1.1 Results of VGG16 Model for Stage 1

The VGG-16 model is a convolutional neural network model, as explained in detail in Sect. 3. It was trained in the previous Sect. 4, on a dataset of over 20 thousand images that was divided into three stages, each of which contained 8000 images, and achieved an impressive top-five accuracy of 91.5%. Figures 14 and 15 visualize the results of the training model using two methods. One is to plot metrics such that, when comparing the training phase to the validation phase, the bar chart shows the loss, AUC, precision, recall, and accuracy of the model using an early-stop function. The plots demonstrate that the model stops learning after 53 iterations, indicating that there is no further progress after iteration 300. The previous table shows the epochs of a given process, and the specific column of the epochs indicates the value of the best epoch when the best value appears. This number can be used to identify the best epoch and determine the value associated with it. And two is the confusion matrix, which is a 2-dimensional array comparing predicted category labels to the true label. For binary classification, these are the True Positive, True Negative, False Positive, and False Negative categories (which we can use to calculate the F1-score, recall, precision, and specificity). It also displays the various types of prediction

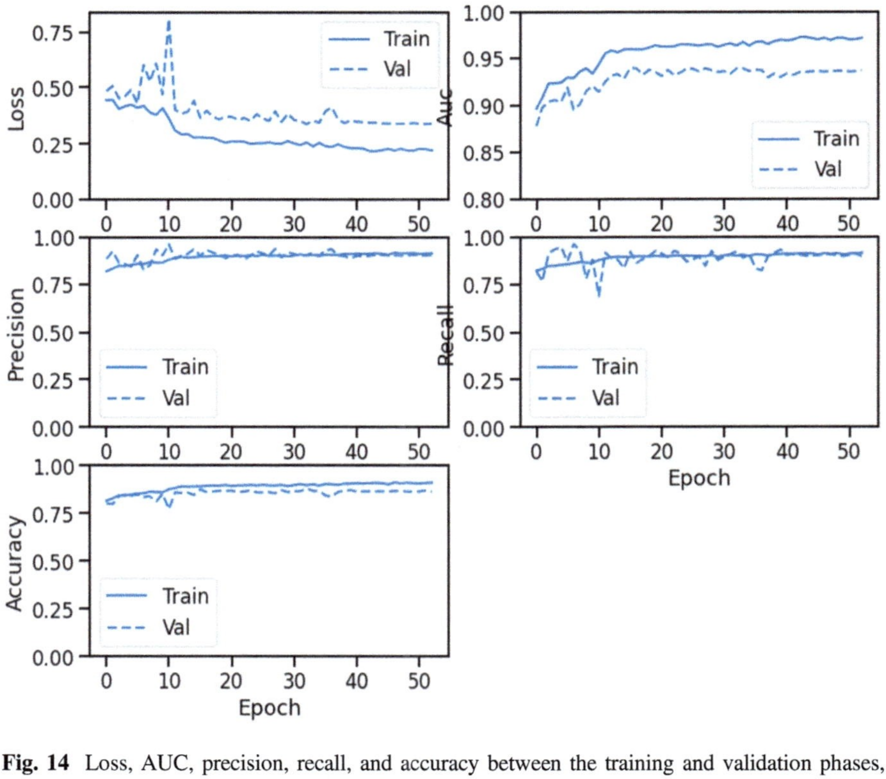

Fig. 14 Loss, AUC, precision, recall, and accuracy between the training and validation phases, with a number of epochs for the VGG16 Stage 1 model

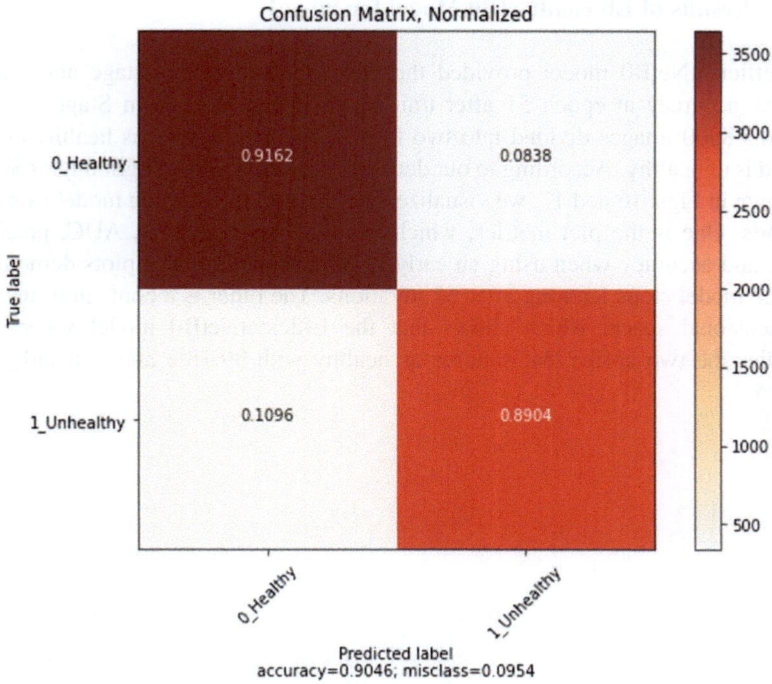

Fig. 15 Confusion matrix of the proposed VGG16 model in Stage 1

errors produced by the model, which can be used to identify any model flaws. Additionally, it serves as a valuable tool for improving the model's performance and comprehension of its behavior. In summary, this is like a table that shows the types of correct predictions and incorrect predictions. This table demonstrated the accuracy of both healthy and unhealthy coffee leaves, the accuracy of healthy coffee leaves was 91.62%, while the accuracy of unhealthy coffee leaves was 89.04%. This showed that the accuracy of healthy leaves was higher than that of unhealthy leaves. This model is widely used in computer vision tasks such as image classification (we used it in the classification of coffee leaf diseases), object detection, and semantic segmentation. We may conclude that VGG 16 is simple to comprehend and use. Only the convolution and pooling layers are present. The VGG 16 architecture resembles the Stack architecture in appearance. This model's primary objective is to build a deep neural network by stacking the convolutional and pooling layers. By adjusting the last softmax dense layer unit to any number we choose depending on the classes we need to categorize, we can adapt this model to work for any number of classes.

4.1.2 Results of EfficientNetB0 Model for Stage1

The EfficientNetB0 model provided the best accuracy at this stage and reached 99.91% accuracy at epoch 51 after training it on a dataset from Stage 1, which contains 8000 images divided into two files, where the first file is healthy and the second is unhealthy. According to our detailed discussion at the beginning of Sect. 4, as shown in Figs. 16 and 17, we visualize the results of the training model using two methods. One is the plot metrics, which show the model's loss, AUC, precision, recall, and accuracy when using an early stopping function. The plots demonstrate that the model stops learning after 52 iterations. The other is a confusion matrix in 2-dimensional space, which shows that the EfficientNetB0 model successfully classifies the two coffee leaf statuses as healthy with 99.78% and unhealthy with 97.60%.

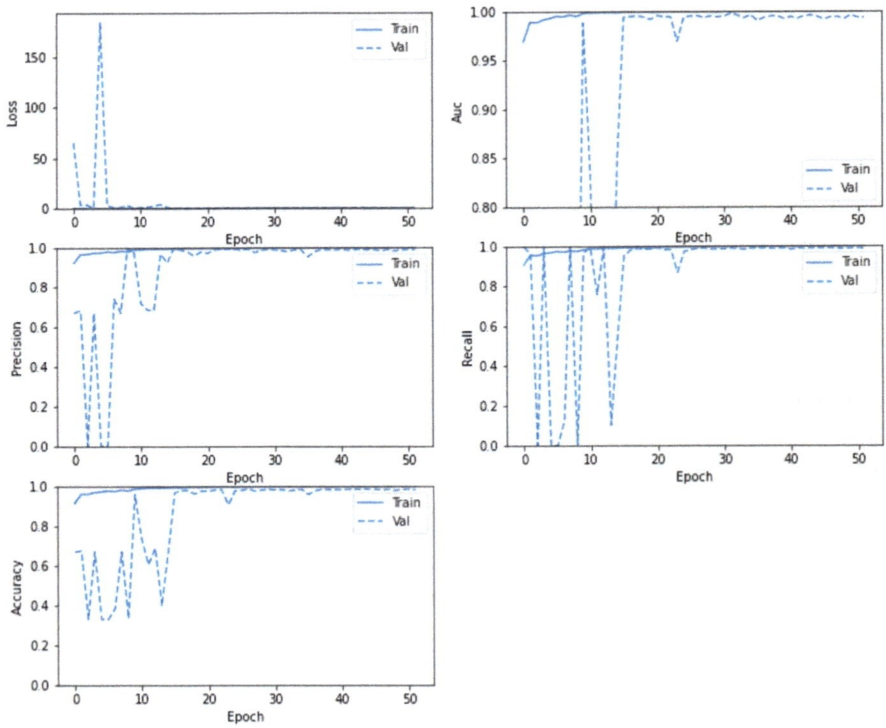

Fig. 16 Loss, AUC, precision, recall, and accuracy between the training and validation phases, with a number of epochs for the EfficientNetB0 model in Stage 1

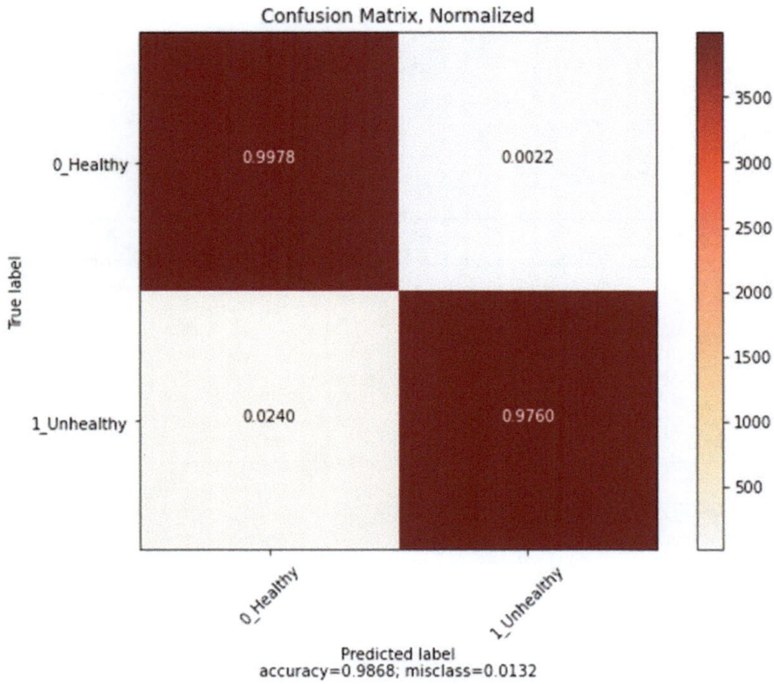

Fig. 17 Confusion matrix of the proposed EfficientNetB0 model in Stage 1

4.1.3 Results of DenseNet121 Model for Stage 1

After training the model, as we discussed in detail in Sect. 4, we present the results of the DenseNet121 model after training in the first stage, where it was trained on 8,000 images, which were divided into healthy and unhealthy. The model achieves an impressive top accuracy of 97.8% in Stage 1. We visualize the results of the training model using two methods: one is to plot metrics, and the other is the confusion matrix. The plot metrics used when comparing the training phase to the validation phase are shown in Fig. 18, which shows the loss, AUC, precision, recall, and accuracy of the model using an early-stop function, as we mentioned earlier. The plots demonstrate that the model stops learning after 52 iterations, indicating that there is no further progress after iteration 300, as shown in Fig. 18. The correctness of the model may be evaluated using the confusion matrix, which shows the number of correct and incorrect predictions made by the model for each class, as shown in Fig. 19. It is explained here that the accuracy of the healthy coffee leaves reached 94.04% and the unhealthy ones reached 97.56%.

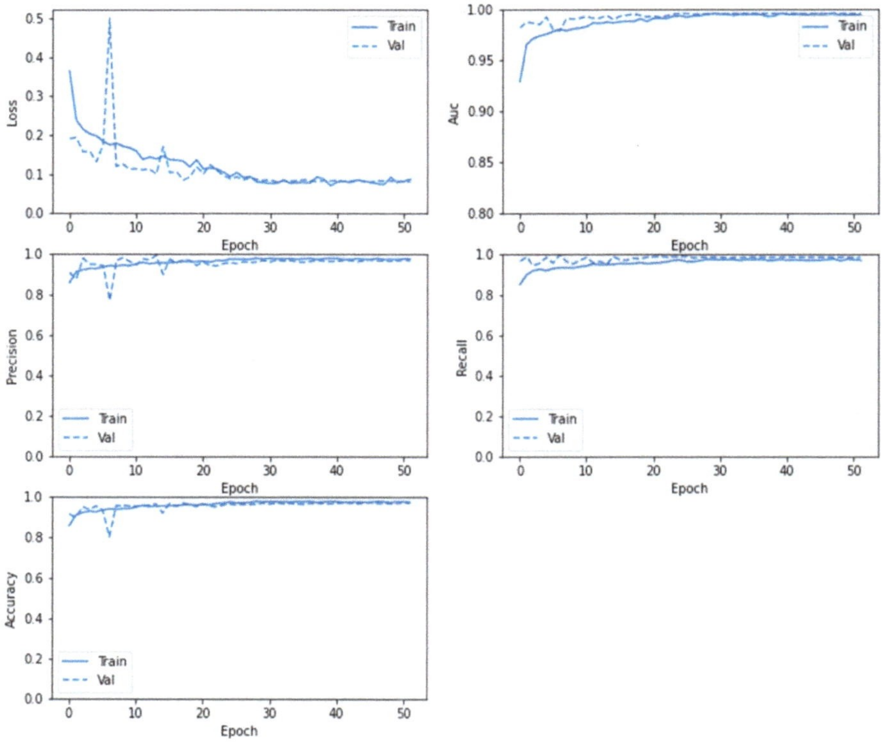

Fig. 18 Loss, AUC, precision, recall, and accuracy between the training and validation phases, with a number of epochs for the DenseNet121 model in Stage 1

4.1.4 Results of ResNet152-V2 Model for Stage 1

The ResNet152-V2 model has been trained on two datasets, healthy and unhealthy leaves, where each group contains 4000 images. In this first stage, the model achieved the best accuracy of 98.85% at epoch 50. In order to visualize the data, we used two methods: a plot matrix and a confusion matrix, as shown in Figs. 21 and 20. In 20, the plot matrix used when comparing the training phase to the validation phase showed the model's loss, AUC, accuracy, recall, and accuracy using the early stopping function. Also, the picture shows that it stopped learning after 56 epochs. Because the colors are bold when they meet in the confusion matrix, the model was able to distinguish them with 99.32% accuracy in detecting the healthy leaf and 98.39% accuracy in detecting the unhealthy leaf, as shown in Fig. 21.

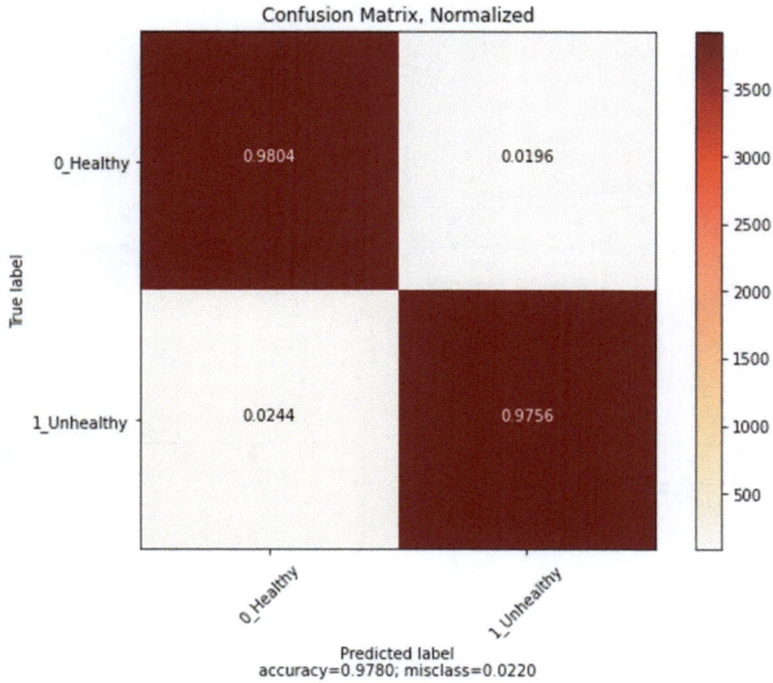

Fig. 19 Confusion matrix of the proposed DenseNet121 model in Stage 1

4.1.5 Comparison of Performance Between Models in the First Stage of CLQC

Previously, we reviewed the results of four models in their first stage. To further illustrate our findings, we have created a bar chart, as shown in Fig. 22, that compares the four models in terms of accuracy, loss, F1-score, recall, precision, and specificity. This evaluation was discussed in Sect. 3, and this provides further insight into the performance of each model and can be a useful way to determine which model is best suited for this project. Comparing results between four models can be a challenging task. Each model has its own strengths and weaknesses, and depending on the application, one model may be more suitable than the others. By comparing the results of each model, we can be a useful way to determine which model is best suited for this project. When comparing the results of these four models in Stage 1, we see that EfficientNetB0 and then ResNet152V2 have the highest accuracy. DenseNet121 and ResNet152V2 have similar accuracy, but they are slightly lower than EfficientNetB0, while VGG16 shows less accuracy than the previous models. Overall, the results show that Efficient-NetB0 and then ResNet152V2 are the best-performing models in the first phase. By analyzing the performance of each model on the same dataset, it can be determined which model is the most accurate, efficient, and reliable. Additionally, the comparison of results

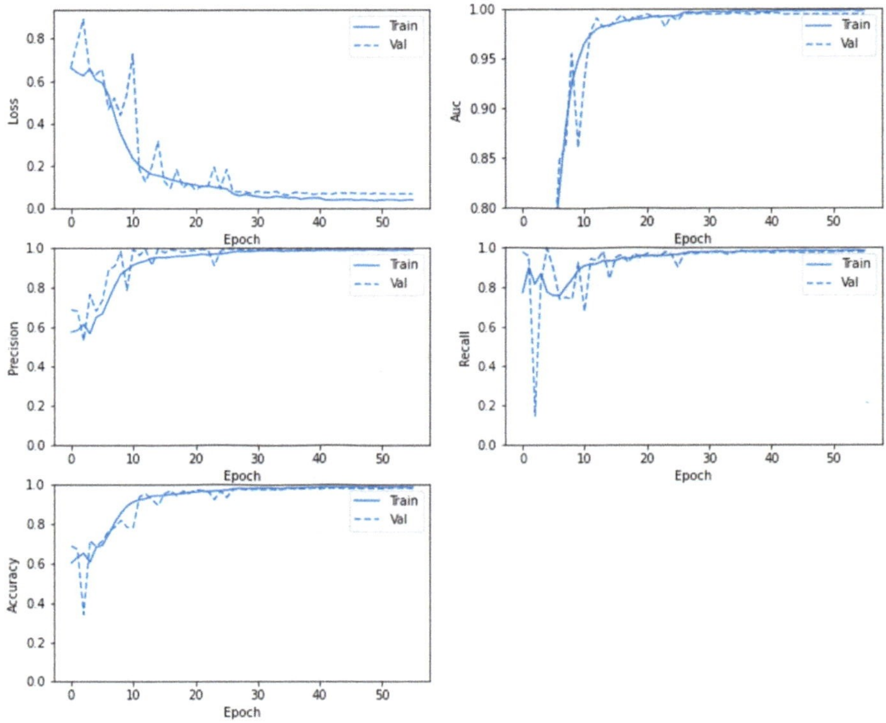

Fig. 20 Loss, AUC, precision, recall, and accuracy between the training and validation phases, with a number of epochs for the ResNet152-V2 model in Stage 1

between the four models can provide insight into which model is most effective in certain situations, allowing for more informed decisions when selecting a model for a given task.

4.2 Results of CLQC Stage 2

The results of Stage 2 using deep learning models were impressive, as the following Table 12 shows; the previous section provides detailed information related to the table. It explains the various elements of the table, such as the columns, rows, and data points. The models were able to accurately identify objects in images with a high degree of accuracy. The team was able to observe the effects of differences between Stages 1 and 2, where multiple categories were added instead of being a binary classification only, so note that the accuracy has increased in some models for the detection of coffee leaf diseases. This indicates that the greater the number of

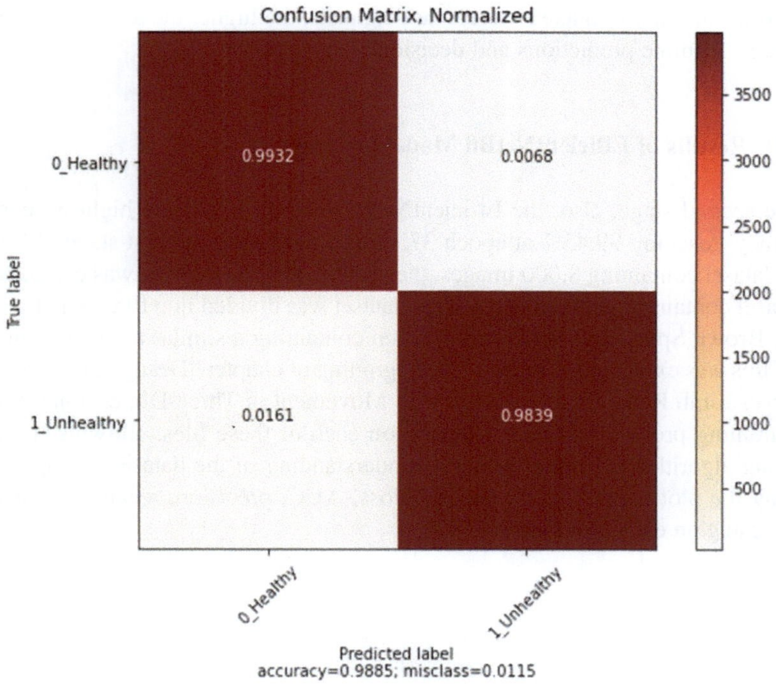

Fig. 21 Confusion matrix of the proposed ResNet152-V2 model in Stage 1

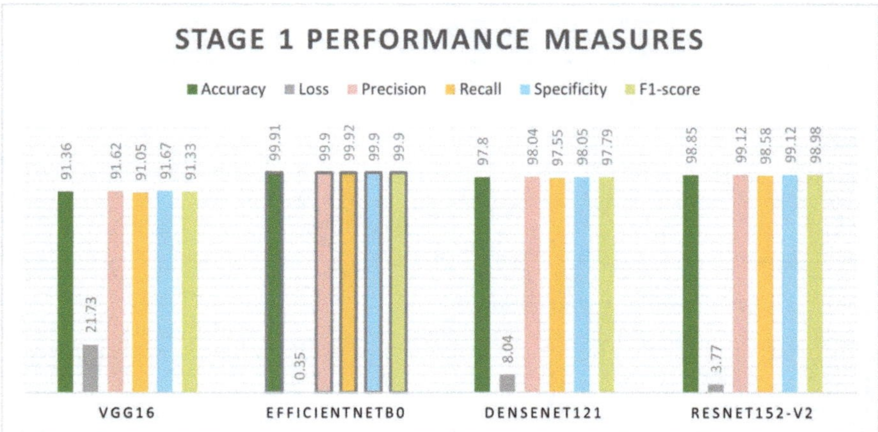

Fig. 22 Comparison of VGG16, EfficientNetB0, DenseNet121, and ResNet152V2 performance measures in Stage 1, where EfficientNetB0 achieved the best results

categories in order to make an accurate diagnosis, the higher the accuracy, allowing for more accurate predictions and decisions.

4.2.1 Results of EfficientNetB0 Model for Stage 2

In the second stage, also, the EfficientNetB0 model offered the highest degree of accuracy, reaching 99.45% at epoch 37. After completing the first stage of training on a dataset containing 8,000 images, the second stage of training was conducted on a dataset containing 8,000 images. This dataset was divided into three separate files: Rust, Brown Spots, and Sooty molds, each containing a similar number of images; all of this was explained in detail at the beginning of chapter "Design and Analysis of a Lower Limb Rehabilitation Robot with Movement in Three-Dimensional Space". The training process was then repeated on each of these files, allowing the deep-learning algorithm to further refine its understanding of the dataset. In Fig. 23, we display the plot metrics of the model's loss, AUC, precision, recall, and accuracy when using an early stopping function.

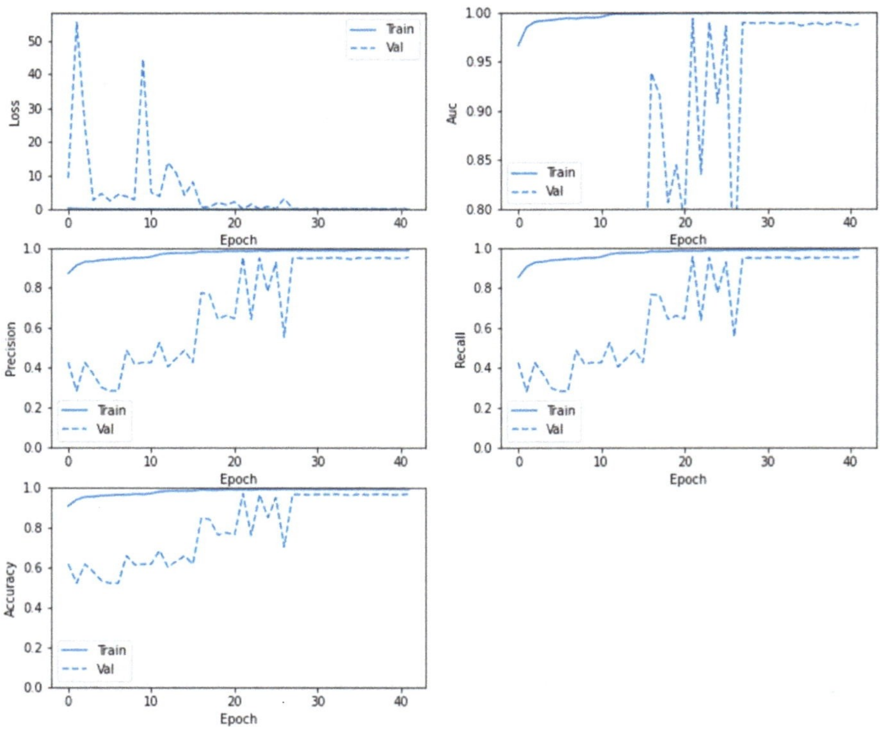

Fig. 23 Loss, AUC, precision, recall, and accuracy between the training and validation phases, with a number of epochs for the EfficientNetB0 model in Stage 1

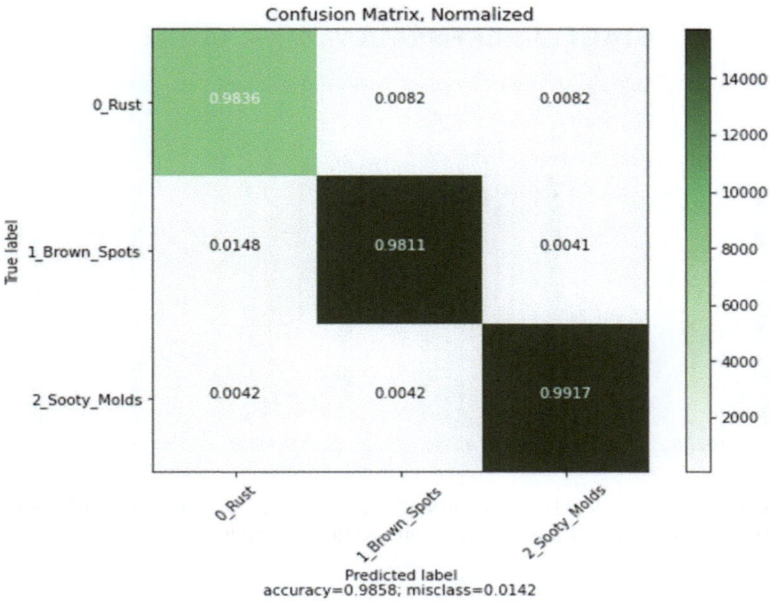

Fig. 24 Confusion matrix of the proposed EfficientNetB0 model in Stage 2

The plots demonstrate that the model stops learning after 42 iterations. The confusion matrix for multiple classes in Fig. 24 shows that the EfficientNetB0 model successfully classifies the three classes of coffee leaf diseases as Rust with 98.36%, Brown Spots with 98.11%, and Sooty Molds with 99.17%, respectively. In Stage 2, we discovered that the EfficientNetB0 model outperformed the three other proposed models and had the highest accuracy.

4.2.2 Comparison of Performance Between Models in the Second Stage of CLQC

In the previous part of this section, we showed the results of the second stage of our experiment, but only for the EfficientNetB0 model. We chose to focus on this model because it had the highest accuracy among the four models we tested. To illustrate this, we created a bar chart to compare the four models in the second stage. As shown in Fig. 25, comparing the results of the four different models in Stage 2 is an important step in understanding the performance of each model. The bar chart revealed that the EfficientNetB0 model had the highest accuracy among the four models. After that, it comes in second place in terms of accuracy, followed by DenseNet121 and ResNet152V2, as they have close accuracy, and VGG16 comes in the last position in terms of accuracy. This made it clear that the EfficientNetB0

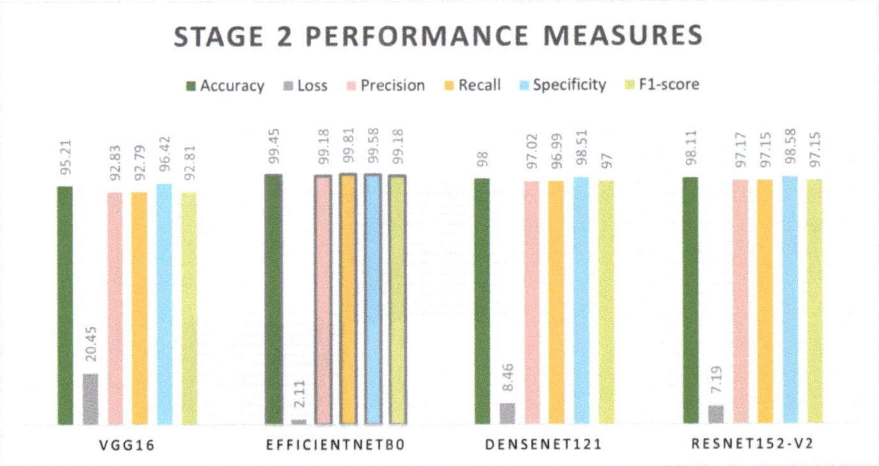

Fig. 25 Comparison of VGG16, EfficientNetB0, DenseNet121, and ResNet152V2 performance measures in Stage 2, where EfficientNetB0 achieved the best results

model was the best choice for our experiment. Therefore, we chose to review only the results of the EfficientNetB0 model in the previous part (Table 12).

Overall, the bar chart was a useful tool to compare the four models and to identify the best one for our experiment. By using the bar chart, we were able to determine which model had the highest accuracy, making it the best choice for our experiment. As a result, we chose to review only the results of the EfficientNetB0 model in the previous part. Comparing the results of these four models in Stage 2 can provide valuable information for understanding the performance of each model and can help us make decisions about which model to use in the future. Overall, all four models performed very well in Stage 2.

4.3 Results of CLQC Stage 3

The results of Stage 3 using deep learning models were impressive. Table 13 illustrates how we may distinguish between Stages 1, 2, and 3 using a variety of criteria, just as we did in the previous stage. Additionally, the previous part contains comprehensive details in this table. It explains the various elements of the table, such as the columns, rows, and data points. The models were able to accurately classify images with an accuracy of over 98%.

Furthermore, the models were able to identify objects in images with a precision of over 90%. This demonstrates the power of deep learning models to accurately identify and classify images. Overall, the results of Stage 3 using deep learning models were highly successful.

Table 12 Performance measures of CLQC Stage 2 using different evaluation metrics

CLQC Stage 2								
			Performance Measures/Metrics					
Models	Activation Function	Epochs	Accuracy(%)	Loss(%)	Precision(%)	Recall(%)	Specificity(%)	F1-score(%)
VGG16	3/Softmax	74/300	95.21	20.45	92.83	92.79	96.42	92.81
EfficientNetB0	3/Softmax	37/300	99.45	2.11	99.18	99.81	99.58	99.18
DenseNet121	3/Softmax	29/300	98	8.46	97.02	96.99	98.51	97
ResNet152-V2	3/Softmax	34/100	98.11	7.19	97.17	97.15	98.58	97.15

Table 13 Performance measures of CLQC Stage 3 using different evaluation metrics

CLQC Stage 3									
			Performance Measures/Metrics						
Models	Activation Function	Epochs	Accuracy(%)	Loss(%)	Precision(%)	Recall(%)	Specificity(%)	F1-score(%)
VGG16	4/Softmax	26/300	95.53	26.99	91.36	90.71	97.14	91.03
EfficientNetB0	4/Softmax	45/300	99.95	0.29	99.9	99.9	99.9	99.9
DenseNet121	4/Softmax	24/300	98.62	8.22	97.38	97.9	99.12	97.23
ResNet152-V2	4/Softmax	36/100	99.43	3.51	98.92	98.81	99.64	98.86

4.3.1 Results of the EfficientNetB0 Model for Stage 3

The EfficientNetB0 model provided the highest level of accuracy in the third stage, reaching 99.95% at epoch 43. Having undergone training on a Stage 3 dataset comprising 8000 images, segregated into four files—Cercospora, Phoma, Miner, and Red Spider Mite—each file containing an equal number of images, the particulars of which are thoroughly elucidated in Sect. 3. The confusion matrix for multiple classes in Fig. 26 shows that the EfficientNetB0 model successfully classifies the fourth class of coffee leaves. Notably, it achieves high accuracy for Cercospora (99.70%), Phoma (99.90%), Miner (99.90%), and Red Spider Mite (99.90%). In Fig. 27, we display the plot metrics of the model's loss, AUC, precision, recall, and accuracy when using an early stopping function. The plots demonstrate that the model stops learning after 47 iterations. Furthermore, the EfficientNetB0 model outperformed the three other proposed models and gave the highest accuracy in Stage 3.

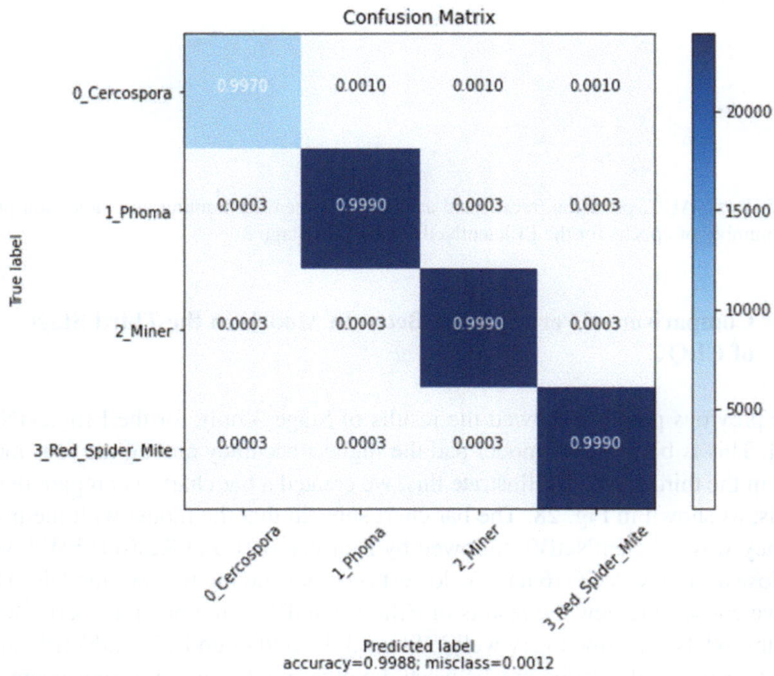

Fig. 26 Confusion matrix of the proposed EfficientNetB0 model in Stage 3

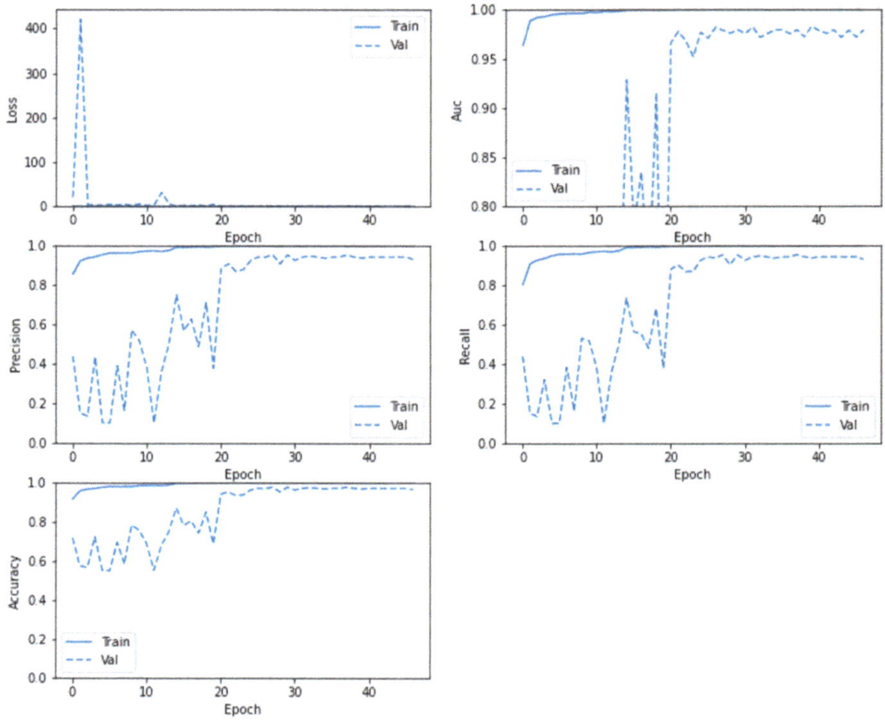

Fig. 27 Loss, AUC, precision, recall, and accuracy between the training and validation phases, with a number of epochs for the EfficientNetB0 model in Stage 3

4.3.2 Comparison of Performance Between Models in the Third Stage of CLQC

In the previous part, we showed the results of Stage 3 only for the EfficientNetB0 model. This is because the model had the highest accuracy among the four models tested in the third stage. To illustrate this, we created a bar chart to compare the four models, as shown in Fig. 28. The bar chart showed that the model with the highest accuracy was EfficientNetB0, followed by DenseNet121 and ResNet152V2, which had close accuracy. VGG16 had the lowest accuracy among the four models. This is why we chose to review the results of EfficientNetB0 in the previous part. Overall, all four models performed very well in Stage 3. Even though EfficientNetB0 had the highest accuracy, the difference between the other models was not significant. This shows that all the models are suitable for the task and can be used depending on our needs.

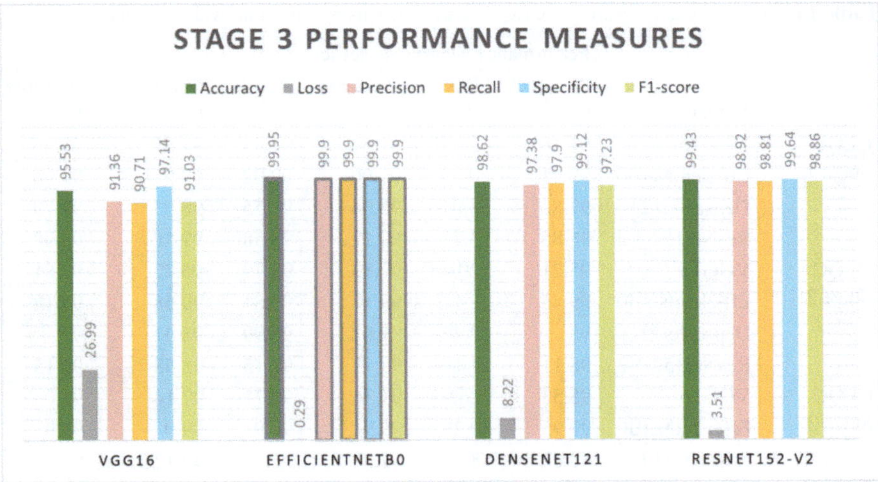

Fig. 28 Comparison of VGG16, EfficientNetB0, DenseNet121, and ResNet152V2 performance measures in Stage 3, where EfficientNetB0 achieved the best results

4.4 Discussion and Comparison of Coffee Leaf Quadruple Classifier (CLQC) Framework

This research proposes three stages, in each of which there are four models: VGG16, EfficientNetB0, DenseNet121, and ResNet152-V2, in order to detect and classify the most common coffee leaf diseases.

Comparing results between the three stages, each of which has four models, can be a challenging task. Each model has its own strengths and weaknesses and depending on the application, one model may be more suitable than the others. For example, VGG16 is a popular model for image classification due to its simplicity, while EfficientNetB0 is more suitable for object detection tasks. DenseNet121 is a good choice for segmentation tasks, while ResNet152V2 is a powerful model for image generation. Comparing the results of each model can be a useful way to determine which model is best suited for this project.

The results of the training and testing experiments for this project showed that the EfficientNetB0 model is the best and has the highest accuracy, as shown in Table 14 in all three stages, as its accuracy reached approximately 99% in all three stages. It comes second in terms of accuracy; ResNet152-V2 has achieved an accuracy of approximately 99% But EfficientNetB0 was the best, followed by model DenseNet121 which achieved approximately 98%. Lastly, in terms of accuracy, there is VGG16, as its accuracy reached approximately 93%. The gray row was used to select the highest value in each column. This table allowed us to compare the different parameter methods and determine which was the most effective.

Table 14 Performance measures of CLQC at all stages using different evaluation metrics

	Models	Performance Measures/Metrics Accuracy (%)	Loss (%)	Precision (%)	Recall (%)	Specificity (%)	F1-score (%)
CLQC Stage 1	VGG16	91.36	21.73	91.62	91.05	91.67	91.33
	EfficientNetB0	99.91	0.35	99.90	99.92	99.9	99.90
	DenseNet121	97.8	8.04	98.04	97.55	98.05	97.79
	ResNet152-V2	98.85	3.77	99.12	98.58	99.12	98.98
CLQC Stage 2	VGG16	95.21	20.45	92.83	92.79	96.42	92.81
	EfficientNetB0	99.45	2.11	99.18	99.81	99.58	99.18
	DenseNet121	98	8.46	97.02	96.99	98.51	97
	ResNet152-V2	98.11	7.19	97.17	97.15	98.58	97.15
CLQC Stage 3	VGG16	95.53	26.99	91.36	90.71	97.14	91.03
	EfficientNetB0	99.95	0.29	99.9	99.9	99.9	99.9
	DenseNet121	98.62	8.22	97.38	97.9	99.12	97.23
	ResNet152-V2	99.43	3.51	98.92	98.81	99.64	98.86

In order to compare the performances of the models, we made a table and a bar chart to compare the three stages, as shown in Table 14. The table and bar chart compare the models in terms of six measures: accuracy, F1-score, loss, precision, and specificity. The results show that the EfficientNetB0 model is the best performer, as it has the highest accuracy, F1-score, and lowest loss. The ResNet152-V2 and DenseNet121 models also perform well, as their accuracy, F1-score, and loss are close to those of the EfficientNetB0 model. Lastly, the VGG16 model has the lowest accuracy and F1-score and the highest loss.

Overall, the results of this research show that the EfficientNetB0 model is the best performer in terms of accuracy, F1-score, and loss. It is followed by ResNet152-V2 and DenseNet121, which have accuracy and F1-scores similar to those of the EfficientNetB0 model. Finally, the VGG16 model has the lowest accuracy, F1-score, and loss.

Table 15 lists a comparison of our study with previous studies related to coffee leaf diseases. Some studies trained on a single dataset, such as [1] and [18] used BrACoL, and [6] used LiCoLe, whereas [9] used a combination of three datasets: BrACoL, RoCoLe, and LiCoLe. As for our study, we used five datasets: BrACoL, RoCoLe, LiCoLe, JMuBEN, and JMuBEN2. So, our dataset is the highest image number with a total of 24000 images. Also, some studies trained a single deep learning model, such as [1] trained InceptionV3. While other studies trained multiple deep learning models, such as [6] trained VGG16, Xception, and ResNetV2-152, and [9] trained EfficientNetB0, DenseNet121, and VGG16. Regarding this study, we trained four deep learning models: VGG16, EfficientNetB0, DenseNet121, and ResNet152-V2. Figure 29 demonstrates that previous studies used accuracy as a performance metric, and our proposed CLQC outperformed them with an overall accuracy of 99.77% for the EfficientNetB0 model. In addition to accuracy, [18] and [9] also used other performance metrics such as F1-Score, Recall, and Precision. Our

Table 15 Comparison of coffee leaf quadruple classifier (CLQC) model with previous studies on coffee leaf disease classification

Ref	DL models	Dataset	Number of images	Performance Measures/Metrics				
				Accuracy (%)	F1-score (%)	Specificity (%)	Recall (%)	Precision (%)
[1]	Inception V3	BrACoL	1747	97.61	–	–	–	–
[23]	LBP VGG16 GLCM	BrACoL	1747	98.04	98.33	–	98	98.66
[6]	VGG16 Xception ResNetV2-152	LiCoLe	4667	97	–	–	–	–
[9]	EfficientNetB0 DenseNet121 VGG16	BrACoL RoCoLe LiCoLe	4675	95.91	95.87	–	95.91	96.05
	VGG16	BrACoL						
Our	EfficientNetB0	RoCoLe						
Proposed	DenseNet121	LiCoLe	24,000	99.77	99.66	99.79	99.87	99.667
CLQC	ResNet152-V2	JMuBEN						
		JMuBEN2						

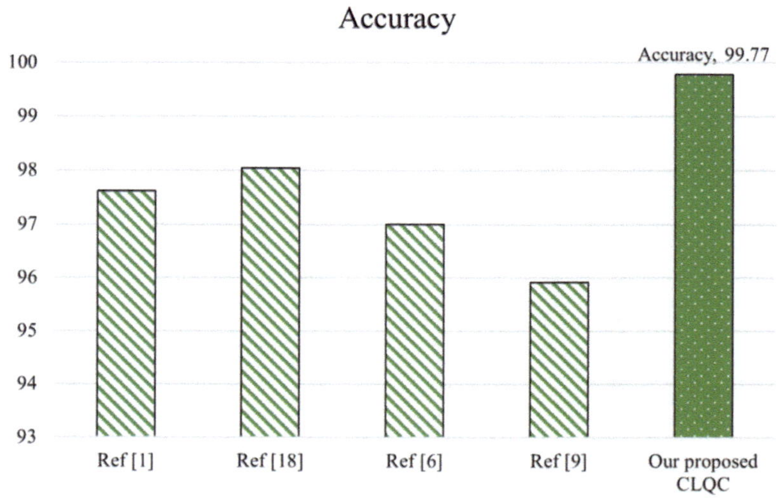

Fig. 29 Accuracy measure comparison between the proposed coffee leaf quadruple classifier (CLQC) framework and previous studies on coffee leaf disease classification

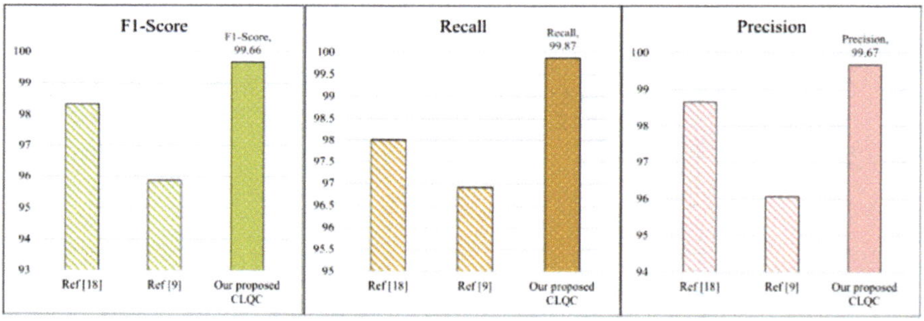

Fig. 30 F1-Score, Recall, and Precision measures comparison between the proposed coffee leaf quadruple classifier (CLQC) framework and previous studies on coffee leaf disease classification.

proposed CLQC achieved the best of these metrics, with an F1-Score of 99.99%, Recall of 99.87%, and Precision of 99.67% for the EfficientNetB0 model, as shown in Fig. 30. This further demonstrates that our proposed CLQC is the best solution for the problem. Overall, our proposed CLQC outperformed the previous studies in accuracy and other performance metrics for coffee leaf disease classification. This makes it the best choice for solving the problem at hand and it is likely to be the preferred solution in the future

5 Concluding Remarks

In this research, we designed and evaluated Coffee Leaf Quadruple Classifier (CLQC) that classifies six coffee leaf diseases that have been widely spread due to their detrimental effect on coffee crops. Early diagnosis is crucial to determining the best course of therapy and preserving coffee leaf survival. Farmers and professionals often use their eyes and their knowledge to observe the plants, but they are prone to misdiagnose crop lesions in general. This study presented four DCNN model architectures: EfficientNetB0, VGG16, DenseNet121, and ResNet152-V2. Each one of them was trained individually to observe which one of these models had the best accuracy in classifying the diseases and we discovered that the EfficientNetB0 model outperformed all the three other proposed models and gave the highest accuracy in all three stages. Developing our models in the field of leaf disease diagnosis is full of possibilities. To ensure that our project is developed in a systematic and organized manner, we have planned the steps of our project in sequential order. The first step is to increase the accuracy, diagnosis, and classification of our system. To do this, we intend to train the system on multiple models and combine the six best models. The next step is to transform the system into two forms: a web-like user interface and an app, as well as a device that can make a preliminary diagnosis of whether or not a leaf is infected. An example of this kind of device is a thermometer, which is the closest example to our idea. Also, to further develop our system, we need to expand it to detect more diseases. We should also think about adding an option called "possibility of infection," which would give a preliminary diagnosis or warning of the possibility of the presence of a disease in the case where the system does not recognize the disease. Still, there appear to be suspicious or strange symptoms, such as a plant or leaf showing traces or strange spots. We are excited to see what further research and development can be done in the future. We plan to continue to improve the accuracy and reliability of our system, as well as to develop the user interface and device. We are confident that with these plans, we can make a huge difference in the field of leaf disease diagnosis. We look forward to seeing the results of our efforts in the near future.

References

1. M. Kumar, P. Gupta, P. Madhav, Disease detection in coffee plants using convolutional neural network, in *2020 5th International Conference on Communication and Electronics Systems (ICCES)*, (IEEE, 2020), pp. 755–760
2. List of countries by coffee production. Wikimedia Foundation (2022), Available at: https://en.wikipedia.org/wiki/List_of_countries_by_coffee_production (Accessed: 17 May 2023)
3. A. Al-Shatti, Coffee travel. Available at: https://www.goodreads.com/ar/book/show/60829843 (Accessed: 17 May 2023)
4. Information about the coffee tree; Learn about the origin of coffee beans, their cultivation and types. Available at: https://almalomat.com/85624 (Accessed: 17 May 2023)

5. L.M. Shita, P.M. Shashi, Ethiopian coffee leaf diseases identification using deep learning features. Int. J. Emerg. Technol. Innov. Res. (www.jetir.org) **8**, b498–b508 (2021). https://doi.org/10.1729/Journal.28324
6. M. Yebasse, B. Shimelis, H. Warku, J. Ko, K.J. Cheoi, Coffee disease visualization and classification. Plants **10**(6), 1257 (2021)
7. F.J.P. Montalbo, A.A. Hernandez, Classifying Barako coffee leaf diseases using deep convolutional models. Int. J. Adv. Intell. Inform. **6**(2), 197–209 (2020)
8. J.G. Esgario, R.A. Krohling, J.A. Ventura, Deep learning for classification and severity estimation of coffee leaf biotic stress. Comput. Electron. Agric. **169**, 105162 (2020)
9. Facts about coffee roya in Central America Salt Spring Coffee. Available at: https://www.saltspringcoffee.com/blogs/news/facts- about-coffee-roya-in-central-america (Accessed: 17 May 2023)
10. F.J.P. Montalbo, A.A. Hernandez, An optimized classification model for Coffea Liberica Disease using Deep Convolutional Neural Networks, in *2020 16th IEEE International Colloquium on Signal Processing & Its Applications (CSPA)*, (IEEE, 2020), pp. 213–218
11. S. Suda, C.N. Sujatha, *Leaf Disease Detection and Classification, National Conference On Advances in Electronics Signal Processing and Communications (AESPC-2020)* (2020), Available at: https://www.researchgate.net/publication/358282588_Leaf_Diseas
12. E. Lisboa, G. Lima, F. Queiroz, Coffee leaf diseases identification and severity classification using deep learning, in *Anais Estendidos do XXXIV Conference on Graphics, Patterns and Images*, (SBC, 2021), pp. 201–205
13. F.J.P. Montalbo, Automated diagnosis of diverse coffee leaf images through a stage-wise aggregated triple deep convolutional neural network. Mach. Vis. Appl. **33**(1), 19 (2022). Available at: https://link.springer.com/article/10.1007/s00138-022-01277-y
14. mikias legesse, Deep learning for coffee disease classification. Int. J. Comput. Trends Technol. **69**(9), 27–32 (2021). Available at: https://doi.org/10.14445/22312803/IJCTT-V69I9P105
15. S. Bhandarkar, R. Prasad, V. Agarwal, R. Hebbar, D. Uma, Y.B. Venkata Reddy, Y. Raghuramulu, K. Ganesha Raj, Deep learning and stastical models for detection of white stem borer disease in Arabica Coffee, in *International Archives of the Photogrammetry, Remote Sensing & Spatial Information Sciences*, (2019)
16. A. Afifi, A. Alhumam, A. Abdelwahab, Convolutional neural network for automatic identification of plant diseases with limited data. Plants **10**, 28 (2021). https://doi.org/10.3390/plants10010028
17. L.X.B. Sorte, C.T. Ferraz, F. Fambrini, R. dos Reis Goulart, J.H. Saito, Coffee leaf disease recognition based on deep learning and texture attributes. Procedia Comput. Sci. **159**, 135–144 (2019)
18. H. Taddese, Coffee disease detection using convolutional neural network an image processing approach, AAU. Addis Ababa University (2021), Available at: http://etd.aau.edu.et/handle/123456789/29686 (Accessed: 17 May 2023)
19. CropScience, B. Cercospora leaf spot, Bayer Crop Science UK. Available at: https://cropscience.bayer.co.uk/threats/diseases/sugar-beet-diseases/cercospora-leaf-spot-beet (Accessed: 17 May 2023)
20. CropScience, B. Phoma and stem canker identification and management, Bayer Crop Science UK. Available at: https://cropscience.bayer.co.uk/threats/diseases/oilseed-rape-diseases/phoma-stem-canker/ (Accessed: 17 May 2023)
21. J. Jepkoech, D.M. Mugo, B.K. Kenduiywo, E.C. Too, Arabica coffee leaf images dataset for coffee leaf disease detection and classification. Data Brief **36**, 107142 (2021)
22. Coffee—Diseases and Pests, Description, Uses, Propagation. Available at: https://plantvillage.psu.edu/topics/coffee/infos (Accessed: 17 May 2023)
23. P.A. Arneson, Coffee rust (2020), Available at: https://www.apsnet.org/edcenter/disandpath/fungalbasidio/pdlessons/Pages/CoffeeRust.aspx (Accessed: 17 May 2023)
24. P. Chomnunti, S. Hongsanan, B. Aguirre-Hudson, Q. Tian, D. Peršoh, M.K. Dhami, A.S. Alias, J. Xu, X. Liu, M. Stadler, K.D. Hyde, The sooty moulds. Fungal Diver. **66**, 1–36 (2014)

25. E. Albuquerque, Coffee Leaf Miner Leucoptera coffeella. Encyclopedia. Available online: https://encyclopedia.pub/entry/17403 (Accessed: 17 May 2023)
26. The Gardener, Red Spider Mites—Pests and Diseases. [online] (n.d.), Available at: https://www.thegardener.co.za/the-gardener/maintenance/red-spider-mites/ (Accessed: 17 May 2023)
27. Nature and Garden, Red spider mite—treating and fighting them, easy organic solutions. [online] (2020), Available at:https://www.nature- and-garden.com/gardening/red-spider-mite.html (Accessed: 17 May 2023)
28. R.A. Krohling, J. Esgario, J.A. Ventura, BRACOL–a Brazilian Arabica Coffee Leaf images dataset to identification and quantification of coffee diseases and pests. Mendeley Data **1** (2019)
29. J. Jepkoech, B. Kenduiywo, D. Mugo, E. Chebet, JMuBEN. (2021) data.mendeley.com , [online] 1. https://doi.org/10.17632/t2r6rszp5c.1
30. J. Jepkoech, D. Mugo, B. Kenduiywo, E. Chebet, JMuBEN2. (2021) data.mendeley.com, [online] 1. https://doi.org/10.17632/tgv3zb82nd.1
31. J. Parraga-Alava, K. Cusme, A. Loor, E. Santander, RoCoLe: (2019) A robusta coffee leaf images dataset. data.mendeley.com [online] 2. https://doi.org/10.17632/c5yvn32dzg.2
32. freeCodeCamp.org. Preprocessing for deep learning: From covariance matrix to image whitening. [online] (2018), Available at: https://www.freecodecamp.org/news/https-medium-com-hadrienj-preprocessing-for-deep-learning- 9e2b9c75165c/amp/ (Accessed: 17 May 2023)
33. I.H. Sarker, Deep learning: A comprehensive overview on techniques, taxonomy, applications and research directions. SN Comput. Sci. **2**(6), 420 (2021)
34. N. Lang, Using Convolutional Neural Network for Image Classification. [online] Medium. (2022), Available at https://towardsdatascience.com/using-convolutional-neural-network-for-image-classification-5997bfd0ede4 (Accessed: 17 May 2023)
35. HappiestMinds. Convolutional Neural Networks (CNN) with Deep Learning. [online] (n.d.), Available at: https://www.happiestminds.com/insights/convolutional- neural-networks-cnns/ (Accessed: 17 May 2023)
36. www.isahit.com. What is the purpose of image preprocessing in deep learning? [online] (n.d.), Available at: https://www.isahit.com/blog/what-is-the- purpose-of-image-preprocessing-in-deep-learning (Accessed: 17 May 2023)
37. A. Arora, DenseNet Architecture Explained with PyTorch Implementation from TorchVision. amaarora github [online] (2020), Available at:https://amaarora.github.io/posts/2020-08-02-densenets.html (Accessed: 17 May 2023)
38. H. Zulkifli, Understanding Learning Rates and How It Improves Performance in Deep Learning. [online] Medium. (2018), Available at: https://towardsdatascience.com/understanding-learning-rates-and-how-it-improves-performance-in- deep-learning-d0d4059c1c10 (Accessed: 17 May 2023)
39. Stack Overflow, Keras - Why models often benefit from reducing the learning rate during training. [online] (n.d.), Available at: https://stackoverflow.com/questions/65869114/why-models-often-benefit-from-reducing-the-learning-rate-during-training (Accessed: 17 May 2023)
40. K. Team, Keras documentation: ReduceLROnPlateau. keras.io. [online] (n.d.), Available at: https://keras.io/api/callbacks/reduce_lron_plateau/ (Accessed: 17 May 2023)
41. www.v7labs.com. What is Overfitting in Deep Learning [+10 Ways to Avoid It]. [online] (n.d.), Available at: https://www.v7labs.com/blog/overfitting#h1 (Accessed: 17 May 2023)
42. J. Brownlee, *A Gentle Introduction to Early Stopping to Avoid Overtraining Neural Networks. [online]* (Machine Learning Mastery, 2018). Available at: https://machinelearningmastery.com/early-stopping-to-avoid-overtraining-neural-network-models/ (Accessed: 17 May 2023)
43. K. Team, Keras documentation: EarlyStopping. [online] keras.io. (n.d.), Available at: https://keras.io/api/callbacks/early stopping/ (Accessed: 17 May 2023)
44. www.dominodatalab.com. What is Model Evaluation? — Domino Data Science Dictionary. [online] (n.d.), Available at: https://www.dominodatalab.com/data-science-dictionary/model-evaluation (Accessed: 17 May 2023)

45. M. Hossin, M.N. Sulaiman, A review on evaluation metrics for data classification evaluations. Int. J. Data Min. Knowl. Manag. Process **5**(2), 1 (2015)
46. LenaAlenazi, GitHub, Coffee Leaf Diseases Classification-using Deep Learning Models. https://github.com/lenaAlenazi/Coffee-Leaf-Diseases-Classification-using-Deep-Learning-Models.git (2023)
47. Google Drive, Dataset. https://drive.google.com/drive/folders/1gu-ghdHmOv5irBs7r8fsZuLEfswR4fY?usp=share link
48. Google, Colaboratory–Google. [online] research.google.com. (n.d.), Available at: https://research.google.com/colaboratory/faq.html (Accessed: 17 May 2023)
49. Simplilearn, What is Keras and Why it so Popular in 2021. [online] Simplilearn.com. (2022), Available at: https://www.simplilearn.com/tutorials/deep- learning-tutorial/what-is-keras (Accessed: 17 May 2023)

Teleworking in Service of Work–Family Balance

Fatima El Amrany, Basma El Ouaghlidi, and Abderrahmane Ouddasser

1 Introduction

Due to the growing importance of work–life balance, the issue of work–life balance has gained considerable momentum in recent decades, becoming a key topic for organizations seeking to attract and retain the best employees [1]. This HR practice has become one of the most popular, impacting attractiveness and staff retention.

In today's world, marked by accelerating work rhythms, job instability, restructuring, pressure to meet deadlines, and the emergence of new technologies, human resources have become essential to companies. Companies increasingly rely on the human factor to strengthen their competitiveness and sustainability. Companies, therefore, have a social responsibility to create a healthy working environment for their employees, who are no longer willing to sacrifice their private lives for their professional careers. Organizations must now be more attentive to their employees' needs and consider their difficulties in reconciling their personal and professional lives.

According to Tremblay [1] and Fusulier, Tremblay, and Di Loreto [2], these organizations today play an intermediary role that can be positive or negative in implementing and applying articulation schemes.

Moroccan organizations are increasingly aware of the negative consequences of work–family conflict. They are seeking to adopt human resource management practices that can reduce gender inequalities, promote work–life balance, and improve their attractiveness.

F. El Amrany (✉) · B. El Ouaghlidi · A. Ouddasser
Industrial Technologies and Services Laboratory, Sidi Mohamed Ben Abdellah University, Fez, Morocco
e-mail: Fatima.elamrany@usmba.ac.ma; basma.elouaghlidi@usmba.ac.ma; abderrahmane.ouddasser@usmba.ac.ma

© The Author(s), under exclusive license to Springer Nature Switzerland AG 2024

A. Shaikh et al. (eds.), *Advances in Emerging Information and Communication Technology*, Signals and Communication Technology,
https://doi.org/10.1007/978-3-031-53237-5_15

Telecommuting has therefore emerged as an appropriate response to the issue of work–life balance. This form of organization corresponds to work activity carried out away from the organization's premises using Information and Communication Technologies, on an occasional, regular, or permanent basis, within the framework of an agreement with one's employer [3, 4].

Based on these findings, we raise the following issue, which may represent a new line of research:

To what extent does teleworking contribute to maintaining work–family balance?

To address this question, we will first briefly present the conceptual framework of our subject by examining the different definitions of telework from the literature and the different types, advantages, and disadvantages associated with this practice. We will then use semi-directive interviews with several public organizations to explore telecommuting in Morocco and its impact on work–life balance.

2 Literature Review

We will begin by examining the fundamental concepts associated with teleworking, which will serve as the theoretical basis for our empirical analysis. In particular, we will focus on the different forms of teleworking and the advantages and disadvantages of this practice. We will then look at the current telework situation in Morocco and how it is used in this context.

2.1 The Genesis of Telework

In the 1970s, J. NILLE [5] introduced the innovative concept of telecommuting. It is based on using telecommunications to replace the need for physical travel in the workplace.

Many people can work remotely thanks to the rise of flexibility in the workplace and the emergence of information and communication technologies since the 2000s. One of the main reasons for this is the growing availability of the equipment needed for telecommuting, such as Internet access, computers, and telephones. As a result, many companies have begun to see telecommuting as an effective tool for economic and social development. Research by DI MARTINO and WIRTH [6] has highlighted the benefits of teleworking in terms of improving the quality of life at work and work–life balance for employees.

Technological advances and changing attitudes are major factors in the growth of teleworking. According to several authors studying the subject, it is clear that telework is defined in various ways. Indeed, each author perceives it from his or her point of view. Although telework is a complex concept, examining several definitions, it is possible to identify the key elements that characterize it by examining several definitions to identify the key elements that characterize it.

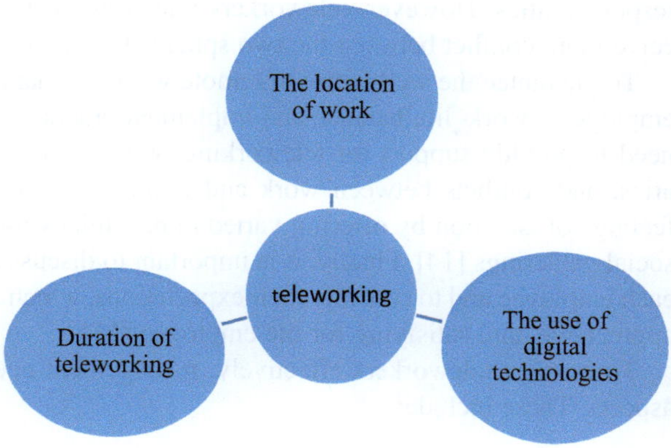

Fig. 1 The components of teleworking. (Source: authors)

According to the definition given by W. B. KORTE and R. WYNNE [7], telework consists of performing work tasks, in whole or in part, at a distance from the company, using communication and information technology tools.

As a result, the existence of telework depends on the combination of these three elements (Fig. 1):

2.2 *What Impact Does Teleworking Have on Work–Life Balance?*

2.2.1 Positive Effects of Teleworking on Work–Life Balance

Telecommuting is a topic that has attracted the interest of several authors, who have studied its role in maintaining work–life balance [8]. Studies have shown that teleworking can positively affect this balance, enabling individuals to reconcile their professional and personal lives better. For example, flexible working hours and reduced commuting can reduce stress and give teleworkers more time for family and personal activities. Indeed, working from home enables teleworkers, particularly women, to devote the time saved by reducing travel to caring for their home and family and to sporting and leisure activities.

Teleworking also offers the possibility of reducing family pressures and diminishing the overloaded roles felt by teleworkers [9], while increasing motivation at work [10].

This issue of work–life balance is of particular interest to working women. According to D-G Tremblay [11], the search for work–life balance is the main reason women telecommute.

However, the positive impact of telecommuting on work–life balance depends on several factors, including the extent of family responsibilities. According to a SOLIS [12] study, work–life balance is more successful among teleworkers with few family responsibilities. However, teleworkers with significant family responsibilities perceive more conflict between the two spheres than non-teleworking employees.

To guarantee the well-being of remote workers, managers must take care of their employees' work–life balance and implement practices to foster it. To do this, they need to provide support for teleworking, with advice on managing stress, boundaries, and conflicts between work and family [13]. It is also important to reduce feelings of isolation by offering varied opportunities for interaction and organizing social gatherings [14]. Finally, it is important to discuss the professional future with each employee and to consider their expectations, which are often highly motivating, appreciable, and satisfying for the employee [15].

To manage teleworkers effectively, management needs to focus on several key aspects. These include:

- Social support, which aims to help employees manage the stresses associated with teleworking [13, 16].
- Participation in decision-making, which helps employees feel more involved in their work [13].
- Empathy and benevolence, which aim to create a positive work environment [17].
- Exemplarity involves setting an example regarding time management and work–life balance [18].

So, to maintain strong social cohesion and a solid social bond, it is important to implement flexible practices that are based on mutual trust [19]. These practices must foster collaboration, exchange, empowerment, and a sense of belonging to maintain the motivation and commitment of all remote workers [20].

2.2.2 The Risks of Work–Family Conflict in the Context of Teleworking

It is important to note that telecommuting also has its drawbacks. Research has highlighted the risks associated with this practice, including the feeling of being unable to disconnect and working without time limits [21]. This can lead to confusion between work and private life, particularly when work takes place at home [13, 22].

Working remotely and without limits can be a source of work overload, occupational stress, and, consequently, burnout [16].

Added to this is the pressure and mental load [23] felt by teleworkers as a result of excessive electronic control by managers [24, 25].

Teleworking leads to job dissatisfaction [16] or difficulties in self-motivation [21, 26].

Studies have also shown that excessive use of digital tools causes cognitive, nervous, and visual fatigue [27], musculoskeletal disorders, or stress linked to hyperconnection and feelings of isolation (INRS, 2020).

According to Golden [28], the higher the intensity of telecommuting, the more committed employees are to their organization and the less emotionally drained they are.

Previous studies show, therefore, that telecommuting is a phenomenon that can both enhance and detract from the well-being of its practitioners. In this respect, the role of management seems particularly important in the introduction, development, and management of teleworking [14, 15].

3 Current Status of Teleworking in Morocco

Morocco has implemented government programs to digitalize and dematerialize procedures to accelerate economic and social development. These efforts have been reinforced by adopting teleworking practices in Moroccan organizations in response to the current health crisis. Thus, to stem the spread of the virus and ensure the continuity of public services, and following a circular from the Ministry of the Economy, Finance and Administrative Reform n°03/2020 on teleworking in public administrations, a methodological guide to teleworking has been drawn up for the first time in Morocco. This crisis period is considered the real beginning of teleworking in Morocco.

It is important to note that studies on teleworking in Morocco are still limited. To analyze the current situation of teleworking in Morocco, we will rely on semi-directive interviews conducted with public organizations that adopted teleworking during the period of confinement.

3.1 Analysis of Factors Favoring the Development of Telecommuting

Analysis of the factors favoring the development of telecommuting involves studying the various elements that have contributed to the expansion of this form of work organization. This may include technological advances, socio-economic trends, corporate policies, and employee preferences.

- Technological advances: The development of information and communication technologies (ICTs) has made it possible to carry out professional tasks remotely, thanks in particular to the use of tools such as computers, telephones, online communication software, etc.
- Flexibility: Teleworking enables employees to work at times and places that suit them best, making reconciling their professional and personal lives easier [29]. Moreover, on the other hand, it saves more time thanks to reduced travel [3, 30].
- Societal changes: Societal evolutions, notably changes in family structures and social norms, have led individuals to seek more flexible forms of work organization to reconcile better their professional and personal lives [8].

- Benefits for employers: Teleworking can also benefit employers, such as cost savings, increased productivity, better employee retention, and reduced sickness absences [21].
- Changes in employee attitudes and preferences are also important. Several studies show that employees increasingly consider teleworking an important criterion when looking for a job. Telecommuting is, therefore, an attractive factor for companies [31].
- Societal trends: Trends in today's society, such as increased mobility, individualized lifestyle choices, growing concern for the environment, and quality.

3.2 Legal Framework for Teleworking

From our analysis of the Labor Code, we found that the legislator has addressed the issue of home working (Article 8 of the Labor Code), which should be distinct from the notion of telecommuting; home working is simply a form of telecommuting.

To help public administrations implement telecommuting during this crisis, the Ministry of the Economy, Finance, and Administrative Reform has drawn up a methodological guide published on April 15, 2020, to provide a framework for telecommuting within the civil service. This guide is organized into six axes, including the procedures to be followed to set up this new work mode and the obligations to be met by civil servants and administrations.

Although telecommuting has become a common practice in the current health crisis, it is not regulated in Morocco. It is, therefore, important to put in place a law that clearly defines the standards for telecommuting, particularly regarding working arrangements, remuneration issues, and potential risks and abuses [32].

There has yet to be a legislative framework for telecommuting in Morocco, which can expose public organizations to various risks. For this reason, a draft decree (n° 2.20.343) on teleworking in public administrations is currently being prepared. It aims to regulate and frame the concept of telecommuting and integrate it into the public sector. This decision has been taken in the context of the digitalization of Moroccan public administration to make administrative life easier for Moroccan citizens.

4 Research Methodology and Discussion of Results

Our empirical survey was conducted among 27 teleworkers belonging to Moroccan public organizations.

Several authors have dealt with the issue of telework in the private sector. However, only some have focused on telework in the public sector, so we chose to address the issue of telework in public organizations.

4.1 *Research Methodology*

Our research aims to analyze the role of telework in maintaining work–family balance. To fully understand this new phenomenon, we used a qualitative approach; the choice of this approach is explained by the aim of our research, which is to explain and understand the particularities of a lived phenomenon. WACHEUX [33] shows that adopting a qualitative approach is justified by "wanting to understand the how and why of events in concrete situations."

For data collection, our qualitative research relies on semi-structured interviews with public servants who work remotely. As Thietart [34] argues, the semi-structured interview remains the most appropriate data collection method for qualitative research. "The interview is a technique designed to collect, in the perspective of their analysis, discourse data reflecting, in particular, the conscious or unconscious mental universe of individuals" [34].

In order to conduct our survey, we prepared an interview guide consisting of open-ended questions. We used a convenience sampling method by targeting 27 public employees from different organizations.

4.2 *Analysis of Results and Discussion*

Based on a manual analysis of the semi-structured interviews, we were able to identify the following results:

4.2.1 Frequency of Teleworking (Fig. 2)

According to the surveys, Moroccan teleworkers worked partially a few days a week. This is due to certain activities requiring a physical presence in the organization's offices, such as managing machinery or using specific equipment. In addition, there may be constraints linked to remote work, such as communication or collaboration problems between employees, difficulties in maintaining productivity, or security problems linked to access to sensitive data.

4.2.2 Work–Life Balance (Fig. 3)

Numerous studies have examined work–family balance in the context of teleworking. Some see job–family reconciliation as an advantage of teleworking, while others highlight the potential conflict between family and work. In our survey, the reduction in commuting time is perceived as a factor facilitating the reconciliation of work and family activities.

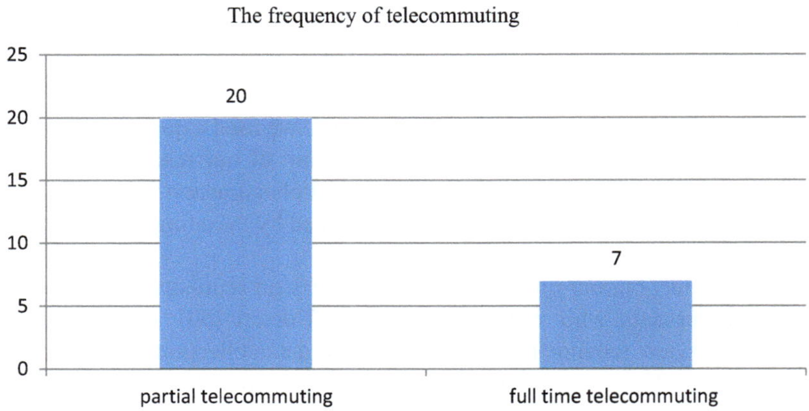

Fig. 2 Frequency of teleworking

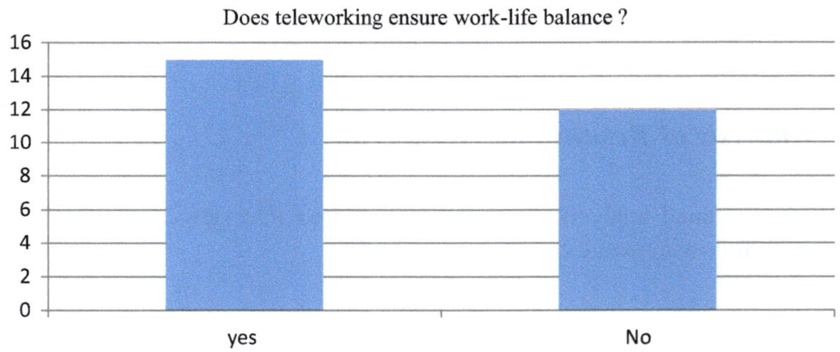

Fig. 3 Work–life balance

According to the responses obtained during the survey, most respondents indicated that they had succeeded in establishing rules and setting up a home workspace that was respected by their families, despite the difficulties encountered at the start of the telecommuting process. Teleworkers, in particular, reported difficulties in making their families understand that they were working from home and unavailable to their households during working hours.

Interestingly, the benefits of teleworking are numerous, with most respondents citing reduced commuting time as one of the main advantages of teleworking. Indeed, the time saved on travel and flexible working hours means that people can devote more time to their family or personal activities. Moreover, teleworkers and their supervisors report that teleworking reduces stress and fatigue, improving productivity and the quality of the work delivered. "Thanks to teleworking, I spend enough time with my children and family; I feel more comfortable working from home," according to a civil servant in the tax department.

Although teleworking offers advantages in reconciling family and professional obligations, thanks to flexible working hours and reduced commuting times, the latter being particularly relevant for women, it is vital that the organization takes into consideration the potential conflicts that may arise, especially for women teleworkers.

4.2.3 Organizational Support in a Teleworking Context

Managers are crucial in maintaining a strong social bond and cohesion between teleworkers. Hence, they must choose the right managerial practices to succeed in this new form of work. Most respondents confirmed the need to implement flexible practices that stimulate collaboration, exchange, and empowerment to maintain the motivation and involvement of all teleworkers [20]. Remote management was, therefore, based primarily on mutual trust between managers and employees [19].

"The management style adopted within our organization is more or less based on trust, empowerment, collaboration, sharing, and exchange, so it is more of a participative style," according to an official at the tax department.

Moreover, an organization that supports and considers the issue of work–life balance fosters a sense of satisfaction and self-worth among its teleworkers, which can be a source of motivation and improved productivity.

4.2.4 Autonomy

For most respondents, the autonomy and independence acquired through teleworking are a source of creativity, satisfaction, and confidence. The fact of not being too attached to the hierarchy gives teleworkers certain flexibility and a considerable margin of freedom to articulate their private and professional lives.

"When I worked remotely, I was freer and more autonomous: I carry out my tasks in my way, my superior simply guiding and directing me if necessary," according to an official at the general tax office.

5 Conclusion

Organizations, conscious of their image, especially in the "war for talent," have integrated work–life balance as a powerful asset in attracting employees. It is the first reason they give for implementing policies to promote work–life harmony. How employees' personal lives fit in with their professional lives directly impacts their behavior and attitudes at work. These issues are also important for the organization regarding recruitment, motivation, retention, career management, professional equality, and diversity.

Whether they like it or not, employers face the challenges and opportunities presented by the gradual disappearance of boundaries between their employees' private and professional lives. This phenomenon, known as "work–life blending," affects employees and poses challenges for human resource managers [35].

Organizations are introducing innovative management practices such as telecommuting to meet the challenges of work–life balance. This attracts talent, builds employee loyalty, and motivates them to become more involved with the company. To successfully implement telecommuting, managers must ensure regular motivation and monitoring of employees working remotely to maintain their sense of belonging to the company. In addition, it is important to provide training for the employees concerned so they can make the most of the benefits of telecommuting. An adequate digital infrastructure is also necessary to enable telecommuting to achieve the objectives set effectively.

Acknowledgments I would like to thank all the authors for their collaboration and their excellent articles and for their invaluable help, patience, proofreading, and corrections.

References

1. D.G. Tremblay, *Articuler famille emploi, Le rôle du soutien organisationnel au cœur de trois professions* (Presses de l'Université du Québec, 2012)
2. B. Fusulier, D.-G. Tremblay, M. Di Loreto, La conciliation emploi-famille dans le secteur du travail social au Québec: une différence de genre ? Quelques éléments de réponse. Politiques Sociales **68**(1), 63–81 (2008)
3. L. Taskin, Les Enjeux du télétravail pour l'organisation. Reflets et perspectives de la vie économique **1**(1), 81–94 (2003)
4. S. Craipeau, Télétravail: le travail fluide. Quaderni **71**, 107–120 (2010). https://doi.org/10.4000/quaderni.534
5. J.M. Nilles, Autres., *The Telecommunications-Transportation Tradeoff: Options for Tomorrow* (A Willey Interscience Publication, John Wiley & Sons, 1976)
6. V. Di Martino, L. Wirth, Le télétravail: un nouveau mode de travail et de vie. Rev. Int. Trav. **129**(5), 27 (1990)
7. W.B. Korte, R. Wyne, *Telework. Penetration, Potential and Practice in Europe* (European Commission, DG XIII-8, Ios Pr Inc, 1996)
8. R.S. Gajendra, D.A. Harrison, The good, the bad and the unknown about telecommuting: Meta-analysis of psychological mediators and individual consequences. J. Appl. Psychol. **92**(6), 1524–1541 (2007). https://doi.org/10.1037/0021-9010.92.6.1524
9. L. Duxbury, M. Halinski, When more is less: An examination of the relationship between hours in telework and role overload. Work **48**(1), 91–103 (2014). https://doi.org/10.3233/WOR-141858
10. N.B. Kurland, D.E. Bailey, Telework: The advantages and challenges of working here, there, anywhere and anytime. Organ. Dyn. **28**(2), 53–67 (1999)
11. D.-G. Tremblay, Telework: A new mode of gendered segmentation ? Results from a study in Canada, Canadian Journal of Communication **28**(4), 461–478 (2003c)
12. M. Solís, Moderators of telework effects on the work-family conflict and on worker performance. Eur. J. Manag. Bus. Econ. **26**(1), 21–34 (2017). https://doi.org/10.1108/EJMBE-07-2017-002

13. S. Mann, L. Holdsworth, The psychological impact of teleworking: Stress, emotions and health. N. Technol. Work. Employ. **18**, 196–211 (2003)
14. E. Brunelle, E-Leadership: L'art de gérer les distances psychologiques. Revue internationale de Gestion **34**, 10–20 (2009)
15. E. Brunelle, Télétravail et leadership: déterminants des pratiques efficaces de direction. Manag. Int. **14**, 23–35 (2010)
16. T.A. Bentley, S.T. Teo, L. McLeod, F. Tan, R. Bosua, M. Gloet, The role of organizational support in teleworker wellbeing: A socio-technical systems approach. Appl. Ergon. **52**, 207–215 (2016)
17. O. Bachelard, Optimiser le bien-être au travail et la performance globale: enjeux et perspectives. Regards **51**, 169–179 (2017)
18. E. Abord de Chatillon, C. Desmarais, Espace de discussion, management et épuisement professionnel. @GRH, De Boeck Supérieur **23**, 13–36 (2017)
19. S. Frimousse, J.-M. Peretti, Concilier distance et proximité, nouveau défi managérial. Question (s) de management **30**(4), 107 (2020). https://doi.org/10.3917/qdm.204.0107
20. D. Ollivier, *Manager le travail à distance et le télétravail . . . Librairie Eyrolles*, (2019)
21. D.G. Tremblay, C. Chevrier, M. Di Loreto, Le télétravail à domicile: Meilleure conciliation emploi-famille ou source d'envahissement de la vie privée? Intervention économiques **34**, 26 (2006)
22. V. Aguilera, A.R. Lethiais, L. Proulhac, *Le télétravail, un objet sans désir ?* (Revue d'Économie Régionale & Urbaine, vol. Février, no 1, 2016), pp. 245–266. https://doi.org/10.3917/reru.161.0245
23. L. Taskin, V. Devos, Paradoxes from the individualization of human resource management: The case of telework. J. Bus. Ethics **62**(1), 13–24 (2005). https://doi.org/10.1007/s10551-005-8710-0
24. L. Taskin, Télétravail: les enjeux de la déspatialisation pour le management humain. Interventions Economiques **34** (2006)
25. B.A.C. Groen, S.P. Van Triest, M. Coers, N. Wtenweerde, Managing flexible work arrangements: Teleworking and output controls. Eur. Manag. J. **36**, 727–735 (2018)
26. E. Vayre, Les incidents du télétravail sur le travailleur dans les domaines professionnel, familial et social. Le travail humain **82**, 1–39 (2019)
27. L. Fosslien, M.W. Duffy, How to combat zoom fatigue. Harv. Bus. Rev. **29**, 8–13 (2020)
28. T.D. Golden, Applying technology to work: Toward a better understanding of telework. Organization Management Journal **6**(4), 241–250 (2009). https://doi.org/10.1057/omj.2009.33
29. D. Tremblay, E. Genin, *IT Self-Employed Workers Between Constraint and Flexibility* (New Technology, Work and Employment, 2010)
30. A. Scaillerez, D.-G. Tremblay, Teleworking, as a new mode of regulation of the flexibilization and organization of work: Analysis and impact of the European and North American legalframework. Account. Organ. Rev. **11**(1), 21–31 (2016)
31. D.-G. Tremblay, L. Thomsin, Telework and mobile working: analysis of its benefits and drawbacks. Int. J. Work Innovation **1**, 100–113
32. H. Sqalli, Le Télétravail Au Maroc: Nouveaux espaces, nouvelles temporalités: Nouveaux rapports au travail?, (2020)
33. F. Wacheux, Méthodes qualitatives et recherche en gestion. Economica (1996)
34. R. Thietart, *Méthodes de recherche en Management* (Dunod, 3ème édition, 2007)
35. S. Robin, Revue: vision R.H, faculté de droit, Economie et gestion à l'UCO, (2019)

An Investigation of Broken Access Control Types, Vulnerabilities, Protection, and Security

Elaf Almushiti, Raseel Zaki, Nora Thamer, and Rima Alshaya

1 Introduction

Access control is a process that mediates resource access based on identity. Most web applications are concerned with it as the primary security service, with most other security services supporting it. Access control decisions, for example, are typically enforced by a user-specific authorization, and authentication is the method used to identify the user in question [1]. In addition, access control is the confidentiality of the ability to read data. Because confidentiality is frequently associated with encryption in computer security, it is used to enforce an access control policy [2].

Web vulnerabilities enable the attacker to exploit them and carry out the attack, resulting in data exposure, modification, exploitation, denial of service, and so on. When there are no restrictions on access to the authorized person, only sensitive data such as user accounts and data can be accessed, sensitive data modification, and so on [3]. Access restrictions should specify who can access this data according to their needs. However, some security vulnerabilities may allow a hacker to bypass authorization and perform tasks that would typically restrict functions to users with higher privileges, such as an administrator. An attacker might be able to add or remove plugins and users, for instance, by taking advantage of this vulnerability [4].

In the last version of OWASP Top 10, the Broken Access Control became the first position, which means that recently, it has been the most exploited vulnerability by attackers. Some authors examined websites for access control vulnerabilities using methods such as Dorks, URL Crawling, and Forced Browsing [4, 5]. On the other

E. Almushiti · R. Zaki (✉) · N. Thamer · R. Alshaya
Department of Information Technology, College of Computer, Qassim University, Buraydah, Saudi Arabia
e-mail: 441212471@qu.edu.sa; Elaf.cyber@gmail.com; 441212470@qu.edu.sa; raseel.zaki@gmail.com; 441212456@qu.edu.sa; NoraTH@outlook.sa; 441212459@qu.edu.sa; 3rima3@gmail.com

© The Author(s), under exclusive license to Springer Nature Switzerland AG 2024
A. Shaikh et al. (eds.), *Advances in Emerging Information and Communication Technology*, Signals and Communication Technology,
https://doi.org/10.1007/978-3-031-53237-5_16

hand, some other researchers were concerned with the detection of access control flows [6, 7]. Also, there was some research fux on the protection against Broken Access Control by applying some policies [8, 9]. After our study of the most important research in access control, we find that static detection is mostly used in research. Dynamic detection is preferable to static detection; however, because static detection relies on source codes, it is difficult for a third party to identify access control problems because no one will share their source codes. More studies on dynamic detection would be preferable.

The contributions of our research are:

- Represent the history of Broken Access Control.
- Outline the vulnerabilities of Broken Access Control.
- Clarify how to detect the Broken Access Control and the protection against it.
- Review the most important research on this topic.

The rest of this paper is organized as follows. Section 2 is the history of Broken Access Control. Section 3 is an overview. Section 4 is about the most important related work. Section 5 about vulnerabilities. Section 6 is about protection and security. Section 7 is our finding. Finally, Sect. 8 is the paper's conclusion.

2 History-Related Work

This section will discuss how the Broken Access Control appeared on different versions of OWASP's Top 10. In the first version of OWASP's Top 10 in 2003, there was risk under the name of (Remote Administration Flaws) Its order was A2. In version 2004, the (Remote Administration Flaws) changed to (Broken Access Control) in the same order as A2 and with the same meaning. In version 2007, the (Broken Access Control) split into A4. (Insecure Direct Object Reference) and A10. (Failure to Restrict URL Access). Version 2010, only (Failure to Restrict URL Access) is moved from A10 to A8, and the (Insecure Direct Object Reference) stayed in the same order and name [1, 4]. In version 2013, (Failure to Restrict URL Access) broadened into (Missing Function Level Access Control) and moved from A8 to A7, and the (Insecure Direct Object Reference) stayed in the same order and name. In version 2017, A4, (Insecure Direct Object Reference), and A7, (Missing Function Level Access Control) merged into A5, (Broken Access Control). Finally, in version 2021, the (Broken Access Control) moved from A5 to A1 [10, 11].

3 Overview

Businesses have altered how they deliver services to their customers to meet rising expectations and quickly evolving behavioral patterns due to the pace of the current technology revolution. In practically every industry today, web applications are the go-to tool for businesses to manage their business processes, such as supply chain management, customer relationship management, staff management, and the like. A web application can quickly and securely respond to different service requests from its authorized users by utilizing the session management functionality.

Typically, a session of the application has been started by verifying the user's identity using specific factors, like username and password. As a result, the application may give users a comfortable, personalized environment where they feel at ease and enjoy just accessing resources that have been granted permission. Access control features guarantee restricted access to web resources like web pages, database tables, and the like. They are also a security setting that prevents unauthorized access from attackers.

While the usage of web apps to manage corporate activities signaled a significant turning point in the history of modern companies, there is also a greater danger of loss if the application's vulnerability persists due to sloppy design and coding throughout its creation. The Broken Access Control (BAC) weakness, which is considered a high-priority issue due to its prevalence in web applications and the significant harm it may cause, is prominently featured in the latest OWASP list [12].

The web application architecture's access control section may contain design errors, including inadequate input validation, leakage of sensitive data, misconfiguration of sessions, and directory readability, among others. These vulnerabilities might potentially grant elevated rights to both regular users and malicious actors. The exploitation of the BAC vulnerability can have significant consequences for applications, such as illegal access to regions with administrator credentials or complete compromise of a web application, among other potential harms.

The discovery of authentication and access control vulnerabilities in widely recognized software applications such as IIS and WordPress, which have been identified as prominent weaknesses on the OWASP 2021 list, is not surprising [13]. Instances of vulnerabilities such as structured query language injection (SQLi), cross-site request forgery (CSRF), cross-site scripting (XSS), local file inclusion (LFI), local file disclosure (LFD), and remote file inclusion (RFI) are commonly identified due to inadequate user authentication and active session management. This issue ranks among the highest risks, as indicated by the Open Web Application Security Project (OWASP) [1]. Numerous scholarly investigations have examined the issue of user identification and access control in the context of preventive measures [2].

Access control, also referred to as authorization, encompasses the process of granting or denying specific requests made by an individual, software application, or system procedure. In order to effectively manage access, it is important to both bestow and withdraw such credentials. It is important to acknowledge that there exists a distinction between permission and authentication in the context of confirming access to specific features or resources, particularly in terms of proving identity.

The access control functionality sometimes spans multiple software domains, contingent upon the intricacy of the access control system. Other features that must be developed or managed in an access control system include tasks such as handling access control information and implementing caching mechanisms for scalability purposes [14].

3.1 Access Control Types

From the viewpoint of the user, access controls are divided into the following categories:

Vertical access controls: Users who lack access to other forms of functionality can access sensitive functionality only through access controls (see Fig. 1). According to vertical access controls, various user categories can access distinct functions of an application. It may be possible for an administrator to change or delete an individual user's account, which is different from the ability of a regular user. Vertical access controls may be more successful as a more granular way of enforcing corporate regulations like least privilege and separation of roles [3, 15].

Horizontal access controls: As the name implies, horizontal access control prevents users from using resources (see Fig. 2). Users can access different resources if they use vertical access controls. For instance, using a banking application, a user cannot examine transactions or make payments from another user's account [3].

Context-dependent access controls: Context-dependent access controls allow the current state of application's or the user's interactions to decide the extent to which resources and capabilities can be accessed by other users (see Fig. 3). Thanks to context-dependent access constraints, a user cannot conduct activities

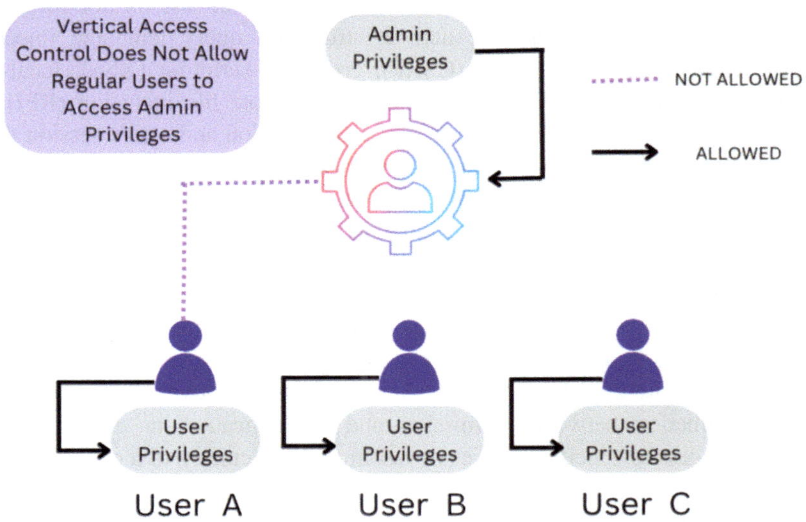

Fig. 1 Vertical access controls [15]

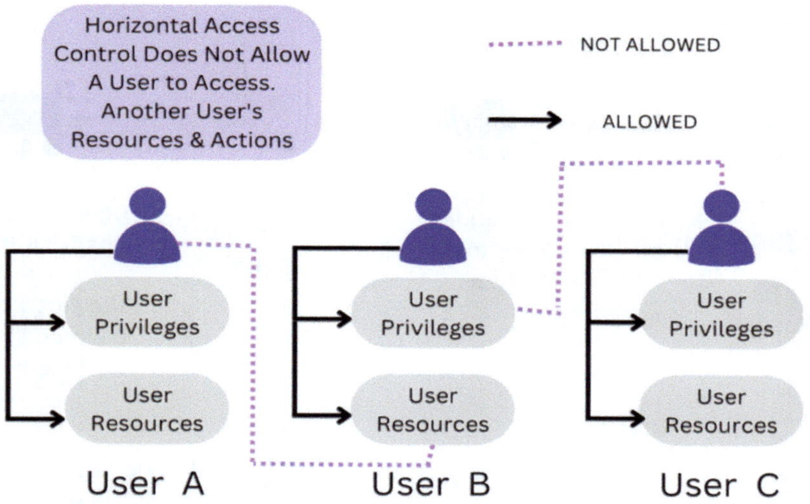

Fig. 2 Horizontal access controls [15]

in the incorrect sequence. For instance, a retail website can forbid users from changing the items in their shopping basket once they have paid [3, 16].

3.1.1 Access Control Systems

Secure your home or business by allowing only the right people in and keeping others out. Controlling access can increase security at home or in the workplace. In addition, they provide information about who has access to which areas at any given time and allow you to control their movements.

Access control systems are divided into three types: discretionary access control (DAC), role-based access control (RBAC), and mandatory access control (MAC). The various types of access control systems each function in their own unique way to provide you command over your premises.

Discretionary Access Control

DAC leaves some access control to the owner of the item or anybody else allowed to govern the object's access. It is commonly used, for example, to restrict a user's access; the owner manages other users' permission to the file. Only the users selected by the owner may have read, write, execute, and other permissions on the file. The DAC policy is extremely adaptable and frequently employed in both the business and government sectors. DAC, on the other hand, is recognized to be intrinsically weak [17]. For homes or small businesses that require security for one or two doors,

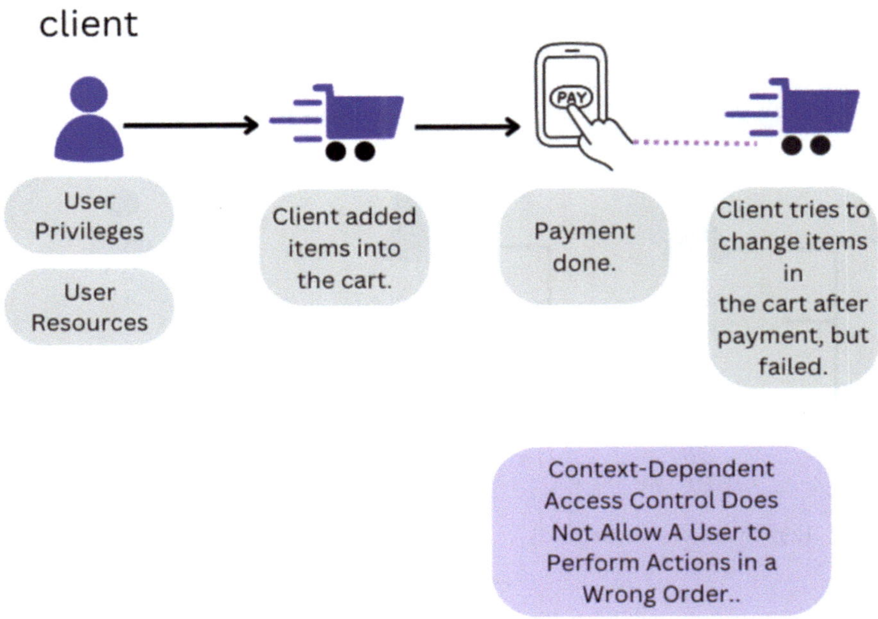

Fig. 3 Context-dependent access controls [15]

discretionary access control is a good option. The business owner or the homeowner can have full authority regarding who enters and exits compared to other types of access control systems. A person can quickly offer someone their admission key or the secret code for the entrance, and they may customize levels of safety and permissions for others. The access control system is standalone, localizing security at the door.

Mandatory Access Control

Compared to DAC systems, mandatory access control systems provide a greater level of security. Mandatory access control contributes to better security and data protection by preventing the end user from controlling who has access to the facility. Owners and system administrators are responsible for access control. An administrator can classify end users using such a system and allow certain areas to be accessible and not others [18]. Even if this solution might fulfill significantly more complex security demands, small to medium organizations are still the ideal candidates for using it. New labels must be assigned manually each time someone needs access to relevant areas, or permissions must be removed when someone leaves. A large number of employees would make this impractical.

Role-Based Access Control

Permissions are doled out to end users of a system that uses role-based access control for doors in accordance with the role that they play in the business or household in question. System administrators control access through this type of access control system, and their responsibilities determine their security allowances. Rather than assigning permissions individually, they will be automatically given the required permissions since their roles are associated with the appropriate security levels [18]. This access control system is most suited for large companies with a high number of employees; however, it can also be useful for smaller companies or families who are trying to increase their level of safety.

3.1.2 Access Control Software

A modern security system is more technologically advanced than a lock and a metal key. Authentication and access privilege control software are also involved. There are several types of access control software (see Fig. 4).

4 Related Work Vulnerabilities

In Hassan, Md Maruf, et al.'s [2] paper, the authors search for websites that have access control vulnerabilities. They discover 330 websites by using dorks and then examining them based on some vulnerabilities, i.e., the topics considered include the disclosure of private information, uncommon redirections, misconfigurations in session management, the programming languages utilized, the operating systems employed, and the servers/platforms involved. The findings showed that Broken Access Control vulnerabilities impacted 39.09% of websites.

In Pan K. and Wang Q. [6] research: To adapt to VUE, they adopted the constructor of grammar without context technique, enhanced it, and connected the extraction process. Two different vulnerability prototypes and patch versions of those two different vulnerability prototypes were created. The AccVulPrototype1 prototype is employed to conduct a comparative analysis of access control breaches between administrator positions and normal user roles. The access control breaches between normal user roles and guest roles are contrasted using a prototype known as AccVulPrototype2.

In Song, Ho-Gil, et al. [8], the authors examine some websites for access control vulnerabilities using URL Crawling and Forced Browsing.

URL Crawling is the practice of identifying and gathering all of the accessible hypertext links on a specific website. Additionally, they employed an algorithm to find privilege escalation. In [19], they offer a method for rapidly identifying access control weaknesses in web applications with little manual work. They employ a static analyzer that checks for known and undiscovered access control flaws.

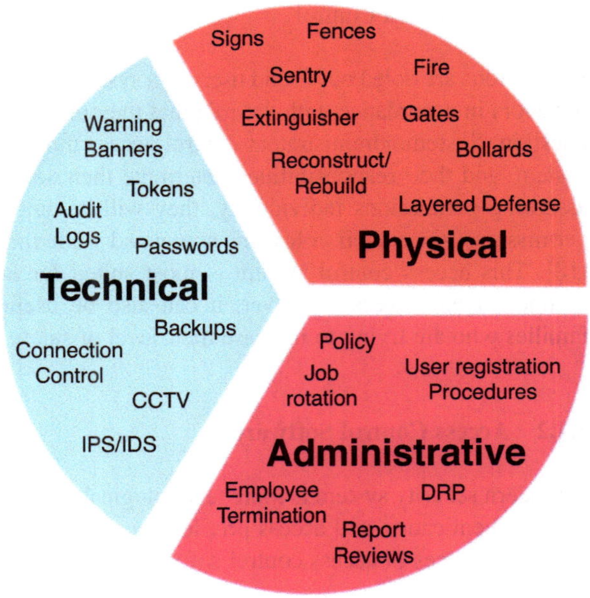

Fig. 4 Access control software types [16]

Research in [20] focuses on access control vulnerability detection using a black box technique. The term "black box" refers to a system that analyzes a live web application by externally interacting with it, without relying on or accessing any internal data, including source code or authorization configurations.

The approach has the potential to detect access control vulnerabilities at both the function level and object level within the scope of HTTP GET queries. The primary objective of the current focus on GET queries is to mitigate the potential occurrence of unfavorable side effects and alterations to an application's status that may arise from PATCH, POST, DELETE, or PUT requests. These requests often lead to alterations in the data as/or the state of the application, potentially impacting its visual appearance and functionality. Consequently, during the testing phase, the program's state would undergo frequent changes, thereby posing significant difficulties in comparing the application's behavior across different users.

In Son S. et al.'s [12] research, they fix access control flaws by enforcing their "FIXMEUP" policy. The process starts with the specification of a high-level access control policy that establishes one or more access control checks on execution routes leading to sensitive actions, including database queries, links to privileged pages, operations that rewrite cookies, and file deletion. Sensitive operations need to be stated in advance. The program does not carry out the delicate procedures if the checks are unsuccessful. Since user roles and entry points within the same application have varying access control logic, various paths could require different checks—or none at all. Access control logic is commonly interprocedural and context-sensitive in web applications.

In Vincent C. Hu et al.'s [4] research, to assist agencies in securing their computer applications, this article presents an overview of access control rules, models, and techniques. Each access control policy is discussed in the document, along with its capabilities, limitations, and qualities.

5 Vulnerabilities

A security flaw defined as a Broken Access Control vulnerability allows for the possibility of limited resources being accessed by a third party who is not allowed to do so. Attackers are able to evade typical security processes and obtain unauthorized access to private information or systems when they take advantage of this flaw and exploit it. Access control vulnerabilities frequently result from shoddy authentication and permission procedures, giving attackers access to unauthorized rights. The elimination of such vulnerabilities is essential for the safety of your systems and data [21].

5.1 Access Control Common Vulnerabilities

Even while there are occasions when users should only have access to a limited number of methods, this is frequently the case. This could be the result of the developer not completely understanding which subset of proxy methods should be accessible and giving access to all of them, or it is because the HTTP server's mapping API defaults to granting all methods access (see Fig. 5).

In theory, requests that specify the execution of a server-side API need not be any less secure than ones that specify the running of a script on the server side or the use of additional resources. However, in actual use, this kind of device typically has flaws. Frequently, the client communicates straight with server-side API calls by getting beyond the application's customary access rules or unanticipated input vectors.

There are some common access control vulnerabilities [11], including the following:

Completely Unprotected Functionality
Functionality and resources that are particularly sensitive can frequently be accessible by anyone with the necessary URL, thanks to failed access controls. For instance, in many programs, the administrative features are fully accessible to anyone who accesses a particular URL.

In this instance, the application usually only implements access control to the level that it only shows administrators—and not other users—a link to this Site on the user interface.

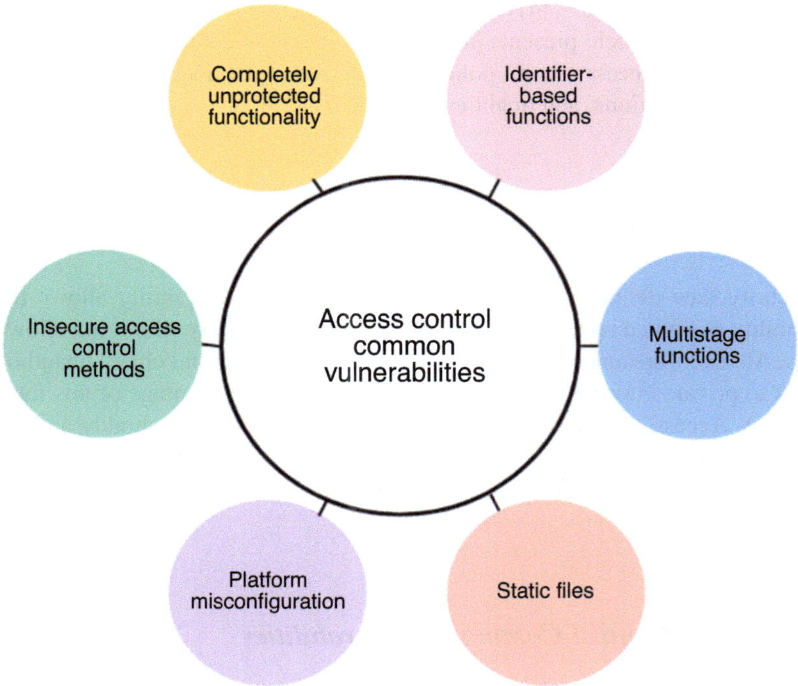

Fig. 5 Common vulnerabilities of access control

Identifier-Based Functions

It is not uncommon to see an identification for the resource being requested being supplied to the server as a requested parameter when an application procedure is used to acquire access to a particular resource. This identifier can be found either within the query string of the URL or within the body of a POST request. When the user who owns the document is logged in, a link to this URL will appear on the page that displays the user's My Documents. The link is hidden from other users.

On the other hand, if the access controls are broken, any user who queries the relevant URL may be able to read the material in the same manner as the authorized user. The issue is significantly worse and easier to exploit when the IDs are readily predictable.

Multistage Functions

An application's implementation of many functions occurs throughout multiple phases, necessitating numerous requests from the client to the server. To add a new user, you must first select this choice from a user service menu, then select the department and role of the user from drop-down menus, and finally enter the new username, first password, and any other information that is required. Programs typically contain delicate functionality that has been secured against unauthorized access. However, the access controls employed are inadequate due to false expectations about how the functionality would be used.

Static Files
By making queries to dynamic pages running on the server, users can frequently access forbidden resources and capabilities. Each page must accomplish the required access control to checking and confirming that this user has the required rights to perform the intended action. The static resources, which are found in the server's web root, are sometimes the target of requests for protected resources.

Platform Misconfiguration
Some apps use controls to regulate access at the web server or application platform layers. A user's job within the application typically determines whether they have access to specific URL paths. For instance, users not members of the Administrators group might not be allowed access to the /admin path. This is a perfectly acceptable method of restricting access, in theory. However, incorrectly configured platform-level protections frequently make it possible for illegal access to occur. The configuration of platform-level frequently consists of rules like the policy rules of firewall, which allow or deny access depending on the following criteria: the method for the HTTP request, the path of the URL, and the role of the user.

Information retrieval is the primary target of the GET method, whereas actions that modify the application's data or state are carried out using the POST method. Only authorized access may happen if care is taken to build rules that appropriately provide access based on the proper HTTP methods and URL destinations.

Insecure Access Control Methods
Access control vulnerabilities fall under the heading of insecure access control methods. It happens when a program reads input from the user in order to directly access objects and an attacker is able to modify the input in order to get unauthorized access. Some apps use an access control architecture that is fundamentally insecure because it bases decisions on the client's request parameters or other factors that an attacker can manipulate. Many insecure access control methods [12, 22] are as follows:

Parameter-Based Access Control

In certain iterations of this paradigm, the determination of a user's role or access level is carried out by the application during the login process. Subsequently, the aforementioned information is transmitted through the client by means of a concealed form field, cookie, or pre-established query string parameter. Users' regular URLs can have a different parameter or none at all. Anyone with knowledge of the option designated for administrators can easily set it in their requests to access organizational features. Without actually accessing the application as a user with high privileges and observing the requests made, this access restriction may occasionally be challenging to spot.

Referer-Based Access Control

Few websites based access limitations on the HTTP request's Referer header. In order to identify the website from which a request was made, browsers often append the Referer header to requests. Consider a scenario in which an application imposes access restriction firmly on the central administration page at /admin but not on sub-pages like /admin/delete. Only the Referer header is examined by the user. The request is accepted if the principal/admin URL is in the Referer header. In this scenario, an attacker may fake direct requests to sensitive sub-pages to give the required Referer header and gain unauthorized access because the attacker may have complete control over the Referer header.

Location-Based Access Control

Depending on the user's location, some websites impose access limits on resources. This may be the case for media services or financial applications subject to local laws or commercial constraints. Getting around these access limitations is frequently possible by using web proxies, VPNs, or geolocation tricks on the client side.

Here are a few typical workarounds for them:

- Making use of a web proxy with a local server
- Making use of a VPN that finishes in the necessary place
- Making use of a mobile device with data roaming capabilities
- Direct interference with client-side geolocation systems

5.2 Broken Access Control Vulnerability

An application that does not require users to authenticate before viewing or updating sensitive data is an example of vulnerability for a failed access control. An attacker could use this vulnerability to access private data or alter it without authorization to clarify more (see Fig. 6).

Another example of a problem with access control is an application that inappropriately limits access to particular functions based on a user's role. For example, an administrator account should be allowed to add new users to the system, but a regular user account should not. If the application does not limit access to the feature, a regular user could create extra users in the system and possibly grant them administrator rights. Attackers could exploit these vulnerabilities to access sensitive data without authorization or change data without permission. Organizations should implement sufficient security controls to reduce the risk posed by these vulnerabilities.

Fig. 6 Example of an attacker exploiting the vulnerability [15]

5.3 Broken Access Control Vulnerability Identify

The presence of Broken Access Control vulnerabilities is associated with a multitude of attack avenues. However, some of the most popular ways to exploit these weaknesses are as follows:

Injection Flaws

When an application is subjected to the injection of untrusted input, injection defects happen, leading to unexpected behavior, this can be used to change application data or obtain unwanted access to sensitive data.

Cross-Site Scripting (XSS)

Untrusted input incorporated into a web page's output might cause XSS issues. Attackers can utilize this to launch destructive scripts in the user's browser, which may result in cookie theft, session hijacking, or other undesirable actions.

Broken Item Authentication and Session Management

Applications that improperly validate or safeguard data related to user authentication and sessions have broken authentication and session management issues. This vulnerability can be exploited by an unauthorized individual to get unauthorized access to resources or sensitive data.

Implementing security controls like input validation, appropriate session management, and permission controls is essential to preventing exploits of Broken Access Control vulnerabilities [10].

6 Protections and Security

The application's access controls, which are in charge of making important decisions, serve as a crucial protection mechanism. When the access control is broken, the attacker can frequently compromise the entire application, seizing control of administrative features and obtaining private information belonging to every other user. Access control has been compromised if any user can access resources or functionality for which they are not permitted. According to the three categories of controls, there are three primary types of attacks on access controls vertical privilege escalation, horizontal privilege escalation, and business logic exploitation. An illustration of vertical privilege escalation is the ability to access other user privileges on different levels. In contrast, horizontal privilege escalation is the access to other user privileges at the same level, and business logic exploitation depends on the situation and what the attacker can access or what he finds. At the same time, he tries to break the access control. This section will present how to protect web applications from Broken Access Control and detect if the web application is infected.

6.1 Protection from Vulnerabilities

To secure web applications, we need to build effective access control. The developers should build effective access control by taking care of all vulnerabilities that may open the door to attackers to exploit them. The developers also should take care of the following:

- The user's knowledge and not assume the users are unaware of the URLs, account numbers, and document IDs.
- Assume that not all user-submitted parameters are trusted.
- The user may access pages in an inappropriate order.

Also, there is some good practice the developers can follow, such as based on the user's session for all access control actions and checking access restrictions using a central application component. In addition, if the functionality is sensitive, like administrative pages, the developers can further limit access by the utilization of IP addresses which serves as a means of verification to restrict access to the functionality just to users within a designated network range. There are too many good practices that can be followed. However, the most efficient method for creating multiple layers of security as part of the defense-in-depth approach is to use a multilayered privilege model that implements access controls at each of these layers. We can use a matrix similar to the one in (see Fig. 7) that shows users' roles within the application and the privileges that should be granted to each role at each tier. Finally, we can say protection as several layers are adequate because if the attacker succeeds in breaching one layer, he cannot go through the other layers [5].

Application Server		Application Roles	Database Privileges													
User type	URL path	User role	Search	Create Application	Edit Application	Purge Application	View Applications	Policy Updates	Rate Adjustment	View User Accounts	Create Users	View Company Ac	Edit Company Ac	Create Company	View Audit Log	Delegate privilege
Administrator	/*	Site Administrator	✓	✓	✓	✓	✓	✓	✓	✓	✓	✓	✓	✓	✓	✓
		Support	✓		✓		✓	✓		✓	✓	✓	✓	✓		
Site Supervisor	/admin/* /myQuotes/* /help/*	Back Office – New business		✓			✓									
		Back Office – Referrals		✓	✓	✓		✓	✓							
		Back Office – Helpdesk	✓				✓			✓		✓			✓	✓
Company Administrator	/myQuotes/* /help/*	Customer – Administrator		✓	✓	✓	✓				✓	✓	✓			✓
		Customer – New Business		✓		✓	✓									
		Customer – Support	✓				✓			✓						
Normal User	/myQuotes/dash.jsp /myQuotes/apply.jsp /myQuotes/search.jsp /help/*	User – Applications	✓	✓			✓									
		User – Referrals														
		User – Helpdesk														
		Unregistered (Read Only)	✓				✓									
Audit	(none)	Syslog Server Account													✓	

Fig. 7 An example of privileges matrix [11]

6.2 *Detection of Vulnerabilities*

Static and dynamic analyses are the two methods available for detecting Broken Access Control vulnerabilities. Dynamic analysis is used to find bugs in programs that are already running. The opposite strategy is used for static precision analysis, implemented in a non-runtime context.

These four types of static analysis techniques [6]:

Forcing a Browser
By bypassing links presented on the page and instead using some brute force techniques, force browsing refers to gaining immediate access to a webpage. It is common practice to replicate the actions of low-level users who acquire access to high-level users' resources using force browsing methods [8].

Context-Based Detection
The program's security-sensitive operations must first be identified via context-based detection technique [12]. In order to achieve this, researchers commonly employ without-context grammar and other methodologies. According to researchers, it is argued that security-sensitive operations ought to possess consistent contextual elements and redirects. If there is consistency between the two, it becomes feasible to determine whether the software exhibits access control concerns.

Model-Based Detection
In order to construct a comprehensive framework, it is imperative for researchers to initially analyze the architecture of an access control implementation [6]. This analysis serves as the foundation for developing a model that accurately describes

the aforementioned framework. The model incorporates components for the purpose of access control. Utilize the model to assess the accuracy of the web application's implementation of the access control mechanism, with the objective of identifying possible weaknesses.

Graph-Based Detection

The application of this detection method is very flexible. Web applications are described by some academics using control flow graphs, site maps, and permission verification graphs. Nodes in a graph frequently stand in for pages or other resources in a web application. Researchers frequently employ algorithm restrictions, flow analysis, and other techniques to determine the reachability between nodes. A privileged node has access control flaws if a non-privileged user can access it [6].

This technology is very scalable, and when nodes are detected, appropriate solutions can be chosen based on the features of the web service. Hence, the bulk of researchers have employed this technique.

7 Findings

Most of the research depends on static detection. Static detection has some limitations, such as depending on source codes, making it difficult for the third party to detect Broken Access Control because no one will give source codes to the third party to examine it. Also, static detection cannot detect the new hacking techniques. It would be better if we had more dynamic detection research.

8 Conclusion

This paper discusses the most used access mechanisms, and it is possible to expand on the different mechanisms. All mechanisms and access control designs have trade-offs and limitations, so it is the user's responsibility to determine which mechanism provides the best fit for their business functions and requirements. There are also several commonly used access control policies included in this document.

As with access control mechanisms, users must select the most appropriate policies for their needs. It is discussed how to achieve safety and the complicated issues involved with it.

References

1. THE OWASP®. OWASP Foundation, the Open Source Foundation for Application Security—OWASP Foundation. (n.d.), https://owasp.org/. Last accessed 20 Sept 2023

2. A. Alotaibi, L. Alghufaili, D.M. Ibrahim, Cross site scripting attack review. ISC Int. J. Inf. Secur. **13**(3), 21–30 (2021)
3. Access control vulnerabilities and privilege escalation. Web Security Academy. (n.d.), https://portswigger.net/web-security/access-controlvertical-access-controls. Last accessed 20 Sept 2023
4. M.S. Albulayhi, D.M. Ibrahim, Open web application security project components with known vulnerabilities: a comprehensive study. ISC Int. J. Inf. Secur. **13**(3), 59–67 (2021)
5. S. Alotaibi, K. Alharbi, B. Abaalkhail, D.M. Ibrahim, Sensitive data exposure: data forwarding and storage on cloud environment. Int. J. Online Biomed. Eng. **17**(14), 4–18 (2021)
6. A.H. Alrobaish, W.F. Al-mutairi, H.A. Alsuqayhi, D.M. Ibrahim, Common attacks on near field communication technology, in *2nd International Conference on Computing and Information Technology (ICCIT)*, (IEEE, Tabuk, 2022), pp. 110–114
7. M.M. Hassan, M.A. Ali, T. Bhuiyan, S. Biswas, Quantitative assessment on broken access control vulnerability in web applications, in *International Conference on Cyber Security and Computer Science 2018*, (2018)
8. V.C. Hu, D.F. Ferraiolo, D. Rick Kuhn, Assessment of Access Control Systems – csrc.nist.rip. (2006), https://csrc.nist.rip/external/nvlpubs.nist.gov /nist-pubs/Legacy/IR/nistir7316.pdf. Last accessed 20 Sept 2023
9. H. Aljoaey, K. Almutawa, R. Alabdali, D.M. Ibrahim, Broken authentication and session management vulnerabilities. ISC Int. J. Inf. Secur. **13**(3), 11–19 (2021)
10. D. Stuttard, P. Marcus, *The Web Application Hacker's Handbook: Finding and Exploiting Security Flaws* (Wiley, 2011)
11. K. Pan, amp; Wang, Q., Static detection of access control vulnerabilities in vue applications. J. Phys. Conf. Ser. **1646**(1), 012021 (2020)
12. M. Alsalamah, H. Alwabli, H. Alqwifli, D.M. Ibrahim, A review study on SQL injection attacks, prevention, and detection. ISC Int. J. Inf. Secur. **13**(3), 1–10 (2021)
13. H.G. Song, Y. Kim, K.G. Doh, Automatic detection of access control vulnerabilities in web applications by URL crawling and forced browsing. Inf. Sci. Technol. **3**(2), 482–486 (2012)
14. F. Sun, L. Xu, Z. Su, Static detection of access control vulnerabilities in web applications, in *20th USENIX Security Symposium* (USENIX Security 11), 64(1) (2011)
15. Broken Access Control, https://owasp.org/www-community/BrokenAccessControl. Last accessed 20 Sept 2023
16. What Is Broken Access Control Vulnerability, https://www.eccouncil.org/cybersecurity-exchange/web-application-hacking/broken-access-control-vulnerability. Last accessed 20 Sept 2023
17. D. Stuttard, M. Pinto, J.J. Pauli, *The web application hacker's Handbook: Finding and exploiting security flaws* (Wiley, 2012)
18. Access control vulnerabilities and privilege escalation, https://portswigger.net/web-security/access-control. Last accessed 20 Sept 2023
19. M. Rennhard, M. Kushnir, O. Favre, D. Esposito, V. Zahnd, Automating the detection of access control vulnerabilities in web applications. SN Comput. Sci. **3**(5), 376 (2022)
20. Son, S., Shmatikov, V., McKinley, K. S., FixMeUp: repairing access-control bugs in web applications. Microsoft Research, NDSS. https://www.microsoft.com/en-us/research/publication/fixmeup-repairing-access-control-bugs-in-web-applications/. Last accessed 20 Sept 2023
21. A comprehensive guide to broken access control. PurpleBox (2022), https://www.prplbx.com/broken-access-control/. Last accessed 20 Sept 2023
22. Wallarm, A5: Broken Access Control! – top 10 owasp 2022. RSS (n.d.), http://www.wallarm.com/what/a5-broken-access-control-2017-owasp. Last accessed 20 Sept 2023

Considering Uncertainty Expression in Sentiment Analysis and Tweet Classification

Zendaoui Fairouz and Hidouci Walid Khaled

1 Introduction

Automatic information extraction from noisy sources has opened up new possibilities for querying and analyzing data, particularly in the current era of social media dominance, where microblogs like Twitter are likely the most popular source of information. Several studies have focused on information extraction from microblogs for various purposes such as opinion investigation [1], rumor detection [2], information retrieval [3], user behavior tracking [4], prediction [5], sentiment analysis [6] and data modeling [7]. Microblogging platforms provide a wealth of information where users share their opinions and exchange information on various topics, but these data are often subjective and uncertain. Many social media-based applications require the detection of uncertainties in user's statements, as they increasingly rely on these platforms as a source of information to generate or deduce interpretations based on them [8]. Identifying uncertainty is a fundamental task in semantic processing for many natural language processing techniques and applications, such as question answering, information extraction, etc., as it determines the quality of information in terms of factualness. Since writings on social media are generally produced in a highly inconsistent manner, factualness becomes a significant issue [9]. Detection techniques for uncertain text that rely on lexical cues are limited in social media, where cue phrases are often missing or poorly formulated. Users express their uncertainty in a manner different from natural language. For example, instead of using cue phrases like "I'm not sure" or "It is possible that," users may often ask questions, refer to external information, or provide additional

---

Z. Fairouz (✉) · H. W. Khaled
Laboratoire de la Communication dans les Systèmes Informatiques, Ecole Nationale Supérieure d'Informatique, Oued-Smar, Alger, Algeria
e-mail: f_zendaoui@esi.dz; http://www.esi.dz; w_hidouci@esi.dz; http://www.esi.dz

© The Author(s), under exclusive license to Springer Nature Switzerland AG 2024
A. Shaikh et al. (eds.), *Advances in Emerging Information and Communication Technology*, Signals and Communication Technology,
https://doi.org/10.1007/978-3-031-53237-5_17

details that indicate their uncertainty. Therefore, uncertain text detection techniques must be capable of recognizing these different expressions and incorporating them into their analysis. Sentiment analysis can be particularly affected by this aspect, as opinions and attitudes expressed in microblogs can often be more nuanced and complex than in more formal texts. Expressions of uncertainty can also have a significant impact on the polarity or intensity of the expressed opinion, making it challenging to accurately interpret sentiment analysis results. For this reason, sentiment analysis techniques need to consider the semantic uncertainties expressed in microblogs in order to generate more realistic results. The research questions addressed in this article are as follows: How to represent uncertainty expressed in a microblog? How to classify microblogs into semantic uncertainty categories? How to conduct sentiment analysis while considering these uncertainties in tweets? To address these research questions, we conduct an experimental study to explore a form of uncertainty present in historical knowledge, specifically the intrinsic uncertainty expressed in microblogs. Our work builds upon the findings presented in our state-of-the-art article on uncertainty identification in microblogs [10]. Additionally, we employ our model for representing uncertain historical knowledge proposed in previous works [11, 12] to implement our experiments. We analyze the sentiment expressed on Twitter while considering the semantic uncertainties present in the tweets. We propose an algorithm for classifying tweets based on their degree of uncertainty and sentiment, and examine sentiments related to the Covid-19 pandemic. Furthermore, we explore whether the expression of uncertainty could affect the outcome of the sentiment analysis and analyze relevant uncertainty categories to consider in our case study. This chapter is organized as follows. In Sect. 2, we discuss the identification of uncertainties in microblogs. In Sect. 3, we present related works that have classified microblogs into semantic uncertainty categories. In Sect. 4, we present our experimental methodology, starting with problem definition, introducing the data collection, establishing the classification algorithm, and outlining the performance evaluation method. Section 5 is dedicated to the presentation, analysis, and interpretation of our experimental results. We conclude the chapter in Sect. 6.

2 Uncertainty Identification in Microblogs

The majority of current uncertainty annotation works are based on the CoNLL-2010 uncertainty identification task [13], in which uncertainty texts were annotated using cues typically found in uncertain statements [9]. However, existing techniques that rely on lexical cues suffer significantly from the informal or conversational nature of social media, where indicator phrases are frequently presented in a poor or even absent form within sentences. Compared to the current uncertainty corpus, the expression of doubt in social media is quite different from that in natural language, in the sense that when users make uncertain remarks, they often ask questions or refer to external information [8]. Szarvas et al. [14] proposed an interesting classification of semantic uncertainty phenomena, which has now been expanded in some research to include uncertainty in microblogs.

There are now a large number of social networks available on the Internet, and they continue to expand and evolve, both widely and deeply. The number of social media users worldwide exceeded 3.8 billion in early 2020, representing a 9% growth compared to 2019, and this number will continue to increase, driven by individuals actively transitioning to using social media on mobile devices [15]. With the rise of social media, there is a growing amount of textual content that includes a large number of informal or conversational terms. In terms of factualness, the quality of information on social media has become a major concern [8]. In general, only true information with a high level of credibility holds value. However, the majority of social media expressions are posted with assumptions and epistemes that cannot be determined as true or false at the moment.

Previous research has examined the existence of uncertain statements and found them to be common on social networking sites such as Facebook, Twitter, and Sina Weibo. According to a statistic from the Twitter dataset [8], at least 18.91% of tweets exhibit uncertainty characteristics. As a result, the quality of information on social media has become a genuine concern for various social media-related activities, such as rumor detection and credibility assessments. Users need to be able to identify uncertainty in order to synthesize data and generate reliable interpretations [9]. However, the identification of uncertainty in the context of social media is rarely studied. Furthermore, unlike biological studies and Wikipedia entries, social media posts are typically brief and informal. Many cue phrases are presented in a substandard form or even omitted from sentences due to word count constraints and casual discourse. Ambiguous semantics will be implicitly provided by the entire sentence rather than explicitly by cue words in this situation [9]. Therefore, the application of current out-of-domain corpora to the context of social media is ineffective. Moreover, compared to the existing uncertainty corpus, the expression of uncertainty in social media differs from that in formal language in the sense that when users make uncertain statements, they frequently ask questions or refer to external information. However, none of the uncertainty expressions can be described using the known categories of uncertainty from the literature. Therefore, in the context of social media, a distinct method of categorizing uncertainties is required [8].

Wei et al. [8] conducted an empirical study on the identification of uncertainty in social media texts, taking into account features other than raw text, such as the number of tweets and their relationships. Vincze [16] proposed the use of lexical, morphological, syntactic, semantic, and discourse-based features in a supervised classifier. Based on supervised sequence labeling algorithms, Al-Sabbagh et al. [17] provided a unified framework for identifying and extracting uncertainty cues, holders, and scopes in Arabic tweets. Deep learning has also been used in uncertainty detection. On the CoNLL-2010 reference datasets, Adel & Schutze [18] proposed an attention-based architecture for uncertainty detection that used a CNN or RNN with an attention mechanism to achieve state-of-the-art results. An external lexicon of starting cue words or phrases was also incorporated into word embedding, and their model performed well on English datasets with this external information.

However, due to the long sentences that frequently appear in English and Chinese social media, and since the use of a CNN or RNN leads to a loss of semantics, this model is not satisfactory for social media texts [17]. Han et al. [19] proposed an approach with three major differences compared to previous work: (1) their model solely uses word embedding without additional knowledge, external systems, or cue words; (2) their proposed neural networks employ LSTM and attention mechanism to generate semantic focus, effectively representing long sentences and typical low-quality expressions found in social media texts; and (3) they are the first to construct a unified experimental dataset.

3 Related Work

This section of the study focuses on similar investigations that used the semantic categorization of uncertainty proposed by Szarvas et al. [14] to detect uncertainties in microblogs. First, we will review the work of Wei et al. [8]. The latter investigated the effectiveness of several categories of social media context characteristics in an empirical study of identifying uncertainty on a dataset of tweets. They made three main observations. Firstly, they found that no tweets of the Investigation type were identified. They noticed that when users post tweets, they rarely use terms such as "review" or "test," which are indicators of the Investigation category. When they have, the assertion should be considered extremely certain. Secondly, the researchers observed that individuals regularly ask clarifying questions on specific topics, demonstrating their hesitation or uncertainty. This suggests that people are not always confident in their knowledge or understanding of certain subjects, leading them to seek clarification and additional information. Lastly, the team noticed that people are more likely to share messages containing external information, such as a story from a friend. This behavior implies a level of doubt, as individuals rely on outside sources to support or validate their claims. The tendency to seek external validation suggests a lack of complete confidence in the information they are sharing. These three observations shed light on the patterns and dynamics of user behavior on social media platforms when it comes to expressing certainty or doubt in their posts.

Wei et al. [8] developed a variation of uncertainty types in the context of social media based on these findings, removing the Investigation category and adding the Question and External categories under hypothetical. Their suggested technique is based on the work of Kiefer [20], which was previously developed by Szarvas et al. [14] to normalize uncertainty corpora in other genres. However, they did not test these expanded schemes for specific genres, as even the basic one [20] proved inadequate in the domain of social media. As a result, Wei et al. [8] proposed a new classification method to identify uncertainty in social media and created the first uncertainty corpus using tweets. They performed uncertainty identification tests on the created dataset to assess the performance of different types of features, such as content-based features, user-based features, and Twitter-specific features. The results demonstrate that all three types of social media-specific features can improve uncertainty detection. Furthermore, among the three, content-based features show the greatest improvement, and the presence of uncertain cues contributes the most to content-based features.

Another study was conducted by Li et al. [9], who focused on annotating Chinese social media texts to identify uncertainty. They were the first to annotate a Chinese social media corpus for an uncertainty detection challenge, the UIR Uncertainty Corpus (UUC). The annotation technique used in this study is based on the uncertainty categories established by Wei et al. [8]. However, due to differences in linguistic habits or usage between English and Chinese, several modifications were made to their annotation scheme. They noticed that the epistemic subclasses of Possible and Probable in Chinese uncertainty expressions are very similar compared to Chinese uncertainty corpora. These two subclasses were merged into a single class called Possible. Additionally, they found that in Chinese social media, there were no expressions belonging to the Dynamic subclass in any of the microblogs. Therefore, the Dynamic subclass was also removed. On the contrary, they observed that there were many words in Sina microblogs conveying the statement of uncertainty with a predicted semantics. They created a variant of the uncertainty subclass in the Chinese social media corpus based on the above data, removing the Dynamic type and adding the Hope subclass. The uncertainty schema has been redefined into six subclasses: Question, External, Doxastic, Hope, Possible, and Condition. These subclasses categorize different expressions of uncertainty, such as asking for information or testing someone's knowledge, repeating others' statements, expressing beliefs and assumptions, desires for something to happen or be true, possibilities, or conditions that may modify a proposition. This updated schema provides a more comprehensive understanding of how uncertainty is conveyed in language. Each tweet needs to be labeled as uncertain or certain in order to fulfill the annotation objective. Furthermore, rather than the existence of uncertain cues, uncertainty statements should be recognized in terms of judgments about the intended meaning by the author or implicit semantics. The subclasses should also be designated according to the annotation system. The study found that the aforementioned classification enhanced the corpus details and made it more valuable for future natural language processing (NLP) applications.

By analyzing the available works that have classified microblogs based on semantic uncertainty types, we observe that this research domain is highly context-dependent, specifically in terms of the community, their way of thinking and speaking, and their own language. Even between just two different languages (English and Chinese), there were numerous differences in the expression of uncertainty that led to these two distinct adaptations of the same semantic uncertainty classification model given by Szarvas et al. [14], maintaining and/or eliminating certain types from the model.

4 Experimental Methodology

4.1 Problem Definition

We aim to examine the sentiments expressed in tweets related to the Covid-19 pandemic while considering the uncertainty that users may express. To achieve this, we have developed a tweet classification algorithm based on their sentiment and level of uncertainty. We collect tweets in real-time using specific keywords and subject them to a series of processing steps before classifying them. As a result, each tweet is categorized as positive, negative, or neutral, and also as certain or uncertain, with a specified uncertainty category (question, condition, hope, or belief). Based on the semantic uncertainty categories adopted by Wei et al. [8] and Li et al. [9] presented in the previous section, we propose a modified classification variant suited to our case study. We proceeded as follows:

- We eliminated the "External" category as we noticed that the majority of Twitter users use the retweet option to quote others. Therefore, this category is effectively captured by retweets. We excluded retweets from the classification and considered only original tweets.
- We also removed the "Dynamic" category as it is not relevant to our context. Deontic, dispositional, or circumstantial modalities do not provide useful information for our sentiment analysis.
- We adopted the term "Belief" for the "Doxastic" category.
- We merged the "Possible" and "Probable" categories with the "Belief" category since all three categories provide information about beliefs and hypotheses.

In our classification (see Fig. 1), we distinguish two main categories:

- Uncertain: includes tweets that exhibit one or more forms of uncertainty, such as questions, conditional expressions, expressions of hope, and explicit expressions of uncertain belief.
- Certain: encompasses tweets that do not exhibit any forms of uncertainty.

We consider subcategories for the "uncertain" category, which are:

- Question: represents tweets containing requests for information, confirmation, or knowledge tests.

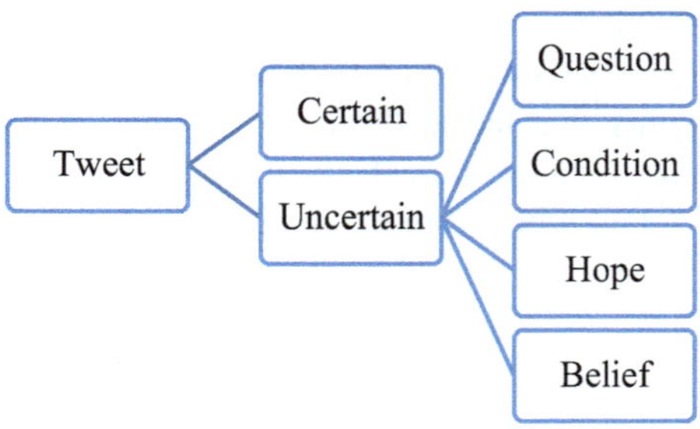

Fig. 1 Categories of tweet classification

- Condition: represents tweets containing propositions under conditions.
- Hope: represents tweets in which the user desires something to happen or be true, or promises that it will happen.
- Belief: groups tweets expressing the user's beliefs or hypotheses.

It should be noted that our model is designed for single-label annotation for the same tweet. Furthermore, when a tweet exhibits multiple forms of uncertainty, it will be classified into a category based on the following priority order: uncertain (question, condition, hope, belief), certain.

4.2 Data Collection

The set of tweets we annotated was collected from Twitter using the Tweepy streaming API over a period from February 27th to March 18th, 2022, resulting in a total of 110,701 tweets. The search criteria included the following keywords: corona, covid, coronavirus, covid-19. The tweets were stored in a MySQL database. For each tweet, we recorded the following information: tweet ID, date and time, tweet text, language, the user who posted the tweet, the user's location, and the number of followers the user has. Our classification algorithm collects and categorizes tweets in real-time (through streaming).

4.3 Classification Algorithm

We outline our classification algorithm in the following steps:

1. Importing the necessary libraries.
2. Authenticating with the Twitter API.
3. Connecting to the MySQL database.
4. Streaming tweet collection based on keywords (corona, covid, coronavirus, covid-19) using the Tweepy Streaming API (excluding retweets).
   For each tweet:
   (a) Cleaning and preprocessing of the tweet (removing usernames, links, hashtags, etc.).
   (b) Sentiment analysis of the tweet using TextBlob: positive, negative, neutral.
   (c) Classification of the tweet into its corresponding belief category (certain or uncertain (question, condition, hope, belief)) using the following methods:
      - Lexical analysis and searching for specific words and characters for the types: question and condition.
      - Semantic similarity analysis between the tweet text and uncertainty cue words for the categories: hope and belief.

(d) Saving the tweet into the database.

The algorithm was developed in Python, and the data were stored in a MySQL database.

4.4 Performance Evaluation Method

We evaluated the performance of our classification model using cross-validation and considering various metrics, such as accuracy, recall, F1 score, and Cohen's kappa.

- Precision: The precision of a model corresponds to the measure of the proportion of positive predictions that have been correctly made. When precision is high, it means that the model minimizes the number of false positives. In other words, when precision is high, the model makes fewer errors in identifying positive predictions.
- Recall: The recall of a model corresponds to the measure of the proportion of positive predictions that have been correctly identified. When recall is high, it means that the model maximizes the number of true positives. In other words, when recall is high, the model effectively identifies a large number of positives.
- F1-Score: The F1-Score is a harmonic mean of the two metrics, precision and recall, and it evaluates the overall performance of a model. The higher the F1-Score, the more performant the model is.
- Cohen's Kappa: Cohen's kappa coefficient is used to measure the agreement between two evaluators who classify N items into C mutually exclusive categories.

5 Experimental Results

In this section, we present the classification results generated by our program and discuss the performance achieved.

5.1 Presentation of Results

The results presented in Table 1 were obtained by executing SQL queries directly in the database. In the following section, we will proceed to analyze various aspects of these results.

1. *Classification of Tweets by Belief*

Figure 2 presents the result of classifying tweets into two belief categories: "certain" and "uncertain." In the presented case, the results show that certainty is

Table 1 Classification results statistics

Tweet distribution				Sentiment		
				Positive	Negative	Neutral
Belief types	Indifferent		110 701	41143	22469	47089
	Binary	Certain	74827	27436	14441	32950
		Uncertain	35874	13707	8028	14139
	Uncertainty categories	Question	11785			
		Condition	4865			
		Hope	10332			
		Belief	8892	3502	2115	3275

Fig. 2 Distribution of tweets by belief (Certain, Uncertain)

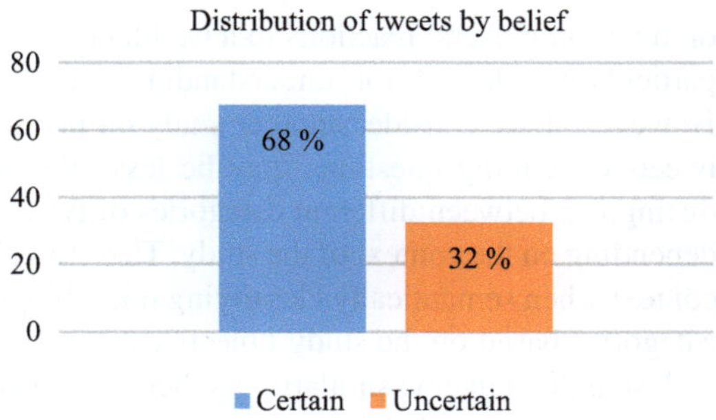

the predominant belief expressed in the tweets, representing 68% of the tweets. This may indicate that microblog users tend to express strong and affirmative beliefs in their messages. However, it is also interesting to note that tweets expressing uncertain beliefs account for 32% of the tweets. In other words, this means that statements devoid of questions, conditions, hopes, and beliefs make up two-thirds of the tweets, while the remaining one-third contains elements of uncertainty. The classification of tweets into two belief categories ("certain" and "uncertain") is an interesting approach to analyzing the nature of beliefs expressed in tweets and can have implications for understanding how microblog users express their opinions and beliefs. However, it is important to note that these results should be interpreted with caution, as the semantic classification of tweets can be complex and context-dependent.

Figure 3 illustrates the distribution of tweets in the "uncertain" category across subcategories of uncertainty, namely "question," "condition," "hope," and "belief." The "question" category recorded the highest percentage at 33%, followed by the "hope" category at 29%, and the "belief" category at 23%. Tweets containing conditions constituted the smallest proportion, accounting for only 14%. Therefore, we can deduce that questions dominate the content of uncertain tweets in our collection. Questions can be a common way for microblog users to express their uncertainty or their need for additional information regarding the subject of the pandemic. However, it is important to note that the relevance of different categories of tweets may vary depending on the context of the study. For example, in our study

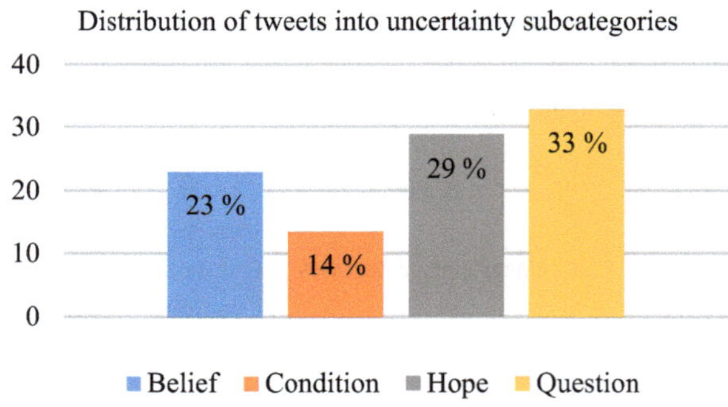

Fig. 3 Distribution of tweets across subcategories of uncertainty: question, condition, hope, belief

on microblog users' reactions to a health crisis, tweets containing questions may be particularly relevant for understanding users' concerns and information needs. However, if we consider another study on public opinion regarding a sports event, tweets containing questions may be less relevant. Our classification allows us to distinguish between different categories of tweets that may be more or less relevant depending on the context of the study. Therefore, it is important to consider the study context when semantically classifying microblogs and select the most relevant tweet categories based on the study objectives.

Using the semantic similarity method, we obtained accurate results even when the expression was implicit. For example, the tweet "Covid a thing of the past," which means that Covid has become a thing of the past, was classified in the "Hope" category because it implies that the Covid-19 pandemic is receding, which is a source of hope for many people. Similarly, tweets containing promises were also classified in the "Hope" category, even if the promise itself was not an explicit expression of hope, but it implies a positive expectation regarding the fulfillment of something. Semantic similarity method is a useful technique for the semantic classification of microblogs as it enables the analysis of semantic relationships between words and expressions, even when they are implicit. It allows for the detection of these semantic nuances and accurate classification of microblogs, even when the language is implicit or subtle. Therefore, it can be a valuable tool for the semantic classification of microblogs in contexts where the meaning is complex or ambiguous.

Table 2 presents and compares our classification results with the annotation results obtained by Wei et al. [8] and Li et al. [9], respectively. We observe that the semantic classification of microblogs is a complex challenge as it heavily depends on the context and the community to which the microblogs are targeted. Microblogs are often short messages, quickly written, and lacking significant context, which can make understanding difficult for natural language processing algorithms. Additionally, communities often have their own jargon, language, and way of thinking, further complicating the classification process. To successfully classify microblogs effectively, it is crucial to consider the context and target community and adapt the classification accordingly. It may be necessary to readjust the categories to

Table 2 Comparison of our results with the annotation results obtained by Wei et al. [8] and Li et al. [9], respectively

Annotation results	Wei et al. [8]	Li et al. [9]	Our results
Data source	Twitter	Sina Weibo	Twitter
Annotation process	Two annotators trained to independently annotate for certain types/uncertainties and one annotator to resolve tag conflicts.		A single annotator that collects tweets in real time and classifies them as uncertain in the presence of forms of uncertainty, otherwise classifies them as certain.
Uncertainty assertions	Based on semantics rather than cue phrases.		
Multiplicity of labeling	Multi-label allowed for subclasses.	Single label annotation.	Single label annotation.
Total number of microblogs	4743	40168	110 701
Uncertain microblogs	19,52 %	27,56 %	32%
Statistics of uncertain categories	Question (52,69%) External (22,46%) Probable (13,93%) Condition (7,66%) Doxastic (5,18%) Dynamic (2,26%) Possible (1,72%)	Question (53,02%) Possible (26,85%) Condition (9,29%) Hope (5,52%) Doxastic (4,24%) External (1,05%)	Question (33%) Hope (29%) Belief (23%) Condition (14%)

include and exclude based on the specificities of the relevant community and gather additional data to enhance the accuracy of the classification.

2. *Analysis of Tweet Sentiments*

Figure 4 presents the results of the classification of tweets related to the Covid-19 pandemic based on their sentiment. It indicates that the dominant sentiment in the studied collection of tweets is neutrality, representing 43% of all tweets. Positive sentiment is the second most common sentiment, accounting for 37% of all tweets. The remaining distribution is attributed to negative sentiment. This distribution pertains to all tweets from both the certain and uncertain belief categories combined. It is important to note that this distribution only applies to tweets related to the Covid-19 pandemic in our collection and cannot be generalized to other topics or contexts. The results of this analysis can be used to guide the decisions of public

Fig. 4 Distribution of tweets by sentiment (positive, negative, neutral)

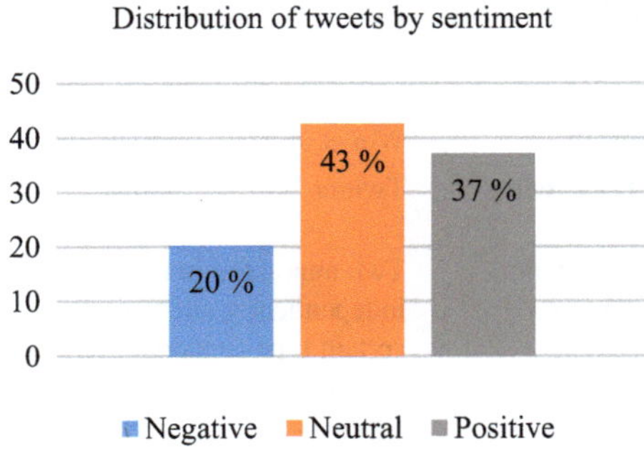

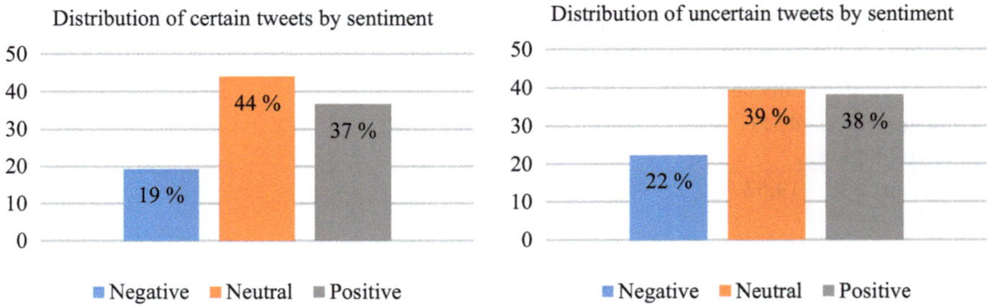

Fig. 5 Distribution of tweets by sentiment and belief

health professionals, policymakers, and organizations regarding communication strategies and measures to be taken in response to the Covid-19 pandemic, taking into account the public's perceptions and sentiments on social media. This can contribute to a more effective and tailored real-time response to the situation.

3. *Considering Uncertainty in Sentiment Analysis*

The sentiment of tweets was analyzed separately for certain and uncertain tweets. We found a predominance of neutrality, followed by positive sentiment, with a very close distribution in both categories (see Fig. 5). It is important to note that the relevance of different subcategories of uncertainty may vary depending on the context of sentiment analysis. In our case, we determined that the Belief category was the most relevant as it could help refine sentiment analysis by focusing on tweets expressing affirmative yet uncertain beliefs and could have implications for understanding public perceptions and attitudes towards the Covid-19 pandemic.

If we consider only the Belief category for sentiment analysis, we observe a shift in the dominant sentiment from neutrality toward positivity (see Fig. 6). This suggests that the predominant sentiment, albeit slightly, in tweets expressing beliefs is positivity. This finding is highly relevant and could be of significant importance for investigative surveys or opinion studies as it allows for targeting specific categories of uncertainty in sentiment analysis and excluding other categories based on the study's context. Indeed, by focusing on the Belief category for

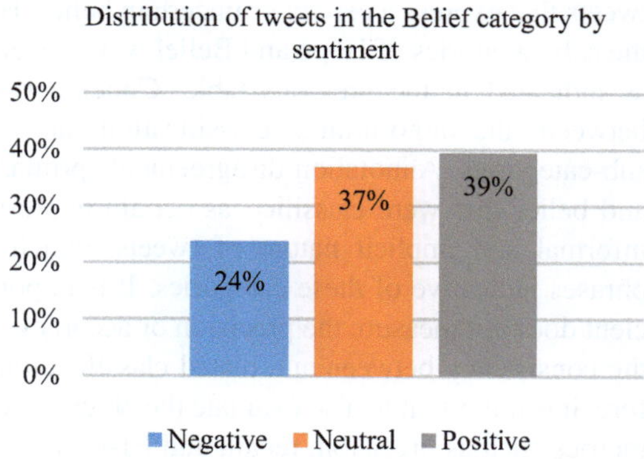

Fig. 6 Distribution of Belief category tweets by sentiment

Table 3 Performance statistics of classification

Classification performance		Precision	Recall	F1-score	Cohen's kappa
Classification (certain / Uncertain)		0,74	1	0,85	0,79
Classification into subcategories of uncertainty	Question	0,81	0,93	0,87	0,84
	Condition	1	1	1	1
	Hope	0,54	1	0,70	0,68
	Belief	0,6	1	0,75	0,74

sentiment analysis, researchers can better understand public attitudes and opinions on a specific subject. This method can be particularly useful for opinion surveys, where respondents may express affirmative yet uncertain beliefs about a topic. By using this approach, investigators can gain a better understanding of public preferences and attitudes and adapt their communication strategies accordingly. Furthermore, this approach can also assist organizations and policymakers in better understanding public concerns and needs in real-time, which can contribute to a more effective response to the Covid-19 pandemic and other public health issues.

5.2 *Performance Evaluation*

To obtain all the performance measures of our classification, we manually annotated a random set of 1000 tweets. All the statistics regarding the identification of the binary belief certain/uncertain and the sub-categories of uncertainty are presented in Table 3.

The table presented above indicates that the majority of annotations were consistent between automated classification and manual classification. The Kappa coefficient, which measures the agreement between these two classifications, is 79% for the binary classification of certain/uncertain, indicating that our results are significantly reliable. This suggests that our model successfully and accurately identified tweets that express a certain or uncertain belief. However, the Kappa coefficients for the sub-categories of Hope and Belief were lower compared to other sub-categories, as indicated in the previous table. Cohen's Kappa showed more disagreements between the algorithm‘s classification and manual classification for these sub-categories. Annotation disagreements primarily stem from expressions of hope and belief that were classified as certain expressions by our algorithm due to the informal and implicit nature of tweets, which may not always contain explicit phrases indicative of these categories. It is important to note that the Kappa coefficient does not measure the precision or accuracy of the classification itself but rather the consistency between automated classification and manual classification. Therefore, it is important to also evaluate the precision of our model using other evaluation metrics such as precision, recall, and F1-score.

The results of our model for the binary classification of certain/uncertain show a precision of 0.74, recall of 1, and F1-score of 0.85, suggesting that our model has good capability in identifying tweets that express a certain or uncertain belief. For the classification into sub-categories of uncertainty, the results vary depending on the sub-category. The model showed a precision of 0.81, recall of 0.93, and F1-score of 0.87 for the Question sub-category, suggesting overall good performance for this sub-category. For the Condition sub-category, the model demonstrated a precision of 1, recall of 1, and F1-score of 1, indicating perfect performance. However, for the Hope and Belief sub-categories, the model exhibited lower performance with precisions of 0.54 and 0.6, respectively. Although the recall is 1 for both sub-categories, the F1-score of 0.70 for Hope and 0.75 for Belief suggests a lesser ability of the model to accurately identify tweets expressing uncertainty of these types.

Overall, these results indicate that our model is capable of classifying tweets expressing certain or uncertain beliefs with good precision, but there is still room for improvement in the precision for some of the sub-categories of uncertainty. It may be useful to analyze the classification errors to identify underlying causes and enhance the performance of our model. These findings highlight the importance of considering the nuances and subtleties of natural language when classifying tweets into sub-categories of uncertainty, especially when cue phrases are not always present or easily identifiable. Exploring more sophisticated approaches, such as the use of neural networks or other natural language processing algorithms, could be beneficial in identifying these types of uncertainty.

5.3 Discussion

5.3.1 Summary of Results

Our study aimed to develop an algorithm for classifying microblogs, particularly tweets, based on belief and sentiment. We utilized a collection of tweets related to Covid-19 and classified them into two binary categories: certain and uncertain. The uncertain tweets were further categorized into more specific sub-categories:

question, condition, hope, and belief. Additionally, we conducted sentiment analysis on the entire set of tweets, distinguishing between different types of uncertain beliefs. The results of our classification demonstrated good reliability. Regarding beliefs, it is interesting to note that over two-thirds of the tweets expressed certainty, while one-third showed forms of uncertainty. Within this latter category, our analysis revealed that over one-third of the uncertain tweets were predominantly questions (in the form of inquiries, requests, etc.). The remaining uncertain tweets were distributed among the sub-categories of hope, belief, and condition. Comparing our results to similar studies, we found that the semantic classification of microblogs is highly dependent on the context, including the community, its way of thinking, speaking, and its own language. Therefore, it requires adaptation and adjustment in terms of the categories to be represented and those to be excluded. In our collection of tweets, neutrality was the dominant sentiment, followed by positivity, both for the certain category, the uncertain category, and the combined categories. However, for the belief sub-category, the sentiment leaned towards positivity. It is therefore important to consider the relevance of including sub-categories of uncertainty in sentiment analysis, as well as those that may introduce bias in the results and should be excluded during analysis. For our study, only the belief category was considered.

5.3.2 Advantages of Classifying Tweets into Uncertainty Categories

This study highlights the importance of considering the expression of uncertainty in sentiment analysis, as it can significantly influence opinion. Therefore, it is essential to only consider categories that are relevant to a given study. This rule also applies to other domains, where it is important to target specific categories based on the case at hand. For example, the question category can be used to target uninformed users, the hope category can be used to identify users' wishes and expectations regarding a product, invention, technology, or brand, and the belief category can help uncover knowledge and truth in relation to facts, among others. In general, classifying tweets or microblogs into uncertainty categories allows for:

- Better understanding the reliability of information contained in microblogs and avoiding the spread of rumors or false information.
- Facilitating the task of researchers and analysts by allowing them to focus on the most relevant and reliable microblogs.
- Gaining a better understanding of the evolution of public opinion on a given topic.
- Assisting organizations in identifying the most concerning topics for users of microblogging platforms and adapting their communication accordingly to better address the needs and concerns of their target audience.
- Lastly, classifying microblogs based on their level of uncertainty can have significant implications for historical studies, helping researchers gain insights into people's attitudes and opinions during a specific period and how these attitudes have evolved over time.

Classifying microblogs can, therefore, help evaluate the quality of information in terms of truthfulness and reliability, assisting users in making informed choices regarding how they use that information. However, if we consider history, we can also see how this approach can be useful in the context of historical research. Most historical information is based on primary sources such as documents, journals, and archives. However, these sources can be imperfect or biased, making it challenging to determine historical truth. By classifying historical information into uncertainty categories, historians can better assess the quality of this information and use it to make informed decisions regarding how they utilize these sources in their research. This approach can also help identify gaps in historical documentation, guiding future data collection efforts. Furthermore, the use of uncertainty categories can demystify the process of historical research for non-specialists by providing a clear and concise explanation of the quality of available historical information.

5.3.3 Other Possible Exploitations of the Model

Our classification model can be used in various ways, offering intriguing research prospects in specific domains. Here are a few ideas summarized below:

(a) Real-time analysis of the evolution of uncertainty sentiment in tweets:

This enables prompt and effective responses to situations that may require quick decision-making. By monitoring uncertainty sentiment in real-time tweets, organizations can quickly detect potential issues and react accordingly, adjusting their communication or taking measures to address a problematic situation. This approach can help prevent communication crises and maintain user trust in the organization. Additionally, it can assist organizations in gaining better insights into trends and patterns in tweets, enabling them to adjust their communication more targeted and effectively in the future.

(b) Considering additional criteria such as hashtags, user location, and user follower count during tweet filtering

This can enhance the relevance and accuracy of tweet classification. Hashtags can provide additional information about the context of tweets, aiding in better understanding the user's intent and improving tweet classification. For example, a tweet containing the hashtag #covid19 may be classified differently from an unrelated topic tweet. User location can also provide supplementary information about the context of tweets, as well as cultural and geographical differences that can influence how uncertainty is expressed. Lastly, the user's follower count can serve as an indicator of the user's credibility and, therefore, the relevance and reliability of their tweet. Users with a large number of followers may be considered influencers or experts in their field, and their tweets may carry more weight in classification. By combining these criteria with the content analysis of tweets, classification methods can become more precise and effective, helping users obtain relevant and reliable information.

(c) Studying opinion changes in a specific population

Studying opinion changes can be highly valuable for numerous reasons. Firstly, it can help understand how opinions and attitudes are formed and evolve over time, which is important for researchers and policymakers. Additionally, it can aid in predicting future behaviors and designing effective strategies to achieve specific goals, such as persuading people to adopt a new behavior or support a specific policy. Lastly, it can help identify factors that can cause sudden or dramatic opinion changes, such as political events, natural disasters, or disinformation campaigns, which can guide actions to minimize their negative impact.

5.3.4 Extension of the Proposed Classification Model

Refining the classification could be highly valuable for certain research domains, particularly by exploring more nuanced types of uncertainty within the "Belief" sub-category, considering gradual semantic uncertainty such as doubt, hesitation, or presumption. Currently, this sub-category only includes tweets where uncertainty has been explicitly expressed. Research domains focusing on truth discovery, investigation, or inquiry could particularly benefit from this finer classification. In these domains, uncertainty is often implicit and subtle. Therefore, understanding how people express their uncertainty is crucial to better grasp the nature of their uncertainty and utilize it to guide research towards truth discovery. Introducing more refined categories of uncertainty would enable researchers to better comprehend the nuances and subtleties of uncertainty in the language of Twitter users, potentially leading to more effective and targeted investigations. Furthermore, understanding different types of uncertainty can help researchers gain insights into the communication, motivations, and beliefs of Twitter users, which could be particularly valuable in fields where uncertainty is often implicit, such as psychology, sociology, politics, and economics. By understanding the nuances of uncertainty, researchers may be better able to comprehend why users express doubts or hesitations, helping identify trends and patterns in Twitter user behavior. In summary, refining the classification could bring significant benefits across various research domains, enabling a better understanding of the behaviors and beliefs of Twitter users.

6 Conclusion

Uncertainty is expressed differently on social media than in formal language. Social media authors have the freedom to write in their preferred style. This issue should be addressed in every investigative activity, including research and information extraction. Indeed, the author's confidence in their statement can be demonstrated through their manner of expression. While uncertainty corpora exist in several domains, there is no standard corpus created specifically for the aspect of uncertainty in social media texts, and this aspect of microblogging has received little attention. We have conducted these experimental studies to explore the intrinsic form of uncertainty expressed in statements. We studied the sentiments expressed on Twitter, taking into account the semantic uncertainties present in the tweets. For this purpose, we developed a tweet classification algorithm based on their degree of uncertainty and sentiment, and studied the sentiments related to the Covid-19 pandemic. The tweets were classified into two binary categories: "certain" and "uncertain." The uncertain tweets were further classified into more specific sub-categories of uncertainty: "question," "condition," "hope," and "belief." The results showed good performance of the classification algorithm. We concluded that the expression of uncertainty can impact sentiment analysis results, and it is important to target relevant categories of uncertainty based on the specific case study. Through this study, we conclude that to adapt to the peculiarities of social media, it is necessary to recognize uncertainty based on contextual semantic uncertainty rather than conventional cues, and considering sub-classes could provide additional insights for studies on managing uncertainty in social media texts. Our study has highlighted the need to carefully consider the semantic categories of uncertainty to be represented and included in sentiment analysis in terms of relevance, as well as categories that could introduce bias and should, therefore, be excluded during analysis. In perspective, we propose the two direct extensions to this research work. An improvement that could be considered is refining the semantic classification for the "belief" category. It would be useful to explore different types of uncertainty, such as doubt, hesitation, and presumption, in a gradual manner, as this sub-category currently only includes tweets explicitly expressing uncertainty. This enhancement would be particularly interesting for research domains focusing on truth discovery, investigation, or inquiry. Another extension worth exploring would be studying the correlation between the expression of uncertainty in a statement and its truth value. This approach would involve merging these two forms of intrinsic and extrinsic uncertainty to infer the plausibility of statements. This approach would provide a better understanding of how the expression of uncertainty in statements affects their degree of truth.

References

1. S. Rouhani, E. Abedin, Crypto-currencies narrated on tweets: a sentiment analysis approach. Int. J. Ethics Syst. **36**(1), 58–72 (2020). https://doi.org/10.1108/IJOES-12-2018-0185
2. R.J. Feng, H.J. Zhang, W.M. Pan, Z.Y. Zhou, Y.J. Li, A new method of microblog rumor detection based on transformer model, in *Artificial Intelligence in China: Proceedings of the 2nd International Conference on Artificial Intelligence in China*, (Springer, Singapore, 2021), pp. 531–537. https://doi.org/10.1007/978-981-15-8599-9_61
3. M. Basu, K. Ghosh, S. Ghosh, Information retrieval from microblogs during disasters: In the light of irmidis task. SN Comput. Sci. **1**, 1–10 (2020). https://doi.org/10.1007/s42979-020-0065-1

4. S. Giglio, F. Bertacchini, E. Bilotta, P. Pantano, Using social media to identify tourism attractiveness in six Italian cities. Tourism Manage. **72**, 306–312 (2019). https://doi.org/10.1016/j.tourman.2018.12.007
5. S.E. Jordan, S.E. Hovet, I.C.H. Fung, H. Liang, K.W. Fu, Z.T.H. Tse, Using Twitter for public health surveillance from monitoring and prediction to public response. Data **4**(1), 6 (2018). https://doi.org/10.3390/data4010006
6. J. Kothandan, P. Murugesan, ML based social media data emotion analyzer and sentiment classifier with enriched preprocessor. J. Inf. Technol. Manage. **13**(Special Issue: Big Data Analytics and Management in Internet of Things), 6–20 (2021). https://doi.org/10.22059/jitm.2021.80614
7. F. Rajabi, A. Saghaei, S. Sadinejad, Monitoring of social network and change detection by applying statistical process: ERGM. J. Optim. Ind. Eng. **13**(1), 131–143 (2020). https://doi.org/10.22094/joie.2019.581174.1615
8. Z. Wei, J. Chen, W. Gao, B. Li, L. Zhou, Y. He, K.F. Wong, An empirical study on uncertainty identification in social media context, in *Social Media Content Analysis: Natural Language Processing and Beyond*, (2018), pp. 79–88. https://aclanthology.org/P13-2011
9. B. Li, J. Xiang, L. Chen, X. Han, X. Yu, R. Xu, et al., The UIR uncertainty corpus for Chinese: annotating Chinese microblog corpus for uncertainty identification from social media, in *Proceedings of the Eleventh International Conference on Language Resources and Evaluation (LREC 2018)*, (2018) https://aclanthology.org/L18-1078
10. F. Zendaoui, W.K. Hidouci, S. Rouhani, Uncertainty identification in microblogs. J. Optim. Ind. Eng. **15**(1), 301–309 (2022). https://doi.org/10.22094/joie.2021.1945486.1913
11. F. Zendaoui, W.K. Hidouci, Multi-version representation of historical event. Periodicals Eng. Nat. Sci. **7**(1), 141–147 (2019). https://doi.org/10.21533/pen.v7i1.329.g247
12. F. Zendaoui, W.K. Hidouci, Considering uncertainty in modeling historical knowledge. ISeCure **11**(3), 10.22042/isecure.2019.11.0.8 (2019)
13. R. Farkas, V. Vincze, G. Móra, J. Csirik, G. Szarvas, The CoNLL-2010 shared task: learning to detect hedges and their scope in natural language text, in *Proceedings of the Fourteenth Conference on Computational Natural Language Learning–Shared Task*, (2010), pp. 1–12. https://aclanthology.org/W10-3001
14. G. Szarvas, V. Vincze, R. Farkas, G. Móra, I. Gurevych, Cross-genre and cross-domain detection of semantic uncertainty. Comput. Linguist. **38**(2), 335–367 (2012). https://doi.org/10.1162/COLI_a_00098
15. A. Bessarab, O. Mitchuk, A. Baranetska, N. Kodatska, O. Kvasnytsia, G. Mykytiv, Social networks as a phenomenon of the information society. J. Optim. Ind. Eng. **14**(Special Issue), 17–24 (2021). https://doi.org/10.22094/joie.2020.677811
16. V. Vincze, Uncertainty detection in natural language texts. PhD, University of Szeged, 141. https://doi.org/10.14232/phd.2291
17. R. Al-Sabbagh, R. Girju, J. Diesner, A unified framework to identify and extract uncertainty cues, holders, and scopes in one fell-swoop, in *Computational Linguistics and Intelligent Text Processing: 16th International Conference, CICLing 2015, Cairo, Egypt, April 14-20, 2015, Proceedings, Part I 16*, (Springer International Publishing, 2015), pp. 310–334. https://doi.org/10.1007/978-3-319-18111-0_24
18. H. Adel, H. Schütze, Exploring different dimensions of attention for uncertainty detection. arXiv preprint arXiv:1612.06549 (2016). https://doi.org/10.48550/arXiv.1612.06549
19. R.J. Feng, H.J. Zhang, W.M. Pan, Z.Y. Zhou, Y.J. Li, A new method of microblog rumor detection based on transformer model, in *Artificial Intelligence in China: Proceedings of the 2nd International Conference on Artificial Intelligence in China*, (Springer, Singapore, 2021), pp. 531–537. https://doi.org/10.26599/TST.2019.9010022
20. [Review of Lehetőség és szükségszerűség: Tanulmányok a nyelvi modalitás köréből [Possibility and necessity: Papers on linguistic modality], by F. Kiefer]. Acta Linguist. Hung. **52**(2–3), 339–340 (2005). http://www.jstor.org/stable/26190076

Design Together: Uncovering the Impact of Co-design and Design Thinking on Designing for People with Dementia

Fatemehalsadat Shojaei

1 Introduction

Most thought-provoking in our thought-provoking time is that we are still not thinking. [1]

Among the approaches in human-computer interaction (HCI), co-design has recently received considerable attention [2]. The co-design workshop is used by designers with the aim of improving innovation and finding the most suitable solution to increase user satisfaction [3].

Martin Heidegger once wrote, "... We are all on the way together, and are not reproving each other. To learn means to make everything we do answer to whatever essentials address themselves to us at a given time. Depending on the kind of essentials, depending on the realm from which they address us, the answer and with it, the kind of learning differs" [1]. Taking Heidegger's quote into consideration, the importance of co-design workshops, where designers meet users, will grow. In other words, designers may believe they know what the user wants, but in reality, they do not!

The purpose of this chapter is to explain the relation between two important topics in design, "Design thinking" and "Co-design," as they might help People with Dementia (PWD). To that end, based on the instinct of these two approaches, I provide an outline of the process and how they, depending on how they are interpreted, can facilitate or hinder this population.

According to Ferdinand Braudel's concept in "Civilization and Capitalism," the notion that "everything is technology" implies that technology can be viewed as an active process [4]. Building upon this perspective, it becomes essential to consider technology as an active component within co-design workshops. Raymond Williams

F. Shojaei (✉)
Computer Science, State University of New York, Oswego, NY, USA
e-mail: fatemehalsadat.shojaei@gmail.com

© The Author(s), under exclusive license to Springer Nature Switzerland AG 2024

A. Shaikh et al. (eds.), *Advances in Emerging Information and Communication Technology*, Signals and Communication Technology,
https://doi.org/10.1007/978-3-031-53237-5_18

emphasized the significance of comprehending the "right-knowing and right-doing" aspects [5]. Hence, this chapter aims to explore any activities endeavored within co-design workshops, aligning with Williams' insight.

I will first define Co-design and identify the advantages of using this method. After that, I will define the design thinking process and how it could help. In the third section, I will give a definition of Dementia and identify the concerns of PWD to better understand the difficulties that PWD face. In the last section of this chapter, I will give suggestions for applying co-design and the design thinking process to designing for people with dementia and highlight how it could help this population.

2 Co-design

2.1 What Is Co-design?

Co-design is an HCI method that has recently received considerable attention [2]. Co-design workshops are used by designers with the aim of finding the most suitable solution to increase user satisfaction while approaching innovation in their solutions [3]. Co-design, according to Steen, can refer to open innovation processes in which people from various organizations share and integrate ideas, or with more emphasis on user involvement, it can refer to a process that involves users as participants in the design process. Diverse methodologies are used in co-design, from approaches that users are approached by researchers and designers (e.g., usability testing) to approaches that encourage users to engage with researchers and designers (e.g., participatory design). Co-design can also be defined as "the process in which actors from different disciplines share their knowledge about both the design process and the design content... in order to create shared understanding on both aspects... and to achieve the larger common objective: the new product to be designed" [6].

2.2 Why Do We Need Co-design?

Users are considered "expert[s] of their experience" [7], which is why co-design approaches are increasingly embraced by researchers and design teams. The recognition of the value that users bring to the design process has led to the utilization of co-design workshops as a means to gather additional ideas and insights. In these workshops, ideas flow collaboratively, with the users' thoughts serving as the starting point. Co-design workshops create an environment where the combination of user input and designer expertise generates outcomes that surpass mere additive contributions. The inclusion of end-users in the design process not only benefits the designers but also empowers the users to express their feelings and ensure that the final product aligns with their specific needs and preferences.

The essence of co-design lies in recognizing that users play an active role as co-creators of value and co-developers of their experiences with a product [3]. However, understanding the actual needs of users can be challenging during the co-design process, as designers must decipher users' needs from their words and expressions within the co-design sessions.

3 Design Thinking

3.1 What Is Design Thinking

Design thinking lacks a universal, single definition, although its contemporary form owes much to the pioneering efforts of IDEO, a prominent design agency. Tim Brown, the CEO of IDEO, defines design thinking as "a human-centered approach to innovation that draws from the designer's toolkit to integrate the needs of people, the possibilities of technology, and the requirements for business success" [8]. The contributions of IDEO have significantly influenced how design thinking is now understood and used in the fields of innovation and problem-solving. It is important to keep in mind that there may be other viewpoints and variants on design thinking, but this particular definition offers an informative overview of its fundamental ideas and objectives (see Fig. 1).

3.2 How Can Design Thinking Help?

According to IDEO, a prominent design firm, the designer's mindset in the design process should embody a "beginner's mind," characterized by openness, curiosity, and the absence of assumptions, thereby perceiving ambiguity as an opportunity for

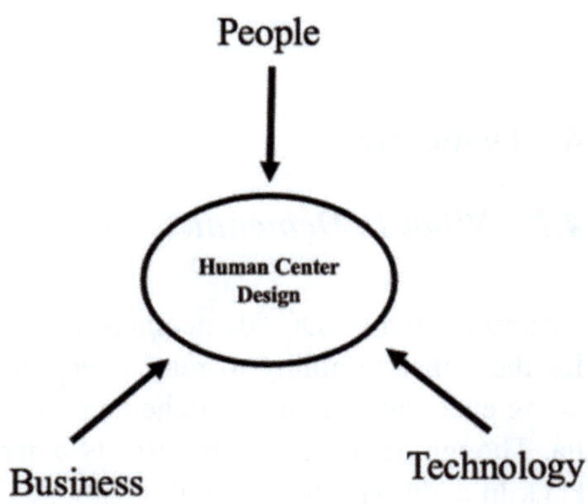

Fig. 1 Unpacking the designer's toolkit in Human-Centered Design (HCD)

Table 1 IDEO's perspective on design's reflective core values

Values	Enhancing design through diverse teams and prototyping
Many eyes	Diverse design teams find hidden insights with unique perspectives.
Customer perspective	Design teams visit customers, interview them, and observe them to understand their experiences deeply.
Tangibility	Design teams use prototypes for iterative, user-centered improvements in designs.

exploration and discovery [9]. In a broader context, IDEO emphasizes that design is a collaborative endeavor, reflecting three core values:

Many eyes: Design teams comprise individuals from diverse disciplines such as engineering, human factors, communication, graphics, ethnography, and sociology. Each team member brings a unique perspective, enabling the collective group to perceive insights and nuances that might have otherwise gone unnoticed.

Customer perspective: Design teams actively engage with customers by visiting their homes, conducting interviews, and observing their real-life activities. This immersive approach allows teams to gain firsthand insights into customer behaviors, especially in high-stress situations, thereby incorporating a deep understanding of customer experiences into the design process.

Tangibility: Design teams employ prototyping and mockup techniques to create tangible representations of their ideas. By testing these prototypes and learning from the results, teams iteratively refine and enhance their designs, ensuring a user-centered and iterative approach [10] (Table. 1).

The concepts of co-design and design thinking are inherently interconnected, making it challenging to discuss one without acknowledging the other. Co-design recognizes users as active participants who contribute to the creation of value and the development of their experiences with a product [3]. However, identifying users' authentic needs poses a significant challenge within the co-design process, as designers must discern these needs from users' verbalizations and expressions during co-design sessions. This is where the significance of design thinking in the co-design process becomes evident.

4 Dementia

4.1 What Is Dementia?

According to the CDC "dementia is not a specific disease but is rather a general term for the impaired ability to remember, think, or make decisions that interfere with doing everyday activities. Alzheimer's disease is the most common type of dementia. Though dementia mostly affects older adults, it is not a part of normal aging" [11]. In addition, Shojaei et al. mentioned in their article that dementia affects how people perceive objects, interact with others, and act, as well as how they process information in their brain [12]. Therefore, Dementia is a broad term used to describe a group of symptoms that affect memory, thinking, behavior, and the ability to perform everyday activities.

4.2 How Do People with Dementia Apply the Limitation to Co-design?

PWD exhibit diverse needs that may evolve with the progression of their condition. However, identifying and understanding the specific needs of PWD can be challenging due to their emotional fragility and sensitivity. Dementia affects various cognitive functions such as memory, executive functioning, communication, and decision-making abilities, leading to a complex set of challenges for PWD [2]. Consequently, engaging in co-design activities with PWD presents certain limitations and considerations. It becomes crucial to prioritize the exploration and comprehension of the needs unique to this user group in order to develop designs that are meaningful, supportive, and responsive to the experiences of PWD.

Shojaei et al. introduced a "Model of Needs of People with Dementia" that encompasses both physical and emotional needs, further subdivided into specific domains [13]. Drawing upon this model, in conjunction with Ekbia's assertion that emotions influence judgment and judgment influences emotions [14], engaging in co-design workshops with PWD becomes a cyclical process. If PWD do not have a positive experience or struggle to comprehend the co-design workshop, they are likely to disengage and perceive it as a futile exercise. Such negative experiences can trigger feelings of frustration, anxiety, or anger, resulting in lasting memories associated with the design process and the co-design workshop, which may adversely affect their emotional well-being.

5 Discussion

Maslow's hierarchy of needs, which includes psychological, safety, and love/belonging needs, should be taken into consideration while designing for PWD. Greater focus can be placed on this mentally fragile demographic by utilizing design thinking techniques and keeping the user at the center of the co-design workshop. Since each target audience in the design process is unique, it is essential to build the user's conceptual, mental, and emotional model early on.

Framing the bond between PWD as distinct users with unique behaviors and the world they live in becomes an important part of establishing the design process model. The user demographics (age, gender, and cultural variety), the activities being carried out, and the location and timing of interactions must all be taken into account while framing a co-design workshop. Additionally, when building the framework for a co-design workshop with PWD, consideration must be given to

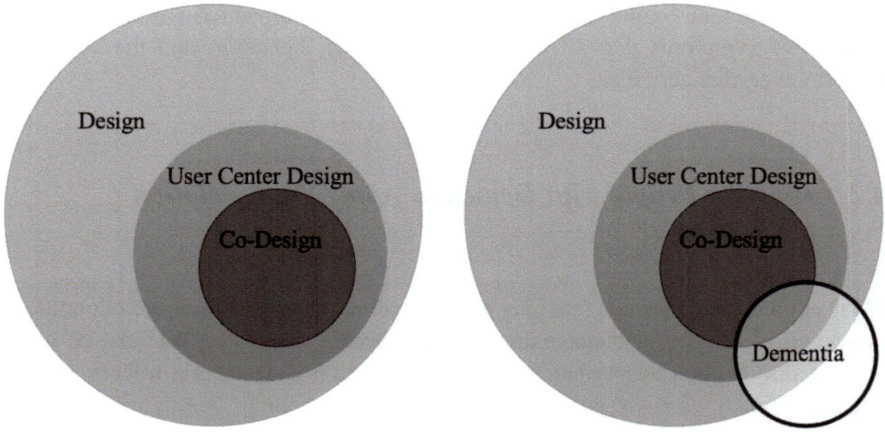

Fig. 2 Interconnection of design, user-centered design, co-design, and dementia

their unique needs and constraints, such as physical impairments, mobility problems, and cognitive challenges [13]. In order to comprehend how PWD communicate with the outside world (question about intelligibility) and with one another (question about mutual intelligibility), it is also crucial to address the issues of mutual and internal intelligibility.

By taking into account these elements, human-centered design can efficiently address the mental, physical, and emotional labor of users in co-design workshops, ensuring a reciprocal and meaningful exchange of value between the designers and PWD while designing based on the data collected from the users (see Fig. 2).

6 Conclusion

In conclusion, by actively incorporating PWD in the design process, co-design offers a potent method for addressing their demands. Co-design workshops give users and designers a place to work collaboratively by recognizing individuals as experts in their own experiences. Co-design workshops can provide creative and user-centered solutions that address the psychological, safety, and love/belonging requirements of PWD by putting design thinking principles like customer-centricity, interactive responsiveness, and physical metaphors to use.

It is essential to frame the co-design workshops precisely for PWD, taking into account their particular cognitive impairments and challenges. Designers may develop empowering and inclusive design solutions by comprehending the conceptual, mental, and emotional models of the users. Determining and comprehending the demands of PWD, however, can be difficult and requires careful study and integration of numerous variables, including user demographics, activity context, and interaction settings. Additionally, the design approach for co-design workshops with PWD must take into account their emotional needs in addition to their physical and cognitive limitations. Aiming for a mutual exchange of value, a human-centered design should make sure that participants' contributions are valued and recognized. Co-design workshops can prevent heteromation, or the extraction of value from participants' unseen labor, by actively involving users and developing using the data they provide.

The study's findings emphasize the value of co-design and design thinking in empowering PWD, improving their well-being, and developing products and services that are tailored to their specific needs. Designers and healthcare professionals may build meaningful collaboration, encourage user ownership, and eventually collaborate to develop more inclusive and supportive environments for PWD by embracing co-design methodologies. Future research can develop the area of inclusive design and improve the quality of life for PWD by exploring and improving co-design frameworks and methodologies that are specifically targeted to the needs and challenges of this vulnerable population.

References

1. M. Heidegger, *What Is Called Thinking?* (1968), p. 275
2. S. Foley, D. Welsh, N. Pantidi, K. Morrissey, T. Nappey, J. McCarthy, Printer Pals: experience-centered design to support agency for people with dementia, in *Proceedings of the 2019 CHI Conference on Human Factors in Computing Systems*, (2019), pp. 1–13. https://doi.org/10.1145/3290605.3300634
3. T. Tiainen, T. Jouppila, Use of virtual environment and virtual prototypes in co-design: the case of hospital design. Computers **8**(2), 44 (2019). https://doi.org/10.3390/computers8020044
4. F. Braudel, *Civilization and Capitalism, 15th–18th Century: The Structures of Everyday Life* (University of California Press, 1981)
5. R. Williams, *Culture and Society 1780–1950* (Anchor Books, New York, 1960)
6. M. Steen, Co-Design as a Process of Joint Inquiry and Imagination. Des. Issues **29**(2), 16–28 (2013). https://doi.org/10.1162/DESI_a_00207
7. E.B.-N. Sanders, P.j. Stappers., Co-creation and the new landscapes of design. CoDesign **4**(1), 5–18 (2008). https://doi.org/10.1080/15710880701875068
8. J. Kolko, The divisiveness of design thinking. Interactions **25**(3), 28–34 (2018). https://doi.org/10.1145/3194313
9. IDEO, IDEO Design Thinking (2021). IDEO.org. Retrieved from https://designthinking.ideo.com/
10. S. Denning, *The Leader's Guide to Radical Management: Reinventing the Workplace for the 21st Century* (Jossey-Bass, 2013)
11. Center for Disease Control and Prevention, What is Dementia? (2021). Retrieved from https://www.cdc.gov/aging/dementia/index.html
12. F. Shojaei, F. Shojaei, E. Bergvist Stolterman, P.C. Shih. (2024). Exploring the Impact of Digital Art Therapy on People with Dementia: A Framework and Research- based Discussion. In *Advances in Emerging Information and Communication Technology* (ISBN:978-3-031-53236-8). Springer Nature.
13. F. Shojaei, F. Shojaei, J. Velez. (2024). Does Game Help People with Dementia? A Guideline to Design Games for People with Dementia's Needs. In *HCist 2023 Book of Industry Papers and Poster Papers*.
14. H.R. Ekbia, B.A. Nardi, *Heteromation, and Other Stories of Computing and Capitalism* (MIT Press, Cambridge, MA, 2017)

Roller Design of the Delivery Mechanism of the Novel Endovascular Intervention Robot

Lingwu Meng, Shiqi Liu, Xiaoliang Xie, Zengguang Hou, and Xiaohu Zhou

1 Introduction

1.1 Research Background

Percutaneous coronary intervention (PCI) operation has been the most popular approach to treat vascular stenosis lesions because of its lower cost, shorter hospitalization stays, and reduced procedural morbidity. However, because of prolonged exposure to X-ray radiation, many interventional doctors suffer from diseases such as cataracts or brain tumors [1]. To avoid radiation, doctors need to wear protective lead clothing during the operation in the catheter lab, but heavy lead clothing may lead to occupational diseases such as cervical spine and lumbar disc herniation [2].

Many scholars and medical companies have proposed and released various intervention surgery robots to solve the above problems based on different working principles and clinical demands. This subsection introduces several typical robots or devices. The remote navigation system (RNS) was developed in Israel's Haifa Hospital in 2006 [3]. Based on RNS, the Corindus company developed the CorPath 200 vascular intervention robot system and became the first robotic system approved by the Food and Drug Administration (FDA) for coronary intervention in 2012

L. Meng
School of Artificial Intelligence, University of Chinese Academy of Science, State Key Laboratory of Management and Control for Complex Systems, Institute of Automation, Chinese Academy of Sciences, Beijing, China
e-mail: menglingwu2019@ia.ac.cn

S. Liu · X. Xie · Z. Hou (✉) · X. Zhou
State Key Laboratory of Management and Control for Complex Systems, Institute of Automation, Chinese Academy of Sciences, Beijing, China
e-mail: shiqi.liu@ia.ac.cn; xiaoliang.xie@ia.ac.cn; zengguang.hou@ia.ac.cn; xiaohu.zhou@ia.ac.cn

© The Author(s), under exclusive license to Springer Nature Switzerland AG 2024
A. Shaikh et al. (eds.), *Advances in Emerging Information and Communication Technology*, Signals and Communication Technology,
https://doi.org/10.1007/978-3-031-53237-5_19

[4]. Its safety and effectiveness have been verified [5]. The R-One vascular intervention robot system developed by French company Robocath obtained Conformite Europeenne (CE) certification in 2019 [6]. The slave delivery device of the vascular intervention robot designed by Woo and others at Hanyang University in South Korea has 5 degrees of freedom and can realize the delivery and rotation of guidewires or catheters [7]. Guo et al. from Kagawa University in Japan designed a vascular intervention surgical robot [8], which used a magnetorheological fluid structure to provide force feedback to doctors. Guangzhong Yang et al. from Imperial College London proposed a master-slave system platform for operating standard vascular intervention devices in different imaging environments [9]. Hongbo Wang et al. from Yanshan University invented a vascular intervention surgical robot system with force feedback and clamping force adjustment [10]. Kundong Wang et al. from Shanghai Jiao Tong University developed an open vascular intervention robot system, which can cooperate with a common guidewire, catheter, balloon, and other instruments to perform surgery [11]. A real-time image-guided master-slave robot system was developed by Shanghai Jiao Tong University and Shenzhen Institutes of Advanced Technology, Chinese Academy of Sciences [12]. The feasibility of the robot system was successfully verified through live animal experiments [13]. A team from Xiamen University developed a vascular intervention surgical robot system with two force-sensing functions [14]. For the above robots, the complex mechanical structure and the enormous size limit their application in clinical surgery. Even the current state-of-the-art Siemens CorPath GRX commercial vascular intervention surgical robot still has some unresolved disadvantages, which greatly limits its clinical application potential [15].

The main disadvantage is that the translation and rotation of the guidewire are separate, which causes two problems. Firstly, the overall structure of the robot is too large to install and repair in the catheter lab, which is not suitable for clinical practice. Secondly, the excessive structure is not conducive to achieving functional expansion for robots. Current robots can only deliver a single guidewire, suitable for simple lesions. Complex lesions including tortuous lesions, opening lesions, and bifurcation lesions in coronary heart disease account for a large proportion of the clinical practice [16], all of which require several guidewires or balloons. The existing surgery robots can only push one guidewire or balloon at a time. This brings longer X-ray exposure time, greater work intensity, and lower surgical efficiency. To deal with more common complex lesions in clinical practice, additional new structures need to be designed to deliver two or more guidewires, which makes the robot structure more complex and even unable to achieve functional expansion.

To overcome these disadvantages, we propose a novel guidewire delivery mechanism that can realize the translation and rotation motion simultaneously. It can realize the translation and rotation respectively and simultaneously, which greatly expands the motion modes, improves the delivery flexibility, makes the guidewire easier to pass through tortuous vessels, improves the efficiency of surgery, and reduces the working time and intensity. Meanwhile, it provides convenience for the expansion of functions.

1.2 Contributions

According to the principle of the delivery mechanism, the guidewire is driven by friction between the rollers. Therefore, the friction force is significant for the delivery of the intervention instrument. The innovations and contributions of this study are as follows. Firstly, we design a novel delivery mechanism that can realize the translation and rotation motion simultaneously. It increases the operation modes of the guidewire and improves the robot's work efficiency. Secondly, this study explores the evaluation method of the surgical robot's guidewire delivery ability, providing a reference for the surgical robot's performance evaluation. Thirdly, most research on intervention surgical robots focuses on the mechanical structure, the control policy, the algorithm design, the surgical safety, and the clinical application. This study can expand the whole robot study range and research direction. Finally, for the specific simulation analysis, the software settings and methods of the finite element simulation of the contact and friction behavior of the rubber, a nonlinear material, provide a valuable reference for others to perform similar finite element analysis. Other researchers in the robot community can benefit and take inspiration from this paper both in the surgical robot's performance evaluation and research methods.

The main structure of this paper is as follows. Section II describes the mechanical structure and working principle of the delivery mechanism of the novel endovascular intervention robot. Section III analyzes the roller design and the delivery process. Section IV introduces the constitutive rubber model and the boundary conditions. Section V shows and discusses the simulation results. Finally, conclusions are presented in Sect. 6.

2 Delivery Mechanism

A common PCI treatment is mainly composed of the following steps: gaining arterial access, guiding catheter delivery, guidewire delivery, balloon catheter delivery and expansion, and stent catheter delivery and deployment. The delivery of the intervention instruments is the key to a PCI operation. The delivery mechanism is the key to an invention robot. The intervention instruments in clinical practice include the guidewire, the guiding catheter, the balloon catheter, and others. The guidewire is taken as an example in this paper.

2.1 Mechanical Structure

The novel intervention robot consists of a master console and a slave delivery mechanism. It can control the intervention instruments to complete translation, rotation, and simultaneous movement in the blood vessel, enabling it to cross

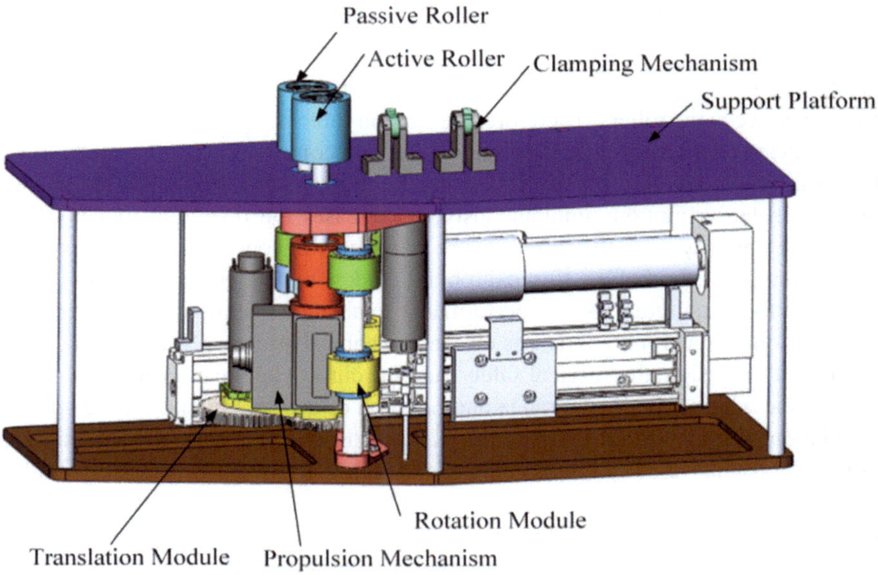

Fig. 1 Mechanical structure of the slave delivery mechanism

complex lesions better. More details can be seen in other articles of our research group [17]. The slave delivery mechanism includes the propulsion mechanism, the clamping mechanism, the support platform, and the rollers, as shown in Fig. 1. The guidewire is driven by the friction between the active and passive rollers directly. The roller structure and the outer material property are very important for the delivery performance of the robot.

2.2 Working Principle

The "delivery and twisting" scheme is adopted. As shown in Fig. 2, the axial propulsion operations, including the forward and backward of the guidewire, are realized through precise cooperation and the friction-based driving force between the two rollers. The two rollers rotate in opposite directions at the same angular velocity.

As shown in Fig. 3, the working procedure of axial rotation of the guidewire is as follows: when the active roller and the passive roller are in contact, the passive roller moves up and down along the axial direction to realize the relative movement of the active and passive rollers, driving the guidewire to rotate. Upward motion corresponds to counterclockwise rotation, and downward motion corresponds to clockwise rotation.

By combining translation and rotation motion, the original four motion modes can be increased to eight. The original four motion modes are forward, backward,

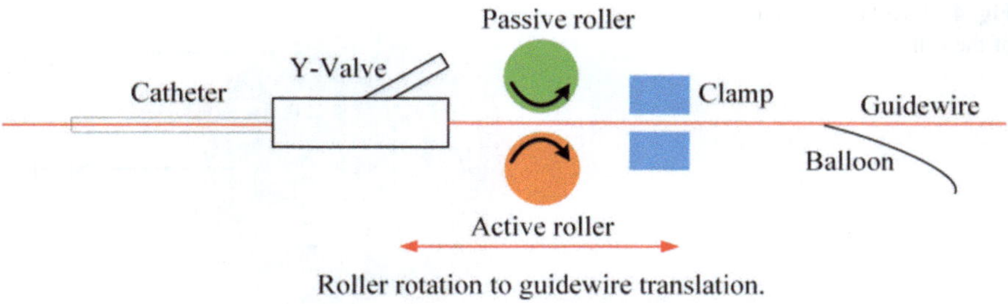

Fig. 2 Schematic diagram of delivery motion

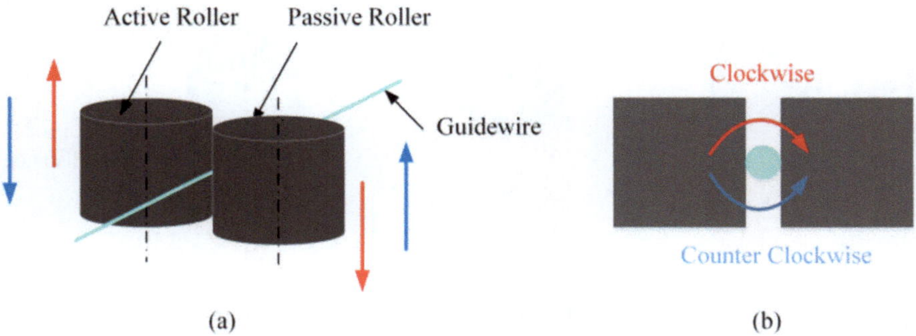

Fig. 3 Schematic diagram of rotation motion

rotate clockwise, and rotate counterclockwise. The new eight motion modes are forward, backward, rotate clockwise, rotate counterclockwise, forward + clockwise, forward + counterclockwise, backward + clockwise, and backward + counterclockwise. The new eight modes can better simulate the actual delivery motions of the doctor's or operator's fingers.

3 Roller Design and Delivery Process

3.1 Roller Structure and Material

The roller design includes the structure design and the material selection. As shown in Fig. 4, the active and passive rollers are all two-layer structures. The two layers are fixed to ensure that the outer rubber coating and steel core do not slip. The diameter of the cylinder is 35 mm, and the thickness of the rubber is 1 mm. The inner cylinders of the roller are made of 304 stainless steel, and the outer coatings of the roller are covered with thermoplastic polyurethane (TPU) rubber.

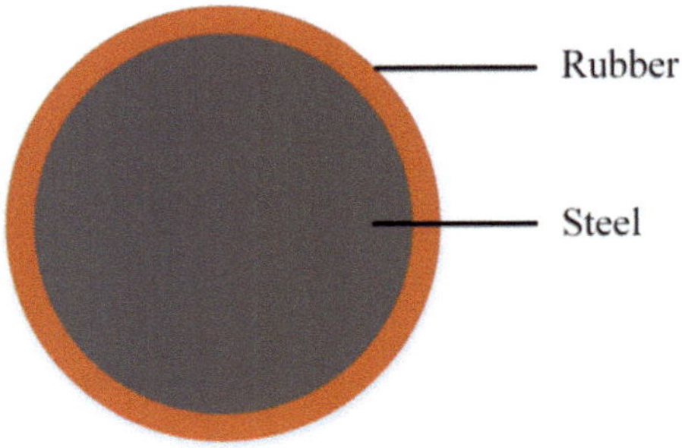

Fig. 4 Two-layer structure of the roller

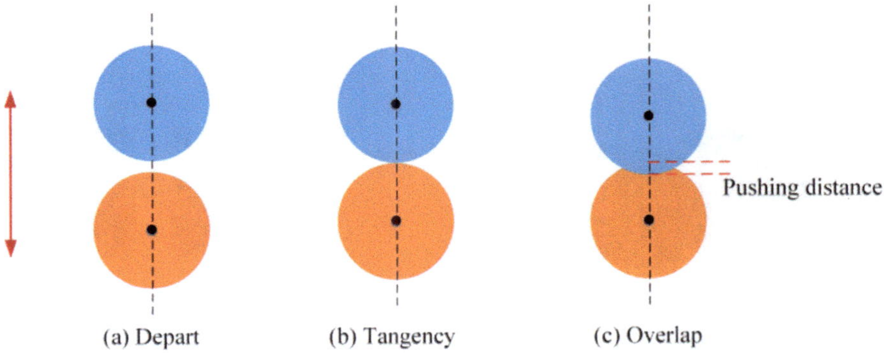

Fig. 5 Different states of the rollers in the contact motion

3.2 Delivery Process

The whole delivery process of the guidewire is divided into contact behavior and delivery behavior. As shown in Fig. 5, the passive roller can be controlled to move along the red line by the roller pushing distance adjusting mechanism. The roller pushing distance is used to indicate the contact pressure. The greater the pushing distance, the greater the contact pressure between rollers. We can control the contact press between the rollers by adjusting the pushing distance. There are three different states between rollers in the contact motion successively: depart, tangency, and overlap.

The friction between the rollers converts the rollers' rotating motion into the guidewires' linear motion to realize the transmission of motion and power. There are many factors affecting the delivery process, such as contact pressure and roller surface friction coefficient. But this paper mainly analyzes the impact of material hardness on the delivery process.

4 Finite Element Analysis

The simulation purpose is to select the proper rubber materials, reduce the manufacturing and experimental cost, and predict the results under different rubber materials.

4.1 Finite Element Model

As is shown in Fig. 6, the finite element model is built based on the software HyperMesh 2019. The finite element analysis is carried out based on the software Abaqus 2020. Due to the slight deformation of the inner metal cylinder of the rollers, the finite element model of the rollers is simplified. The central axes of the rollers are parallel. Because the external coating of the guidewire is too thin, the guidewire is simplified to a slender rod. We conducted a mesh independence test to determine the most appropriate mesh size, which can both consider accuracy and computation. The mesh size of rollers is 0.1 mm. The mesh size of the guidewire is 0.05 mm, and there are 1000 layers in the length direction. The mesh size of the middle part of the roller that contacts the guidewire is smaller to make the analysis more accurate.

4.2 Constitutive Model of the TPU Rubber

As a kind of hyperelastic material, the elastic modulus and Poisson's ratio are generally not applicable in the TPU rubber deformation analysis by finite element method. The constitutive model is needed to be established. Many scholars have proposed a variety of constitutive models suitable for different rubber materials and working conditions. In this study, the Mooney-Rivlin model [18] based on strain energy function is selected for the finite element analysis, which can simulate the mechanical behaviors of almost all rubber materials.

Fig. 6 Finite element model

$$\begin{cases} W = W(I_1, I_2, I_3) \\ I_1 = \lambda_1^2 + \lambda_2^2 + \lambda_3^2 \\ I_2 = \lambda_1^2\lambda_2^2 + \lambda_2^2\lambda_3^2 + \lambda_1^2\lambda_3^2 \\ I_3 = \lambda_1^2\lambda_2^2\lambda_3^2 \end{cases} \tag{1}$$

where, I_1, I_2, and I_3 are the invariant of the Cauchy-Green deformation tensor, respectively. λ_1, λ_2, and λ_3 are the principal stretches.

$$W = \sum_{i,j=0}^{n} C_{ij}(I_1 - 3)^i (I_2 - 3)^j + \sum_{k=1}^{n} \frac{1}{D_k}\left(I_3^2 - 1\right)^{2k} \tag{2}$$

where, W is the strain energy density function, C_{ij} are material constants related to the distortional response, and D_k are material constants related to the volumetric response. For impossible rubber material, $I_3 = \lambda_1^2\lambda_2^2\lambda_3^2 = 1$:

$$W = \sum_{i,j=0}^{n} C_{ij}(I_1 - 3)^i (I_2 - 3)^j. \tag{3}$$

If only one material constant is used to describe the Mooney-Rivlin model, it is called the Neo-Hookean model:

$$W = C_{10}(I_1 - 3) \tag{4}$$

The material constant C_{10} of the Mooney-Rivlin model is obtained by the following equation [19]:

$$C_{10} = \frac{15.75 + 2.15 \times H_A}{6 \times (100 - H_A)} \tag{5}$$

where, H_A is the hardness of TPU rubber.

4.3 Boundary Conditions

The contact between the rubber and the guidewire is set as friction contact. The whole finite element analysis includes two steps: the contact analysis and the friction-based driving force analysis. The time for step one is 0.005 s. In this step, the mechanical changes in the contact motion of the rubber rollers are mainly analyzed. The time for step two is 0.004 s. In this step, the friction-based driving force caused by the rotation motion of rollers is analyzed. The boundary condition is that the roller rotates at a certain angle around the center. In this analysis, the roller pushing distance is set to 0.8 mm, and the friction coefficient is set to 0.35. The analysis type is set to the Explicit 3D stress based on the Abaqus 2020 (Tables 1 and 2).

Table 1 Material properties of 304 stainless steel

Material properties	Density (kg/m^3)	Young's modulus (GPa)	Poisson's ratio
Value	7830	204	0.285

Table 2 Material properties of TPU rubber

Hardness (Shore A)	C_{10} (MPa)	C_{01} (MPa)	D
50	0.4168	0	0
55	0.4963	0	0
60	0.6031	0	0
65	0.7405	0	0
70	0.9236	0	0
75	1.18	0	0
80	1.5646	0	0

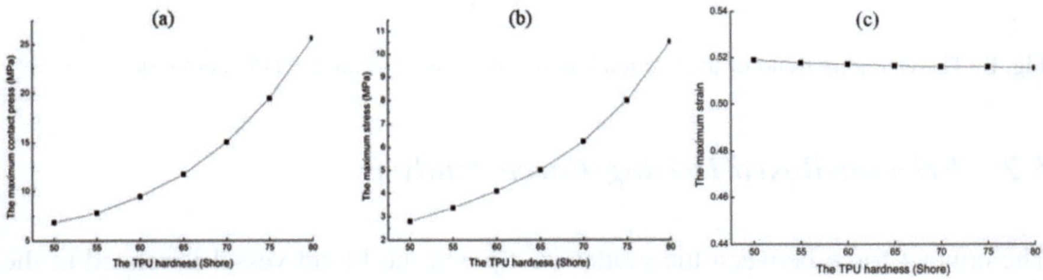

Fig. 7 The changing trend of the contact indexes with the TPU hardness. (**a**) The maximum contact press, (**b**) The maximum stress, (**c**) The maximum strain

5 Simulation Results and Discussion

5.1 Contact Motion Analysis

Three indicators were identified to quantify the contact behavior in step 1, namely, the maximum contact press, the maximum stress, and the maximum strain. The changing trend of the three indicators with the TPU hardness is shown in Fig. 7. When the TPU hardness increases, the maximum contact press and the maximum stress both increase obviously, but the maximum strain is almost constant. To reduce the roller deformation effectively, TPU with high hardness is recommended. However, if the maximum contact press is too high, the guidewire may be destroyed at the same pushing distance and friction coefficient. We need to select the material of the roller coating with proper hardness carefully.

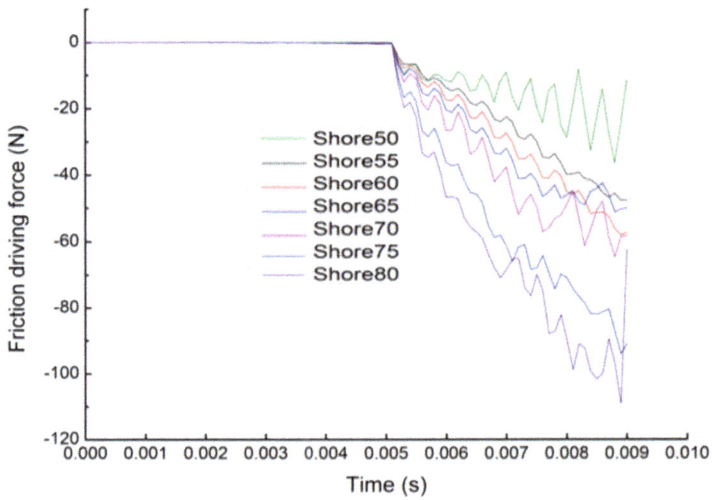

Fig. 8 The changing trend of the friction-based driving force with the TPU hardness

5.2 *Friction-Based Driving Force Analysis*

The contact force between the guidewire tip and the blood vessel is related to the friction-based driving force. If the driving force is too large, the blood vessel may be damaged when it collides with the guidewire tip. It is necessary to analyze the changes in the friction-based driving force with the TPU hardness to ensure the surgery's safety. The changing trend of the friction-based driving force with time under the TPU hardness is shown in Fig. 8. As we can see, at the same time, the friction-based driving force increases as the TPU hardness increases. The changing trend of friction-based driving force under different TPU hardness is similar.

6 Conclusions

This study introduces a novel endovascular intervention surgery robot, especially the delivery mechanism. The working principle and advantages of the novel robot are presented. The contact and friction between rollers are very significant to realize the function of the delivery mechanism of the robot. We explore the performance evaluation method. Three indicators were identified to quantify the contact behavior: the maximum contact press, maximum stress, and maximum strain. The indicator of friction behavior is the friction-based driving force. The finite element method simulates and analyzes the influence of the rubber material property on the contact behavior and friction-based driving force. For contact behavior, the TPU hardness obviously affects the first two indicators but not the maximum strain. For the friction-based driving force, the changing trends of friction-based driving force with time under the different TPU hardness are consistent, just with different values. These finite element analyses provide a valuable reference for selecting an appropriate outer material and designing a better roller.

Acknowledgement This work is supported in part by the National Key Research and Development Plan of China under Grant 2019YFB1311700, in part by the National Natural Science Foundation of China under Grant U1913210, Grant 62073325, and Grant U20A20224, in part by the Youth Innovation Promotion Association of CAS under Grant 2020140.

References

1. J. Goyes, N.J. Gomez, V. Restrepo, H. Contreras, J. Villegas, W. Giraldo, J. Bedoya, J. Morales, E. Arias, Radiation exposure in a group of interventional cardiologists in a high volume pci center. Interv Cardio **10**(5), 73–79 (2018)
2. R. Dixon, V. Khiatani, J. Statler, E. Walser, M. Midia, D. Miller, G. Bartal, J. Collins, K. Gross, M. Stecker, Society of interventional radiology: Occupational back and neck pain and the interventional radiologist. J. Vasc. Interv. Radiol. **28**(2), 195–199 (2017)
3. R. Beyar, L. Gruberg, D. Deleanu, A. Roguin, Y. Almagor, S. Cohen, G. Kumar, T. Wenderow, Remote-control percutaneous coronary interventions: Concept, validation, and first-in-humans pilot clinical trial. J. Am. Coll. Cardiol. **47**(2), 296–300 (2006)
4. J.F. Granada, J.A. Delgado, M.P. Uribe, A. Fernandez, G. Blanco, M.B. Leon, G. Weisz, First-in-human evaluation of a novel robotic-assisted coronary angioplasty system. J. Am. Coll. Cardiol. Intv. **4**(4), 460–465 (2011)
5. G. Weisz, D. Metzger, R. Caputo, J. Delgado, J. Marshall, G. Vetrovec, M. Reisman, R. Waksman, J. Granada, V. Novack, Safety and feasibility of robotic percutaneous coronary intervention: Precise (percutaneous robotically-enhanced coronary intervention) study. J. Am. Coll. Cardiol. **61**(15) (2013)
6. Robocath, Robocath successfully completes first robotic coronary angioplasties in germany with r-one™. https://www.robocath.com/robocath-successfully-completes-first-robotic-coronary-angioplasties-in-germany-with-r-one/ (2020)
7. J. Woo, H.-S. Song, H.-J. Cha, B.-J. Yi, Advantage of steerable catheter and haptic feedback for a 5-dof vascular intervention robot system. Appl. Sci. **9**(20), 4305 (2019)
8. S. Guo, Y. Song, X. Yin, L. Zhang, T. Tamiya, H. Hirata, H. Ishihara, A novel robot-assisted endovascular catheterization system with haptic force feedback. IEEE Trans. Robot. **35**(3), 685–696 (2019)
9. D. Kundrat, G. Dagnino, T.M. Kwok, M.E. Abdelaziz, W. Chi, A. Nguyen, C. Riga, G.-Z. Yang, An mr-safe endovascular robotic platform: Design, control, and ex-vivo evaluation. IEEE Trans. Biomed. Eng. **68**(10), 3110–3121 (2021)
10. H. Yu, H. Wang, J. Chang, J. Niu, F. Wang, Y. Yan, H. Tian, J. Fang, H. Lu, A novel vascular intervention surgical robot based on force feedback and flexible clamping. Appl. Sci. **11**(2), 611 (2021)
11. Q. Lu, Y. Shen, S. Xia, B. Chen, K. Wang, A novel universal endovascular robot for peripheral arterial stent–assisted angioplasty: Initial experimental results. Vasc. Endovasc. Surg. **54**(7), 598–604 (2020)
12. H. Shen, C. Wang, L. Xie, S. Zhou, L. Gu, H. Xie, A novel remotecontrolled robotic system for cerebrovascular intervention. Int. J. Med. Robot. Comput. Assisted Surg. **14**(6), e1943 (2018)

13. H. Shen, C. Wang, L. Xie, S. Zhou, L. Gu, H. Xie, A novel robotic system for vascular intervention: Principles, performances, and applications. Int. J. Comput. Assist. Radiol. Surg. **14**(4), 671–683 (2019)
14. J. Zhou, Z. Mei, J. Miao, J. Mao, L. Wang, D. Wu, D. Sun, Y. Zhao, A remote-controlled robotic system with safety protection strategy based on force-sensing and bending feedback for transcatheter arterial chemoembolization. Micromachines **11**(9), 805 (2020)
15. J. Chakravartti, S. Rao, Robotic assisted percutaneous coronary intervention: Hype or hope? J. Am. Heart Assoc. **8**(13), e012743 (2019)
16. G. Hyeon-Cheol, Understanding the coronary bifurcation stenting. Korean Circ. J. **48**(6), 481–491 (2018)
17. H.-L. Zhao, S.-Q. Liu, X.-H. Zhou, X.-L. Xie, Z.-G. Hou, Y.-J. Zhou, L.-S. Zhang, M.-J. Gui, J.-L. Wang, *Design and performance evaluation of a novel vascular robotic system for complex percutaneous coronary interventions* (2021 43rd Annual International Conference of the IEEE Engineering in Medicine Biology Society (EMBC), 2021), pp. 4679–4682
18. R. Rivlin, Large elastic deformations of isotropic materials. i. Fundamental concepts. Philos. Transact. R. Soc. London. Series A, Math. Phys. Sci. **240**(822), 459–490 (1948)
19. W. Wang, T. Deng, S. Zhao, Determination for material constants of rubber mooney-rivlin model. Spec. Purp. Rubber Prod. **25**(4), 8–10 (2004)

A Machine Learning Framework for Enhancing Security of Transaction in Saudi Banks Based on User Behavior

Haneen Almayouf, Shoaa Almudhibri, Wejdan Alsayegh, Meshaiel Alsheail, Salam Almneiy, Arwa Albelaihi, and Haya Duhisan

1 Introduction

In light of the digital transformation witnessed by the Kingdom of Saudi Arabia and with the launch of Vision 2030, with the transformation of many government sectors and institutions to a business model that relies on digital technologies to innovate products and deliver services and for the banking sector, the number of branches decreased significantly as a result of digital transformation, as the number of operating branches in the Kingdom reached 1420, i.e., 62 branches less compared to the previous year, according to a report the Saudi Central Bank (SAMA) for the year 1442/2021, and one of the reasons for this decline is due to the users' demand for spending through electronic payment and e-commerce, which exposed many users of Saudi banks to many financial frauds, as the protection system used in banks is less than what is expected [1]. Among the financial fraud methods described by SAMA, include unknown parties calling your mobile phone, sending you text messages, or emailing you while pretending to be official institutions such as banks or government institutions such as the Saudi Central Bank and asking you to disclose your credit card information, account numbers secret banking, or a verification code that the financial authority sends to your mobile phone on the pretext of updating your data. The exploitation of the psychological demands of people, like their desire to obtain quick wealth, earn cash or rewards, deal with unlicensed parties and people who work illegally and who provide services related to

H. Almayouf (✉) · S. Almudhibri · W. Alsayegh · M. Alsheail · S. Almneiy · A. Albelaihi · H. Duhisan
Department of Information Technology, Collage of Computer, Qassim University, Buraydah, Saudi Arabia
e-mail: hanenalmayouf@gmail.com; shoaarageh@gmail.com; wejdanalsayegh2@gmail.com; m.alsheail@qu.edu.sa; salamaalnuaimi11@gmail.com; a.albelaihi@qu.ed.sa; haya6ju@gmail.com

© The Author(s), under exclusive license to Springer Nature Switzerland AG 2024
A. Shaikh et al. (eds.), *Advances in Emerging Information and Communication Technology*, Signals and Communication Technology,
https://doi.org/10.1007/978-3-031-53237-5_20

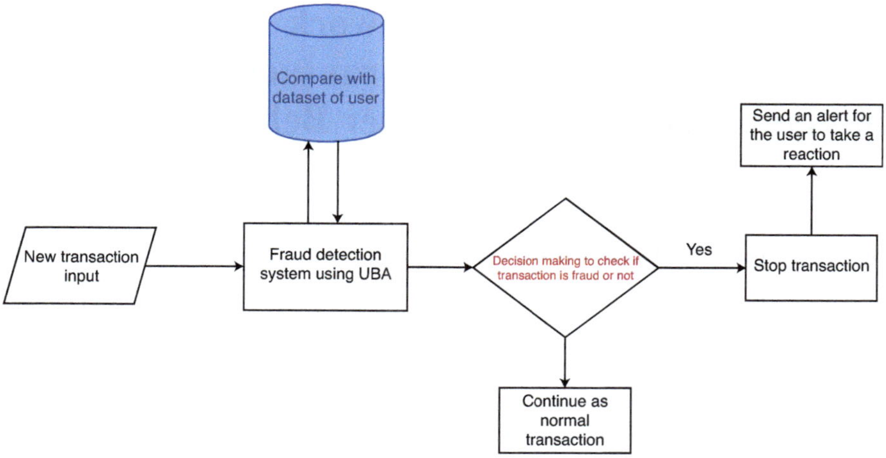

Fig. 1 Proposed model framework

granting loans with better terms and prices, or repay loans and financial difficulties without reasonable conditions to protect against financial fraud, is ignoring any calls or text messages asking you to provide details about your credit card and bank accounts or give the verification code that the financial institution sends to your cell phone, dealing with approved parties only. Install reliable software and security measures on your computer and mobile device to prevent hacking and make Internet use safer. Review the financial transactions by looking at your bank notifications for your bank transactions and your credit card statement to report any fraud to the bank immediately. Change your username and change your online banking password frequently. Unfortunately, all of these ways to try to prevent fraud do not prevent it enough, and this is what made us turn to this research to find an optimal way to protect the user from fraud before it occurs [2].

Recently, many bank users have fallen into electronic financial fraud, coinciding with the increase in the use of credit cards by Saudis and the increase in the number of fraudulent and untrusted websites, noting that more than 20% of Saudis have fallen into fraud cases through these applications and websites, which create a strong motivation for us to know the reasons for the persistence of the problem with the presence of many warnings and preventive measures [3]. The main objective of this research is to enhance safety in Saudi banks by studying user behavior, where we study the user's behavior pattern to discover abnormal behaviors that may indicate fraudulent behavior. The purpose of this research:

(1) User protection
(2) Reducing financial fraud
(3) Developing security in bank system

Based on previous studies and data in the field of fraud detection and user behavior analysis, the proposed system provides protection for users from financial fraud by studying user behavior and discovering patterns of use of bank accounts and credit cards based on artificial intelligence algorithms and machine learning. Figure 1 shows proposed model framework.

The steps of the proposed framework are as follows:

- Input a new transaction into the account of the user.
- Process the new transaction by entering the fraud detection system.
- Compare the new transaction with the user behavior analysis through the user dataset in the bank.
- If the result is matched with a true positive alarm, it will be blocked for the transaction, if not will continue as a normal transaction.

2 Background

User behavior analysis (UBA) is a tool that helps study and predict user behavior by collecting, studying, and analyzing user information by looking at patterns of human behavior. In other words, it is a scientific analysis of human behavior that starts with the knowledge and isolation of the parts of an event to determine the characteristics and the dimensions of the occasion where the behavior occurs and to define the changes that are produced by the activities in response to the environment, space, time, and opportunities [4, 5]. Due to the complexity of the task of analyzing user behavior and the high amount of various actions performed by the user, it is necessary to use an automated approach to identify suspicious user behavior [6]. For example, credit cards are offered by financial institutions, and these cards should be secured using UBA solutions. These solutions gather typical user behavior patterns using data from credit card transactions. Consider a user who consistently makes similar purchases in the same store. The credit card company may raise an alert and ask the user for transaction verification if this user attempts to spend a larger amount of money in another store since it is unusual behavior [7]. User activity using behavioral features:

- *Used System:* Users frequently have specialized systems for carrying out their tasks. Consequently, these systems are most frequently linked to the good intentions of users. If a user account is used on another system, it's possible that the user credentials are being used maliciously by an attacker. The use of a different system is obviously not a guarantee of suspicious user activity, as it is also possible that the user is working remotely or has a new system [6].
- *Location:* The system and logon method can determine where the account credentials were used. Therefore, the location from which the actual user logs on should be pretty consistent. Still, if the user, for instance, travels on business, then the location of the related user activities also shifts. However, this could

again be a sign of suspicious activity if there are numerous user activities in a short period from distant locations [6].

- *Date and Time:* If there are any actions the user deviates from the pattern associated with the time and date, this may also be a red flag for questionable behavior [6].
- *Session Length:* Similar to date and time features, deviations from the pattern of session length used to access client accounts can be used as evidence to detect suspicious user activity. However, sessions that take less time may be the result of beneficial user behaviors [6].

Behavioral analysis of user behavior is according to the following five steps:

(1) Define the target of the behavior to be analyzed.
(2) Observe the behavior.
(3) Record the frequency of the behavior in the database of user behavior together with other measurements of the behavior that happened throughout the procedure.
(4) Introduce the test variable, the test variable used to measure the system efficiency.
(5) Compare the frequency of behavior before and after the test variable or the occurrence of a response [8].

Cybersecurity Cybersecurity started to spread and became more important after people increased the use of the Internet in all aspects of their lives. In addition, the need to protect business and personal information on the network has led to an increased demand for the field of cybersecurity, which is a subfield of information security [9]. Cybersecurity focuses on the digital world and the goal of reducing cyberattack threats across the network, greatly affecting nonexpert end users connected to the Internet, to reduce risks, stop, and identify cyberattacks and detect fraud. User behavior analysis (UBA) has been used alongside cybersecurity and is a specialized field that focuses on modeling user behavior [7].

Fraud Detection It is about online fraud, a type of cybercrime that takes advantage of victims to carry out fraudulent transactions to gain financial gain. It has increased with people's dependence on the Internet in their daily lives. The key element shared by all of the works in this category is the use of UBA solutions to identify typical behavior patterns that could be used to spot fraudulent transactions. UBA solutions for identifying telecommunications fraud and credit card fraud are discussed technical fraud, contractual fraud, hacking fraud, and procedural fraud are the four subcategories of telecommunications fraud, on the other hand. All of them can potentially be found and avoided using UBA [7].

Pattern Recognition It is one of the inductive, indicative, mathematical, and statistical methods used to analyze and study user behavior, which aims to categorize objects into one category or group of categories and is part of artificial intelligence systems designed for decision-making. Moreover, pattern recognition algorithms can accurately, correctly, and quickly distinguish between similar patterns and any different and unusual behavior in the data. Over many years, pattern recognition has undergone significant developments and includes many methods, such as the following:

- Statistical pattern recognition: This classic method of identifying patterns that were discovered through a long process of development has been used extensively for a long period of time. This method depends on the distribution of the features vector on which a statistical model and probability are made.
- Data clustering: It aims to extract a similar set of data from the data block. The data can be divided into two parts: divide sections or hierarchical groups.
- The application of fuzzy sets: Fuzzy group theory can effectively explain the intent and extension of a fuzzy concept. The application of fuzzy groups in the process of identifying patterns began through two main processes: generalization and abstraction.
- Neural networks: These networks were invented based on simulating the human brain and deducing knowledge of its physiology and functions. These biological concepts are applied to machines in order to identify patterns.
- Structural pattern recognition: How some sub-patterns are formed into a single pattern. Pattern architecture can deal with statistics, classification, and neural networks and is one of the ways it can deal with complex problems that do not contain structural patterns.

Artificial Intelligence Branch of science and engineering concerned both with the development of intelligent artifacts and the computational explanation of what is often referred to as intelligent behavior [10]. There are two methods of structural pattern recognition syntactic analysis, which is the formal language, and structure matching, which is a mathematical technique based on sub-types [11].

Machine Learning (ML) Machine learning is one of the artificial intelligence fields in which data is used to benefit from it through algorithms that mimic human learning. The main task of machine learning is to extract sample conclusions using statistical theories to build a mathematical model. The performance of the models is improving through previous experiences. In some machine learning applications, the efficiency of inference or learning algorithms is no less critical than their predictive accuracy. Supervised, unsupervised, and reinforcement learning are the three machine learning types.

The two most popular types of machine learning (ML) are as follows: supervised learning: different algorithms that aim to map from input to output and the correct values are provided by the supervisor. Examples of supervised learning algorithms are neural networks, naïve Bayes, and linear regression. Unsupervised learning: there is only evidence of input, and the goal is to determine the regularity of the input. Examples of an unsupervised learning algorithm are KNN (k-nearest neighbors), hierarchal clustering, and anomaly detection. There are other types of machine learning, such as semi-supervised machine learning and reinforcement learning [12].

Artificial Neural Network Artificial Neural Network is a model inspired by the way the nervous systems in the brain process information. It consists of several layers called neurons. These cells perform two functions: collecting inputs and generating outputs [13].

Fuzzy Systems Using mathematical calculus, one can translate one's subjective understanding of real-world processes [14].

Deep Neuro-Fuzzy Refers to all methods combining neural networks and fuzzy logic [14].

3 Literature Review

This section will review some of the previous studies related to the study of user behavior analysis and fraud detection, then will show comparison between the techniques that enable us to analyze user behaviors and detect fraud with high quality and best cost. As shown in Tables 1, 2, and 3. The studies were arranged on the basis of the oldest and most recent and were divided into two tables, the first representing the period from 2007 to 2010 and the second representing the period from 2020 to 2022. Studies generally rely on different solution methods; this study by Quah and M. Sriganesh (2007) assesses a multilayer approach called self-organizing map (SOM), which is based on neural networks and consists of the initial authentication, screening layer, risk scoring layer, behavior analysis layer (core layer), and decision-making layer, to identify a customer's spending patterns in a credit card database. The main goals of the SOM approach are to categorize and cluster input data, find and identify hidden patterns in the data, and derive those patterns; this study proved that the SOM approach is an effective and simple approach compared to traditional approaches for pattern detection and clustering [15].

Table 1 Literature analysis comparison of the prior researches

Study	Technique	Dataset	Outcomes
[15]	Multilayer approach called SOM based on neural network	Extraction from the actual banking database	SOM approach is efficient and simple; it follows unsupervised learning without prior knowledge of fraud cases
[16]	Rule-based filtering, dumpster-Shafer theory, and Bayesian learning	Virtual dataset that was done by MS-SQL Server 2000	Simulation yielded up to 98% TP and less than 10
[17]	Neuro-fuzzy system	Not mentioned	Ambiguous data and prior knowledge is present, but applicability was limited number of flaws

Table 2 Literature analysis comparison of the recent researches

Study	Technique	Dataset	Outcomes
[18]	Integrating both k-means and ABC clustering algorithms and KNN for testing	Actual dataset of 99 customers' online purchases	Comparative studies have shown significant improvements in accuracy, with the simulation producing a true positive rate of 100% and a false positive rate of less than 2% (FP)
[19]	BiLSTM-MaxPooling and BIGRU-MaxPooling	IEEE-CIS Fraud detection	Model performed better with a score of 91.37%
[20]	Asexual reproduction optimization (ARO)	A Brazilian bank' dataset	ARO has increased the AUC, sensitivity and precision, and specificity; in comparison to AIS, it increased accuracy by 13%, 25% 56%, 3%, and 3%, respectively. Reduces training time by 75%
[21]	First approach: Deep convolutional neural network and feature engineering Long short-term memory network is the second method Deep convolutional long short-term memory network is the third technique	Actual data from one of Iran's biggest commercial banks	High performance in contrast to the most popular and conventional techniques. Deep neural network (DNN) based merger classifier Increase the random forest findings by roughly 38% to match the error rate (EER) and 0.58 for the distance degree. Using GAN files data significantly enhances the performance of every model utilized, particularly the final models formed at 50%
[22]	DNFN	UCSD network telescope aggregated DDoS metadata	With a maximum precision of 93.3 percent, recall of 95.7%, and F-measure of 92.1 percent, the proposed MSSO-based DNFN provided greater efficiency
[23]	ML, DL algorithm (CKHA)	An open-source credit card dataset	The OCSODL-CCFD technology was demonstrated by the experimental results obtained with the OCSO-selected attributes
[24]	Convolution neural network (CNN)	53 studies used Alexa's website or downloaded legitimate datasets, 53 for the phishing dataset, and 53 for the PhishTank website	The methods based on machine learning have been used the most in the selected

Table 3 Table of comparison of current methodologies

Techniques	Advantages	Disadvantages
Artificial neural network (ANN)	Ability to learn from the past and not require reprogramming. Capability to derive rules and forecast actions based on the current context. The ability to generate code for real-time systems, portability, high detection speed, ease of construction and operation, high accuracy, effectiveness in handling noisy data, identifying patterns resolving difficult issues, and processing new instances. Knowledge discovery and data mining, as well as adaptability and sustainability	Confirmation of the structure is challenging, large neural networks require a lot of processing time and excessive training, explanation capabilities are poor, setup and operation are challenging, costs are high, and nonnumerical data must be converted and normalized responsiveness to data format
Artificial immune system (AIS)	High ability to identify patterns. Strong learning and memory self-organization, simplicity in system integration, dynamically changing coverage, multilayered self-identification, noise and fault tolerance in predator-prey dynamics, low cost, and no requirement for a training phase in DCA are just a few of its many advantages	Lack of capability to deal with missing data in ClonalG and NSA and need for extensive training
Genetic algorithm	Functions well with noisy data, is easy to build, is simple to integrate with other systems, and is frequently used in conjunction with other techniques to improve their performance and optimize their parameters regarding expenses, detection speed, flexibility, upkeep, information finding, and data manipulation	Difficult to understand and requires extensive tool knowledge to set up and operate
Hidden Markov model (HMM)	Fast in detection	High cost, poor accuracy, and incapacity for handling large datasets
Support vector machines (SVM)	Given the convex nature of the optimality problem and the ability to be robust even in the presence of biased training samples, SVMs offer a novel solution	Large datasets that are difficult to process expensive have slow detection times, have medium accuracy levels, and lack results transparency
Bayesian network	High accuracy and processing speed for detection	Excessive training need and expensive
Expert system	High degree of accuracy, facilities for explanation, good performance, simplicity in handling complexity or missing information, and ease of system development and construction, rules from other techniques, such as	Knowledge representation languages are inadequate at handling missing data or unexpected data values processing various data types and approaching human flexibility poorly designed, operated, or integrated

(continued)

Table 3 (continued)

Techniques	Advantages	Disadvantages
	NN and DT, can be extracted, modified and stored in the KB	
Inductive logic programming (ILP)	Powerful when processing various data types, powerful when modeling complex relationships, and powerful when handling missing data	Low predictive accuracy high noise sensitivity and rapid performance degradation in the presence of erroneous data
Case-based reasoning (CBR)	Effective, flexible, and easy to update and maintain, can be used in a hybrid approach, works with incomplete or noisy data, and is useful in domains with a lot of examples	May experience the issue of imperfect on noisy data
Decision tree (DT)	High flexibility, soundness explainability, ease of implementation, simplicity in display and comprehension	Requirements for individually examining each condition. The transaction is a requirement for fraud detection
Fuzzy logic-based system Fuzzy neural (network)	Extremely quick detection and good accuracy	Expensive
Fuzzy logic-based system Fuzzy Darwinian (system)	High level of accuracy and maintainability	Has a very slow detection speed and is very expensive
Rule induction (RI)	Easy to use, easy to explain results, easy to interpret rules, and simple to implement applications. Scalable High predictive accuracy	The size of the dataset, the number of splits for each attribute, the order in which to choose the attributes, the difficulty of handling missing data, the possibility of over-fitting, and the difficulty of handling continuous data all have an impact on how well the tree is built
ARO (Asexual Reproduction Optimization)	Individual-based algorithm does not need any parameter setting, the selection mechanism is not necessary, and a model-free method can be utilized for various issues including optimization and decision-making	One potential long-term weakness of ARO could be its willingness to adapt to an unruly environment
Regression	Simple to comprehend construct and use with two class classifications For small datasets, logistic regression can provide more accuracy than some learning techniques	Poor with noise or outliers data unsuitable for complicated applications, ineffective with nonnumerical data, good but not high accuracy, unsuitable for large datasets

While in another study they discuss another technique by Suvasini Panigrahi et al. (2009), filtering by rules, Bayesian learning [PS1], and Dempster-Shafer theory are the three techniques that comprise the proposed fraud detection system. The Dempster rule is used to compare customer's current behavior to prior behavior, the rule-based filtering strategy determines the suspicious level of each incoming transaction, and the Bayesian learning approach updates the suspicious score of the transaction using the customer history database [16].

But this study (lens Brosso et al., 2010) conducted research on a continuous authentication mechanism based on known users and a security system that is based on user behavioral analysis and trust. The user is recognized upon first entry into the application, the initial data is recorded, and the user's behavior is started to be analyzed using trust. They choose the trust based on the user's propensity to repeat the behavior in situations that are repeated. According to the user's behavior analysis and the trust restrictions they have created, the proposed system establishes trust levels. With the help of the fuzzification mechanism and the neuro-fuzzy system, the user behavioral database can be updated continuously to reflect changes in user behavior and maintain trust levels. The system interacts with the user and gathers evidence to determine whether to increase or decrease the user's level of trust as time goes on and in accordance with behavior analysis. The system has two stages for allocating trust: first, the system accounts for the behavior in the database of behaviors and assigns a minimum trust before accounting and storing the captured data in the event that there is insufficient information about the behavior. Second, when a user interacts more with the system in subsequent accesses, a database's trust restrictions are checked first. If there are no restrictions, the system compares the user's actual behavior to that information; otherwise, an alarm is set off, the new behavior is stored, the trust is recalculated, and security mechanisms are triggered. Finally this system has some advantages over traditional methods, particularly when ambiguous data or prior knowledge is present, but their applicability was limited by a number of flaws in the individual models [17].

However, after 10 years of stagnation in research specialized in financial fraud through the analysis of user behavior, this study by Saad M. Darwish (2020) provides a system to improve correct fraud detection and reduce the percentage of false alarms. The proposed model was divided into two stages: training phase—k-means and ABC-clustering algorithms are integrated to record the user's historical spending patterns in the database and cluster transactions according to these patterns. Testing phase: using the KNN classifier, this builds the customer profile for the new transaction and compares it to the cardholder's stored spending profile. Adaptive decision integration, a type of data integration that integrates various business decisions into a single decision about the activity that occurred, is one of the reasons why this proposed model is superior to other approaches (i.e., input decisions for better or new decisions). For each individual classifier, this kind of fusion is required to increase accuracy and decrease uncertainty regarding the credit card fraud detection (CCFD) decision. The proposed method uses an integrated rule engine to filter the dataset's features based on many customer behavior (profile) parameters, including usage frequency, book balance, and geographic regions, to evaluate if an operation is normal or suspect. In order to identify unusual e-commerce transactions, they employed an actual dataset of 99 customers' e-commerce transactions. There are 927 transactions in the training dataset and 100 transactions in the test data. Up to 100% true positives (TP) and less than 2% false positives (FP) were produced by the simulation. Comparative research has demonstrated considerable increases in accuracy. Based on the findings, it can be said that learning and the fusion of multiple evidences are the best ways to handle this kind of real-world issue, where the behavior patterns are complex, and there might be little to no understanding of the application domain's semantics. To improve this model in the future and make it more accurate, more rules can be added to the rule engine. Furthermore, collaborative intrusion detection scenarios are implemented on real datasets to verify their effectiveness [18].

And another study by Hassan Najadat et al. (2020) works on the model BiLSTM-MaxPooling-BiGRU. The work done on the IEEE-CIS Fraud Detection dataset, which is available on Kaggle, forms the basis of this model, which uses machine learning and deep learning to address the problem of credit card fraud. Bidirectional long short-term memory (BiLSTM) and bidirectional gated recurrent units (BiGRU) that present method for predicting whether transactions on the IEEE will be legitimate or fraudulent. Additionally, used six machine learning classifiers: logistic regression, naïve Bayes, voting, Ada boosting, random forest, and decision tree. When results from machine learning classifiers are compared to those from this model, the results show that this model performed better with a score of 91.37%. IEEE-CIS Fraud Detection is the name of the dataset that was used for this study. They applied three techniques—SMOTE, random oversampling, and random undersampling—to deal with the imbalanced dataset. Additionally, they applied deep learning models to the dataset; they experimented with various deep learning models and selected BiLSTM-MaxPooling-BiGRU-MaxPooling as the top candidate [19].

While in this study (Farhang et al., 2021) try to find algorithms that can distinguish between legitimate and fraudulent transactions with a high level of detection accuracy in the least amount of time and at the lowest possible cost is the main goal of fraud detection studies. The evolutionary single-objective optimization (ARO) algorithm has many benefits that make it appropriate for use in fraud detection issues. The ARO (Asexual Reproduction Optimization) in credit card fraud detection is an individually based method. The best techniques used didn't produce as good of results as this efficient approach. They contrasted the outcomes with those of the AIS one of the best techniques ever used on the dataset [20].

In order to find the best method for detecting fraud, the most recent study (Baratzadeh et al., 2022) compares three widely used models—LSTM, CNN, and C-LSTM—with the three data mining techniques that are most frequently used for fraud detection: random forest (RF), support vector machine (SVM), and a conventional shallow-structured artificial neural network. The study offers solutions based on deep learning. The combination model of short-term memory and deep convolutional networks, which is trained using the combined data obtained from the generative adversarial network, exhibits the best performance, as per the observed findings. The goal of this study is to give data that will enable more effective approaches to the problem of uneven class distribution than those used in the past. Furthermore, it enhances performance through the integration of the long short-term memory network with the deep convolutional network. In this application, the distance score and the equal error rate are used to evaluate models in a more transparent and accurate way due to the inefficiency of assessment criteria like accuracy. The effectiveness of the experiment is assessed by contrasting the novel method with conventional approaches. They used a real-world dataset from one of the leading commercial banks in Iran for this investigation. Just 617 (0.02%) of the 268,465 transaction records in the sample are fraudulent transactions used at random. Twenty percent of the dataset was used for system assessment, and 80% was used for model training. In addition, they created some fictitious synthetic data using the generative adversarial network (GAN) technique to address the unbalancing problem. Two different approaches to improving modeling precision and taking temporal dependencies into consideration are proposed in this paper and are applied in three different modeling procedures. The first method is neural networks with deep convolution and feature engineering. In legitimate transactions, the deep convolutional neural network is taught to detect fraudulent activity. The second method is called long short-term memory network. Because recursive networks store memory in their cells, these networks may theoretically understand temporal relationships.. They are also capable of understanding the short- and long-term dependencies of time series data because of the unique architecture of this network's short-term persistent memory cells. As a result, this technique transforms raw data into time series of constant length and sends it to the long short-term memory network after undergoing little preprocessing. The third method deep convolutional long-term memory network: This model combines the best features of long-term and convolutional memory networks to understand 2D patterns and their temporal correlations. Their connections are profitable to both of them. This article describes how feature engineering extracted data is divided into time series with predetermined lengths and fed into a network for training. The evaluation results show that the recommended model performs better than the most widely used and accepted methods in this sector, including support vector machines and random forests. This study's DNN-based classifier, which is based on combinatory deep neural networks, improved the random forest's performance by more than 38% at an equal error rate (EER) and 0.58 for the distance score. All the models' performance, including their final model, improved by 50% when using the GAN-generated data. To summarize, the utilization of GAN-generated data significantly enhances the performance of any model, and its combination with LSTM and CLSM classifiers exhibits potential [21].

In this study, Suman Thapaliya et al. (2023) developed a generalized model using MapReduce as a basis for identifying cybercrimes. The goal of this form is to provide an automated form that can be used to show device misbehavior. As a result, attacks that take advantage of vulnerability can be detected by the newly developed automatic model, in which a partial routing trajectory search algorithm is used in order to send information between nodes. A deep neural fuzzy network (DNFN) is taken into account when using MapReduce to detect cybercrime in a base station (BS) [22].

In this study by N. Prabhakaran et al. (2023), e-commerce and communication technology advancements have led to a significant rise in credit card fraud in recent years. Fraudsters are always looking for new ways to commit fraud, and fraudulent credit card transactions frequently have an impact on the financial status of both businesses and customers. To maintain the trustworthiness of electronic payments, it is essential to recognize credit card fraud. Therefore, developing effective and precise credit card fraud detection (CCFD) methods is crucial. Due to the characteristics of building an effective model to identify fraudulent transactions, CCFD can make use of the recently established machine learning (ML) and deep learning (DL) methods. A deep learning model for CCFD is combined with a novel oppositional cat swarm optimization-based feature selection model in the OCSODL-CCFD approach. The OCSODL-CCFD method's main goal is to recognize and classify credit card fraud. The OCSODL-CCFD method is used to build a novel OCSO-based feature selection algorithm for choosing the optimal subset of features. Furthermore, the bidirectional gated recurrent unit (BiGRU) model is used in conjunction with the chaotic krill herd algorithm (CKHA) to perform hyperparameter tuning for the categorization of credit card frauds. Numerous simulation analyses were carried out to illustrate the better outcomes that the OCSODL-CCFD model produced. Based on a variety of evaluation metrics, the extensive comparative analysis revealed that the OCSODL-CCFD model produced superior results to its competitors [23].

In this study by Asadullah Safi et al. (2023), phishing is a kind of fraud when a perpetrator poses as someone they know and trust in order to obtain private information from a user on the Internet. A variety of phishing detection techniques, such as list-based, visual similarity, heuristic, machine learning, and deep learning-based techniques, are compared and examined in this systematic literature review (SLR). For this objective, several algorithms, datasets, and techniques for phishing website identification are revealed by the presented study topics. A thorough literature study covering the last 5 years included 80 scientific publications from renowned websites, conferences, leading workshops, research theses, book chapters, and research journals. This research updates earlier systematic literature surveys by emphasizing the latest advancements in phishing detection strategies. This study will provide readers with a better knowledge of the various phishing website detection techniques, datasets, and algorithms' relative performance. With 57 research covered, the SLR indicates that machine learning approaches have been applied the most. Furthermore, the survey indicated that researchers mostly used two sources to obtain the datasets: while 53 students entered the PhishTank website (53 for the phishing dataset), 29 students downloaded authentic datasets from Alexa's website. Furthermore, the majority of studies used machine learning techniques, with 31 using the random forest classifier, according to the literature study. The convolution neural network (CNN) has the best accuracy, 99.98%, according to numerous studies when it came to identifying phishing websites [24].

In this study, Kai Zheng et al. (2023) look for sequential usage patterns in the way physicians interact with the software user interface of an electronic health records (EHR) system. Clinicians exhibited consistent behavior while engaging with the EHR, some of which the system's creators or the clinic management did not foresee. By helping clinicians standardize daily clinical practice through consistent interaction with EHRs, awareness of such behaviors would help EHR design become more effective. Additionally, it would assist doctors in anticipating certain unanticipated outcomes of IT use in relation to workflow integration and software user interface design [25].

In this research by Achmad Tarmizi et al. (2023), this study aims to look into the possibilities of financial reporting fraud using the fraud hexagon hypothesis. A number of factors influence fraudulent financial reporting, including stimulus (financial target, financial stability, and external pressure), capability (change in director), collusion (total number of positions on the board of commissioners), opportunity (ineffective monitoring), rationalization (auditor switching), elegance (frequency of the number of photos of the CEO in the annual financial statements), and these other factors. The 215 banking firms that were listed on the Indonesia Stock Exchange (IDX) between 2017 and 2021 make up the study's sample. SPSS 25 software was employed for data processing. The results of this study indicate that external pressure and conceit are factors that contribute to dishonest financial reporting. However, collusion, director changes, auditor switching, financial targets, and stability all have no bearing on misleading financial reporting. Therefore, in order to have a system in place to avoid embezzlement, a corporation needs to build a human resources (HR) detection, monitoring, and systems review policy system [26].

In this study by Kibreab Adane et al. (2023), Adane and Beyene aim to investigate key variables affecting users' online security behavior in the context of email and website-based phishing attacks using the health belief model because the effectiveness of phishing attacks depends on human behavior. The model was validated using data from 138 faculty members. The aim of the study was to better understand email phishing and website-based phishing attacks in order to design online and offline content for cybersecurity awareness programs [27].

Although much research has been done in the field of fraud detection, no effective technology or algorithm in and of itself outperforms all others that can learn transaction data streams as they are in progress and can provide effective and reliable fraud detection systems. We could not find clear and explicit information about Saudi banks and how they protect users from financial fraud. Unfortunately, we also did not find a comprehensive dataset for all the elements that we need to test in the research, and we did not have access to the largest number of specialized research in this field, cybersecurity, and financial fraud, based on domain specificity.

3.1 Saudi Banks Customers

Data collection process was conducted to measure the extent to which Saudi banks protect users to obtain general knowledge about users using a survey (questionnaire).

Questionnaires:

In order to study and analyze these responses to demonstrate the importance and seriousness of this topic, we must first collect the views of the beneficiaries to help us determine the degree to which the beneficiary is exposed to fraud and whether the banks have a response. It has 132 responses collected by Google Forms, which included seven questions:

(1) What is your gender?
(2) What is your age?
(3) Do you think online banking is safe?
(4) Do you use electronic payment?
(5) Have you been subjected to financial fraud?
(6) Do you know someone who has been financially fraud?
(7) Did you find any interaction from the competent authority (bank)?

Analyze the results of the survey: The first and second questions were about age group and gender, and most of the age groups responding to the questionnaire were 40 years and over. The proportion of female respondents was 81.1%, and the proportion of males was 18.9%. The third and fourth questions in Figs. 2 and 3 are about if they think online banking is safe, also if they use electronic payment; we noticed that the responses differed regarding the security of electronic banking transactions, and most of them use electronic payment. The fifth, sixth, and last questions are about whether they have been financially defrauded; most of the answers are no; however, most of them know people who have been financially defrauded, and the ratios were close on the interaction of the competent authority (the bank) on the fraud issues they faced.

3.2 *System Design Procedure*

This section describes the diagrams and tools that were used to explain the functionality of the suggested system based on the requirements discussed in the previous level to aid in figuring out the overall structure of the system. By using a flow chart which is a diagram used to simplify and display the structuring of interactions in execution between system programs. Analysts mostly use system flowcharts to describe in detail a step-by-step chronology of functions and procedures [28].

First step: The customer logs in. Second step: This customer's identity is verified via the user verification page. Third step: If the customer is not registered in the system, he is returned to the user's login page. Fourth step: The main screen of the customer that contains the account data appears. Fifth step: The customer chooses to perform one of the activities available in the system, such as transfer, withdrawal, deposit, and other services provided by the systems of banks. Sixth step: All operations performed by the customer are saved in the database. Seventh step: The client's behavior in this session is compared with the client's behavior in previous sessions using a deep neural fuzzy. Eighth step: The data is returned after the comparison to be stored in the database. Ninth step: Verify whether this process is legitimate or fraudulent. Tenth step: If it is fraudulent, it will suspend the account and send an alert to the customer. Eleventh step: If his project will carry out this process. Twelfth step: Logout the system.

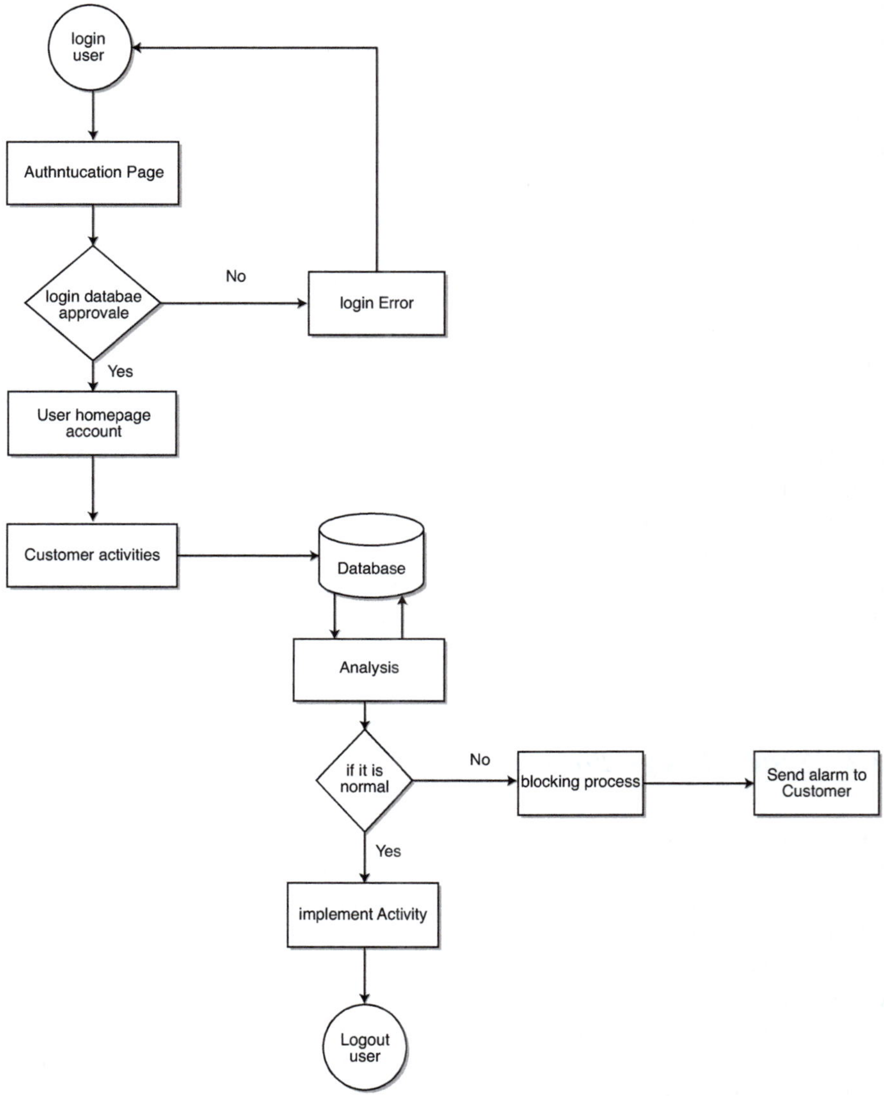

Fig. 2 Flowchart diagram of the proposed system

3.3 *Proposed Model*

This section will describe the steps in the implementation process used in the proposed system and recommend the requirements for that system, as well as the way the algorithm works. Figure 3 describes the overview of our proposed model implementation process.

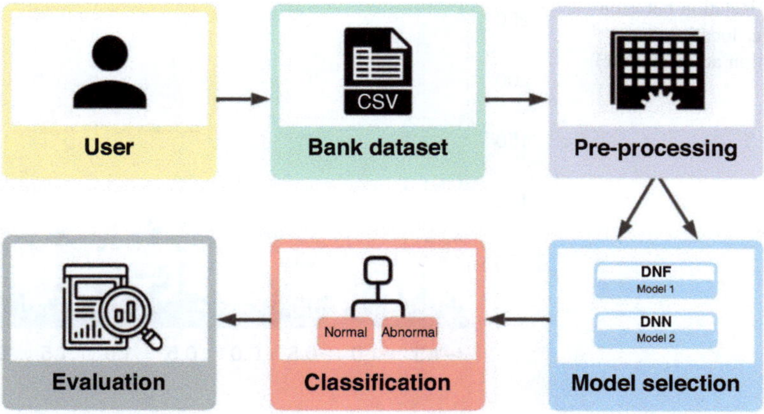

Fig. 3 Proposed model implementation process

Table 4 Description of bank transaction dataset features

Feature	Description
CustomerID	Unique value for each customer
TransactionID	Unique value for each transaction
CustomerDOB	Date of birth of the customer
CustGender	Male/female
CustLocation	The physical location of the customer
CustAccountBalance	The current total in account
TransactionDate	The date at which a trade occurs
TransactionTime	The value of the transaction being submitted

3.4 Dataset

We used two datasets for two different models to be able to test the effectiveness of the model if it were applied to an actual dataset.

3.4.1 Dataset Used with DNF Model

Almost all banks serve a large variety of customers based on their age, income, values, lifestyle, and other factors. Customer segmentation is the process of clustering a dataset of customers based on their common characteristics. Therefore, we decided to use this dataset to study user behavior.

In order to detect fraud and analyze user behavior, we used a dataset containing more than one million transactions made by over 800,000 customers for an Indian bank. This data includes information as shown in Table 4, such as a transaction ID, a customer ID, and customer age, which represents a customer's date of birth, physical

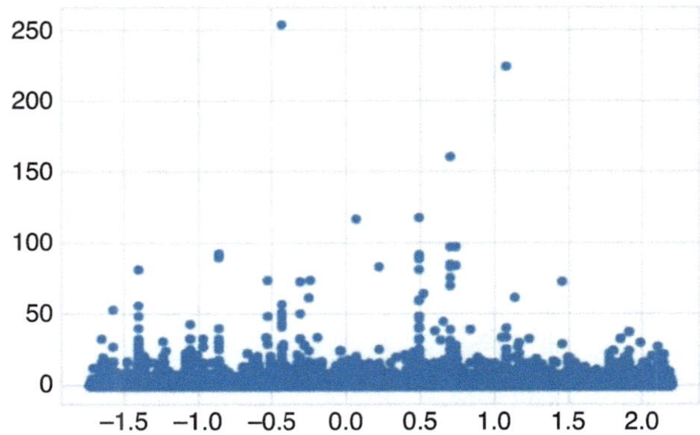

Fig. 4 Relation between customer location and transaction amount (INR)

location, and gender. There is also the account balance, which shows the total amount currently in the customer's account, as well as the transaction date and time and the amount of the transaction. Data preprocessing is required before we begin implementing, such as:

- Cleaning data: To clean data, we eliminate inaccurate information from a dataset, such as the fact that a person's birth year should not be less than 2022 or that it is corrupted, such as 1/1/1800.
- Feature extraction from other features: We extracted the age feature from the date of birth by calling the function to_datetime() from the Pandas library, subtracting it from the date of birth, and then dividing it by the function timedelta64 that was called from the Numpy library.
- Drop all columns that we will not use: Remove specific labels from columns using the DataFrame.drop() function from the Pandas library.
- The data has been changed into float numbers: Import preprocessing from the SkLearn library to use the function LabelEncoder().

When we searched for a suitable relationship to calculate the extreme values that indicate abnormal behavior, we tried to represent the location of the customer in relation to the amount of the transaction and also the age of the customer in relation to the amount of the transaction, and we observed that the values based on the age of the customer indicate that there is a strong relationship between the amount of the transaction and the age of the customer, as shown in Figs. 4 and 5.

3.4.2 Dataset Used with DNN Model

We utilized a virtual dataset created with the PaySim simulator because there aren't many datasets on financial services that are available to the general public, especially in the newly emerging industry of mobile money transfers. PaySim uses aggregated data from the private dataset to construct a synthetic dataset that contains malicious

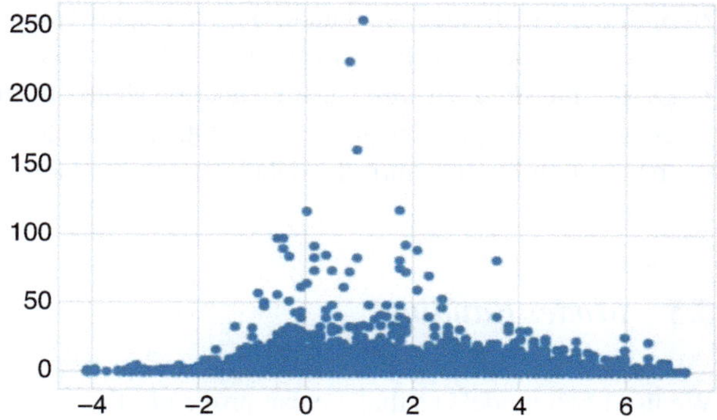

Fig. 5 Relation between customer age and transaction amount (INR)

Table 5 Synthetic financial dataset features

Feature	Description
Step	Represent a time unit in the real world. In this instance, 1 step equals 1 h. 744 steps total (30 days simulation
Type	CASH-IN, CASH-OUT, DEBIT, PAYMENT and TRANSFER
Amount	Amount of the transaction in local currency.
NameOrig	Customer who started the transaction
NewbalanceOrig	New balance after the transaction
OldbalanceOrg	Initial balance before the transaction
NameDest	Customer who is the recipient of the transaction
NewbalauceDest	New balance recipient after the transaction
OldbalanceDest	Initial balance recipient before the transaction
Is fraud	The transactions made by the fraudulent agents inside the simulation

behavior and replicates the typical operation of transactions in order to assess the efficacy of fraud detection algorithms.

The dataset contains values that represent a real-world time unit, such as means and steps. Here, one step is equivalent to 1 h. A total of 744 steps total—30 days of simulation—include types of transactions like cash in, cash out, debit, payment, and transfer, the amount of the transaction in local currency, the customer who initiated the transaction and was represented by nameOrig along with his initial balance prior to the transaction, the customer who received the transaction and was represented by nameDes along with his initial balance prior to the transaction, a new balance recipient following the transaction, no information for customers beginning with M (Merchants), and, lastly, the transactions carried out by the fraudulent agents during the simulation. In this particular dataset, the agents' dishonest actions are motivated by a desire to make money. Table 5 shows synthetic financial dataset features. They attempt to access clients' accounts, move money to another account, and then cash out of the system. There are a few steps in the dataset's preprocessing before we begin implementing:

- *Split dataset* divided the data into three stages: validation, evaluation, and testing by calling function split() from the Numpy library.
- *Divide the data for input* using function data() from the TensorFlow library to create an input pipeline and using batch(), prefetch(), and shuffle() functions from the random library, until the data is entered randomly for every 256 datasets.

3.5 Model Building

We used two models so that we can prove our theory by working on the two models according to the availability of the dataset. The models were created in Python 3 and tested in the Google Colab environment.

DNF is divided into three phases. The first phase was a preprocessing of data features to clean the data, such as removing empty records and extracting features from other features, such as age from the date of birth. Then the second phase was to collect the data to find the interrelationship between the data in order to label it and then save it in a (CSV) file. The final phase was to classify the data that had already been classified by deep ANN for regression. All these phases will be used to predict the results of the data. Figure 6 explores DNF implementation.

Dense layers are layers where neurons reside; these neurons define what these layers are with a specific activation function that specifies where the activation takes place, which means the possibility of passing data or not. There are many dense layers: the layer containing 500 and 300 neurons has a sigmoid activation function, the layer containing 100 neurons has a TAN activation function, and the layer containing 50 neurons has a ReLU() activation function. Table 6 explores the model extraction code for DNF model.

The DNN model was divided into two phases, with the first phase preprocessing the data to divide it into three parts: testing, validation, and evaluation, and then giving the data a volume and a way to enter it into the model, and the second phase

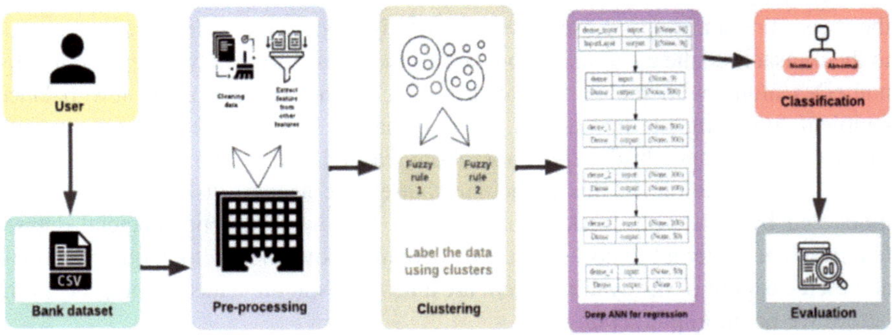

Fig. 6 Proposed model implementation for DNF

Table 6 Summary of the model extracted from the code

DNF model		
Laver (type)	Output Shape	#Param
Dense (Danse)	(none, 500)	5000
Dense_1 (Dense)	(none, 300)	150,300
Dense_2 (Dense)	(none, 100)	30,100
Dense_3 (Dense)	(none, 50)	5050
Dense_4 (Dense)	(none, 1)	51
Total params 190,501		
Trainable params 190,501		
Non-trainable params 0		

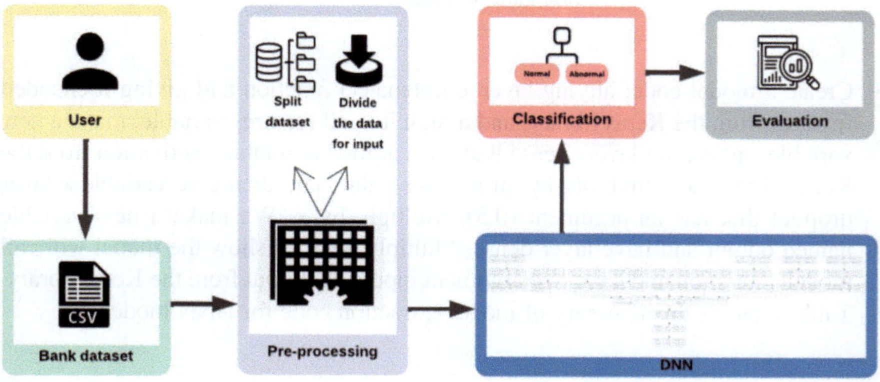

Fig. 7 Proposed model implementation for DNN

classifying the results into normal and abnormal. Figure 7 explores DNN implementation.

The model was trained on two types of data: numerical and categorical. The numerical data includes the amounts, OldBalanceOrig, NewBalanceOrig, NewBalanceDest, and OldBalanceDest. The categorical data includes step, type, numOrig, and numDest. In categorical data, we pass the data into an integer lookup layer, and then the data will go to the encoding layer. But in numerical data, the data will pass through the normalization layer. Then the model will concatenate all the data after processing.

- Integer features code: calling the input() function from the Keras library and giving it a shape, name, and data type and then encoding layer by calling the function get_category_encoding_layer_() and giving it a name, dataset, data type, and maximum number of tokens then assign encoding layer() to encoded step call, then will append step call, then will append encoded step call.
- Categorical features: make a set that has type, nameOrig and nameDest, and then make a loop that has a header to move on the set and encode the layer.

Table 7 Summary of the model extracted from the code

DNN model		
Layer (type)	Output Shape	#Param
Dense (dense)	(None, 1000)	10,000
Dense 1 (dense)	(None, 500)	500,500
Dense 2 (dense)	(None, 250)	125,250
Dense 3 (dense)	(None, 100)	25,100
Dropout (dropout)	(None, 50)	5050
Dropout2 (dropout)	(None, 1)	51
Dense 4 (dense)	(None, 1)	51
Total params 631,752		
Trainable params 631,752		
Non-trainable params 0		

- Create a model code: allying layer_cocatenate() function and giving it encoded features from the Keras library, and assign it to all features variable; make a new variable and calling layers dense that have argument unit and activation from the Keras library, and multiply by all features value, and then give variable x layer dropout that has an argument (0.5). Multiply by x. We make a new variable named output and have layer dense. Multiply by x, to show the model will call model() function and have two argument input and output from the Keras library. Table 7 shows the summary of model extraction code for DNN model.

4 Results and Discussions

In this research, we applied two different models with two different datasets; one of the datasets is real, while the second is simulated. This research evaluated the performance of two different models: the first is a hybrid model based on a combination of deep neural networks and fuzzy logic algorithms (DNF), where the data is classified using clustering in fuzzy logic, and the outputs of this clustering become inputs for neural networks to be classified into normal or abnormal results, while the second is a deep neural network (DNN) model that used a clustered dataset divided into fraud and not fraud and then trained the model to give us the results of whether this transaction is normal or abnormal.

As shown in Figs. 8 and 9, we demonstrate the accuracy of the two models of detection of anomalies in transactions, as Fig. 8 proves the efficiency of DNN in detecting anomalies in transactions in terms of the false positive and the true positive.

At this point, since each model has been trained with the same epoch size, we can compare the training time and accuracy results obtained to determine which model performs the best. We picked 300 epochs to ensure that the values will not change even when trained over large epochs, and we compared the results, looking at the training accuracy results and real-time results with the epoch number. The following

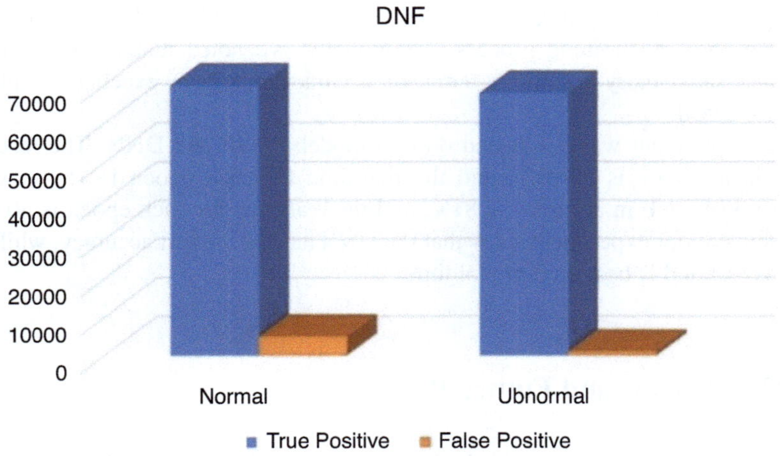

Fig. 8 DNF model

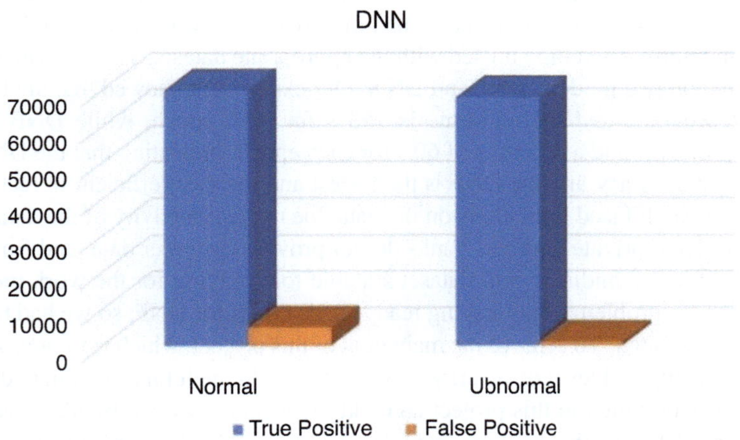

Fig. 9 DNN model

Table 8 Comparison for accuracy and real time

Model	Accuracy at 300 epochs	Runtime at 300 epochs
DNF model	81.0%	60 s
DNN model	99.95%	153 s

Table 8 is a summary of the testing performed to evaluate the performance of the models; the highest accuracy achieved is the result listed.

According to Table 8, the DNN model achieved the highest accuracy for our dataset. We have also confirmed that increasing the number of periods beyond

300 epochs will not significantly improve the results, and this is where the early stop function ends the training process. Additionally, we noted that the DNF model provided satisfactory results in terms of accuracy but gave excellent results in terms of speed.

The experiment was conducted on two models, DNF and DNN. The result of DNN in accuracy is 99.95%, and the real-time for each epoch is 153 s, while the result of DNF in accuracy is 81%, and the real-time for each epoch is 60 s, so the results of our experiment show that the DNN has the highest accuracy, while the DNF is fast and is best used in real time.

5 Conclusions and Future Work

The reason for enhancing security in Saudi banks by studying user behavior has been discussed in detail throughout this research. Previous studies have proven the effectiveness of neural networks in detecting fraud, but the time factor was a big problem in detecting anomalies in transactions, and after searching for algorithms that can improve the performance of neural networks, it was concluded that fuzzy logic can improve real time if used with the appropriate dataset. The experiment was conducted on two models, DNF and DNN. Our experiment showed that the DNN is 99.95% accurate, and the real-time is 153 s for each epoch, while DNF has an accuracy of 81% and a real time of 60 s for each epoch, indicating that the DNN has the highest accuracy and the DNF is the fastest and the most efficient in real time.

This research faced limitations on the data due to the sensitivity of the data, since it is considered private data and banks do not provide customer data, and from this, we had difficulty finding a real dataset suitable for applying for the work on it; we also faced the problem of not having ready-made codes for DNF, so we had to build codes from scratch. To achieve the main goal of this project, which is to improve and raise the quality and level of security in Saudi banks. In the future, we aim to develop the services provided in this project and add the location service by identifying the customer's IP with the banking bank, which raises the level of security in the banking bank. This project was presented as a proposal to cooperate with Saudi banks to provide this system that will help banks achieve security.

References

1. Annual report 57. [Online]. (2021). Available: https://www.sama.gov.sa/ar-sa/EconomicReports/pages/annualreport.aspx
2. How to protect yourself against financial fraud (2019) Accessed: 31 Oct 2022. [Online]. Available: https://samacares.sa/en/guides/protect-financial-fraud/
3. 20% of saudi online shoppers defrauded, arab news. Accessed: 04 Nov 2022. [Online]. Available: https://www.arabnews.com/node/768106/amp
4. R. Akerkar, P. Sajja, *Knowledge-Based Systems* (Jones & Bartlett Publishers, 2009)

5. M. F. K. A. Mrs, A. Sheth, S. Bhosale, *Research paper on cyber security* (The International Journal name, 2021) [Online]. Available: https://www.researchgate.net/publication/352477690_Research_Paper_on_Cyber_Security
6. M. Ussath, D. Jaeger, F. Cheng, C. Meinel, Identifying suspicious user behavior with neural networks, in *2017 IEEE 4th International Conference on Cyber Security and Cloud Computing (CSCloud)*, (IEEE, 2017), pp. 255–263
7. A.G. Martín, A. Fernández-Isabel, I. Martín de Diego, M. Beltrán, A survey for user behavior analysis based on machine learning techniques: Current models and applications. Appl. Intell. **51**(8), 6029–6055 (2021)
8. J.M. Helm, A.M. Swiergosz, H.S. Haeberle, J.M. Karnuta, J.L. Schaffer, V.E. Krebs, A.I. Spitzer, P.N. Ramkumar, Machine learning and artificial intelligence: Definitions, applications, and future directions. Curr. Rev. Musculoskelet. Med. **13**(1), 69–76 (2020)
9. S. Parmar, "Computer, Internet and Network Systems Scurity," 2010
10. K.C.M.J.R.D.P.J. Ramesh, A. N, *Artificial Intelligence in Medicine* (Annals of the Royal College of Surgeons of England) Accessed: 04 Nov 2022. [Online]. Available: https://www.ncbi.nlm.nih.gov/pmc/articles/PMC1964229/
11. V. Dutt, V. Chaudhry, I. Khan, Pattern recognition: an overview. *American Journal of Intelligent Systems* **2**, 23–27 (2012)
12. E. Alpaydin, *Introduction to Machine Learning* (MIT Press, 2020)
13. A. Dongare, R. Kharde, A.D. Kachare, et al., Introduction to artificial neural network. IJEIT **2**(1), 189–194 (2012)
14. J. Vieira, F. Morgado-Dias, A. Mota, Neuro-fuzzy systems: A survey. *WSEAS Transactions on Systems* **3**, 414–419 (2004)
15. J.T.S. Quah, M. Sriganesh, Real time credit card fraud detection using computational intelligence. Int. J. Conf. Neural Network **2007**, 863–868 (2007)
16. S. Panigrahi, A. Kundu, S. Sural, A.K. Majumdar, Credit card fraud detection: A fusion approach using dempster–Shafer theory and bayesian learning. Inf. Fusion **10**(4), 354–363 (2009)
17. I. Brosso, A. La Neve, G. Bressan, W.V. Ruggiero, A continuous authentication system based on user behavior analysis, in *2010 International Conference on Availability, Reliability and Security*, (IEEE, 2010), pp. 380–385
18. S.M. Darwish, A bio-inspired credit card fraud detection model based on user behavior analysis suitable for business management in electronic banking. J. Ambient. Intell. Humaniz. Comput. **11**(11), 4873–4887 (2020)
19. H. Najadat, O. Altiti, A.A. Aqouleh, M. Younes, Credit card fraud detection based on machine and deep learning, in *2020 11th International Conference on Information and Communication Systems (ICICS)*, (IEEE, 2020), pp. 204–208
20. A.F. Ghahfarokhi, T. Mansouri, M.R.S. Moghaddam, N. Bahrambeik, R. Yavari, M.F. Sani, Credit card fraud detection using asexual reproduction optimization. Kybernetes (2021)
21. F. Baratzadeh, S.M. Hasheminejad, Customer behavior analysis to improve detection of fraudulent transactions using deep learning. J. AI Data Min. **10**(1), 87–101 (2022)
22. S. Thapaliya, P.K. Sharma, Optimized deep neuro fuzzy network for cyber forensic investigation in big data-based iot infrastructures. IJISP **17**(1), 1–22 (2023)
23. N. Prabhakaran, R. Nedunchelian, et al., Oppositional cat swarm optimization-based feature selection approach for credit card fraud detection. Comput. Intell. Neurosci. **2023**, 2693022 (2023)
24. A. Safi, S. Singh, A systematic literature review on phishing website detection techniques. J. King Saud Univ.-Comput. Inf. Sci. (2023)
25. K. Zheng, *An Interface-Driven Analysis of User Behavior of an Electronic Health Records System* (Ph.D. dissertation, The University of Michigan, 2029)

26. T. Achmad, I. Ghozali, M.R.A. Helmina, D.I. Hapsari, I.D. Pamungkas, Detecting fraudulent financial reporting using the fraud hexagon model: Evidence from the banking sector in Indonesia. Economies **11**(1), 5 (2022)
27. K. Adane, B. Beyene, Email and website-based phishing attack: Examining online users security behavior in cyberspace environment. IJISM **21**(1), 245–262 (2023)
28. Flowchart. [Online]. Available: https://dl.acm.org/doi/abs/10.5555/1074100.1074406

Hybrid HQ Stereo Cameras and RPLIDAR Sensor System Applied to Navigation of the Autonomous Mobile Robots

Luige Vladareanu, Hongbo Wang, Marius Pandelea, Victor Vladareanu, Ionel-Alexandru Gal, and Ştefan Ghibanu

1 Introduction

Artificial intelligence and robots are present in the industrial and domestic fields in the form of manipulators, mobile or anthropomorphic robots, self-driving cars, home systems that support the elderly or disabled, medical equipment for physical treatment and recovery, surgical operations, etc.

The introduction into the informational flow of precise data, in relation to the environment from which they come, from the latest generation equipment, gives the chance of improved control and high precision movements. The movement of the robot is carried out in the unknown working environment only by knowing itself and the surrounding space very well. Sensors are equipment that observes, identifies, tracks, and recognizes environmental entities, practically maps the space, and also helps define the position and orientation of the robot.

At the hardware level of the robots' perception subsystem are found combinations of equipment such as camera and LiDAR sensor, camera and radar sensor, camera and IMU sensor, LiDAR sensor and radar sensor, LiDAR sensor and IMU sensor, and camera with LiDAR sensor and radar sensor.

Among the various factors that influence the movement of these entities, the authors identified the multitude of categories and forms of obstacles (convex/concave, small lakes), the type of camera (active/passive, monocular/stereo, infrared

L. Vladareanu (✉) · M. Pandelea · V. Vladareanu · I.-A. Gal · Ş. Ghibanu
Romanian Academy, Institute of Solid Mechanics, Bucharest, Romania
e-mail: luige.vladareanu@vipro.edu.ro; marius.pandelea@imsar.ro; victor.vladareanu@imsar.ro; alexandru.gal@imsar.ro; stefan.ghibanu@imsar.ro

H. Wang
Academy for Engineering & Technology, Fudan University, Shanghai, China
e-mail: wanghongbo@fudan.edu.cn

© The Author(s), under exclusive license to Springer Nature Switzerland AG 2024
A. Shaikh et al. (eds.), *Advances in Emerging Information and Communication Technology*, Signals and Communication Technology,
https://doi.org/10.1007/978-3-031-53237-5_21

camera, LiDAR, radar, RGB-D, sonar, thermal), the technical specifications of the sensors (angle, range, size, and velocity), the optimal multiple sensor configurations, the environmental conditions in which the movement takes place (day lighting, fog, low visibility, night time, rain, snow), and the principle information fusion method (raw data level, feature level, decision level).

In the article [1], the authors visually approach SLAM as a hybrid method of stereo environmental perception for autonomous robots. The localization of the robotic system in motion in unfriendly environments, such as those with strong illumination fluctuations and without texture, is initially approximately estimated by online photometric calibration and feature matching, and then the proposed hybrid pyramidal algorithm combines direct and indirect methods of depth map error elimination images.

The concern of the authors [2] is the subject of simultaneous localization and mapping (SLAM) analyzed extensively in the scientific article that reviews, for a period of over 30 years, the models, methods, algorithms, difficulties, evolution, and current trends. This overview is particularly useful for the field of mobile robots because it concisely shows the correspondence between factual situations, methods, and solutions with adopted particle filters, the evolution of the revision over time of the visual-SLAM methods specifying the type of sensor, the solution method, the outputs, and the disadvantages. The process of locating a Caterpillar vehicle and tracking the path traveled in an unknown and uncertain environment is studied theoretically and experimentally [3]. Modeling the vehicle, developing the EKF (extended Kalman filter) algorithm for positioning through the information of the SLAM mode and the RPLIDAR model A2 laser sensor, as well as the robust control for tracking the trajectory in a disturbed environment through MIMO (multiple input and multiple output) are the stage of the authors research.

The authors of [4] focused on the real-time exploration of the unknown indoor environment by a mobile robot through the SLAM module connected to a LiDAR sensor. The proposed exploration method, called e-SLAM, builds the space map by filtering the identified entities and predictions in the navigation process and by reporting the movements to the MSC (most significant corner) benefiting from the help of the LQE (linear quadratic estimation) algorithm.

2 Hybrid HQ Stereo Cameras and RPLIDAR Sensor System

Starting from the observation, identification, and recognition of the environment together with its objects and beings, as well as the monitoring of the main qualitative factors, the autonomous bipedal robots placed in the public space collect data and bring a plus by alerting the authorities about major local risk situations, help people with visual impairments in the performance of urban actions, provide important information and social assistance to citizens, and indicate proactive solutions. The connection of several hardware and software entities, such as SLAM (simultaneous localization and mapping), cameras or video cameras in monocular or stereo system, and various sensors, facilitates the production of intelligent movements and the fulfillment of missions. The collaboration between the laser scanner, cameras, and the SLAM module annihilates their weak points and gives power to identify and recognize objects/obstacles, good operation in low light conditions, tracking fluctuations from the environment, etc. [5]. Specialized literature knows numerous situations of singular use of peripheral equipment that transmit data and information about the environment, as well as complex solutions with composite architecture of vision-LiDAR, visual-IMU (inertial measurement unit), vision-SLAM, and LiDAR-SLAM. All of these offer remarkable qualities to robots in adverse conditions, such as autonomous navigation, planning their own movements and routes, determining decisions, and static and dynamic stability [5].

The collaboration between the 2D LiDAR sensor and RGB-D cameras leads to the merging of data that greatly expands the application field and offers advantages for 3D mapping, semantic mapping, and people detection and recognition. Real-time route planning with environment mapping for a mobile indoor wheeled robot is studied with the cost function algorithm for planning the direction to the target point and the DWA (dynamic window algorithm) algorithm for the immediate zonal route planning neighborhood of the robot. The hardware structure used for the mobile robot benefited from the information received from the SLAM module, on the one hand, and the RGB-D depth camera or from the 2D LiDAR sensor, on the other hand. At the same time, three types of mapping algorithms were considered: "GMapping algorithm, HectorSLAM algorithm, and Cartographer algorithm." Good results, simulated and experimental, for the movement of the robot in indoor spaces were obtained by using the mapping algorithms alongside the SLAM module, and in the conditions of a nonuniform environment, the HectorSLAM algorithm proved superior to the others [6].

For route planning and robot navigation, the generation of real-time 3D topographic map of the environment with a high density of points is a crucial necessity and is obtained by fusing data and information from LiDAR sensor and cameras, according to [7]. The SLAM and IMU modules help to develop the map, together with the CNN segmentation of the images captured and reported to the PSPNet-50 base. The experimental testing validated the proposed method by drawing up a precise map that includes objects whose classification was improved but also optimized by the CRF (conditional random fields) method, using the KITTI dataset. The data fusion module in permanent collaboration with the perception subsystem facilitates the existence of firsthand information for mapping the environment, identifying obstacles, and locating and navigating autonomous robots. For environmental mapping, the techniques used in data fusion are landmark shape reconstruction, 3D depth map modeling, SLAM module in combination with 2D LiDAR scanner and stereo camera, LiDAR scanner, and a camera. The detection and avoidance of obstacles were introduced into the scheme systems equipped with LiDAR and camera, whose information merges gradually optimized: fusion of information from multimodal systems, Gaussian algorithms for dynamic environments; CNN (convolutional neural network), RNN (recurrent neural networks), SSD (source to skin distance) techniques [8]. The article [9] proposes the modular approach for real-time localization of mobile ground robots in outdoor environments, regardless of lighting conditions. The information received from the LiDAR sensor is the basis of this research which refers to the regeneration of the environmental map through the LOAM (LiDAR odometry and mapping) method, corrective techniques, and segmentation techniques of the identified point clouds, as well as the precise establishment of the robot's location. The experimental results obtained for localization can be successfully applied in real time to autonomous vehicles, thanks to the NVIDIA Jetson Xavier supercomputer and the good performance recorded by the convergence speed of the estimated position matrix compared to NDT (normal distributions transform) scanning. Article [10] brings as novel elements the NNMPA (neural network-based multipeak analysis) method and the prediction of relatively large distances (4060 m) to landmarks in strong background lighting conditions, through real-time analysis and correlation of the generated histograms, as well as optimization through automatic learning. Using the LiDAR sensor and the FNN (feedforward neural network) extraction of a small number of main features of the objects resulted in two sets of data, one theoretical and one real. Experiments have shown improved distance accuracy and reduced measurement time compared to CDP (classical digital processing). The system proposed in [11] experiences in real time the movements of the robot, located in external spaces, on a solid support surface or on the water surface. The studied component of the system is the subsystem made up of 3D LiDAR and the SLAM module which, by means of matching algorithms for scanning with coplanar extraction in four points, on the one hand, and the odometry functions and informational feedback, on the other hand, locates, through high frequency, and accurately corrects the location, through low frequency, optimizing the position of the robot in a dynamic environment. The evaluation in the virtual environment through the KITTI dataset and in the real environment on an unmanned ground vehicle (UGV) shows that the proposed method has adaptability to the two environments and positioning accuracy below 1% compared to the landmark. For UAV (unmanned aerial vehicle) position estimation and navigation in 3D environments, the authors of [12] used the combination of IMU-LiDAR sensors and designed an algorithm with a loss function for adjustment. At the same time, they built a CNN neural network structure with LSTM (long short-term memory) deep learning function that records the values of the parameters perceived by each sensor. The results obtained on the flight simulator, reported to the KITTI dataset, show that the proposals, the algorithm, and the network generate accurate positions and navigation compared to the version without the IMU sensor.

The mobile robotic platform (MRP) architecture in Fig. 1 is presented. The main hardware components are two HQ IMX477 cameras with 6 mm "fisheye" lenses located at a distance of 0.232 m between them and a RPLIDAR A3 laser scanner. The NVIDIA Jetson AGX Xavier board has the role of quickly processing the large volume of recorded information. The mobile robotic platform (MRP) architecture was used for real research experiments for obstacle recognition.

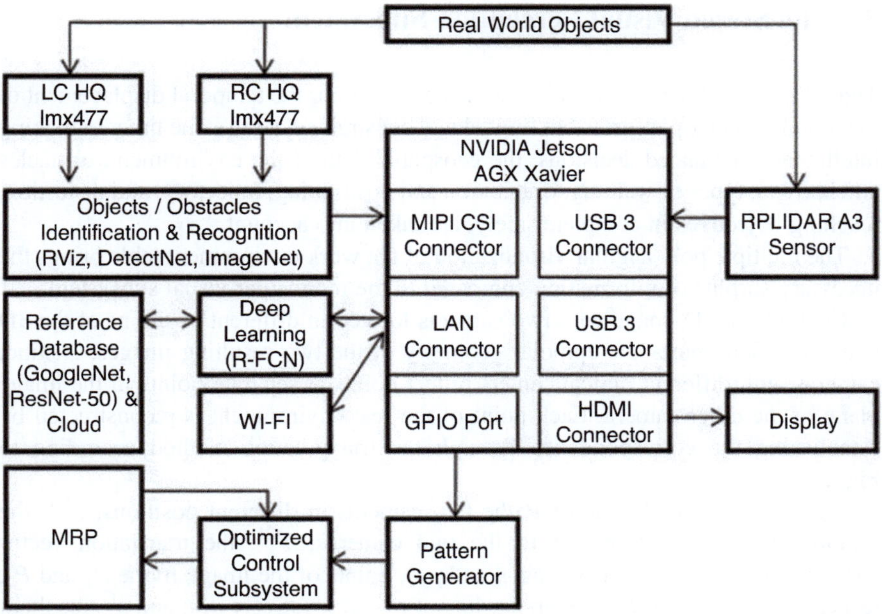

Fig. 1 The mobile robotic platform (MRP) architecture with hybrid technological solution

Communication between the Raspberry Pi HQ stereo cameras with Sony IMX477 sensor and the RPLIDAR A3 laser sensor is realized through the NVIDIA Jetson AGX Xavier board for increased calculation speed and ensures connections with the subsystems and equipment of the robotic system: cameras, sensors, control subsystem, display, Ethernet, and SD card.

Measuring the distance to objects, landmarks, and environmental obstacles is realized through the stereoscopic technique with the help of two cameras positioned differently in space. By capturing two different images that are later processed and superimposed, see obtain depth perception.

For real-time identification of the contours of objects in the robot's environment, as well as the distances to them with stable operation, the 2D telemetry solution with the RPLIDAR A3 low-power laser scanner is used. The increased accuracy of the mapping and obstacle avoidance process is ensured by the rotation speed at 16,000 rpm of the RPLIDAR head.

Part of the perception subsystem, the RPLIDAR model A3 scanner equipment is a rangefinder that reflects the laser beam in its own rotating mirror, based on the principle of time of flight (ToF), providing values of the distance in the plane of the beam to the surface of the objects encountered.

Mobile robotic platform (MRP) has a classic shape with four wheels, each driven by an electric motor. The dimensions of the platform are the length of 0.26 m, the width of 0.34 m, and the height with a Wi-Fi antenna of 0.18 m.

3 The Stereo Visual Perception Subsystem

The stereo visual perception subsystem is essential for the temporal displacement of the mobile robotic platforms but limited and constrained at the same time. In making intelligently advanced decisions, the geospatial data of the environment, obstacles and barriers, types of walking, orientation and positioning, trajectory, and distortion, reaching objectives in complete safety, are taken into account.

The multiple poly angular visualization of the work environment table brings the necessary surplus of information compared to the monocular visual subsystem.

Capturing a 2D image with two cameras located in different positions of the 3D real space determines the epipolar geometry of the two resulting images. Distinct cameras with different optical centers reflect points at separate points in the image plane of the other camera. Each point of the real environment is reconstructed by establishing the correspondence through the triangulation method according to Fig. 2.

Figure 2 contains CL and CR the two cameras in different positions: R^{C1}, the rotation matrix corresponding to the two cameras; T^{C1}, the translation vector corresponding to the two cameras; p and p_{P2}, points of the image frame P_1 and P_2, respectively; Fp_{p1} and Fp_{p2}, epipolar line of p_{P1}, respectively p_{P2}; and p^C, resultant point of the epipolar plane that includes the reference line of the two chambers. The attached fundamental matrix F verifies the epipolar constraints at points p_{p1} and p_{p2} in the two images captured by the cameras [13]. Knowing the position of the cameras, the corresponding image points in several frames are attached the epipolar constraint in the form of the relation in [13]:

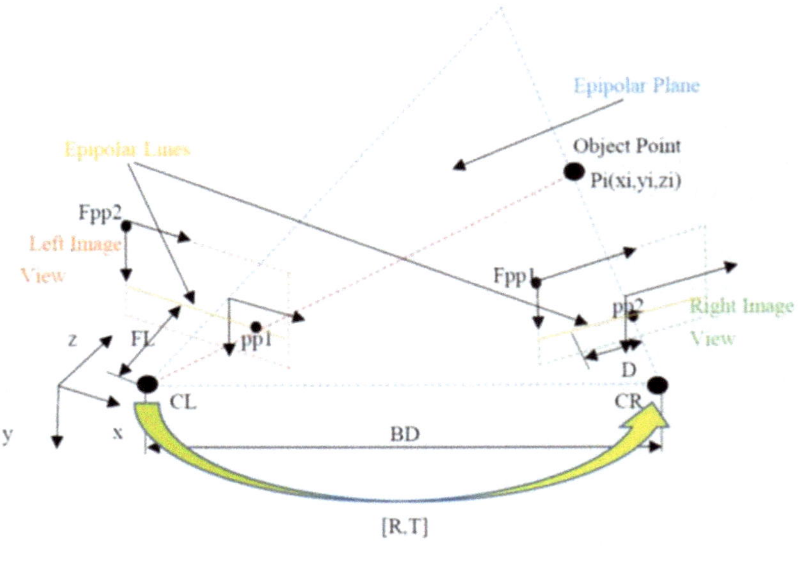

Fig. 2 The main geometric elements of binocular stereo vision

$$p_{p_1}^T \times F \times p_{p_2} = 0. \tag{1}$$

The advantage of using epipolar geometry is given by the fact that the search for the corresponding point is greatly reduced, so that one-dimensional examination is done on the epipolar line. The MATLAB programming and numerical calculation platform can be used for the implementation algorithm. Areas with major discontinuity, as well as regions with high homogeneity, raise the most problems for the correspondence verification process.

The disparity map is based on epipolar geometry; it accurately determines the distance to an object under the conditions of a well-made calibration and a correct match between the pixels of the captured images.

Basically, it is assumed that two points belonging to the two images have the same depth if they have identical intensities of the characteristic pixels. Better information about the match is obtained when the points of interest are analyzed alongside the other points of a predefined area.

The calculation of the disparity values at the level of each pixel is generally performed locally, with the WTA (Winner Takes All) strategy, according to [14]:

$$d_p = \underset{d \in \gamma}{\arg\min}\ (C(p, d)) \tag{2}$$

where γ is the disparity range and $C\ (p, d)$ is the cost function of pixel p. We obtain improved results by reconditioning the disparities, involving in the process a disturbance filter, but also interpolation.

Stereoscopy algorithms, whether local or general, need new hardware solutions that significantly reduce the computation processing time. They aim either to increase the quality of the disparity map or to increase the speed of data processing and the accuracy of calculations. In order to have noticeable results in real time, local algorithms use the following relations for the calculation cost:

Cost function for the AD (absolute intensity differences) algorithm [14]:

$$C_{AD}(x, y, d) = \frac{1}{n} \sum_{i=1-n} \left| I_l(x, y)_i - I_r(x - d, y)_i \right|. \tag{3}$$

Cost function for the SAD algorithm (sum of absolute differences):

$$C_{SAD}(x, y, d) = \frac{1}{n} \sum_{(a, b) \in N(x, y)} C_{AD}(a, b, d). \tag{4}$$

The global approach methods, DP (dynamic programming), GC (graph cuts), and BP (belief propagation), are based on the influence of all pixels. Smoother disparity maps are defined by minimizing the energy by scanning in both the horizontal and vertical directions. Their disadvantage is that induce the streaking effect in images [14]:

$$E(f) = E_{\text{data}}(f) + E_{\text{smooth}}(f) \tag{5}$$

where $E_{\text{data}(f)}$ is the data term corresponding to the values of each pixel and E_{smooth} $_{(f)}$ is the uniformity term that ensures the smoothing of major differences between neighboring pixels.

Summing up the information from the sensors provides the proposed MRP system with increased real-time action capacity and robustness through software and the NVIDIA Jetson AGX Xavier card.

4 Software Architecture

The functional duality of the peripherals incorporated in the proposed perception subsystem is presented in Fig. 3. The two collaborative equipment in the proposed perception subsystem create connections with the NVIDIA Jetson AGX Xavier card. The hardware components of the robotic system collaborate with the following software parts, as shown in Fig. 3.

Object Identification and Recognition It consists of software processing and extraction and segmentation and filtering and interpretation and fusion data and matching. RViz, DetectNet, and ImageNet were used in the current research stage.

RViz is a software utility of the ROS visualization tool, with a simple graphical interface and a rich data library. It allows the 2D/3D visualization of robotic systems integrated in natural environments based on the information provided by existing sensors (cameras, odometric equipment). RViz has the ability to provide multiple data in real time by using the attached plugins. It allows the configuration and observation of the customized parameters of the attached robotic system. Displaying the values of the state parameters simultaneously with the movement of the robot requires the fast processing of the extremely large volume of datasets. It also allows motion planning and the study of the simulation of robotic systems in the virtual environment. The mobile robot is tracked on the display using the RViz software installed under ROS Melodic (http://wiki.ros.org/rviz).

DetectNet is one of many software for detecting and classifying objects identified from captured images, based on DNN (deep neural network). It is the creation of NVIDIA, similar to YOLO (you only look once), RetinaNet, R-CNN (regions with convolutional neural networks), and SSD (single-shot detector) and can be accessed from Python or C++. It is especially used for images with a large number of different objects, providing information about the class and position coordinates of the workbox attached around each one. The information flow provided by cameras in the natural environment is stored by memory on image frames and processed. The DetectNet method includes the analysis of the recorded data and the primary labeling by class, the extraction and filtering of the main features, the anticipation of the class by semantic segmentation of the superficial layers with R-FCN (region-based fully

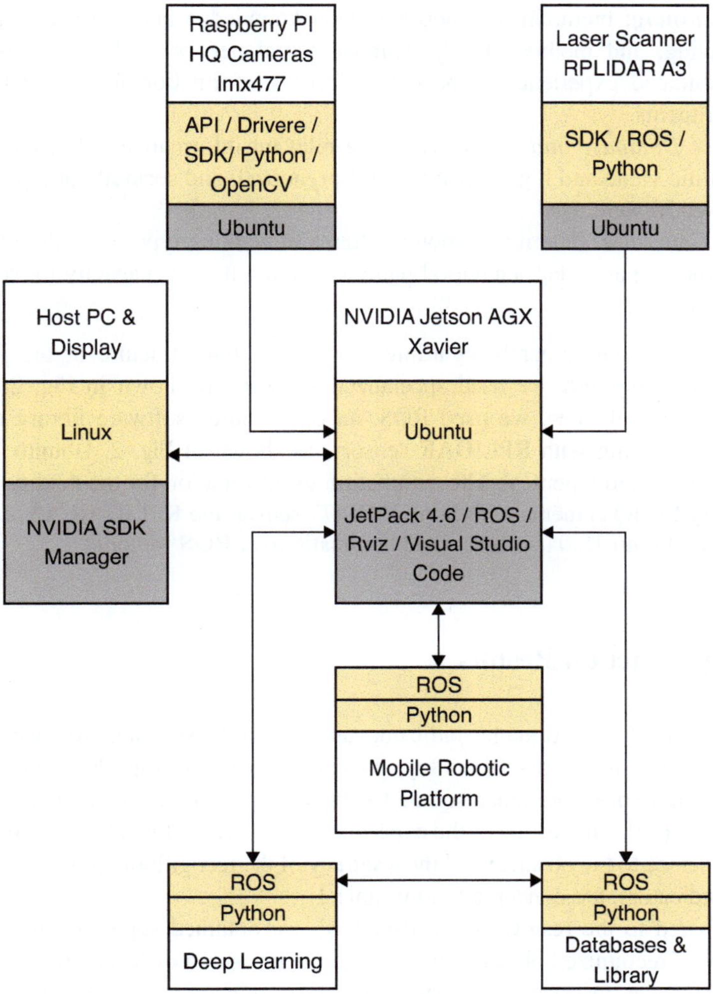

Fig. 3 Software architecture

convolutional networks), the measurement of information losses and the correction of errors regarding the prediction object class, grouping and validation of the attached bounding box, and calculation and indication of mAP (mean average precision). Unlike CNN, R-FCN only targets convolution without fully connected layers (https://raw.githubusercontent.com/NVIDIA/caffe/caffe-0.15/examples/kitti/detectnet_network.prototxt).

ImageNet represents a hierarchically structured database by categories, which is of great help to visual perception and image classification. Each node encapsulates thousands of images described according to the synset (https://www.imagenet.org/).

Deep Learning: hierarchical mode based on knowledge and statistics, predictive modeling, and methods using artificial neural networks; draws lessons from accumulated experience to provide effective information for future tests and experiments.

Reference Database and Cloud: creates a relevant informative collection of great scientific value and high-capacity, well-organized, and secured. GoogLeNet and ResNet-50 were used.

Pattern Generator: determines motor patterns of walking types according to norms and known particular behavioral patterns, having flexible capacity for sequential analysis.

To validate our research regarding the constructive structure of the proposed perception subsystem, we used specialized software as shown in Fig. 2. From a software point of view, we used ROS, an open-source software library that also facilitates working with RPLIDAR sensors, as shown in Fig. 2, Ubuntu 18.04 of ROS Melodic and OpenCV. The summation of information for the duality between Raspberry Pi HQ cameras with Sony IMX477 sensor and RPLIDAR A3 is done by means of Ubuntu 18.04 software and applications of ROS Melodic.

5 Experimental Results

Two categories of performance indicators are applicable for evaluation, which are as follows: metric indicators (accuracy, precision, recall, average localization error, mean squared error, root mean squared error) and qualitative indicators (scalability, robustness) [15]. The results of the experimental tests regarding the measured values of the distance to the obstacle and the accuracy of the recognition process, two of the metric indicators, are concentrated in Table 1.

Compared to the results obtained in [16], we obtained superior values for the accuracy of recognized obstacle classes and the measured distance error.

In Fig. 4, we observe the superiority of HQ-type cameras compared to low-cost cameras in the process of identifying obstacles, compared to the results obtained in the article [16].

Table 1 The result of experimental identification and recognition research with fisheye lens

GoogLeN et database	ResNet50 database	Fisheye lens	Distance (m)	Measured average distance (m)	Distance error (%)	Class accuracy (%)
x		x	1.000	1.031	3.10	12.73
x		x	1.000	0.987	1.30	15.10
	x	x	0.750	0.739	1.47	31.23
x		x	0.750	0.744	0.80	43.90
	x	x	0.500	0.494	1.20	54.44
x		x	0.500	0.496	0.80	68.38

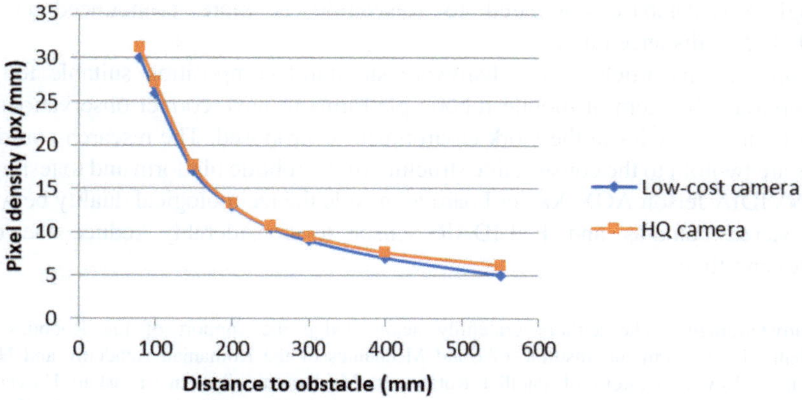

Fig. 4 Comparative graph depending on the cameras type

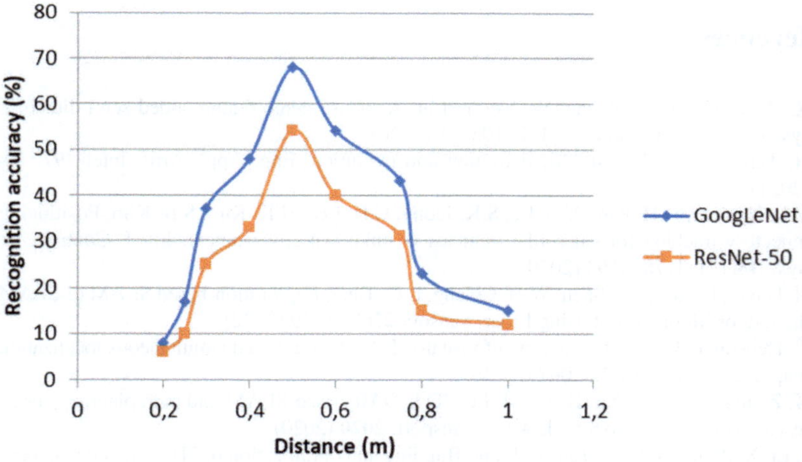

Fig. 5 Comparative graph depending on the database

The differences are more pronounced in the distance range of 0.25–0.55 m, so that the mobile robot has the time and the necessary maneuvering space to adapt its movement strategy to the natural environment.

The distance to the obstacle, measured horizontally in mm, was related to the pixel density, measured in px/mm. Collaborative perception, visual and telemetry tools, Raspberry Pi HQ with Sony IMX477 sensor and RPLIDAR A3 laser sensor are successfully used.

Figure 5 shows the results recorded for the accuracy of obstacle recognition depending on the distance at which they are located from the mobile robot, based on the knowledge accumulated by GoogLeNet and ResNet50. The superiority of the

GoogLeNet database compared to ResNet-50 is more pronounced in the 0.30–0.75 m distance range.

Through this article a new hardware structural composition suitable for the perception subsystem of mobile robotic platforms to have correct observations on the existing obstacles in the work environment is proposed. The research contributions are twofold to the constructive structure of the robotic platform and to testing of the NVIDIA Jetson AGX Xavier board alongside the technological duality between HQ stereo cameras and RPLIDAR sensor to considerably reduce the data processing time.

Acknowledgments The authors gratefully acknowledge the support of the Robotics and Mechatronics Department, Institute of Solid Mechanics of the Romanian Academy and Hebei Provincial Key Laboratory of Parallel Robot and Mechatronic System, Yanshan University, Qinhuangdao, China, and Ningbo University, Faculty of Mechanical Engineering and Mechanics, China.

References

1. X. Zhao, L. Liu, R. Zheng, W. Ye, Y. Liu, A robust stereo feature-aided semi-direct SLAM system. Robot. Auton. Syst. **132**, 103597 (2020)
2. H. Taheri, Z.C. Xia, SLAM; definition and evolution. Eng. Appl. Artif. Intell. **97**, 104032 (2021)
3. V.L. Nguyen, D.H. Kim, V.S. Le, S.K. Jeong, C.H. Lee, H.K. Kim, S.B. Kim, Positioning and trajectory tracking for caterpillar vehicles in unknown environment. Int. J. Control. Autom. Syst. **18**(12), 3178–3193 (2020)
4. H. Ismail, R. Roy, L.J. Sheu, W.H. Chieng, L.C. Tang, Exploration-based SLAM (e-SLAM) for the indoor Mobile robot using Lidar. Sensors **22**(4), 1689 (2022)
5. C. Debeunne, D. Vivet, A review of visual-LiDAR fusion based simultaneous localization and mapping. Sensors **20**(7), 2068 (2020)
6. X. Zhang, J. Lai, D. Xu, H. Li, M. Fu, 2D LiDAR-based SLAM and path planning for indoor rescue using mobile robots. J. Adv. Transport. **2020** (2020)
7. J. Li, X. Zhang, J. Li, Y. Liu, J. Wang, Building and optimization of 3D semantic map based on Lidar and camera fusion. Neurocomputing **409**, 394–407 (2020)
8. P. Kolar, P. Benavidez, M. Jamshidi, Survey of data fusion techniques for laser and vision based sensor integration for autonomous navigation. Sensors **20**(8), 2180 (2020)
9. I. Belkin, A. Abramenko, D. Yudin, Real-time LIDAR-based localization of mobile ground robot. Procedia Comput. Sci. **186**, 440–448 (2021)
10. G. Chen, F. Landmeyer, C. Wiede, R. Kokozinski, Feature extraction and neural network-based multi-peak analysis on time-correlated LiDAR histograms. J. Opt. **24**(3), 034008 (2022)
11. B. Zhou, Y. He, K. Qian, X. Ma, X. Li, S4-SLAM: A real-time 3D LIDAR SLAM system for ground/water surface multi-scene outdoor applications. Auton. Robot. **45**(1), 77–98 (2021)
12. H. Son, B. Lee, S. Sung, Synthetic deep neural network design for Lidar-inertial odometry based on CNN and LSTM. Int. J. Control. Autom. Syst. **19**(8), 2859–2868 (2021)
13. G. Klancar, A. Zdesar, S. Blazic, I. Skrjanc, *Wheeled mobile robotics: From fundamentals towards autonomous systems. Butterworth-Heinemann* (Chapter 5 – Sensors Used in Mobile Systems, 2017). https://doi.org/10.1016/B978-0-12-804204-5.00005-6
14. K.Y. Kok, P. Rajendran, A review on stereo vision algorithm: Challenges and solutions. ECTI-CIT **13**(2), 112–128 (2019)

15. P. Roy, C. Chowdhury, A survey of machine learning techniques for indoor localization and navigation systems. J. Intell. Robot. Syst. **101**(3), 1–34 (2021)
16. C. Dragne, I. Todirite, M. Iliescu, M. Pandelea, Distance assessment by object detection – for visually impaired assistive mechatronic system. Appl. Sci. **2022**(12), 6342 (2022). https://doi.org/10.3390/app12136342

Using Eye Movement Features for Secure Authentication

Esraa Almohaimeed, Daad Albriki, Fatima Abdulkreem, and Abeer Alghulayqah

1 Introduction

1.1 Introduction

One of the security requirements for public authentication systems is to be easy, fast, and secure. People face authentication mechanisms daily and must authenticate themselves using traditional knowledge-based methods. For example, biometrics is a technology that uses human biometrics to recognize an identity, such as a password. Still, these methods are not secure because their malicious monitors use surveillance techniques such as a user monitoring shoulder browsing while typing the password through the keyboard to capture user authentication data. There are also security issues due to poor interactions between systems and users; as a result, the researchers suggested eye-tracking techniques, where users can enter the password by looking at the appropriate codes in the proper order, and thus the user is not prone to shoulder browsing. Commonly used biometric features are the face, iris, etc. Unlike static biometrics, eye tracking is a natural interaction method, and safety systems based on eye motion tracking provide a promising solution for system security. Furthermore, ease of use and behavior are believed to be faster in response to environmental changes. This makes it difficult to falsify eye movement data. Therefore, eye movement recognition is a very safe identification technique and the goal of this. The paper reviews techniques or solutions for eye motion tracking in security systems.

E. Almohaimeed (✉) · D. Albriki · F. Abdulkreem · A. Alghulayqah
Department of Information Technology, College of Computer, Qassim University, Buraydah, Saudi Arabia
e-mail: 382215880@qu.edu.sa; 391214603@qu.edu.sa; amgliekh@qu.edu.sa

© The Author(s), under exclusive license to Springer Nature Switzerland AG 2024
A. Shaikh et al. (eds.), *Advances in Emerging Information and Communication Technology*, Signals and Communication Technology,
https://doi.org/10.1007/978-3-031-53237-5_22

Biometric authentication— identification through a person's distinctive physical or behavioral characteristics—offers a more convenient alternative that requires no memorization. However, biometric systems have failed to gain widespread adoption. However, face authentication is vulnerable to spoofing attacks, with attacks demonstrated as recently as Android Jelly Bean. In the social media era, users' faces are often available online. Therefore, faces should be considered public knowledge, which means face authentication lacks a secrecy component. Secrecy is the strength behind passwords and PINs; to be secure, it may be something that faces authentication must incorporate. So we need to design a secure authentication system with face and eyes for authentication to access sensitive data. As a result, we'll make things simple for:

- Real-time eye movement recognition model using biometric authentication.
- Previous studies comparing them and finding problems.
- Help organizations verify employee or customer data with greater accuracy and flexibility.
- Eye movement security authentication in the shortest possible time.
- An effective system for detecting and recognizing eye movement.
- Increase the security and reliability of the system.

It is the objective of this work to demonstrate how to use an application to verify security authentication by tracking eye movements. All individuals are required to provide identification at airports, the department of the interior, and other government institutions as part of the work. Private organizations, such as companies, monitor employee attendance to ensure discipline. It is necessary to enter an individual's data in each system, link it to their eye movements, and photograph them each time. The computer will display all the information regarding the individual after the eye movements have been verified.

2 Literature Review

2.1 Introduction

A fundamental requirement of a terminal authentication system is to be simple, quick, and secure since users encounter authentication mechanisms daily and must authenticate themselves using traditional knowledge-based techniques such as passwords. As a result, these methods are not protected since they are observed by nefarious individuals who use surveillance methods such as shoulder surfing (watching a person enter a password while they type it onto a keyboard) to record the authentication details of the users. Additionally, there are security issues brought on by subpar user and system interactions. Therefore, the researchers devised eye-tracking devices in which users may input their passwords by looking at the correct symbols in the proper order, allowing them to remain unaffected by shoulder surfing. Furthermore, communication is naturally accomplished through eye tracking. It has been established that eye tracking is an effective method for interacting with systems and that incorporated eye-tracking systems provide a promising means for improving the security and usability of the system. A description of our topic's background, concepts, and definitions will be presented in this work, as well as the type of techniques used, architectures, and history of previous studies relevant to the topic we are discussing. Moreover, the paper provides a comprehensive explanation of prior studies pertinent to our topic before presenting the issues related to the proposed work. Last but not least, we have introduced our proposed hypotheses.

2.2 Background

2.2.1 Concepts and Definitions

- Face detection: This is a computer technique that recognizes human faces in digital images and is used in various applications.
- Pupil detection: This is a fundamental layer in the eye-tracking stack since it generates the majority of the signal used by the other layers.
- Eye tracking: This is the practice of tracking a person's eye movement about their head or point of gaze (where they are looking).
- Electrooculography (EOG) tracking: Measures the potential electrical changes between two electrodes to record eye movements.
- Photo-oculography (POG): This is an eye-tracking method based on the idea of employing straightforward photosensors to gauge how much reflected light—typically infrared light—the eye reflects when it rotates.
- Video-oculography (VOG): This is a noninvasive, video-based technology that uses a mask mounted on the head that has tiny cameras to record the motions of both eyes (eye tracking) in the horizontal, vertical, and torsional planes.
- AOI-based approach: This is a data mining algorithm that is used to extract knowledge from relational data while taking expert knowledge into consideration.
- TRM: It aids talent acquisition teams in developing, tracking, and engaging people to fill unfilled positions as rapidly as possible. By tracking candidate engagement at each touchpoint, all stakeholders comprehensively understand the candidate pipeline, history, and actions.
- Face recognition: It operates by recognizing and quantifying facial features in an image. Facial recognition technology can recognize human faces in images or videos, assess whether a face appears in two different photos of the same person, or look for a face in an extensive database of images that have already been taken.
- Convolutional neural networks (CNNs): These are a specific type of artificial neural network that uses a mathematical operation in at least one of their layers, called convolution, in place of general matrix multiplication in at least one of their layers.

- Gaze tracker: Is a software and hardware system that reflects infrared lights to reflect on users' eyes and tracks pupil movement to determine where the users' eyes are looking.

2.2.2 Architectures

Figure 1, SAFE's architecture is illustrated as highly modular, consisting of the following elements: user interface, gaze tracking technology, facial recognition technology, authentication model, and user icon and profile databases. In Fig. 1, the user interfaces the challenge generated by the SAFE authentication module where the user's eyes were when they were calibrated, displayed as white circles the video stream from the front-facing camera. Through the movement of the pupil, the gaze tracker can calculate the position of the eyes' gaze by reflecting infrared light on the user's eyes. By evaluating the facial features of the current user, the face recognition system determines if they are similar to the facial features of the enrolled

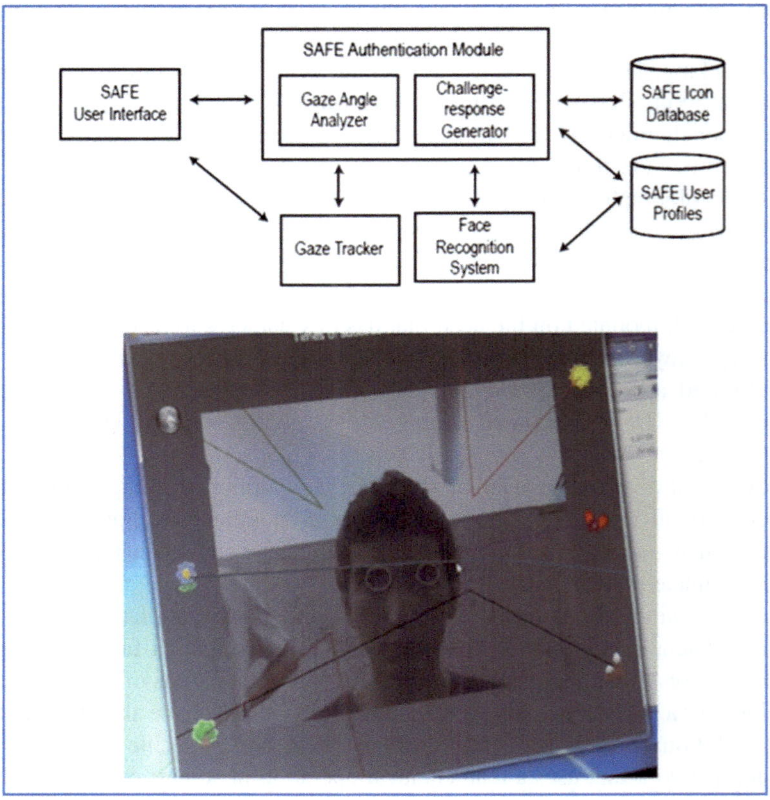

Fig. 1 SAFE architecture [1]

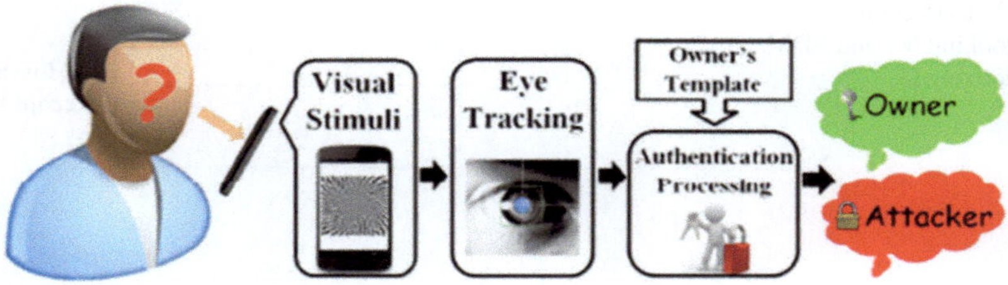

Fig. 2 Implement the eye-tracking module on an Android smartphone [2]

user(s). The face recognition system must identify an enrolled user before the device can be unlocked. The face recognition system communicates its results to the authentication module in the background while unlocking the device. The SAFE authentication module is shown in Fig. 1 [1].

Their research [2] consists of the development of an authentication system in Fig. 1 and the implementation of an eye-tracking module on an Android smartphone in Fig. 2. Volitional and non-volitional eye movements are generally classified as volitional. Movements that are voluntary are those that are undertaken by people voluntarily, while movements that are non-volitional are those that are carried out passively. In order to stimulate various physiological and behavioral aspects of the eyes, the researchers designed and studied four different types of visual stimuli.

As described in the paper [3], the authors propose a method for extracting and classifying features through a network. We have designed two branch networks for the purpose of extracting further features based on MI and SDM, respectively. Figure 3 illustrates all pipelines. Convolutional neural networks (CNNs) are an appropriate network for feature extraction since saccades and fixations do not correlate with eye movement records. There are four 2D-CNNs in the branch network which is used for the processing of SDM. Each 2D-CNN consists of 64 filters, with a kernel size of (5) and a dilation of (1). In order to facilitate the expansion of the perceptual view of each filter, dilated convolutions are used, and MI is processed by a branch network containing 4 1D-CNNs, each with 64 filters and a kernel size and dilation of 7. After batch normalization, the activation function of the ReLU was applied to each convolutional layer. Two fully connected layers are formed by splicing the outputs of both branch networks. Features are merged and classified using the fully connected layers. This is accomplished by flattening the result of the fully connected layer into a one-dimensional array, performing L2 normalization on it, and obtaining an embedding feature having 128 dimensions.

2.2.3 History

In the 1950s, eye-tracking research was conducted by Alfred L. Yarbus, and his 1967 book is often quoted. He showed that the task given to a subject has a very large influence on the subject's eye movement [4]. In addition to exploring the

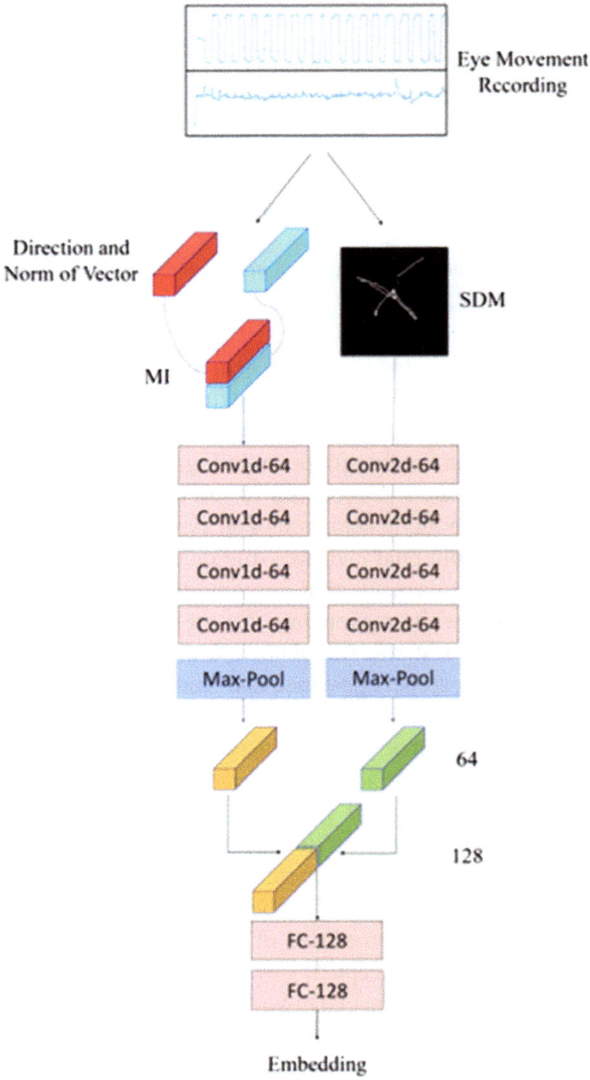

Fig. 3 Extraction overlooking MI and SDM, respectively [3]

connection between fixations and interest, he pointed out that the cyclical pattern in picture analysis was influenced by the size and scope of the issue that the observer was trying to solve, as well as the knowledge he intended to obtain from the image [4]. Research into eye tracking increased rapidly during the 1970s, especially in reading. Rayner's review of research was conducted during this period, and an overview of the research in this field is presented [5]. The strong eye-mind hypothesis was introduced by Just and Carpenter in 1980, in which they proposed that there is no apparent lag between what is fixed and what is processed [6]. Accordingly, when an individual looks at an object or word, they will also think about it for the same time as recording their fixation. The hypothesis is often taken for granted by researchers using eye-tracking. To disentangle overt and covert attention, gaze-contingent techniques provide an intriguing option for distinguishing between the fixation of attention and the processing of information. Human-computer interaction was also revolutionized in the 1980s by the introduction of eye tracking. An investigation was conducted into how users search for commands within computer menus. In addition, computers enabled researchers to analyze eye-tracking results in real time, primarily assisting disabled individuals. The eye-mind concept has generated a lot of debate ever since it was proposed in the 1980s in light of covert attention, which is being aware of what one is not consciously aware of [7–9]. If covert attention is frequently present while being recorded, eye-tracking recordings may be unable to reveal cognitive processing. If this were the case, scan-path and fixation patterns would frequently not depict the object of attention but rather the location of the eye's gaze. Also, during the 1980s, eye tracking was developed to answer questions about human-computer interaction. A particular study examined how users search for commands within computer menus. Furthermore, researchers could interpret eye-tracking results in real time, primarily to assist disabled users [9]. It has become possible to carry out eye-tracking tasks and analyses using artificial intelligence (AI) and artificial neural networks in the twenty-first century [9]. Since convolutional neural networks are designed for tasks requiring image recognition, they are particularly suitable for eye tracking. AI can be used to provide additional information about eye-tracking tasks and studies that were not previously visible to humans. When provided with enough sample data, a deep-learning neural network can also become more proficient at a given task. Obtaining sufficient training data requires a considerable amount of time, however. In eye tracking, AI can be used in various applications, including medical, driver safety, game theory, and even education and training [10].

2.3 *Related Work*

During this section, we will discuss a number of issues related to eye movement validation studies. This system focuses on mobile applications and web applications, so we seek to collect information about these applications. We would appreciate your assistance in resolving and reducing issues with secure access.

2.3.1 Issues Related to the Current Work

There has been a great deal of previous research on eye movement biometrics which has utilized a small number of participants, sometimes utilizing one or two datasets, e.g., [11–13]. Additionally, in most prior studies, eye movements were assessed and trained on datasets accumulated over a short period of time, leaving the question of their permanence unanswered. In most eye movement research studies, the long duration of eye movement recording did not take into account the fact that people are less interested when recorded for such a long period of time. When visual material is located in the fovea, it is a good indicator of whether people are interested in it. It is the retina that captures images of objects, which are then converted into signals by photoreceptor cells before being sent to the brain through the optic nerve.

2.3.2 Previous Studies

In using gaze gestures as an authentication method to counter shoulder surfing attacks, this paper [14] describes a dynamic gaze gesture-based authentication system. In addition, two authentication interfaces were explored: a dynamic interface and a static-dynamic interface which makes use of gaze gestures. As a result of system evaluations, we have determined that there is a more practical solution to the problem by using a static-dynamic interface with an animation speed of two seconds. A knowledge-based authentication method, characterized by the use of passwords, is used in this study to prevent shoulder surfing attacks. This study introduces iType, a novel eye-gazing method that uses widely available mobile platforms to let users enter private data with their eyes [15]. This design faces the following main difficulties: mobile gaze tracking is generally unreliable; input errors are difficult to remedy because there is no way to compare them to the real text value and device noises and motions might affect iType performance. The iType keystroke detection technology, according to numerous evaluations, achieves high keystroke detection accuracy in a respectable amount of time for secure typing. The objective of this paper [3]. This is to achieve authentication utilizing shorter-duration eye movement data. In addition, we advise keeping the eye movement recorded to less than 12 seconds. For the objectives of this study, we employed the spatial distribution elements of the saccade and the temporal motion features of the gaze point to reflect individual identification. For the tests, datasets, including recordings of eye movements for 5 and 12 seconds, are created [16]. Our suggested methods yield equal error rates of 10.62 percent at recording durations of 12 seconds and 12.48 percent at recording durations of 5 seconds in our open-set authentication results on the datasets described in this work. Based on the findings of our closed-set authentication, we have demonstrated that our suggested approaches can achieve an equal error rate of 5.25 percent when recording for 12 seconds and 7.82 percent for 5 seconds. In this way, it proves that it is possible to authenticate identities using eye movement data. Frequency domain approaches are employed to evaluate both horizontal and vertical eye movement data. The statistical method recognizes the fixation and saccade phases of eye movements. The spatial-based method calculates spatial distribution information of eye movements by mapping them into a two-dimensional environment using density maps and trajectory maps. The required length of eye movement records makes it clear in this study that time-based characteristics and artificial design elements used in eye movement-based authentication methods are inapplicable. Given the difference in the degree of divergence between the gaze point and the stimulus point on the screen, a suitable recording period for eye movements should be less than 12 seconds. Two datasets are created, each containing recordings of eye movements for 5 and 12 seconds, in order to execute the tests. The equal error rate of our suggested approaches reaches 10.62 percent when the recording period is 12 seconds and 12.48 percent when it is 5 seconds, according to open-set authentication results from the datasets created in this paper. Our proposed approaches may obtain an equal error rate of 5.25 percent while recording for 12 seconds and 7.82 percent when recording for 5 seconds, according to the results of closed-set authentication. The strategy suggested. Eye movement is a behavioral biometric, as opposed to static biometrics, and eye trackers can capture it. It is demonstrated that the proposed approach offers a standard for evaluating eye movement-based identity identification. According to Komogortsev and A. Abdelwahab, eye movements were recorded for 40 seconds, 60 seconds, and 95 seconds, respectively. The longer the duration of the recording of eye movement, the more information can be gathered, and the more accurate the identification [17–19]. The difficulties of employing eye movements to identify persons in a real context, such as implicit recognition and stimulus independence, are discussed in this work. Thirty-nine volunteers' eye movements were gathered, and 5 sets of eye movements were derived from studies on real-world way-finding. Eye movement biometrics have the benefit of being impossible to spoof [20]. The capacity to continually identify biometric data from eye movements is an added advantage. It should be emphasized that as eye-tracking technology advances, mobile and wearable eye-tracking devices become smaller, more affordable, and more precise; as a result, they may soon replace other types of sensors as commonplace technology. Despite the extensive use of mobile and wearable eye tracking in real-life scenarios for studying human gaze behavior, laboratory-based eye movement biometric research has not yet been carried out in actual settings. MWEMB is a difficult procedure that poses a variety of difficulties in real-world applications in actual situations, in addition to technical difficulties with wearable devices [21, 22]. We subjects were asked to perform navigation in real-world surroundings while using real-time eye movement data. Several studies have been conducted using eye-tracking technology to investigate user behavior when working with a computer. For an overview, see [23, 24]. Through eye-tracking research, it has been widely used to enhance the usability and design of online pages as well as to understand how users are directed by them. An eye-tracking technology is an eye-tracking technique that highlights eye movement. The tools that are utilized to implement eye tracking are called eye trackers. There are several ways to gather information on eye movement, including measuring the eye's position in relation to the head and the "Point of regard" measurement, which is frequently used in interactive applications. The most widely used eye trackers currently in use, according to technology, determine the point of view based on light reflected from the cornea. Video-based systems are one of the four major categories of eye movement measuring technologies, along with eye-attached systems, optical systems, and electrooculography (EOG) tracking.

2.4 Proposed Work

Research from previous studies has been used to support this statement. Safer access to information is provided by the proposed system. The purpose of this process is to verify the identity of individuals. It is simpler to implement than other security measures because reading high-resolution video cameras does not require any eye movement. Through the proposed system, programmers will be able to create websites and applications that can record eye movements using a mobile phone's camera or special cameras set up in government buildings. Eye movement has many benefits, including the capacity to more accurately and flexibly identify people. As seen in Fig. 4, employees and visitors at the Boston airport, for instance, now record their data and track their eye movements, exactly like it is done at the British Post, the Ministry of Defense and Antiquities in America, the Frankfurt Airport, and the airports in Amsterdam and Frankfurt. In addition, it is widely known that the UAE was among the first Arab nations to use this technology for security. In the Kingdom of Saudi Arabia, the same holds true. The National Information Center of the Interior Ministry has started a phase of the identification system. National identity a database containing every eye movement made by inhabitants, visitors, and anybody else entering the Kingdom of Saudi Arabia is one of many objectives being pursued. A number of border crossings started to record eye movements. This message will be distributed to all border crossings in an effort to stop anyone from entering the Kingdom without first taking a motion. All people will have their fingerprints taken when their smart national identity card is renewed or issued.

Fig. 4 Biometric device at the airport [25]

2.4.1 Issues Related to Proposed Work

Eye movement tracking technologies have taken the place of fingerprint identification in some organizations, providing secure data access for institutions and government agencies. The experts in this system have all decided to build it to be more dependable and safer while also making it simpler for every user to utilize. Any organization that wants to increase security can gain from the installation of this eye-tracking device, which offers protection from all sides.

3 Methodology

3.1 System Architecture

There are several components of the system architecture, which is extremely modular, such as the user interface, gaze tracking, face recognition, authentication mechanism, and icon and profile databases. During calibration, white circles indicate the user's eye positions, whereas black circles indicate the current video feed from the front-facing camera. A gaze tracker is a piece of software that determines where the eyes are focused based on pupil movement (Fig. 5).

Through the use of a face recognition system (s), the present user will be compared to the registered user based on their facial characteristics. In order for the facial recognition system to initiate the process of unlocking the device, the current user must be recognized as a registered user. During the process of unlocking, the facial recognition system is continuously running in the background and transmits its findings to the authentication module. This authentication module

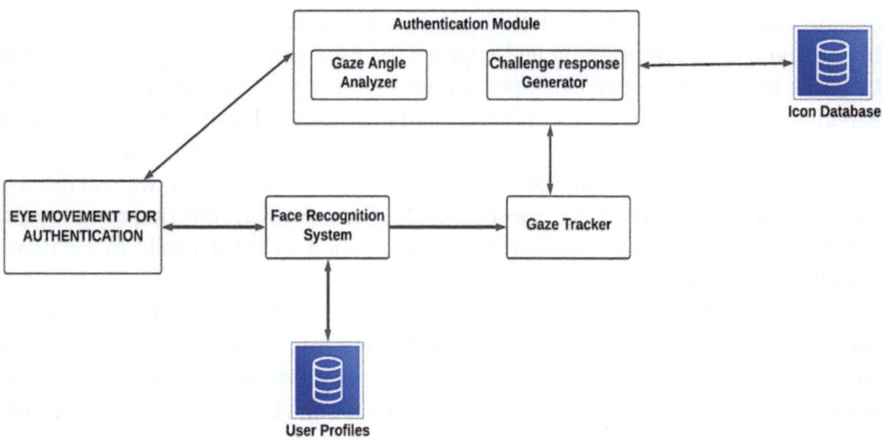

Fig. 5 System architecture

compares the gaze data of a user with a symbol pathway on the screen, as well as generates challenges and evaluates the responses to those challenges. In this case, the challenge response generator gathers a user's secret symbol from the user's profile database. Then it generates a challenge based on the physical screen size, screen resolution, gaze tracking error, and desired level of security. Each challenge consists of a collection of icons, their placement on the screen, and their future trajectory. By using the observed gaze data, the gaze angle analyzer calculates the angle of a user's gaze route.

4 Implementation and Testing

4.1 Implementation Procedure

An application that runs on computers is created. Eye movement tracking and data are two types of user input that are supposed to be recorded in the application and stored in the application's database. The newly registered user's personal data is first requested to log into the application. This is followed by tracking eye movement and storing it in a database. But if the user is already registered, eye movement tracking is the first requirement to enter the application. The system will require scripting on the server side, and then the user's data will be pulled from the database. The system will then compare the eye movement data with the database data. If the data matches, the user can access the application. Otherwise, access will be denied. Designing a user interface is very important to help the user control how he or she uses the application. It should be intuitive and easy to use, allowing the user to quickly understand how to interact with the application. A well-designed design makes the user experience smoother and more enjoyable. User interface design should be tailored to the user's needs. The application must have camera functionality to enable the computer to capture and record eye movement. This data can then be used to analyze the user's interaction with the application and improve the user experience. The data can also be used to identify areas of improvement and optimize the application's user experience. Additionally, camera functionality can be used to further improve user interface design by capturing facial expressions and other behavior. This will provide more insights into the user experience. So in this subsection, we will discuss the materials we use and the purpose of each. Then, we will move on to the steps, breaking them down into smaller parts and explaining what needs to be done. Finally, we will provide an overview of the entire process.

PyCharm and Visual Studio Code programs were used to write the eye motion code. These programs were chosen because they are easy-to-download code libraries and have an easy-to-use interface and error detection. Moreover, open-source software can be modified to suit specific needs. In addition, it is well supported by a wide developer community. It is also very reliable and secure, which makes it ideal for software development systems.

Brief explanations of the libraries used in building the system in this research are listed below:

- Cv2 library: The module import name for OpenCV, a huge open-source library for computer vision, machine learning, and image processing and supports a wide variety of programming languages. It can process images and videos to identify objects or faces.
- NumPy library: Open-source library used to perform mathematical operations on arrays which provide tools for handling the n-dimensional arrays.
- Time library: Python time module allows working with time in Python. It allows functionality like getting the current time.
- MySQL connector: To correlate eye movement with his data, this database was used. This database allowed for quick analysis and visualization of the data. It enabled the researcher to make connections between eye movements and other data points.
- Sys library: The sys model provides information about constants functions and variables of the Python interpreter.
- Math library: A set of built-in math functions, including an extensive math module, which allows you to perform mathematical tasks on numbers.
- PyQt5 library: This library is used to design the application's graphical interfaces. It provides tools to create buttons, menus, text fields, and other components. It also allows developers to customize the look and feel of the UI. The library is easy to use and provides a robust set of features.
- MediaPipe library: This library was used to define the face's details and features the library was used to analyze the facial geometry, such as angles between the eyes and the nose, as well as the shape of the chin and jawline. It was also used to measure the distance between facial features.

4.2 *FaceMech Model*

The FaceMech model is one of the Face Landmark models, which produces an estimated 478 3D facial features on the face. This model was chosen because it is one of the simplest methods. The identification of facial features is an important part of determining eye details. and tracking eye movement is the identification of facial features. The model is accurate and efficient, and it can also be used to detect facial expressions. This makes it ideal for biometric authentication applications (Figs. 6 and 7).

```
if self.vCap.isOpened():
        user_face_mesh = mp.solutions.face_mesh
        with user_face_mesh.FaceMesh(
                max_num_faces=1,
                refine_landmarks=True,
                min_detection_confidence=0.5,
                min_tracking_confidence=0.5
        ) as face_mesh:
            while self.startThread == True:
                ret, frame = self.vCap.read()
                if not ret:
                    break
                img_h, img_w = frame.shape[:2]
                frame = cv.flip(frame, 1)
                rgb_frame = cv.cvtColor(frame, cv.COLOR_BGR2RGB)
                self.face_inside(frame, img_w, img_h)
                results = face_mesh.process(rgb_frame)
                if results.multi_face_landmarks:
                    mesh_points = np.array([np.multiply([p.x, p.y], [img_w, img_h]).astype(int) for p in
                                            results.multi_face_landmarks[0].landmark])
```

Fig. 6 FaceMech code

Fig. 7 FaceMech model

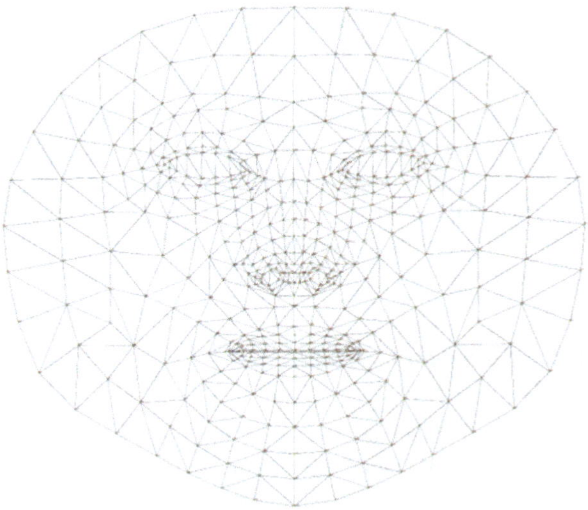

4.3 Function Explanation

K-means clustering algorithm is used to map each data point to a cluster by calculating the distance between it and the centroid. Identifying the number of K groups in the dataset is the objective. As data points are grouped based on similar characteristics, an iterative process is used to assign each data point to a cluster. To reduce the distance between the data points and the block midpoint, the sum of the distances between the data points must be reduced. It is for this purpose that each

```
# clustering eye color by kmeans clusters
def clustering_eye(self, frame, center_right, r_radius):
    iris_frame, width, height = self.extract_iris(frame, center_right, r_radius)
    eye_height, eye_width, _ = np.shape(iris_frame)

    data = np.reshape(iris_frame, (height * width, 3))
    data = np.float32(data)

    number_clusters = 1
    criteria = (cv.TERM_CRITERIA_EPS + cv.TERM_CRITERIA_MAX_ITER, 10, 1.0)
    flags = cv.KMEANS_RANDOM_CENTERS
    compactness, labels, centers = cv.kmeans(data, number_clusters, None, criteria, 10, flags)

    return (centers[0][2], centers[0][1], centers[0][0])
```

Fig. 8 K-mean clustering algorithm

Fig. 9 Calculate distance between two point

```
def distance(self, point1, point2):
    x1, y1 = point1.ravel()
    x2, y2 = point2.ravel()
    dist = math.sqrt((x2 - x1) ** 2 + (y2 - y1) ** 2)
    return dist
```

```
def extract_iris(self, frame, center, radius):
    iris_frame = frame[center.ravel()[1] - int(radius):center.ravel()[1] + int(radius),
                 center.ravel()[0] - int(radius):center.ravel()[0] + int(radius),
                 ]
    width = 2 * int(radius)
    height = 2 * int(radius)
    return iris_frame, width, height
```

Fig. 10 Extract iris

data point is assigned to the appropriate group. The result, the data area has been divided into K groups, and an average value has been assigned to each group, in order to place data points in groups based on their proximity to the average value are grouped together. The distance can be calculated using several different distance scales. It is based on them that the color of the eyes is determined in this function (Fig. 8).

This function is used to find the distance between the two points of the eyes (Fig. 9).

Its functions are to extract the size of the eye's iris and detect the position of iris (Figs. 10, 11 and 12).

The purpose of this function is to check that the eye movement is not repeated in the database. If the motion exists, a message will appear to the user "The movement is exists, Press the Start button" (Fig. 13).

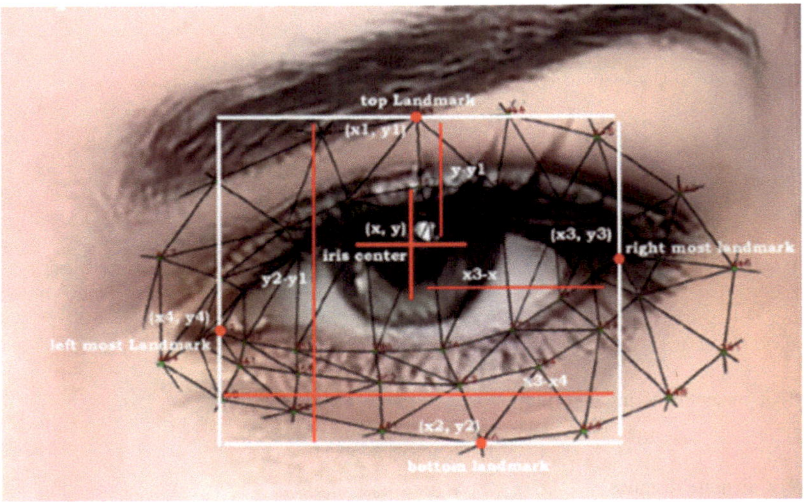

Fig. 11 Extract the size of the eye's iris

```
def iris_position(self, iris_center, right_point, left_point, top_point, botton_point):
    # center_to_right_dist = distance(iris_center, right_point)
    # x3 - x
    center_to_right_dist = math.fabs(iris_center.ravel()[0] - right_point.ravel()[0])
    # totat_distance_horizontal = distance(right_point, left_point)
    # x3 - x4
    totat_distance_horizontal = math.fabs(right_point.ravel()[0] - left_point.ravel()[0])
    ratio_horizontal = center_to_right_dist / totat_distance_horizontal

    # center_to_top_dist = distance(iris_center, top_point)
    # y - y1
    center_to_top_dist = math.fabs(iris_center.ravel()[1] - top_point.ravel()[1])
    # totat_distance_vertical = distance(top_point, botton_point)
    # y2 - y1
    totat_distance_vertical = math.fabs(botton_point.ravel()[1] - top_point.ravel()[1])
    ratio_vertical = center_to_top_dist / totat_distance_vertical

    ratio = ratio_horizontal
```

Fig. 12 Detected the iris position

4.4 Application Design

In this section, we will display a graphic showing the entire movement of the application. This will be followed by a detailed explanation of all application

```python
def isMovementsExist(self):
    dataConn = conn.connect(
        host="localhost",
        user="root",
        password="22630",
        database="eyemoving"
    )
    dataCursor = dataConn.cursor()

    sqlStr = "SELECT eyemoving FROM users WHERE eyemoving = (%s)"
    data = [self.eyeStringMovements]
    dataCursor.execute(sqlStr, data)
    data = dataCursor.fetchone()
    if data is not None and data[0].__eq__(self.eyeStringMovements) == True:
        self.eyeStringMovements = ""
        self.error_msg.setText("The movements is exists Press the Start button")
    dataConn.close()
```

Fig. 13 Check eye movement

Fig. 14 Graphic showing the entire system

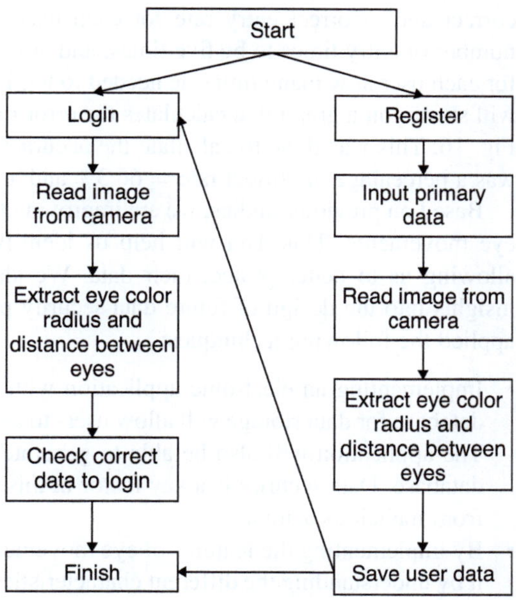

interfaces. We will also provide screenshots of the application as it progresses through the various steps (Fig. 14).

5 Results and Discussions

5.1 Application Result

After we linked the program code of the application, the accurate results of the application became displayed in the form of a table that contains the data of all users, and of course, this data can't be viewed by anyone to preserve the confidentiality of the information. We also added a security feature to the program that prevents unauthorized access to the database. We tested the application and confirmed that the results were accurate and the information was kept confidential.

The most common attacks in our lives are the adversary's shoulder browsing to obtain the user's password or the user's over-the-shoulder observation during the password entry process. So, securing the data with eye tracking as described in the implementation and testing steps succeeded in preventing the attack because its methodology is complex. It is expected that this work will be widely used in the future because it is a simple way to protect data. We plan to keep improving the technology to make it safer. Also, you want to make access to this technology safer to use. Also, we will continue to monitor any potential security risks and update our methods accordingly. Rest assured that our system will become the gold standard for security and privacy protection. We wrote programmatic commands to calculate the correct and incorrect entry rate for each user, as we specified for each user the number of entry times to be five times, and at the time of the dataset, we calculated for each user how many times he needed to log in in order to show him his data. We will show you a graph that calculates the error rate in Fig. 15 and the correct rate in Fig. 16. This was done to calculate the accuracy rate of the application. The result was a percentage of correct rate = 66.3% and a percentage of error rate = 33.7%.

Based on previous studies, we are improving the security of user data by tracking eye movements. This data will help us identify patterns and behaviors of users, allowing us to better protect their data. We also expect this research to provide insights into the design of future data security protocols. To achieve this, we have applied the following techniques:

- Implementing an electronic application with a user interface and linking it to a database for data storage will allow users to access and manipulate data with ease. The application will also be able to generate reports and analyze data from the database. Data security is a key factor in this process, as it needs to be protected from malicious actors.
- By implementing the features of eye movement by studying the characteristics of it by understanding the different characteristics of eye movement, we can develop algorithms to detect and track it.
- Test, debug, and improve application security regularly review security protocols and policies. Ensure data is being stored and accessed securely.

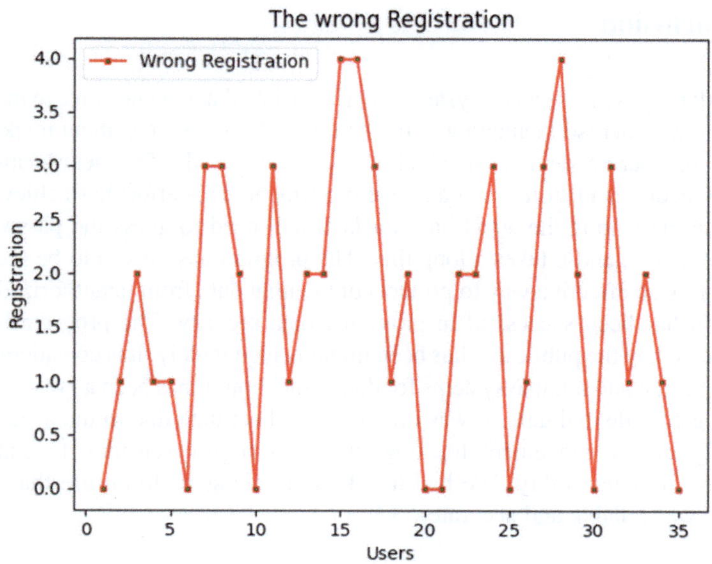

Fig. 15 Graph that calculates the error rate

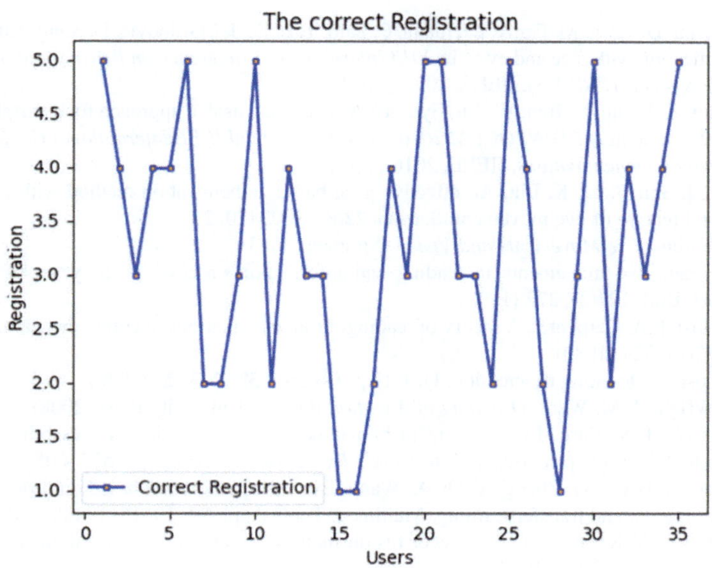

Fig. 16 Graph that calculates the correct rate

6 Conclusion

This work proposed a security system for private data that depends on eye movement tracking. The Microsoft camera was the first model for eye movement tracking, and it is the pioneering device on which the program works for identification. The database records and stores data and eye movement. This effort has achieved great success in preventing the attack, as attackers will need to guess the password and hack the system, and it takes a long time The program has proven to be secure and reliable. It is an effective way to protect confidential data from unauthorized access. Overall, it has been successful in providing data security. The program has been well-received by the public and has been implemented widely. It is considered one of the most secure and reliable systems for data protection. It has been a great success in protecting confidential data. It was difficult to collect data due to the accuracy and sensitivity of eye movement tracking, which is highly sensitive to changes in lighting or user instability. We had to take multiple steps to ensure that the data collected was reliable and accurate.

References

1. A. Boehm, D. Chen, M. Frank, L. Huang, C. Kuo, T. Lolic, I. Martinovic, D. Song, Safe: Secure authentication with face and eyes, in *2013 International Conference on Privacy and Security in Mobile Systems (PRISMS)*, (IEEE, 2013), pp. 1–8
2. C. Song, A. Wang, K. Ren, W. Xu, Eyeveri: A secure and usable approach for smartphone user authentication, in *IEEE INFOCOM 2016 – The 35th Annual IEEE International Conference on Computer Communications*, (IEEE, 2016), pp. 1–9
3. J. Yin, J. Sun, J. Li, K. Liu, An effective gaze-based authentication method with the spatio-temporal feature of eye movement. Sensors **22**(8), 3002 (2022)
4. A.L. Yarbus, *Eye Movements and Vision* (Springer, 2013)
5. K. Rayner, Eye movements in reading and information processing: 20 years of research. Psychol. Bull. **124**(3), 372 (1998)
6. M.A. Just, P.A. Carpenter, A theory of reading: From eye fixations to comprehension. Psychol. Rev. **87**(4), 329 (1980)
7. M.I. Posner, Orienting of attention. Q. J. Exp. Psychol. **32**(1), 3–25 (1980)
8. R.D. Wright, L.M. Ward, *Orienting of Attention* (Oxford University Press, 2008)
9. R.J. Jacob, K.S. Karn, Eye tracking in human-computer interaction and usability research: Ready to deliver the promises, in *The Mind's Eye*, (Elsevier, 2003), pp. 573–605
10. L. Zhao, Z. Wang, G. Zhang, Y. Qi, X. Wang, Eye state recognition based on deep integrated neural network and transfer learning. Multimed. Tools Appl. **77**, 19415–19438 (2018)
11. A. George, A. Routray, A score level fusion method for eye movement biometrics. Pattern Recogn. Lett. **82**, 207–215 (2016)
12. L.A. Jäger, S. Makowski, P. Prasse, S. Liehr, M. Seidler, T. Scheffer, Deep eyedentification: Biometric identification using micro-movements of the eye, in *Machine Learning and Knowledge Discovery in Databases: European Conference*, (ECML PKDD 2019, Würzburg., September 16–20, 2019, Proceedings, Part II. Springer, 2020), pp. 299–314
13. V. Krishna, Y. Ding, A. Xu, T. Höllerer, *Multimodal biometric authentication for vr/ar using eeg and eye tracking* (Adjunct of the 2019 International Conference on Multimodal Interaction, 2019), pp. 1–5

14. V. Rajanna, A.H. Malla, R.A. Bhagat, T. Hammond, Dygazepass: A gaze gesturebased dynamic authentication system to counter shoulder surfing and video analysis attacks, in *In 2018 IEEE 4th International Conference on Identity, Security, and Behavior Analysis (ISBA)*, (IEEE, 2018), pp. 1–8
15. Z. Li, M. Li, P. Mohapatra, J. Han, S. Chen, itype: Using eye gaze to enhance typing privacy, in *IEEE INFOCOM 2017-IEEE Conference on Computer Communications*, (IEEE, 2017), pp. 1–9
16. A. Maffei, A. Angrilli, Spontaneous blink rate as an index of attention and emotion during film clips viewing. Physiol. Behav. **204**, 256–263 (2019)
17. S. Jia, A. Seccia, P. Antonenko, R. Lamb, A. Keil, M. Schneps, M. Pomplun, et al., Biometric recognition through eye movements using a recurrent neural network, in *In 2018 IEEE International Conference on Big Knowledge (ICBK)*, (IEEE, 2018), pp. 57–64
18. O.V. Komogortsev, J.I. Khan, *Eye movement prediction by kalman filter with integrated linear horizontal oculomotor plant mechanical model* (Proceedings of the 2008 Symposium on Eye Tracking Research & Applications, 2008), pp. 229–236
19. A. Abdelwahab, N. Landwehr, Deep distributional sequence embeddings based on a wasserstein loss. Neural. Process. Lett. **54**(5), 3749–3769 (2022)
20. U. Saeed, Eye movements during scene understanding for biometric identification. Pattern Recogn. Lett. **82**, 190–195 (2016)
21. E. Abdulin, I. Rigas, O. Komogortsev, *Eye Movement Biometrics on Wearable Devices: What Are the Limits?* (Proceedings of the 2016 CHI Conference Extended Abstracts on Human Factors in Computing Systems, 2016), pp. 1503–1509
22. M. Tonsen, J. Steil, Y. Sugano, A. Bulling, Invisibleeye: Mobile eye tracking using multiple low-resolution cameras and learning-based gaze estimation. Proceedings of the ACM on Interactive, Mobile, Wearable and Ubiquitous Technologies **1**(3), 1–21 (2017)
23. M. Ivanović, A. Klašnja-Milićević, J. Ivković, M. Porta, *Integration of Eye Tracking Technologies and Methods in an E-Learning System* (Proceedings of the 8th Balkan Conference in Informatics, 2017), pp. 1–4
24. A. Vance, J.L. Jenkins, B.B. Anderson, D.K. Bjornn, C.B. Kirwan, Tuning out security warnings: A longitudinal examination of habituation through fmri, eye tracking, and field experiments. MIS Q. **42**(2), 355–380 (2018)
25. V. Patel, "Airport Passenger Processing Technology: A Biometric Airport Journey," 2018

Implementing Face Recognition Using Deep Learning

Allam Fatima Zohra, Hamami Mitiche Latifa, and Bousbia-Salah Hicham

1 Introduction

The world we live in today has been distinguished by its insecurity and the resurgence of attacks. Security has become a primary concern for government and private institutions. Indeed, for security reasons in order to fight against the risk of crime, the use of the Facial Recognition System is proving increasingly essential. This security system then remains a privilege, especially in public places.

Face Recognition is based on data and the principles of Artificial Intelligence. Increasingly used because of its practical and effective applications in daily life, Facial Recognition is now experiencing considerable growth in surveillance and security. This application continues to attract researchers in the field of computer vision and remains one of the most studied techniques in recent decades [1].

Facial Recognition has been quietly installed in many public places. It is a biometric technology that consists of identifying an individual based on the characteristics of their face. The identification concerns in particular the distance between the different elements that form the face of the face, in this case, the eyes, the ears, the lips, etc.

As a security measure in establishments that receive people, such as airports, Facial Recognition analyzes individuals more quickly. The principle is to look at facial features that do not change significantly as the individual ages or undergoes cosmetic surgery. This technology represents a real potential for application in various fields and should constitute a development factor.

A. F. Zohra (✉) · H. M. Latifa · B.-S. Hicham
Signal and Communication Laboratory, Department of Electronics, National Polytechnic School, Algiers, Algeria
e-mail: fatima_zohra.allam@g.enp.edu.dz; latifa.hamami@g.enp.edu.dz; hicham.bousbia-salah@g.enp.edu.dz

© The Author(s), under exclusive license to Springer Nature Switzerland AG 2024
A. Shaikh et al. (eds.), *Advances in Emerging Information and Communication Technology*, Signals and Communication Technology,
https://doi.org/10.1007/978-3-031-53237-5_23

Facial Recognition technology is, nowadays, evolving, with both practical and secure applications. In general, the system recognizes a person automatically thanks to their features. In concrete terms, this technique makes it possible to identify a given person, authenticate him, or verify his identity [2].

It is in this context that our work takes place; it is a question of developing a Facial Recognition System that meets the needs of security and access control in protected environments. In this article, we are interested in the development of a biometric Facial Recognition System by applying Deep Learning. In the second section, we briefly present general notionsabout Artificial Intelligence. The third section deals with Deep Learning and Artificial Neural Networks.

In our application, we have focused on Artificial Neural Network (*ANN*) and Multi-Layer Perceptron (*MLP*). We will detail the different stages of implementation. The tests carried out and the results obtained.

2 Artificial Intelligence: Principle and Definition

Artificial Intelligence (*AI*) is the science and engineering of making intelligent machines that can learn, reason, and act on their own. These machines can mimic complex tasks performed by human beings. They can make their own decisions when faced with new situations [3, 4].

To achieve this, machines must be trained to replicate their own representation of the world through Machine Learning. The latter is a set of methods that allow a machine to learn from examples. Deep Learning is a subset of Machine Learning based on Artificial Neural Networks, which mimic the human nervous system [5]. Figure 1 illustrates the relationship between *AI*, *ML*, *DL* and *ANN*.

In the following, we will briefly explain these three techniques.

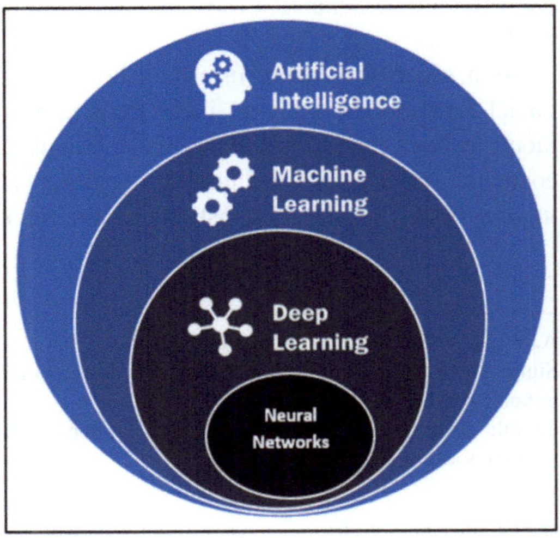

Fig. 1 Relationship between *AI*, *ML*, *DL* and *ANN*

2.1 Machine Learning (ML)

The principle of *ML* consists in teaching a program to perform a task *T* using an experiment *E* and measuring the improvement in its performance *P* in this task. In other words, the program can mimic and generalize behavior for situations similar to experiment *E* [6, 7].

The task *T* in machine learning can be:

- Either a regression, which models a target value based on independent predictors
- Either a classification, which categorizes a set of data into classes
- Either a segmentation, which separates the data into distinct groups

Experiment *E* represents the learning algorithm.

Algorithms can be classified into three main categories according to the type of learning [8]:

1. *Supervised Learning:* In the case of supervised learning, we find labeled or labeled data provided in the form of input (Features) / output (Label) pairs. The goal is to create an algorithm that is able to predict the output of an unlabeled input (maps data *X* to possible predictions *Y*). Supervised learning is divided into two families, namely, classification (separation of input data into several classes) and regression, where the output is a continuous function.
2. *Unsupervised Learning:* From the sample of unlabeled training data (having no known results), we look for general rules in order to group data that share similar characteristics (Clustering), or to find a relation between the data (Association).
3. *Reinforcement Learning:* In a general sense, Reinforcement Learning is the task of learning from successive experiences. What to do in order to find the best solution. In such a problem, an agent (the algorithm) interacts with the "environment" to find the optimal solution.

This type of Machine Learning Algorithm uses the trial-and-error method to produce results based on the highest efficiency of the function. The output is compared to find errors and feedback, which is sent back to the agent to improve or maximize its performance.

Despite the set of advantages provided by *ML*, such as easy identification of relationships, wide application in different fields, and improvement of its performance over time, it encounters some limitations, namely:

- In *ML*, the machine will not work well in real-life scenarios if it has not encountered such scenarios in the past (does not have a code of such inputs).
- *ML* cannot handle high volumes of data.
- In *ML*, the feature engineering process is done manually. It helps structure the data to accurately predict the outcome.

The limitations of Machine Learning have motivated researchers to find other solutions. They developed Deep Learning, which decreases the error rate and gives more accurate results.

2.2 Deep Learning (DL)

DL is a particular subset of *ML*. It is based on an Artificial Neural Network (*ANN*) inspired by the human brain [9]. It presents a solution of a large number of complex problems in different fields, such as image classification and natural language processing.

The principle of *DL* is based on hierarchical learning, layer by layer, in a way that the outputs of each layer represent the inputs of the other, where they become more abstract than before reaching a certain layer. In this way, the system acquires an increasingly precise representation of the data it processes [9].

Deep Learning is based on Artificial Neural Networks:

2.3 Artificial Neural Network (ANN)

Neural Networks represent a computational model with mathematical functions and static concepts with large numbers of coefficients. As illustrated in Fig. 2. The structure of a Neural Network contains a set of interconnected neurons grouped essentially in the following three layers [10]:

- Input Layer: which plays the role of receiving data (inputs).
- Hidden Layer: which contains a transfer function F where all calculations are performed.
- Output Layer: which is obtained after applying transfer function F (the result).

Traditionally, the basic elements of the perceptron are:

- Weight: The weight parameter is directly proportional to the strength of the associated input neuron to decide the output.

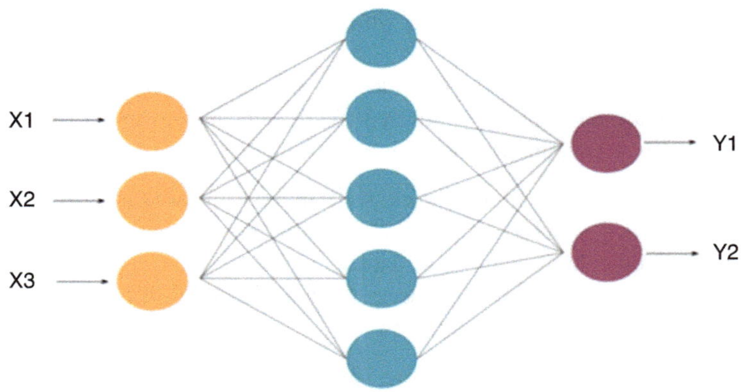

Fig. 2 Layers of the Artificial Neural Network

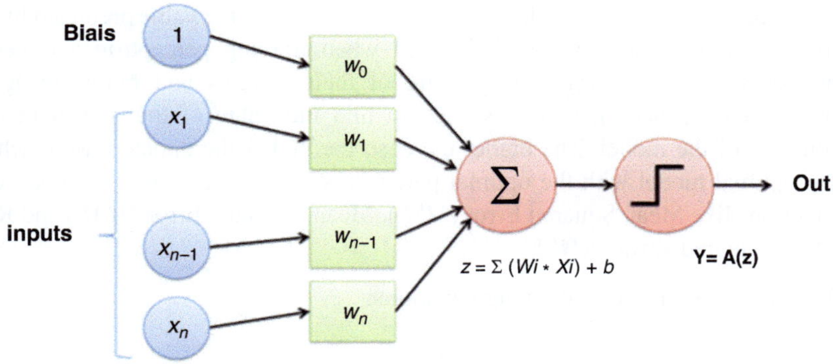

Fig. 3 Elements of the Perceptron

- Bias: It is an additional parameter that is used to adjust the curve of the decision boundary to best fit the data and get a well-accurate output.
- Activation Function: It can be considered, mainly, as a non-linear function to map the weighted sum to the output. This non-linearity makes it possible to spatially modify their representation. The most commonly used activation functions are: Sigmoid, Tanh and ReLU.

The Perceptron can only classify linearly separable data. Otherwise, the Perceptron becomes inoperative and multilayer artificial neural networks are an alternative to solve this problem [6] (Fig. 3).

2.4 *Multi-Layer Perceptrons (MLP)*

MLP is a Feedforward Artificial Neural Network, composed of an input layer, one or more hidden layers and an output layer. When an *MLP* has more than one hidden layer, it is considered a deep *MLP*. Each layer is itself composed of one or more neurons whose role is to apply an activation function to its input [11].

A bias value is also added by the neuron, and each connection between two nodes has a weight according to its importance in the network. We initially give a random value to the weight. Then these values are changed so that the resulting error in the calculation should be minimized. This process is called Backpropagation [11].

In order to develop and train an *MLP*, we repeat the following steps in a loop:

- Forward propagation propagates data from the input layer to the output layer
- Calculation of the cost function
- Backpropagation
- Calculation of Gradient Descent

1. *Cost Function:* It is used to determine the error between the value predicted by the model and the actual value. It is used when training perceptron and neural networks by influencing the updating of their weights and biases using an optimization function, which is used to find the value of the minimum cost function of the model. The higher the cost, the higher the update until reaching the optimal model with the optimal parameters. There are several common cost functions like Mean Squared Error (*MSE*), Mean Absolute Error (*MAE*) and Root Mean Squared Error (*RMSE*).

   They are given by the following equations:

$$\mathrm{MSE} = \frac{1}{N}\sum_{n=1}^{N}\left(y_n^{\mathrm{pred}} - y_n^{\mathrm{label}}\right)^2 \tag{1}$$

$$\mathrm{MAE} = \frac{1}{N}\sum_{n=1}^{N}\left|y_n^{\mathrm{pred}} - y_n^{\mathrm{label}}\right| \tag{2}$$

$$\mathrm{RMSE} = \sqrt{\frac{1}{N}\sum_{n=1}^{N}\left(y_n^{\mathrm{pred}} - y_n^{\mathrm{label}}\right)^2} \tag{3}$$

2. *Backpropagation:* It consists of adjusting the parameters W and b in such a way as to minimize the error function of the model. For this, it is necessary to determine how this function varies with the different parameters, does the function decrease when one of the parameters increases or the reverse. To answer this question, we must calculate a gradient or the derivative of our cost function.
3. *Gradient Descent:* It represents the derivative of the cost function.

   It shows how this function varies:

- If the derivative is negative, we will have to increase W if we want to reduce our errors.
- If the derivative is positive, it will be necessary to decrease W if we want to reduce our errors.

To do this, we use Eq. 4:

$$W_{t+1} = W_t - \alpha \Delta f(W_t) \tag{4}$$

3 Deep Learning Architecture

Convolutional Neural Networks (*CNN*) designed to process images. The architecture of a *CNN* uses four types of processing layers:

- The Convolution Layer,
- The Pooling Layer,
- The ReLU correction Layer
- The Fully-Connected Layer.

Recurrent Neural Networks (*RNN*) works on the principle of saving the output of a previous layer and feeding it back to the input in order to predict the output of the layer.

Short Memory Recurrent Network (*SMRN*) consists of three parts, called gates.

- Forget Gate which decides what information should be discarded or kept.
- Input Gate which is used for cell update.
- Output Gate which has the role of deciding which should be the next hidden state

3.1 Data Preparation Process

We used a Dataset published in the *Kaggle* site to collect and exploit a certain number of images of people. The Dataset is subject to certain conditions, mainly, it must be well organized. Hence the need to pre-process the data.

3.2 Data Preprocessing

Data pre-processing is the stage of preparing the data before providing it to the machine for training. The purpose of this step is to put the data in a more understandable and readable format in order to effectively identify the pattern to improve the overall quality of the results.

The main preprocessing step is the resampling technique. This technique is widely adopted to deal with unbalanced datasets. It consists of removing samples from the majority class (Undersampling) or adding other examples from the minority class (Oversampling).

Figure 4 shows the difference between the two methods.

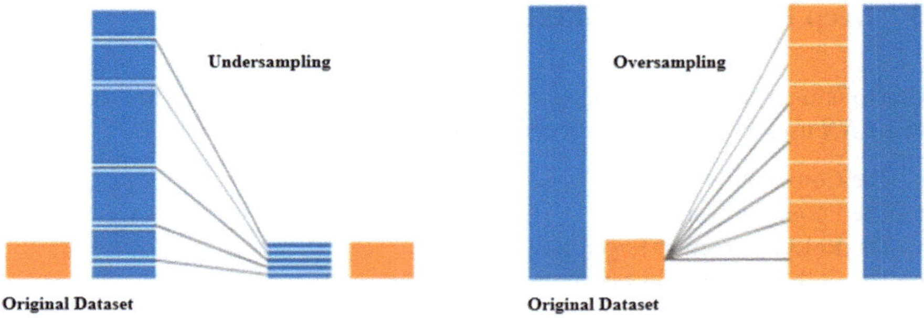

Fig. 4 Difference between *Undersampling* and *Oversampling*

3.3 Data Splitting: Training and Testing Data

In *DL* or *ML*, the performance of the model should not be evaluated using the same data that was used for its training. As mentioned before, we have divided the Dataset into two parts:

- The first is "*Trainset*" (80%) whose data is used to train, train the model and learn its parameters.
- The second part, meanwhile, is "*Testset*" (20%). It is reserved only to evaluate the model. We can thus have an idea of its future performance.

The dividing process is done randomly, i.e., the dataset is shuffled and the samples are chosen arbitrarily and placed in the Trainset and "*Testset*" according to the given percentage. However, this method has a drawback in datasets with an unbalanced class distribution. For this, the researchers are developing another technique, called stratified sampling. The latter distributes the classes in each of the training and test sets [12].

3.4 Performance Indicators

In order to evaluate the performance of the proposed *DL* model, the *Accuracy*, *Precision*, *Recall* indicators are considered performance measures. These measurements are calculated based on True Positive (*TP*), True Negative (*TN*), False Positive (*FP*), and False Negative (*FN*).

1. *Accuracy:* This metric estimates the ratio of correct connection records to total data. The performance of the *DL* model increases with increasing *Accuracy*.

$$\text{Accuracy} = \frac{\text{TP} + \text{TN}}{\text{TP} + \text{TN} + \text{FP} + \text{FN}} \tag{5}$$

2. *Precision:* This metric estimates the ratio of correctly identified records to total predicted positives. The performance of the *DL* model increases with increasing Precision.

$$\text{Precision} = \frac{\text{TP}}{\text{TP} + \text{FP}} \tag{6}$$

3. *Recall:* This indicator essentially estimates the ratio of true positives to all positives in the true class.

$$\text{Recall} = \frac{\text{TP}}{\text{TP} + \text{FN}} \tag{7}$$

4 Implementation

The "Labelled Faces in the Wild" (*LFW*) data set is a collection of 13,233 images of celebrity faces collected from the web, with the individual in each image manually labelled. The dataset is designed for exploring, creating and testing image recognition algorithms.

In this part, we will be attempting to run an algorithm that can successfully identify faces in the dataset after having been trained on similar faces. After doing Library Installation and Data Read-in (Data Cleaning). We will follow the bellow steps:

- Exploratory Data Analysis (*EDA*): exploration, visualizations and splitting the data into training/test/validation sets
- Model Construction
- Results Analysis
- Conclusion and Next Steps: limitations and possible future extensions

4.1 Data Reshaping and Train/Test Split

The training set should incorporate multiple images of some individuals, to equip the model with knowledge of how to deal with the same face in different situations. However, the test set needs to incorporate some individuals that we have seen in the

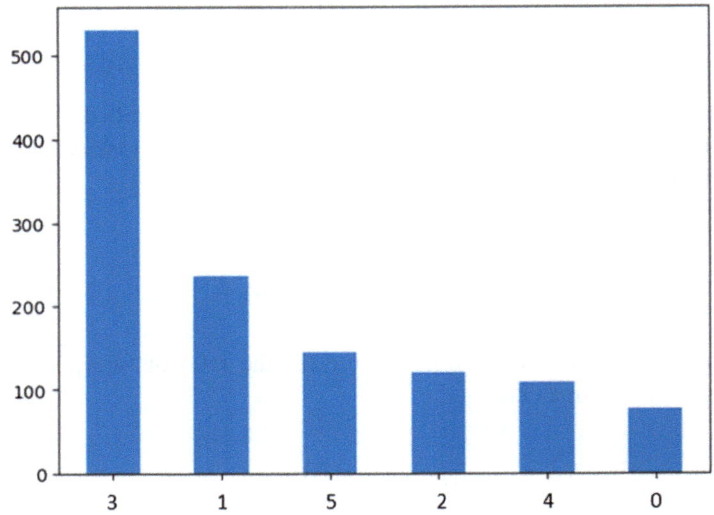

Fig. 5 Resampling method

training set (although not the same images) and some individuals that we haven't, to test the algorithm's accuracy

Therefore, we need to split the data into training and test data.

```
# shape data frame so there is a row per image, matched to relevant jpg file
image_paths = lfw_allnames.loc[lfw_allnames.index.repeat(lfw_allnames['images'])]
image_paths['image_path'] = 1 + image_paths.groupby('name').cumcount()
image_paths['image_path'] = image_paths.image_path.apply(lambda x: '{0:0>4}'.format(x))
image_paths['image_path'] = image_paths.name + "/" + image_paths.name + "_" + image_paths.image_path + ".jpg"
image_paths = image_paths.drop("images",1)

# take a random sample: 80% of the data for the test set
lfw_train, lfw_test = train_test_split(image_paths, test_size=0.2)
lfw_train = lfw_train.reset_index().drop("index",1)
lfw_test = lfw_test.reset_index().drop("index",1)

# verify that there is a mix of seen and unseen individuals in the test set
print(len(set(lfw_train.name).intersection(set(lfw_test.name))))
print(len(set(lfw_test.name) - set(lfw_train.name)))
```

One of the drawbacks of the *LFW* dataset is that if has significant class imbalance. This can be dealt with using selective sampling, but it is useful to get an indication of the scale of the imbalance (Fig. 5):

After the sampling, it become as the bellow (Fig. 6)

As a sanity check, and to check the directories are all correctly connected, we visualize a sample image:

```
im = Image.open("/content/drive/MyDrive/Dataset /lfw-deepfunneled/lfw-deepfunneled/" + str(lfw_train.image_path[0]))
plt.imshow(im)

<matplotlib.image.AxesImage at 0x7f31ac03eda0>
```

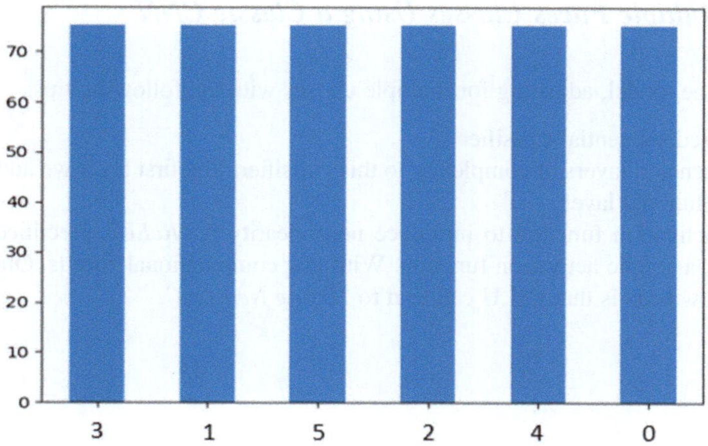

Fig. 6 After sampling

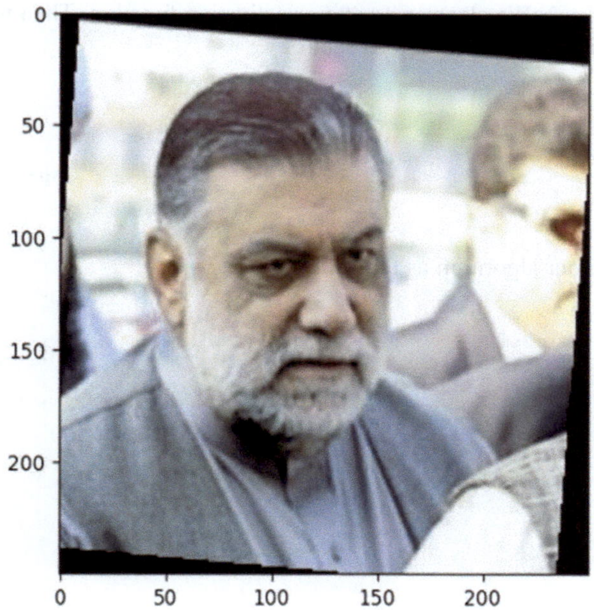

4.2 *Multiple Faces Classes Using a Classic CNN*

We run the model, adjusting for multiple classes with the following tips:

1. We used sequential classifier
2. We then add layers of complexity to this classifier. The first layer we add is a 2D convolutional layer
3. The activation function to introduce non-linearity is "*ReLU*": Rectified Linear Units, a simple activation function. With low computational speeds. One potential downside is that ReLU can lead to "*Dying Neurons*".

```
multi_classifier = Sequential()
multi_classifier.add(Conv2D(32, (3, 3), input_shape = (250, 250, 3), activation = 'relu'))
multi_classifier.add(MaxPooling2D(pool_size = (2, 2)))
multi_classifier.add(Flatten())
multi_classifier.add(Dense(units = 128, activation = 'relu'))
```

4. As we are training on several classes, we need several classification units (one for each class). We also use a *softmax* activation function. Then we change the loss function to categorical_crossentropy.

```
multi_classifier.add(Dense(units = 6, activation = 'softmax'))
multi_classifier.compile(optimizer = 'adam', loss = 'categorical_crossentropy', metrics = ['accuracy'])
```

We can run our algorithm as the bellow

```
multi_history = multi_classifier.fit_generator(multi_training_set,
                                               # set steps per epoch equal to the number of training images
                                               steps_per_epoch=int(304/32),
                                               # as the data is smaller, we can run for more epochs
                                               epochs = 100,
                                               validation_data=multi_val_set,
                                               validation_steps=int(90/32))
```

4.3 *Results Analysis*

Now that the model is operational, we can look at the relationship between the training and validation loss to see if overfitting is likely to have taken place (Figs. 7, 8, and 9):

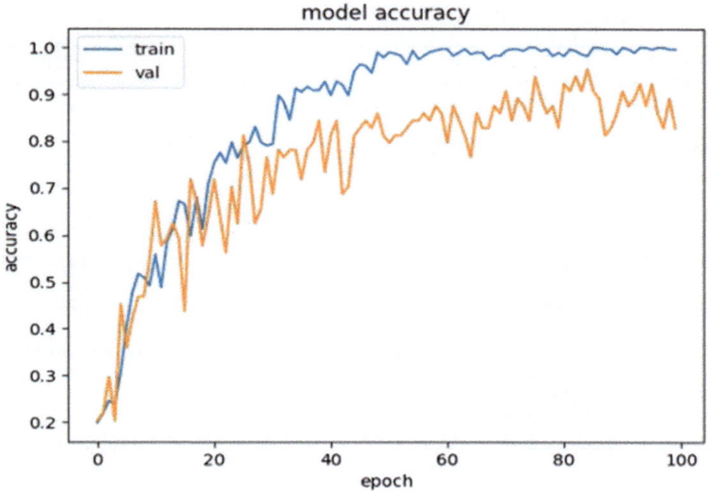

Fig. 7 Visualization of *Accuracy* results

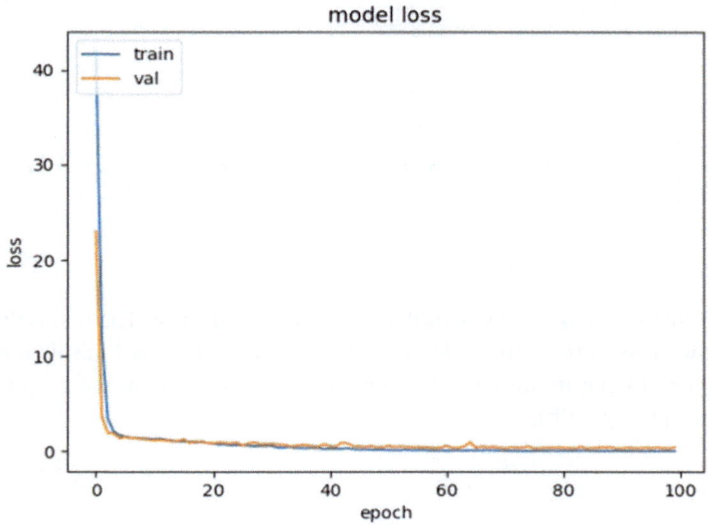

Fig. 8 Visualization of *Loss* results

4.4 Test the Model

For the "Test" phase, we apply a capture through *Google Colab* of an unknown face to verify the accuracy of the algorithm.

```
Model: "sequential_1"
_________________________________________________________________
 Layer (type)                Output Shape              Param #
=================================================================
 conv2d_13 (Conv2D)          (None, 248, 248, 32)      896

 max_pooling2d_7 (MaxPooling  (None, 124, 124, 32)     0
 2D)

 flatten_3 (Flatten)         (None, 492032)            0

 dense_9 (Dense)             (None, 128)               62980224

 dense_10 (Dense)            (None, 6)                 774

=================================================================
Total params: 62,981,894
Trainable params: 62,981,894
Non-trainable params: 0
_________________________________________________________________
```

Fig. 9 Feature of the trained model

```
try:
  filename = take_photo('photo.jpg')
  print('Saved to {}'.format(filename))

  # Show the image which was just taken.
  display(Image(filename))
except Exception as err:
  # Errors will be thrown if the user does not have a webcam or if they do not
  # grant the page permission to access it.
  print(str(err))
```

There are two types of messages that shown during the test. The first is the case of an unknown person (Fig. 10a). The second case is when the machine already trained on this type of data; in this case the camera shows the name of the person (coding with number) (Fig. 10b).

4.5 Next Steps

In this project, we have successfully run *CNN* generalizes reasonably well to multiple classes using images from the *LFW* dataset. However, there are several ways we could both improve the current model or extend to a different model that would likely bring higher accuracy and more reliable results.

There are several immediate improvements that we would make in future work, many of which are listed above: running for more epochs, on more classes, with more complex layers in the model. Another important limitation around the model as

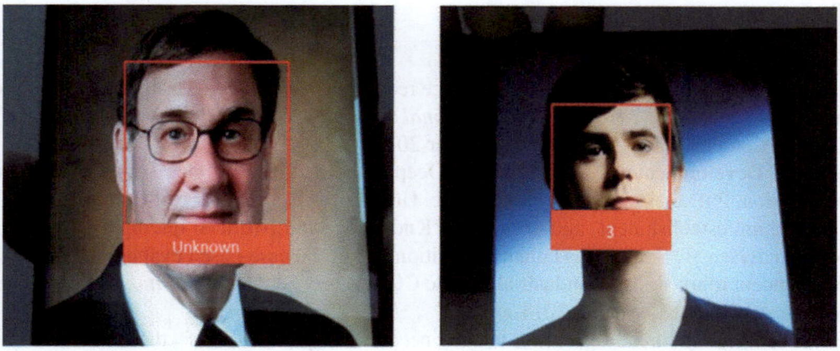

(a) Unknown Person (b) Known Person

Fig. 10 Test phase

it currently stands is its data input. A solution to this could be to incorporate more datasets in the training model, to increase its diversity and generalization (and therefore its impartiality).

5 Conclusion

Facial Recognition is an evolving technology that has many practical and secure applications. It is increasingly used to ensure security in public places and highly protected environments. In the context of this article, we have developed a selfie Facial Recognition System to meet the security and access control needs of individuals.

We are interested in harnessing the power of algorithms based on Artificial Intelligence. Indeed, *AI* is considered to be a very brilliant technology with very great capacities for simulating problems, calculating and learning from very large amounts of data.

This project presented an implementation of Facial Recognition based on the *CNN* architecture. The various tests carried out have shown satisfactory results in terms of recognition rate. These results are encouraged for the further development of this technology and its application in different fields, including surveillance, medicine, etc.

However, it should be noted that the quality of the results strongly depends on the quality of the data used to train the model. Therefore, it is important to have a high quality and large dataset to ensure optimal performance.

In conclusion, the use of machine learning and deep neural networks for image recognition, particularly, facial recognition, is a promising technology that offers many possibilities in many fields. It is important to continue to develop and improve this technology to get the most out of it and benefit society.

References

1. X. Peng, N. Ratha, S. Pankanti, Learning face recognition from limited training data using deep neural networks, in *2016 23rd International Conference on Pattern Recognition (ICPR) Cancún Center*, Cancún, México, December 2016
2. M. Kas, Development of handcrafted and Deep based methods for face and facial expression recognition, PhD thesis in Computer Science, University of Technology of Belfort-Montbéliard (preparation establishment) and Distributed Knowledge and Artificial Intelligence, July 2021
3. R.H. Andersen, T. Solund, J. Hallam, Definition and initial case-based evaluation of hardware-independent robot skills for industrial robotic CO-workers, in *41st International Symposium on Robotics,* Munich, Germany, 2014, pp. 1–7
4. B. Goertzel, Artificial general intelligence: Concept, state of the art, and future prospects. J. Artif. Gen. Intell. **5**(1), 1–46 (2014)
5. C. Hardy, Contribution to the development of Deep Learning in distributed systems, PhD Thesis in Mathematics and Information and Communication Sciences and Technologies, Specialty: Computer Science, University of Rennes I, April 2019
6. O. Biran, C. Cotton, Explanation and justification in Machine Learning: A survey, in *IJCAI-17 Workshop on Explainable Artificial Intelligence (XAI)*, (2017)
7. K. Crowston, F. Bolici, Impact of Machine Learning on work, in *Proceedings of the 52nd Hawaii International Conference on System Sciences*, (2019)
8. L. Deng, D. Yu, Deep Learning: Methods and applications. Found. Trends Signal Process. **7**(3–4), 197–387 (2014)
9. W. Hao, L. Qi, L. Xiaodong, A review of Deep Learning approaches for image classification and object segmentation. Comput. Mater. Contin. **60**(2), 575–597 (2019)
10. T. Guillod, P. Papamanolis, J.W. Kolar, *Artificial Neural Network (ANN) Based Fast and Accurate Inductor Modeling and Design* (Power Electronic Systems Laboratory, ETH Zurich, July 2020)
11. B. Zhang, L. Zhang, The analysis and improvement of Artificial Neural Network models, in *1997 IEEE International Conference on Intelligent Processing Systems*, October 1997
12. M. Almseidin, M. Alzubi, S. Kovacs, Evaluation of Machine Learning algorithms for an intrusion detection system, in *2017 IEEE 15th International Symposium on Intelligent Systems and Informatics (SISY)*, October 2017

A Comprehensive Review of Brain Diseases Classification Using Deep Learning Techniques

Lin M. Saleh Aouto, Leidi M. Saleh Aouto, Rawan Khaled Flifel, and Dina M. Ibrahim

1 Introduction

The human brain is the most sophisticated organ we have ever encountered. The complexity of the human brain results in intricate diseases and disorders. The improving understanding of this complex system has led to potential advancements in treating various illnesses. Damage to the brain during development has a qualitatively different effect than damage throughout maturity. Adult brain damage usually causes significant and selective brain abnormalities, with loss of skill in one function against the context of otherwise preserved cognition. Amnesia (selective difficulty in developing new memories), aphasia (deficiency in interpreting or expressing language), and agnosia (inability to identify persons or objects using one or more of the senses) are some examples. On the other hand, cognitive deficits caused by brain disease acquired during development are often less drastic but broader, impacting a wider scope of cognition. One explanation for these disparities is that early-life focal injury might affect the function of other brain areas related to the affected area during maturation processes [1].

Mental diseases and further neurological illnesses are among the leading causes of injury and infirmity worldwide, frequently causing chronic effects on the person

L. M. S. Aouto (✉) · L. M. S. Aouto · R. K. Flifel
Department of Information Technology, College of Computer, Qassim University, Buraydah, Saudi Arabia
e-mail: 411216176@qu.edu.sa; 411216177@qu.edu.sa; 392216766@qu.edu.sa

D. M. Ibrahim
Department of Information Technology, College of Computer, Qassim University, Buraydah, Saudi Arabia

Department of Computers and Control Engineering, Faculty of Engineering, Tanta University, Tanta, Egypt
e-mail: d.hussein@qu.edu.sa; dina.mahmoud@f-eng.tanta.edu.eg

© The Author(s), under exclusive license to Springer Nature Switzerland AG 2024
A. Shaikh et al. (eds.), *Advances in Emerging Information and Communication Technology*, Signals and Communication Technology,
https://doi.org/10.1007/978-3-031-53237-5_24

suffering from them. The illness processes are complicated, including a variety of hereditary and environmental variables. Although there is considerable inter-individual variation, people with common brain illnesses achieve worse on mental tests, are more likely to fail in performing tasks, and participate in harmful health behaviors more frequently.

One of the most important factors causing and affecting brain disorders is age, as seen by the vast range of illness initiation timings from infancy to senior age. This implies that the underlying pathophysiology relies heavily on the age at which certain trajectories first deviate from the norm. Although attention-deficit/hyperactivity disorder (ADHD) and autism spectrum disorder (ASD) appear during childhood, bipolar disorder(BD) and schizophrenia (SZ) start in adolescence, before the onset of serious symptoms during the adult years. Similarly, multiple sclerosis (MS) most commonly becomes apparent in one's early adult years, but the progression of the disease begins and develops before showing any signs of its presence. Dementia (DEM) and mild cognitive impairment (MCI) are more frequent in older adults, although major depressive disorder (MDD) can develop at every stage of life, from adolescence to elderly age. Age-related deviations from the norm may differ in geographic position, direction, change rate, and size, all of which might challenge the comprehension of recorded effects when attempting to evaluate the underlying brain dysfunction of these conditions [2].

A brain disorder can be caused by a gene variant or mutation and is called an inborn genetic brain disorder. A variation is a variant of a gene. A gene mutation is an alteration. Genetic brain abnormalities have an impact on brain development and function.

Certain genetic brain disorders are caused by random gene mutations or by environmental factors, such as cigarette smoke. Whereas inherited disorders imply that a faulty gene or collection of genes is transmitted from generation to generation. They can also be caused by a mix of genetic alterations and other external causes. Phenylketonuria (PKU), Tay-Sachs syndrome, leukodystrophies, and Wilson syndrome are instances of genetic brain disorders. People with genetic neurological issues generate insufficient amounts of particular proteins that impact brain development and function. Brain abnormalities can have major consequences for the neurological system. There exist some medications to alleviate symptoms. But others are potentially fatal [3]. Numerous brain diseases have been discovered and studied (see Fig. 1). Further reading about each disease can be found in the American Brain Foundation [4].

2 Background

Increased understanding of the connections between the brain's structure, function, and the manifestation of mental illness has resulted from discoveries in neuroscience and genetics; brain disorders are considered to be critical as they have permanent effects on patients' lives and their cognitive and motor skills. The cost of

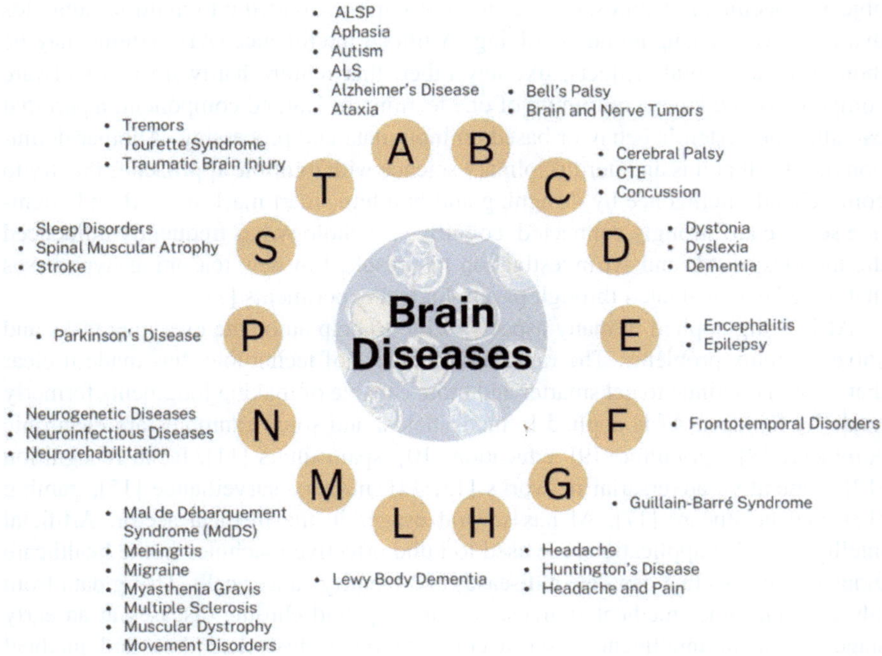

Fig. 1 Brain diseases from A to Z

neurological disorder-related disability and deaths is rapidly being recognized as a worldwide public health concern, and its impact is expected to grow over the next few decades as the population ages. As stated by Pan American Health Organization (PAHO), nervous conditions in Canada, for example, reached 39.49 deaths in 2019, with a 44.1% change relative to the statistics calculated in the year 2000 for the same country. Another example is the numbers found in the Region of the Americas, in 2019, neurological conditions hit 32.90 deaths there, with a percentage change relative to the year 2000 equal to 60.4%. Those statistics are rates per 100,000 populations and for both sexes (male and female) [5]. In order to decrease the dangers of neurological disorders and diseases, we need to find the most accurate and early diagnosis tools possible. This can be achieved by the development and usage of AI medical diagnosis techniques.

2.1 Artificial Intelligence

According to the Cambridge Companion to Artificial Intelligence: Legal and Ethical Considerations from Around the World [6], AI is a complicated, comprehensive, wide-ranging branch that is difficult to describe since it might relate to technical objects, specific procedures, or a scientific discipline divided into multiple subfields that are always changing and evolving. Artificial intelligence (AI) systems may be thought of as digital artifacts, like any other, that require hardware and software components and have a minimum of one learning or learned component, a part that can alter the system's behavior based on input data and processing. Another definition of AI is that it is an interdisciplinary science with multiple approaches that try to comprehend intelligence by designing and building smart machines. AI and cognitive science are strongly connected; cognitive psychology has frequently influenced the theories of AI, and AI investigation has resulted in new reasoning hypotheses that have been evaluated through psychological experiments [7].

AI is being applied in many aspects of life to help automate everyday tasks and solve complex problems. The rapid advancement of technology has made it clear that AI will continue to get smarter and more capable of making judgments formerly made by humans. AI is applied in many fields, and some examples are electronic commerce [8], agriculture [9], education [10], spam filters [11], facial recognition [12], generative adversarial networks [13, 14], military surveillance [15], gaming [16], and healthcare [17]. AI has several usages in the medical sector. Artificial intelligence (AI) applications are used to build effective machines in the healthcare industry, devices that can detect diseases and identify cancer cells. Using data from lab tests and other medical sources, AI can help find chronic diseases at an early stage. Artificial intelligence uses a combination of historical data and medical understanding to develop new drugs. In a more focused scope, we can say that AI techniques should be utilized effectively in brain diseases and disorders diagnosis given the importance of detecting any problem that may affect neurons that are uncoverable once they get damaged. High accuracy of diagnosis is needed since some diseases have familiar symptoms and traits, so misdiagnosis could get common if the tools used are not smart and accurate enough.

AI includes the subfields of machine learning (ML) and deep learning (DL). Nevertheless, DL is a part of ML, which is a part of AI.

2.2 *Machine Learning*

Machine learning (ML) is a subfield of artificial intelligence and computer science that seeks to improve AI's accuracy by modeling human learning strategies with data and algorithms. Machine learning's goal is to learn from experience and create a model that provides effective decision-making by performing prediction tasks (see Fig. 2) [18].

ML methodologies are grouped into four groups based on the learning model: clustering, dimensionality reduction, classification, and association rule mining. Several performance indicators such as precision, accuracy, recall, and so on are used to assess the efficiency of machine learning methods [18].

Machine learning advancements, the availability of digital medical care data, and faster processors have all led to a rise in the number of articles presenting ML

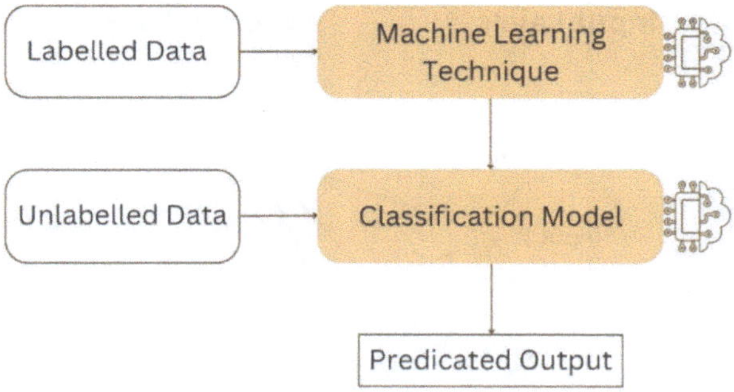

Fig. 2 A basic model of machine learning

applications in medical care. These ML models all have the same aim in mind: to enhance patient care for both doctors and patients [19, 20].

2.3 *Neural Networks*

A neural network model is a subfield of artificial intelligence; it is one form of ML model that has evolved more competitively compared to traditional regression and models statistics in terms of practicality, known as artificial neural networks (ANNs). Its structure and name hinge on the human brain, and they aim to create an information management model that imitative the actual nervous system job of the brain. The brain is made up of two basic components: neurons, which form nodes of ANNs, and synapses, which form connections or weights of ANNs [21, 22]. Instead of designing a computer system to perform certain tasks, ANNs train the system to execute tasks. To complete such tasks, an (AI) system is created. It is a practical model capable of swiftly and precisely locating patterns in data that repeat relevant knowledge. One type of these AI models is the neural network [23].

The most basic version of ANNs comprises of three layers: the first layer is input layer, followed by the second layer which is called hidden layer, and lastly the output layer (see Fig. 3) [22].

ANNs' lightning-fast processing in a wide implementation in parallel has increased the demand for study in this arena. Natural language processing, image recognition, and other products may be constructed using ANNs. ANNs are currently extensively used for universal function approximation in numerical paradigms due to their superior properties of additivity, self-learning, nonlinearity, tolerance to faults, and progress in input-to-output mapping [21].

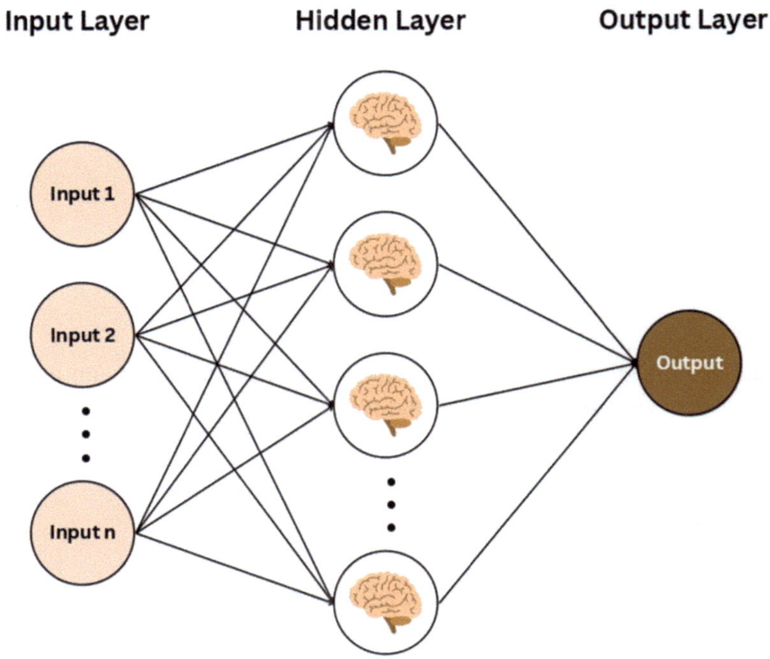

Fig. 3 Basic version of ANN architecture

2.4 Deep Learning

Deep learning DL is a sub-type of AI that performs tasks like feature extraction and transformation by employing a large number of nonlinear processing unit layers [24, 25]. Moreover, DL is another strategy for improving training accuracy. The deep learning system will supply additional training parameters in the high-dimensional feature space when the deep layer neurons are enlarged, and training accuracy will improve. Nevertheless, employing additional training parameters will raise calculation expenses [26]. Each subsequent layer receives the output of the preceding layer as input. Learning algorithms can be unsupervised such as regression or supervised such as classification. Deep learning models are dominated by various versions of ANNs, although there are other types. Unlike ML models that undergo training to execute tasks using features that are manually extracted from unprocessed data, or the features extracted by other ML models, the main feature that DL methods have in common is that they learn useful representations and features instantly from raw data. This is the fundamental difference between DL models and more standard ML approaches. Finding features and completing an assignment are merged into a single challenge and so enhanced simultaneously throughout the training phase. Deep learning frameworks are essential for applications such as image processing, facial recognition, semantic video segmentation, and object identification, among others [27, 28].

3 Related Work

In this project, we will focus on studying the usage of AI techniques in the diagnosis of some of the collective brain diseases and disorders, which are epilepsy, dementia especially Alzheimer's disease stages, Parkinson's disease, and brain tumors (see Fig. 4).

The authors in [29] proposed two deep learning architectures that are independent of the patient and may use data from several participants to train a global function. The suggested models show outstanding results for seizure prediction on the CHB-MIT-EEG dataset, with accuracy of 88.81% and 91.54%, respectively. In summary, the Siamese model trained using the suggested learning technique is capable of identifying patterns in data that are associated with patient variations while foretelling seizures. The same design may be used as a classifier of a specific patient with model adaptation, and their models exhibit higher performance for patient-independent seizure prediction. They demonstrate that they are the first research to employ model interpretation to explain classifier behavior for the goal of seizure prediction, as the MFCC map of features utilized by the models they developed contains biomarkers that are predictive linked with brain states like interictal and preictal. According to the authors in [30], it is amazing how well machine learning models predict seizures with increased sensitivity and specificity. The nonstationary electroencephalogram EEG, however, makes it challenging to construct a functional device for a patient because it necessitates a sizable number of seizures that are unique to a patient and periodic model retraining. In order to lessen this procedure, they suggested a Siamese neural network-based seizure prediction technique that

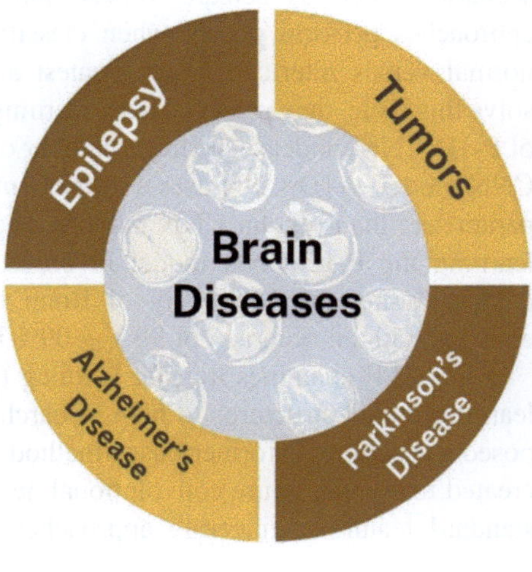

Fig. 4 Brain diseases that we focused on in this study

employs a convolutional neural network (CNN) as its base network to identify change points in EEG, and it takes as input a wavelet-transformed EEG tensor. Their method only takes one seizure for training, which equaled to fewer than 10 minutes of preictal and interictal data. Despite this, their method nevertheless produces results that are equivalent to models that employ numerous seizures for seizure prediction. This is in contrast to explanations in the literature that require days' recordings of the EEG.

The authors in the research [31] presented a one-step semi-supervised epilepsy detection algorithm that completely utilizes unlabeled data to reduce labeling costs. By requiring a more reliable and precise decision, boundary is produced by the proposed neural network learning method due to the stability of double forecasts on the same unlabeled input. For the CHB-MIT and Kaggle datasets, the area under receiver operating characteristic (AUROC) curves of their suggested model are giving values of 10.3% and 4.9% that outperform the supervised techniques, respectively.

The authors in [32] presented multifeature metric learning based on enhanced equidistance embedding (MMLE3) for EEG detection of epilepsy, which is inspired by the equidistance-restricted metric learning technique. The MMLE3 method employs multiple types of EEG features, and the feature weights are weighted adaptively. Without manual adjustment, the feature weight vector can be altered adaptively, which is a significant advantage. The MMLE3 algorithm increases the gap between the cannot-link-restricted samples and normalizes the distances between the samples of the various classes while minimizing the gap between samples forced by the link and compressing samples from the same category to a single point. Under the assumption that the multiple feature classification tasks are consistent, MMLE3 is able to completely excerpt the linked and corresponding information concealed between features. The result of the experiment findings on the CHB-MIT dataset confirms the algorithm's high generalization performance.

According to the authors in [33], many automatic algorithms utilizing conventional methods have been presented to aid neurologists in detecting binary epilepsy scenarios, such as seizure versus non-seizure or normal versus ictal. These approaches perform poorly when classifying ternary cases, such as ictal versus normal versus interictal; their greatest accuracy for this situation is 97.1%. To solve this issue, they present a deep learning-based system consisting of an ensemble of P-1D-CNN models. The huge number of trainable parameters is a bottleneck in a CNN model. P-1D-CNN uses a refinement-based approach and results in 60% less parameters than regular CNN models. They suggested augmentation strategies for learning the P-1D-CNN model in order to overcome the limitations of a short dataset. Using the University of Bonn dataset, the suggested approach detects epilepsy with an accuracy of $99.1 \pm 09\%$ in practically all instances.

Paper [34] examines transfer learning (TL) and model inputs for different deep learning model structures to help researchers construct algorithms. They also proposed a precise transformer-based method. Methods: A hybrid transformer model is created to compare pure convolutional neural network (CNN)-based models to two standard feature engineering approaches and the suggested method, which uses diverse EEG rhythms. Lastly, patient-independent and TL techniques are used to evaluate two model architectures. Results: Their feature engineering strategy improved model performance and was better for transformer-based models on the CHB-MIT scalp EEG dataset. Transformer-based model accuracy improvement using fine-tuning techniques is more robust than CNN-based model performance improvement, and their model attained an ideal sensitivity of 91.7% with 0.00/h FPR. Conclusion: Their epilepsy prediction technique outperforms CNN-based TL structures. The gamma (γ) rhythm also helps predict epilepsy. Significance: Their hybrid transformer is precise.

The research [35] proposed an instance-specific seizure detection via deep learning prediction method for long-term scalp EEG data. The aim is to distinguish the preictal brain stage from the interictal stage as early as possible for real-time application, allowing for more effective treatment. Automatic feature extraction and classification go hand in hand. No preprocessing of the EEG signals is performed by the system. In order to speed up prediction times and enhance classification accuracy, four deep learning algorithms extract the most discriminative characteristics possible. The convolutional neural network (CNN) can extract meaningful spatial input from several scalp locations, and the recurrent neural network can predict seizures with greater accuracy than existing approaches. The situation is enhanced by semi-supervised optimization based on transfer learning. The device is appropriate for real-time use since the algorithm that chooses the channel selects the most relevant EEG channels. Robustness is tested using an effective method. The suggested method is the most efficient in the field due to its 99.6% accuracy.

In the study [36], the authors suggested two classification techniques for automatically identifying epileptic seizures from EEG records. Deep neural networks (DNNs) are used in the suggested approaches for classification and empirical mode decomposition (EMD) for feature extraction. The first classification approach uses a multilayer perceptron to speed up the process of classifying between normal and epileptic instances by lowering the dimension of the features. This approach yields a classification accuracy of 100%. The second classification approach uses deep convolutional neural networks (DCNN) to achieve excellent accuracy in multiclass classification tasks. A classification accuracy of 98.6% was achieved using DCNN to distinguish between normal, interictal, and ictal instances. To assure the system's robustness, the tenfold cross-validation methodology is used to evaluate the proposed methods.

Similarly, in [37], a novel method for detecting epileptic seizures is proposed, one that does not need artificial feature extraction and instead uses recordings of raw EEG data to categorize the seizures. With the help of a one-dimensional convolutional variational autoencoder, the system combines supervised and unsupervised deep learning. Using k-fold cross-validation, the suggested system is evaluated in order to determine the resilience of the system against classifying unknown data. Their technique is one of the most effective among other cutting-edge systems since the categorization results were 100% accurate between normal and ictal patients and 99% correct between normal, interictal, and ictal cases. Reference [38] hypothesizes that scalp EEG data can predict focal seizures. They first wanted to identify interictal and preictal traits. The second objective is to establish a timeframe for forecasting that is accurate and early, two competing goals. Methods: Interictal, preictal, and ictal quantitative signatures are defined and learned using EEG wavelet transformation convolutional filters. The data also determines the appropriate seizure prediction horizon. Results: Computational optimization methods reveal a 10-minute seizure prediction horizon. Kullback-Leibler divergence on automatically generated feature distributions confirms this result. Conclusion: Preictal phase shift occurs approximately 10 minutes before the onset of a seizure, as shown by a dataset of 204 EEG recordings; then testing prediction findings are positive, with 87.8% sensitivity and only 0.142 FP/h false positive rate. Their seizure prediction methods outperform random predictors. Significance: Scalp EEG can identify the preictal condition of focal seizures.

The study in [27] assessed the OASIS database's accuracy in finding dementia biomarkers using machine learning. Dementia is caused by AD globally. It's practically hard to complete anything without help when the challenge level is high. The diagnostic period and population growth are increasing. Testing and medical history are contemporary methods. The imbalance of datasets and their accuracy impact dementia research. A reinforcement learning and neural network system might produce and divide imbalanced categories. High-resolution illness probability maps can be derived from a correct diagnosis that takes into account all four phases of dementia. Dementia risk is expressed clearly and accurately because to the use of deep reinforcement learning. It is important that samples accurately reflect the composition of each class. The MRI image shows imbalance classes. Testing precision improved by 9% points, recall by 13% points, and F score by 9–10% with the deep reinforcement system. Diagnosis efficiency has also increased.

The authors in study [39] used 3D diffusion tensor imaging (DTI) to solve a straightforward four-class classification problem for the first time, i.e., AD, normal control (NC), late mild cognitive impairment (LMCI), and early mild cognitive impairment (EMCI). The fractional anisotropy (FA), echo planar imaging (EPI), and mean diffusivity (MD) concentrations provided by DTI provide information on the structure of the brain. They train CNN individually, more precisely VoxCNNs, using FA, MD, and EPI concentrations on the volumes of the 3D DTI scan. Furthermore, they feed a random forest classifier the normal values of the brain area for FA and MD that were obtained using the Colin27 brain atlas (RFC). For the aforementioned four-class classification task, these four models—one RFC and three separately trained VoxCNNs—are initially used singly. Using a modified rank-averaging method, categorization outcomes for each individual are then combined at the decision level, yielding a classification accuracy of 92.6%. The efficiency of the suggested method has been thoroughly tested using the Alzheimer's Disease National Initiative (ADNI) database, which is freely accessible.

In Reference [40], a deep feature-based method for detecting the stage of Alzheimer's disease is proposed. It does this by deep feature extraction from a CNN and copying over the AlexNet's initialization layers. They employed a KNN, support vector machine (SVM), and random forest, three popular machine learning techniques, to classify the retrieved deep features (RF). An accuracy rate of

99.21% was achieved by a deep feature-based model, which, according to the findings of the evaluation conducted on the proposed plan, surpassed hand-crafted and deep learning techniques. The proposed model also performs better than the most cutting-edge approaches that are currently being used. According to the authors in [41], effective machine learning algorithms were assessed because of their capability to diagnose dementia using the Open Access Series of Imaging Studies (OASIS) dataset. Further tested for the detection of Alzheimer's disease were two CNN algorithms, namely, AlexNet and ResNet-50, in addition to two hybrid deep and machine learning methodologies, namely, AlexNet+SVM and ResNet-50 + SVM. For the OASIS dataset, the t-Distributed Stochastic Neighbor Embedding (T-SNE) algorithm was employed to create a low-dimensional representation of the high-dimensional data after it had been balanced, the missing values had been filled in, and the missing values had been replaced. For detecting dementia, all of the machine learning algorithms—including SVM, random forest, decision tree, and KNN—achieved outstanding performance. The precision, recall, and F1 scores for the random forest method were 93%, 98%, and 96%, respectively, with a total accuracy of 94%. The MRI image dataset, the second dataset, was examined using hybrid AlexNet+SVM and ResNet-50 + SVM models, as well as AlexNet and ResNet-50 models. All models had excellent performance, but the hybrid approaches that combined deep learning and machine learning performed better overall. The hybrid AlexNet+SVM model scored 94.8% for accuracy, 93% for sensitivity, and 97.75% and 99.70%, for both specificity and AUC, respectively. In the paper [42], the authors show how eliminating these feature extraction processes yields comparable performance for the residual and simple 3D CNN designs. They evaluate the effectiveness of the proposed method for Alzheimer's disease classification in relation to that of mild cognitive impairment and normal controls using the ADNI dataset of 3D MRI brain structural scans.

The study [43] assessed a different deep learning algorithm for diagnosing Alzheimer's disease (AD) and jointly predicted South Korean AD patients' MMSE scores. They created functional 3D independent component spatial maps from 331 resting-state functional magnetic resonance imaging data for classification and regression tasks. The classification was done with a 3D CNN. Tree regression with gICA features, logistic regression, support vector regression, bagging-based ensemble regression, and LLSR were all able to predict MMSE scores. SVM-based recursive feature reduction and the least absolute shrinkage and selection operator were utilized to enhance MMSE regression performance. Comparing AD patients to healthy controls, the average balanced test accuracy was 85.27%. The medial visual, the default mode, ventral attention, executive, and acoustic networks were most strongly associated with AD. The highest R2 value (0.63+/−0.02) and the smallest root mean square error (3.27+/−0.58) were achieved with the LLSR approach on gICA and SVM-RFE features to predict clinical MMSE scores. Only rs-fMRI can classify AD and healthy controls and functional independent component characteristics can predict MMSE scores. 3D regression learning and deep learning through rs-fMRI data may predict AD disease status and clinical MMSE scores without physicians. Table 1 presents the previous studies for prediction and classification methods of epilepsy, dementia, and AD.

Table 1 Summary of prediction and classification methods of epilepsy, dementia, and AD in the previous studies

Ref	Disease	Method	Dataset	Metrics*
[29]	Epilepsy	Siamese neural network	CHB-MIT-EEG	Acc = 91.54% Spe = 89.94% ROC-AUC = 96.94% Sen = 92.45%
[30]	Epilepsy	Siamese neural network	CHB-MIT-EEG	Sen = 93.8%
[32]	Epilepsy	Few-shot learning method	CHB-MIT-EEG	AUC-ROC = 85.99%.
			The UPenn and Mayo Clinic's	AUC-ROC = 96.45%
[32]	Epilepsy	Metric learning	CHB-MIT-EEG	Acc = 97.13% Spe = 97.44% Sen = 96.64%
[33]	Epilepsy	Pyramidal one-dimensional (P-1D-CNN) models	Bonn EEG dataset	Acc = 99.1% Spe = 99% Sen = 98%
[34]	Epilepsy	Hybrid transformer model, TL	CHB-MIT EEG	Sen = 91.7%
[35]	Epilepsy	Multilayer perceptron (MLP)	CHB-MIT EEG	Sen = 84.67% Acc = 83.63%
		DCNN + MLP		Sen = 95.41% Acc = 94.10%
		DCAE + Bi-LSTM + CS		Sen = 99.72% Acc = 99.66%
[36]	Epilepsy	Multilayer perceptron (MLP)	Bonn EEG dataset	Acc =100% Spe =100% Sen =100%
		Deep convolutional neural network (DCNN)		Acc =98.6% Spe = 98.2% Sen =99%
[37]	Epilepsy	VAE + latent space size selection + CNN	Bonn EEG dataset	Acc = 99% Spe =99.5% Sen =99% Pre = 99.13% F1 score = 99%
			CHB-MIT EEG	Acc = 99.45% Spe =98.9% Sen =95.2% Pre =98.3% F1 score = 96.7%
[38]	Epilepsy	CNN	MSSM EEG dataset + CHB-MIT EEG	ROC-AUC = 86.6% Sen = 87.8%
[27]	Dementia	CNN + deep reinforcement learning (DRL)		Acc = 90.23% Pre = 92.34%

(continued)

Table 1 (continued)

Ref	Disease	Method	Dataset	Metrics*
		BOOST DRL-XGBOOST	Kaggle-based tiny dataset "Alzheimer's dataset (4 class of Images)"	Sen = 95.45% F1 score = 96.12%
			OASIS-2	Acc = 91.23% Pre = 92.3% Sen = 94.56% F1 score = 95%
			Balanced ADNI dataset "Alzheimers-Disease-5-Class-Dataset-ADNI"	Acc = 93.23% Pre = 94.5% Sen = 96.34% F1 score = 92.12%
[39]	Alzheimer	VoxCNN	ADNI Dataset	Acc = 92.6%
[40]	Alzheimer	AlexNet	OASIS	Acc =92.85%
[41]	Dementia and Alzheimer	Deep learning + hybrid methods	OASIS	Acc =94% Pre = 98% Recall =98% F1 score = 96%,
		AlexNet	Alzheimer's dataset (MRI dataset)	Acc = 92.20% Spe =97.00% Sen =94.50% AUC = 99.45%
		ResNet-50		Acc =93.10% Spe =97.25% Sen =92.25% AUC =98.82%
		Hybrid (AlexNet + ResNet-50)		Acc = 94.10% Spe =97.50% Sen =91.75% AUC =99.31%
[42]	Alzheimer	VoxCNN	Alzheimer's Disease Neuro-imaging Initiative (ADNI)	AUC = 88% Acc = 79%
		ResNet		AUC = 87% Acc = 80%
[43]	Alzheimer	3-D CNN	The Chosun University National Dementia Research Center (South Korea) (Private)	Acc =85.27% Spe =98.21% Sen =67.23%
[44]	Alzheimer	DVE-CNN	GARD (private)	Acc =94.82% Sen =86% F1 score = 91% Pre =97% ROC-AUC = 92.54%

*_Acc_ accuracy, _Sen_ sensitivity, _Pre_ precision, _AUC_ area under curve, _Spe_ specificity, _ROC_ region of interest

The authors in [44] a technique for identifying Alzheimer's disease by analyzing structural MRI (SMRI) data for differences in volume between the left and right hippocampi slice by slice. The suggested technique combines a CNN and DNNs model. A performance with two stages Hough-CNN was able to pinpoint the location of both the left and right hippocampi. 3D patches (80 × 80 × 80 voxels) are extracted from hippocampus locations. Axial, coronal, and sagittal observations separate 2D slices from 3D patches. A discrete volume estimation convolutional neural network (DVE-CNN) model extracts volumetric characteristics from slices using preprocessed 2D patches. Classification network training and testing used volumetric features. The volumetric features that were recovered from both the left and right hippocampi resulted in an average weighted classification accuracy of 94.82% and 94.02%, respectively. Its left and right hippocampal AUC values are 92.54% and 90.62%, respectively. Their technique outperformed others in the same dataset.

The authors in research [45] proposed the prediction of AD using MRI using a deep neural network. Modern image classification networks with transfer learning, such as RESNET, VGG, and others, show encouraging results. By transferring learning, pre-trained versions of these networks perform better. In relation to forecasting different phases of the disease, a RESNET-based design with a lot of layers is proven to perform well. On the Kaggle dataset, the experiments are run.

In [46], the authors presented a Parkinson's disease categorization model using deep neural networks. Their proposed method classifies Parkinson's disease patients faster and more accurately at 94.87%. Parkinsonism can be identified using the patient's dataset. In addition, comparisons were made between naive Bayes, support vector machine (SVM), linear discriminant analysis, K-nearest neighbor (KNN), decision tree (DT), trees of classifications and regressions, random forest (RF), regressions as linear and logistic, and multilayer perceptron.

The work presented by [47] reviews and compares PD detection computational intelligence methods. Classification helps discover PD faster and to improve treatment. The knowledge review shows that various classification techniques have improved performance; however, the challenge is to discover the optimum classifier for PD recognition. Finding the optimal classification algorithm is complicated by the availability of only local datasets. Therefore, this article compares the efficacy and precision of the multilayer perceptron, the support vector machine, and the K-nearest neighbor classifiers on the benchmark (voice) dataset. These classifiers use UCI machine learning repository voice input datasets. The best classifier (95.89%) was ANN using the Levenberg-Marquardt algorithm.

In [48], the authors show that single 1000um regions in sections spanning the frontal brain to the lateral motor region of the vagus nerve can be used for fully automated PD detection. This is accomplished via statistical techniques and image processing, with machine learning showing enhanced accuracy. A deep neural network was used to process digitally preserved stained brain slices to create recolored, or "synthetically stained," images that were processed further before being forwarded to a different network in order to be classified. They offer cutting-edge PD detection (> 90% accuracy on individual 1000um regions), demonstrate the feasibility of doing binary categorization on high-resolution segments in minutes, and mark off regions of interest for the physician to visually verify by hand.

The authors in the article [49] analyze and evaluate the usefulness of existing deep learning-based algorithms to identify neurological disorders using MRI data gathered using several modalities, focus on Alzheimer's, Parkinson's, and schizophrenia using structural and functional MRI. CNN performs better than other approaches in diagnosing neurological illnesses, according to a comparison of the performance of several DL architectures across diverse disorders and imaging modalities. The conclusion identifies a number of ongoing research issues and offers some potential future study possibilities.

Similarly, [50] used machine learning approaches for the early detection of acute neurological illnesses, particularly Alzheimer's disease (AD), which are examined. It is suggested to use a classification algorithm based on machine learning. To obtain the scattered signals, simulations are run using the CST studio suite on accurate numerical brain phantoms. To create the synthetic data needed for ML algorithms, a brand-new method of data augmentation is suggested. To train different ML algorithms, features are extracted using a deep neural network-based autoencoder. The classification outcomes are contrasted with manual feature extraction and raw data. The research demonstrates that the suggested ML-based approach could be utilized to track AD in its early stages.

Study [51] aimed to create a PD gait regression model that can determine the degree of motor dysfunction from provided gait the sequences of image. They avoided the crucial requirements of segmentation and parameter estimation by employing a model-free approach. Additionally, they used linear discriminant analysis (LDA) to boost feature efficiency by optimizing the variances between and within groups. Moreover, categorization was used to evaluate the geographical and temporal information, and then these two new keys were used for linear regression. The trials showed that the results were substantially correlated with the motor examination sections; the accuracy for the training is equal to 0.92, while the accuracy for the testing is 0.85—all these values are analyzed under learning rate of 0.0001. Their technique offered a superior assessment of PD abnormalities when compared to traditional methods. Table 2 presents the previous studies for prediction and classification methods of both Parkinson's and Alzheimer's diseases.

In recent years MRI, which has used such as the identification of brain tumors, has developed into an efficient tool for clinical research. This work, presented by authors on [54], the use of a deep learning classifier in conjunction with multiple kernel-based probabilistic clustering, was recommended as a method for classifying cancers. The three parts of the proposed method are the following: a segmentation module, a feature extraction module, and a classification module. Before the median filter is used in the denoising process, the MRI image is preprocessed to make it appropriate for segmentation. Subsequently, features are retrieved according to the shape, texture, and extent of each individual segment. After features are extracted, we will use linear discriminant analysis (LDA) to select the most important ones for classification. Next, the deep learning classifier is utilized to sort the samples into

Table 2 Summary of prediction and classification methods of Parkinson's and Alzheimer's diseases in the previous studies

Ref	Disease	Method	Dataset	Metrics
[45]	Alzheimer's disease	VGG16, VGG19, ResNet18, ResNet34, ResNet50, ResNet101	Alzheimer's Dataset (4 class of Images) [52]	Acc = 99.51% Sen = 0.98% Pre = 0.99% F1 score = 98%
[46]	Parkinson's disease	PDCD	Kaggle [NAN]	Acc =94% Sen = 92% Pre = 93% F1 score = 92%
[47]	Parkinson's disease	Artificial neural network (ANN)	Parkinson's dataset [53]	Acc = 95.89% Sen = 93.75% Spe = 96.59%
[48]	Parkinson's disease	PDNet (a variation of the EfficientNet-B7 model)	A private dataset of digitized images provided by the Parkinson's UK Brain Bank	Acc = 90% Sen = 93% Spe = 88% Pre = 88% F1 score = 91%

tumor and normal categories. Measures of accuracy, specificity, and sensitivity are employed to evaluate the proposed strategy. The findings of the suggested strategy are also contrasted with those of an existing method that makes use of a feed-forward backpropagation network (FFBN). The proposed method produced average values for sensitivity is 0.88, for specificity is 0.80, and for accuracy 0.83, with the highest values hovering around 1, 0.85, and 0.94, respectively. The results that are better demonstrate how effective the suggested method is.

According to [55], various methods and procedures for applying deep learning to the creation of a model for brain tumor detection will be examined. The objective is to discover a reliable and effective method for MRI-based brain tumor detection to assist brain doctors in quickly, accurately, and conveniently reaching choices. According to a World Health Organization research from February 2018, the greatest death rate from brain cancer or other diseases of the central nervous system (CNS). Early cancer detection is crucial in order to save many of these lives. The model has been developed and put into use, together with a dataset of 10,000 images for the purpose of detecting brain tumors using deep learning methods based on neural networks. These models, Inception, VGG16, MobileNet, and ResNet, have all been tested. These are the F score accuracy results they obtained for each model: InceptionV3 scored 99.88%, VGG16 scored 99.86%, ResNet50 scored 98.14%, and their model scored 98.28.

The authors in [56] discussed how to create a successful business plan. Entropy is employed to optimize these characteristics for precise and quick categorization, and a fused vector is then given to the classifiers. Medical image computing and computer-assisted intervention (MICCAI) challenge models frequently incorporate multimodal brain tumor segmentation (BRATS) databases from 2015, 2016, and 2017. On BRATS 2015, the dice similarity coefficient (DSC) was 0.99, on BRATS 2015 it was 1.00, and on BRATS 2017 it was also 0.99. The work is presented by the authors in [57] utilized brain MRI images to create a CNN to identify a tumor. CNN was the first to use images. The Softmax fully connected layer achieved a classification accuracy of 98.67%. Also, the decision tree (DT) classifier's accuracy is 94.24%, while the radial basis function (RBF) classifier's accuracy is 97.34%. In addition to the accuracy criterion, they used the sensitivity, specificity, and precision standards to evaluate the effectiveness of the network. Data from the classifiers showed that the CNN network accuracy on testing images was highest with the Softmax classifier. This is a new strategy for tumor detection from brain imaging that combines feature extraction methods with CNN. The way proposed an accuracy of 99.12% on the testing data. The doctors' accuracy in diagnosing the patients' tumors and treating them rose as a result of the importance of the diagnosis rendered by the physician.

To aid physicians, a sophisticated method for identifying and categorizing brain tumors is required as illustrated in [58]. The study's innovative aspect is its use of a hierarchical deep learning technique to categorize brain tumors into glioma, meningioma, and pituitary. For a speedy and effective cure, diagnosis and tumor classification are crucial, and medical image processing utilizing a CNN is producing remarkable results in this area. CNN utilizes data training techniques to classify different types of tumors by analyzing image fragments. A novel approach is presented for the classification of brain tumors using a hierarchical deep learning framework, specifically employing convolutional neural networks (CNNs). This method aims to accurately identify and classify different types of brain cancers. Glioma, meningioma, pituitary, and no-tumor are the four categories of tumors that the suggested system divides into. The model under consideration exhibits superior performance compared to current methodologies in the realm of brain tumor identification and segmentation, with a precision rate of 92.13% and a miss rate of 7.87%. In the future, the planned system will be used to provide medical help.

In [59], a brain tumors segmenting based on deep learning technique for from several MRI modalities is given. The suggested hybrid CNN architecture employs a patch-based method that predicts output labels while taking both local and contextual information into consideration. The suggested network uses a dropout regularizer together with batch normalization to address the over-fitting issue, although a two-phase training method is used to address the data imbalance issue. The suggested approach contains a step of preprocessing that normalizes and modifies preference in the images, a feed-forward run via a CNN, and a post-processing step that eliminates minor false positives near the mind section. The BRATS 2013 dataset is used to test the suggested method, achieving scores of 0.86 for the dice score, 0.86 for the sensitivity, 0.91 for the specificity, and for the entire tumor section, which is an enhancement above the results obtained using recent methods.

Similarly, the authors in [60] presented a brain tumor detection model to design and implement the images using a machine learning approach. The study looks at a list of risk factors being monitored by brain tumor surveillance systems. For the identification, classification, and brain tumors segmentation, the suggested technique guarantees to be extremely effective and exact. Automatic or semiautomatic approaches are required to accomplish this accuracy. The research recommends using convolutional neural networks (CNNs) to generate small 3x3 kernels for automatic segmentation. This single method accomplishes both segmentation and classification. CNN one kind of machine learning relies on the classification abilities of neural networks, which are themselves constructed from layers. The recommended procedures include many steps, such as gathering data, cleaning it up, applying an average filter, segmenting it, extracting features, and feeding them into a convolutional neural network for classification and identification. Using data mining (DM) methods, meaningful connections and patterns in the data can be extracted. For the early diagnosis and avoidance of brain tumors, data mining and machine learning approaches are being used successfully.

In [61], the authors proposed a brain tumor classification and detection system based on deep learning. The proposed method has three phases: preprocessing, tumor detection using an extreme learning machine with local receptive fields, and tumor area extraction utilizing image processing. In the beginning, potential sounds were eliminated by utilizing nonlocal ways in addition to local smoothing strategies. The second stage involved classifying cranial MRI images as benign or malignant using ELM-ERF. The cancers were divided into sections in the third stage. In order to reduce the doctor's time, the study solely used cranial MR images, which had a mass. The categorization accurateness of cranial MRI images in experimental trials is about 97.18%. The evaluation results revealed that the proposed strategy performed more effectively than other recent research in the literature. Also, experimental outcomes demonstrated the viability of the suggested approach for computer-aided brain tumor detection. Table 3 presents the previous studies for prediction and classification methods of brain tumor diseases.

4 Datasets

4.1 Open-Access Datasets

The University of Bonn EEG dataset This dataset is made accessible to the public by University of Bonn. It contains of five categories labeled A, B, C, D, and E. Each class comprises 100 segments of single-channel EEG with a total of 4097 samples per channel and a length of 23.6 seconds. Each dataset comes in the form of a zip file containing 100 TXT files (ASCII code). The data was collected using 128-channel electrodes and an average reference. 12-bit analog-to-digital converters with a sampling frequency of 173.61 Hz were employed for data gathering. A band-pass filter with a frequency range of 0.53–40 Hz was used to convert each segment into a time series of 4097 values. As a result, EEG data is five categories X 100 explanations/class X 4097 (23.6 seconds/remark) (see Fig. 5). Five normal subjects with

Table 3 Summary of prediction and classification methods of brain tumor diseases in the previous studies

Ref	Method	Dataset	Metrics %					
			Acc	**Sen**	**Spe**	**Pre**	**Jacc**	**DSC**
[54]	FFBN	Real-time MRI images	83	88	80	–	–	–
[55]	VGG16	Kaggle	99.86	–	–	–	–	–
	ResNet50		98.14	–	–	–	–	–
	MobileNet		88.98	–	–	–	–	–
	InceptionV3		99.88	–	–	–	–	–
	Proposed model		98.28	–	–	–	–	–
[56]	HOG+ LBP	BRATS2015 (original)	99.82	99.59	100	–	99.59	99.79
		BRATS2015 (high/low slices)	90.14	89.80	90.58	–	8381	91.19
	Deep features	BRATS2016 (original)	99.82	99.59	100	–	99.59	99.79
		BRATS2016 (high/low slices)	91.74	95.08	87.33	–	86.76	92.91
	Fusion of HOG +LBP + deep features	BRATS2017 (original)	99.82	100	99.59	–	99.67	99.84
		BRATS2017 (high/low slices)	99.25	99.05	99.46	–	98.59	99.29
[57]	CNN + Softmax	Brain MRI images	98.67	100	94.64	98.26	–	–
	CNN + RBP		97.34	100	89.28	96.59	–	–
	CNN +DT		94.24	100	85.71	95.37	–	–
	Proposed model		99.12	100	96.42	98.83	–	–
[58]	HDL2B-tumor classifier	Kaggle	–	–	–	–	92.13	–
[59]	Two-path: complete	BRATS 2013	–	81	92	–	–	83
	Two-path: core		–	85	90	–	–	85
	Two-path: Enhancing		–	84	92	–	–	87
	Three-path: complete		–	82	91	–	–	81
	Three-path: core		–	86	91	–	–	87
	Three-path: Enhancing		–	85	90	–	–	84
	Hybrid: complete		–	86	91	–	–	86
	Hybrid: core		–	87	93	–	–	86
	Hybrid: enhancing		–	90	94	–	–	88
[60]	CNN	The UCL dataset	91	–	–	–	–	–
[61]	ELM-LRF	DICOM	97.18	96.8	97.12	–	–	–

(continued)

Table 3 (continued)

Ref	Method	Dataset	Metrics %					
			Acc	**Sen**	**Spe**	**Pre**	**Jacc**	**DSC**
	CNN		96.45	96.23	95.92	–	–	–
	Statistical features		96.24	96.15	96.45	–	–	–
	GWF		93.65	92.88	93.15	–	–	–

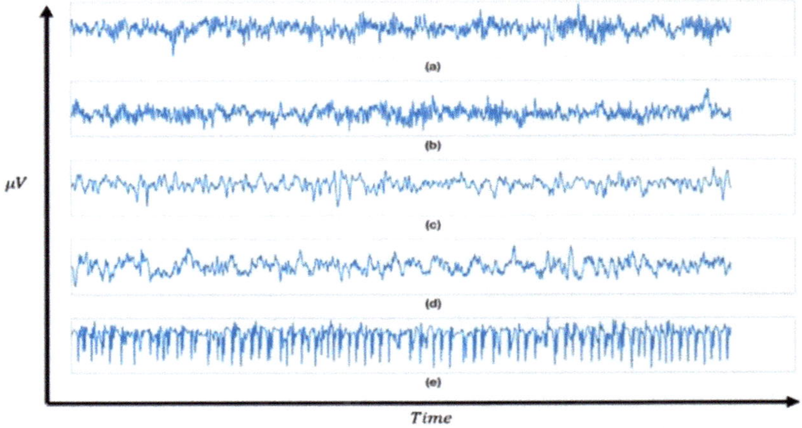

Fig. 5 Samples of Bon EEG dataset

their eyes open made up sets A, and five healthy subjects with their eyes closed made up sets B. Both sets contain typical signals. Five patients that have epilepsy made up sets C, D, and E. Sets C and D are taken in the interictal state, which is when there haven't been any seizures, while set E is taken in the ictal state when there have been seizures [62].

CHB-MIT-EEG The Boston Children's Hospital (CHB-MIT dataset) contains recordings of EEG from pediatric patients suffering from intractable seizures. Subjects were monitored for several days after discontinuing anti-seizure medication to characterize their seizures and determine whether they qualified for surgical treatment. The dataset has a total of 24 instances, which are comprised of 23 patients each (one patient has two recordings, spaced out over a period of 1.5 years). The collection is made up of 969 hours of head EEG recordings and includes 173 seizures. The dataset contains a variety of seizures (clonic, atonic, tonic). The dataset includes a variety of patients between 10 and 22 years old from male and female and various kinds of seizures make them suitable for evaluating the efficacy of automatic seizure detection techniques in actual circumstances [29, 63, 64] (see Fig. 6).

Alzheimer's Dataset (4 class of Images) [52] The competition on Kaggle led to the collection and distribution of the 6400 magnetic resonance pictures that make up

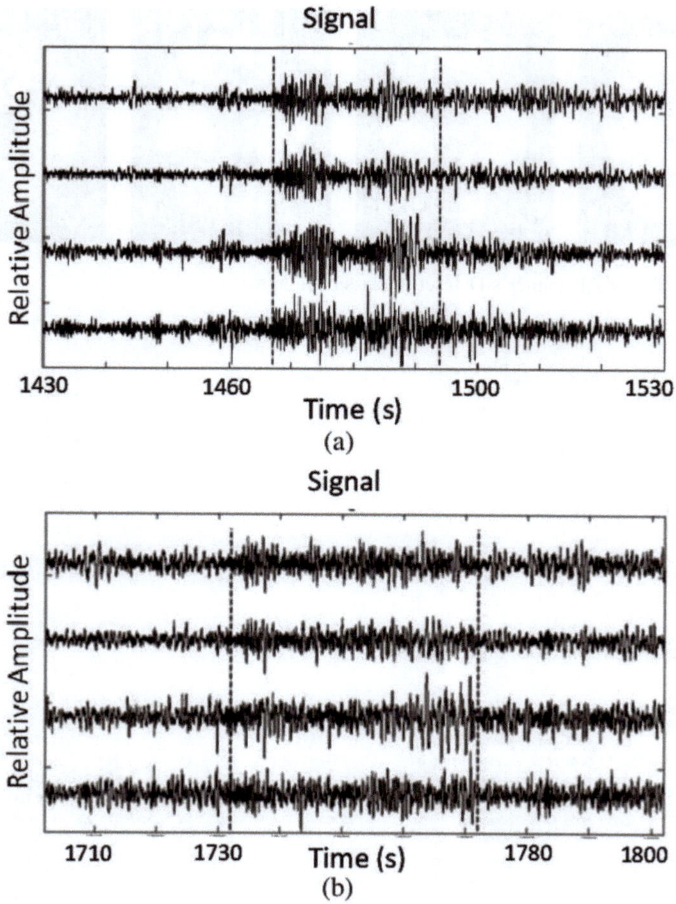

Fig. 6 Samples of CHB-MIT-EEG dataset

the dataset resulted in the dissemination of these images. The whole dataset is separated into two distinct train and test sets, which do not overlap with one another. In comparison, the test set only features 1279 pictures, whereas the training set has 5121 of them. Based on the extent of neurological degeneration, MR images are classified as nondemented (ND), moderate demented (MOD), mild demented (MD), and very mild demented (VMD) [52] (see Fig. 7).

OASIS OASIS stands for the Open Access Series of Imaging Studies. This project intends to make the datasets used for brain neuroimaging open to the scientific community for free and enable future basic and clinical neuroscience discoveries in basic and clinical neuroscience by gathering and openly releasing neuroimaging datasets. At the present time, five datasets are available in OASIS. The first collection of data is referred to as OASIS-1: Cross-Sectional MRI Data in Early Age,

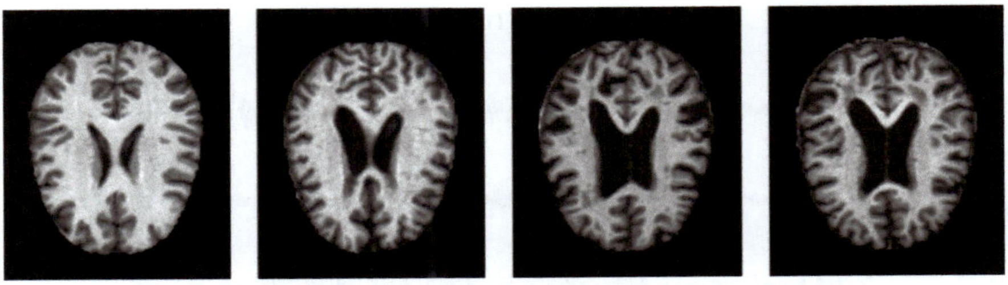

Fig. 7 Samples of Alzheimer's Dataset (4 class of Images)

1	ID	M/F	Hand	Age	Educ	SES	MMSE	CDR	eTIV	nWBV	ASF	Delay
2	OAS1_000	F	R	74	2	3	29	0	1344	0.743	1.306	N/A
3	OAS1_000	F	R	55	4	1	29	0	1147	0.81	1.531	N/A
4	OAS1_000	F	R	73	4	3	27	0.5	1454	0.708	1.207	N/A
5	OAS1_000	M	R	28					1588	0.803	1.105	N/A
6	OAS1_000	M	R	18					1737	0.848	1.01	N/A
7	OAS1_000	F	R	24					1131	0.862	1.551	N/A
8	OAS1_000	M	R	21					1516	0.83	1.157	N/A
9	OAS1_000	F	R	20					1505	0.843	1.166	N/A
10	OAS1_001	M	R	74	5	2	30	0	1636	0.689	1.073	N/A

Fig. 8 A sample from the OASIS-1 dataset

1	Subject ID	MRI ID	Group	Visit	MR Delay	M/F	Hand	Age	EDUC	SES	MMSE	CDR	eTIV	nWBV	ASF
2	OAS2_0001	OAS2_0001_MR1	Nondemented	1	0	M	R	87	14	2	27	0	1987	0.696	0.883
3	OAS2_0001	OAS2_0001_MR2	Nondemented	2	457	M	R	88	14	2	30	0	2004	0.681	0.876
4	OAS2_0002	OAS2_0002_MR1	Demented	1	0	M	R	75	12		23	0.5	1678	0.736	1.046
5	OAS2_0002	OAS2_0002_MR2	Demented	2	560	M	R	76	12		28	0.5	1738	0.713	1.010
6	OAS2_0002	OAS2_0002_MR3	Demented	3	1895	M	R	80	12		22	0.5	1698	0.701	1.034
7	OAS2_0004	OAS2_0004_MR1	Nondemented	1	0	F	R	88	18	3	28	0	1215	0.710	1.444
8	OAS2_0004	OAS2_0004_MR2	Nondemented	2	538	F	R	90	18	3	27	0	1200	0.718	1.462
9	OAS2_0005	OAS2_0005_MR1	Nondemented	1	0	M	R	80	12	4	28	0	1689	0.712	1.039
10	OAS2_0005	OAS2_0005_MR2	Nondemented	2	1010	M	R	83	12	4	29	0.5	1701	0.711	1.032

Fig. 9 A sample from the OASIS-2 dataset

Middle-Aged, Nondemented, and Demented Elder Adults. This dataset consists of 416 right-handed patients ranging in age from 18 to 96 years old and being of both sexes. The second dataset is OASIS-2: Longitudinal MRI Data in Nondemented and Demented Older Adults which contains a longitudinal compilation of 150 right-handed subjects aged 60 to 96 (see Figs. 8 and 9). The third one is OASIS-3: Clinical, Longitudinal Multimodal Neuroimaging, and Cognitive Dataset for standard ages and Alzheimer's disease which comprises a retrospective collection of data for 1378 participants that range in age between 42 and 95 years. The next dataset is OASIS-3 TAU: OASIS-3 Flortaucipir F18 (AV1451) PET; this group consists of OASIS-3 participants who have also participated in TAU (AV1451) PET imaging. Lastly, OASIS-4: Clinical Cohort dataset that includes 663 participants spanning ages from 21 to 94. This clinical cohort was assessed for memory problems

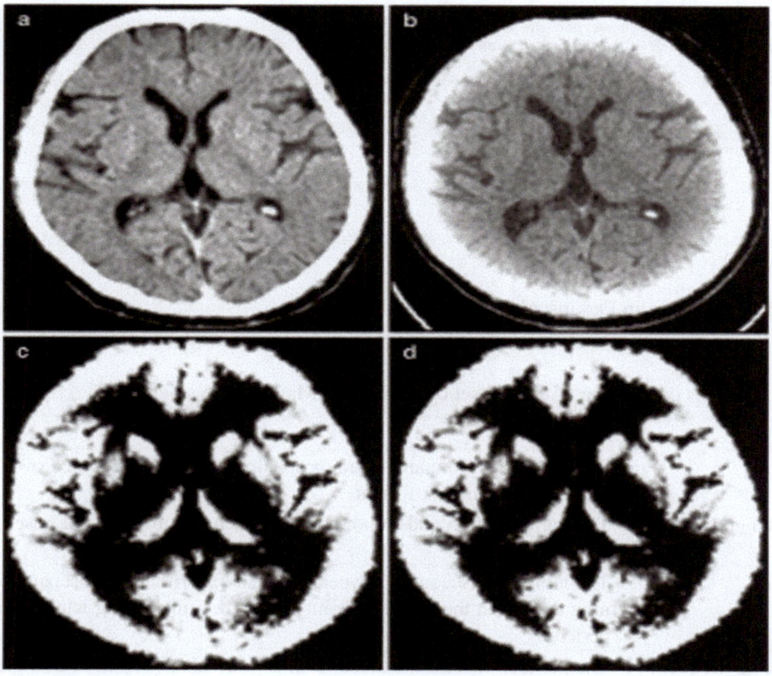

Fig. 10 Samples of ADNI dataset

and dementia through clinical, csf, neuropsychometric, and neuroimaging methods [65].

ADNI The Alzheimer's Disease Neuroimaging Initiative (ADNI) is a multicenter long-term study that aims to create clinical, imaging, genetic, and biochemical indicators for the early diagnosis and monitoring of AD. ADNI consists of ADNI-1 ADNI-GO (200 new patients with young mild cognitive impairment (EMCI)), which included 400 people with MCI, 200 participants with early AD, and 200 old control subjects, in addition to the current ADNI-1 group, ADNI-2 (on top of the ADNI-1 and ADNI-GO tests, they will add 150 older controls, 100 earlier MCI, 150 later MCI, and 150 AD), and ADNI-3 (ADNI-1, ADNI-GO, and ADNI-2 in addition to 133 elderly controls, 151 MCI, and 87 AD) [66] (see Fig. 10).

Parkinson's dataset The voices of 31 people, 23 of whom have Parkinson's disease, are included in this dataset. Voice measurements are shown in each column of the table, and each row leads to one of 195 samples of these patients' voices. The "status" entry, which has a value of 0 for normal health and 1 for PD, is used to tell the difference between health and PD people. The data are all presented in an ASCII CSV format. Each individual voice recording is represented by a single instance

```
name,MDVP:Fo(Hz),MDVP:Fhi(Hz),MDVP:Flo(Hz),MDVP:Jitter(%),MDVP:Jitter(Abs),MDVP:RAP,MDVP:PPQ,Jitter:DDP,MDVP:Shimmer,MDVP:Shimmer(dB),Shimmer:APQ3,Shimmer:A
PQ5,MDVP:APQ,Shimmer:DDA,NHR,HNR,status,RPDE,DFA,spread1,spread2,D2,PPE
phon_R01_S01_
1,119.99200,157.30200,74.99700,0.00784,0.00007,0.00370,0.00554,0.01109,0.04374,0.42600,0.02182,0.03130,0.02971,0.06545,0.02211,21.03300,1,0.414783,0.815285,
-4.813031,0.266482,2.301442,0.284654
phon_R01_S01_
2,122.40000,148.65000,113.81900,0.00968,0.00008,0.00465,0.00696,0.01394,0.06134,0.62600,0.03134,0.04518,0.04368,0.09403,0.01929,19.08500,1,0.458359,0.819521
,-4.075192,0.335590,2.486855,0.368674
phon_R01_S01_
3,116.68200,131.11100,111.55500,0.01050,0.00009,0.00544,0.00781,0.01633,0.05233,0.48200,0.02757,0.03858,0.03590,0.08270,0.01309,20.65100,1,0.429895,0.825288
,-4.443179,0.311173,2.342259,0.332634
phon_R01_S01_
4,116.67600,137.87100,111.36600,0.00997,0.00009,0.00502,0.00698,0.01505,0.05492,0.51700,0.02924,0.04005,0.03772,0.08771,0.01353,20.64400,1,0.434969,0.819235
,-4.117501,0.334147,2.405554,0.368975
```


Fig. 11 Samples from Parkinson's dataset

Table 4 Fields description of the dataset of UPenn and Mayo Clinic's Seizure

Field	Description
Data	Sample values from an electroencephalogram (EEG) recorded over time, presented in a matrix format
Data length sec	The number of seconds represented by each data row (in this case, 1 second)
Latency	The number of seconds between when an expert noticed a seizure beginning and when the first data point appeared. (only during ictal training periods)
Sampling_frequency	The number of readings taken from an individual's brain in a single second of EEG data. (Non-integer values are an average across the entire dataset; this may indicate that some EEG samples are missing)
Channels	Matching electrode names to rows in the data field

inside the rows of the Excel file. Approximately six recordings were recorded per patient, with the patient's name stated in the first column [53] (see Fig. 11).

UPenn and Mayo Clinic's Seizure Detection Challenge Dataset [67] The data pertaining to each human and canine subject is organized in folders that contain data pertaining to both training and testing. The training data is split into 1-second EEG clips designated "Ictal" for seizures and "Interictal" for non-seizure data segments. Training data are ordered consecutively, whereas testing data are ordered randomly. Interictal data sections cover the average seizure duration across all subjects, while ictal data from testing and training segments cover the entire episode. With the restriction that no interictal section should be less than an hour prior to or following a seizure, the beginning locations for the interictal portions of the data were chosen at random from the entire dataset. In the interior folders, data segments are kept in Matlab.mat files, which are organized in the following fields in Table 4.

Human data were collected from individuals with temporal and extratemporal lobe epilepsy who were being evaluated for the epilepsy's surgery. The intracranial electroencephalography (iEEG) records are made using depth electrodes positioned along the hippocampus's anterior-posterior axis, as well as grids of subdural

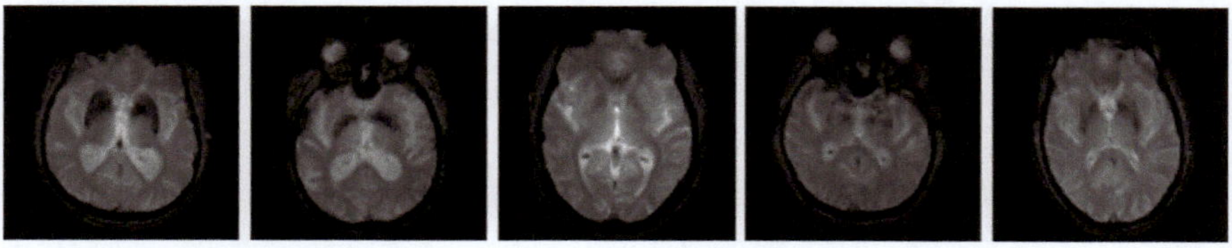

Fig. 12 Samples from Alzheimer's-Disease-5-Class-Dataset-ADNI

electrodes put in various places. The sampling rates of data range from 500 Hz to 5000 Hz.

The canine data is generated by an implanted device that collects information from 16 subdural electrodes. Two four-contact sheets are superimposed over each hemisphere in an anteroposterior configuration. 400 Hz sample frequency allows for continuous data collection and comparison to the mean of the group [67].

Alzheimers-Disease-5-Class-Dataset-ADNI This is an MRI dataset from the ADNI repository representing five phases of Alzheimer's disease. This dataset was gathered from the ADNI website, and all credit goes to the same ADNI; this is a collection of customers to help accelerate Alzheimer's research. The dataset contains 1296 MRI files of five phases of Alzheimer's disease that have previously been divided into two directories for training and testing, which are for training: 204 early mild cognitive impairment (EMCI)/61 late mild cognitive impairment (LMCI)/198 mild cognitive impairment (MCI)/145 AD/439 control (CN), and for testing: 36 EMCI/11 LMCI/35 MCI/26 AD/87 CN images [68] (see Fig. 12).

Parkinson's disease spiral drawings using digitized graphics tablet dataset [69] The control and PD handwriting dataset includes 62 subjects with Parkinson's disease and 15 control subjects who were attracted to the Neurology Department at Istanbul University's Cerrahpasa Medicine Faculty. The dataset contains images of spiral drawings made by persons with Parkinson's disease. As a result, this dataset may be utilized for regression as well (see Fig. 13). Each person had their handwriting recorded in three different ways: static spiral test (SST), the DST, or dynamic spiral test, and STCP (stability test at a constant point). With SST, the patient is given a graphics tablet with three twisted Archimedean spirals on it, and they are encouraged to trace the same loop on as many occasions as they can.

In DST, in contrast to SST, the Archimedean loop appears and disappears (like blinking) at regular intervals. The patient must maintain drawing the pattern in their head. Due to the increased difficulty of repeating the Archimedean spiral under these conditions, this test aims to determine whether or not the patient's drawing speed and pause times have altered. As a consequence of this test, the patients' majority continued to draw, but virtually all of them could not keep the pattern.

Participants in STCP are tasked with not touching the screen while keeping their digital pen on a red dot displayed in the middle of the screen for a set period of time [69].

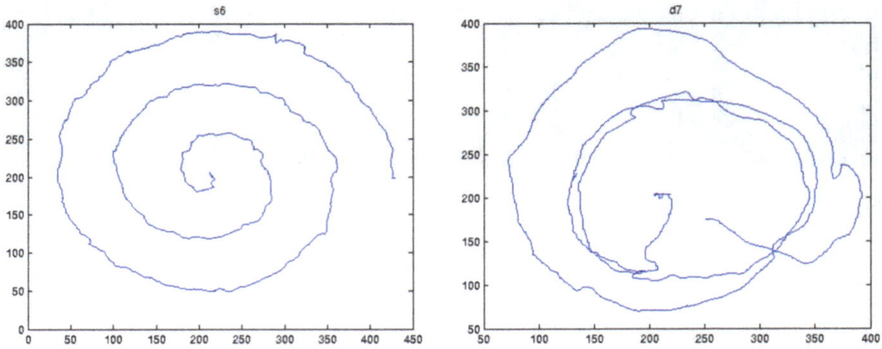

(a) A sample from SST (b) A sample from DST

Fig. 13 Samples of Parkinson's Disease Spiral Drawings Using Digitized Graphics Tablet Dataset

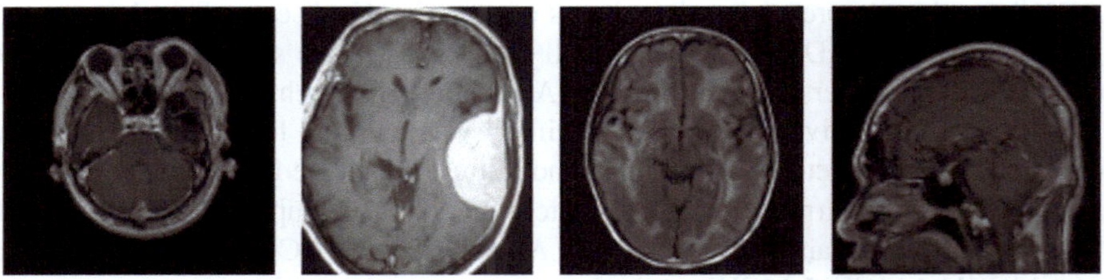

Fig. 14 Samples from Brain Tumor MRI Dataset

Brain Tumor MRI Dataset Three separate datasets—figshare, the SARTAJ dataset, and Br35H—have been combined to form the Brain Tumor MRI Dataset. The 7023 human brain MRI scans in this collection have been labeled as glioma, meningioma, no tumor, or pituitary. The Br35H dataset did not contain any photos of tumor classes (see Fig. 14) [70].

The NTUA Parkinson Dataset This dataset is made up of MRI and DaT scans of a total of 78 people, 55 with Parkinson's disease and 23 healthy controls. Approximately 42,000 images are accessible for academic usage [71] (see Fig. 15).

BraTS 2013 Dataset Segmentation of synthetic and actual pictures of brain tumors, further subdivided into high-grade gliomas (HGG) and low-grade gliomas (LGG), make up this dataset. Twenty-five patients have both fake HGG and LGG scans, whereas ten others have true HGG scans, and ten more have real LGG scans (see Fig. 16) [72].

BraTS 2015 Dataset This dataset is made up of brain tumor segmentation totaling 220 HGG and 54 LGG MRI scans. T1, T1c, T2, and T2FLAIR are the four MRI

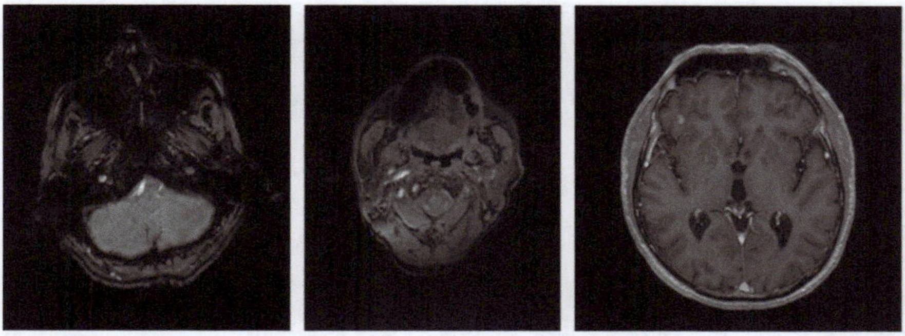

Fig. 15 Samples from NTUA Parkinson Dataset

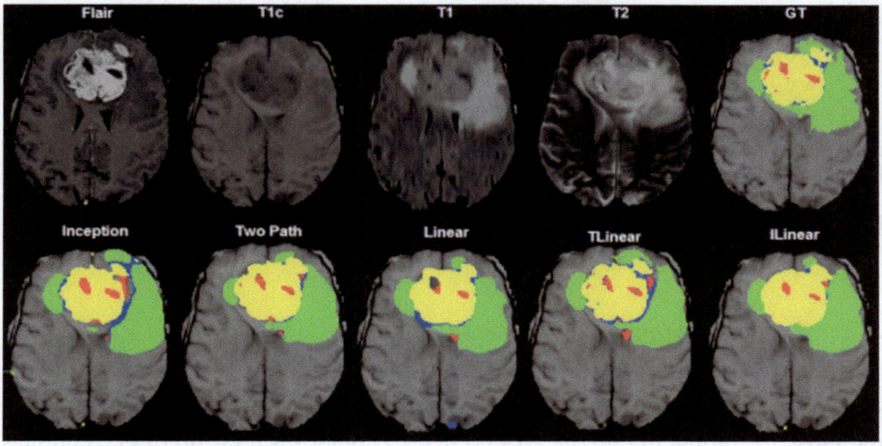

Fig. 16 Samples of BraTS 2013 Dataset

techniques available. Edema, improving tumor, non-improving tumor, and necrosis are the four intra-tumoral classes that can be gleaned from the data that segmented "ground truth" gives [73, 74] (see Fig. 17).

BraTS 2017 Dataset This dataset is made up of brain tumor segmentation includes 285 cases of brain tumors imaged with one of four MRI modalities (T1, T1ce, T2, and Flair). In addition to containing markers for ED, ET, and NET/NCR, the collection also includes full coverings for brain tumors. The effectiveness of segmentation is measured in three ways: classification by WT, TC, and ET [75] (see Fig. 18).

Recent BraTS Datasets Recently there have been new versions of BraTS dataset, BraTS 2020 [76] and Brats 2021 [77]. They are made up of brain tumor segmentation [74, 78] (see Fig. 19).

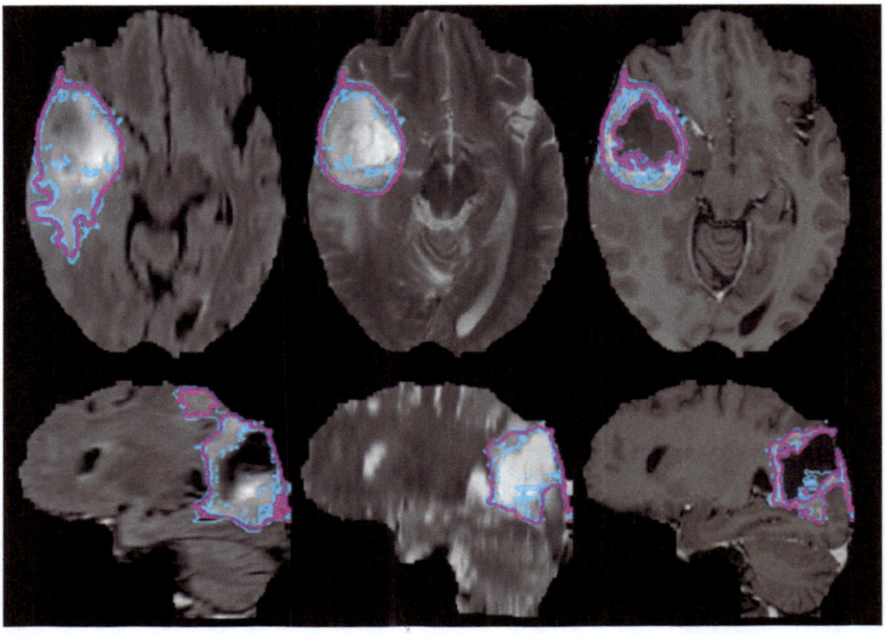

Fig. 17 Samples of BraTS 2015 Dataset

4.2 *Private Datasets*

The EEG dataset collected by Mount Sinai Hospital for epilepsy patients According to [38], the MSSM dataset comprises 86 scalp EEG recordings from 28 epileptic individuals. The records were produced using XLTEK equipment with 22 data inputs and a sample frequency of 256 Hz for continuous EEG research. The ongoing EEG investigation lasted between 2 and 8 days. All patient data were anonymized, and the appropriate EEG activity was converted to EDF files. All of the seizures were concentrated, with different areas of the brain serving as the epileptic focus. All EEG data was selected, analyzed, and anonymized by two trained electroencephalographers. We excluded EEGs where electrode artifacts impacted more than one electrode. However, little effort was made to filter out artifacts associated with sleep and wake states, eye blinks, body movement, or chewing. Without any personally identifiable data, each patient was issued a number varying from 1 to 28. Sixty-one of the 86 recorders included information about the commencement of a seizure, while the others just included information about periods between seizures. In any of the recordings, the electroencephalographers were unable to detect a preictal signal on the raw EEG. Interictal recordings were usually 60 minutes long. 75 minutes of EEG data were collected before a seizure. Seizures collected include both preclinical and symptomatic focal seizures. There were no

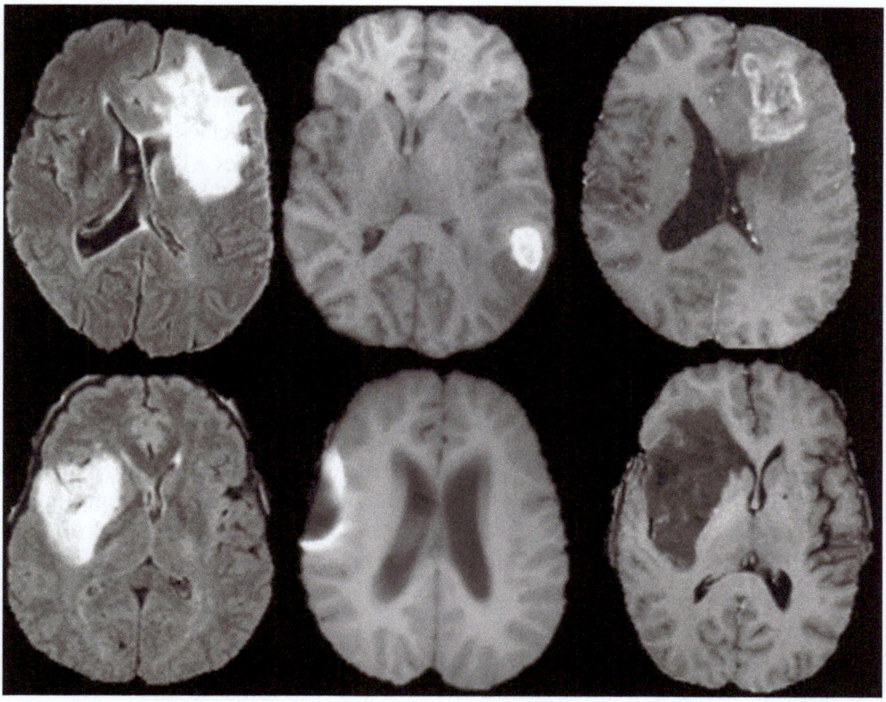

Fig. 18 Samples of BraTS 2017 Dataset

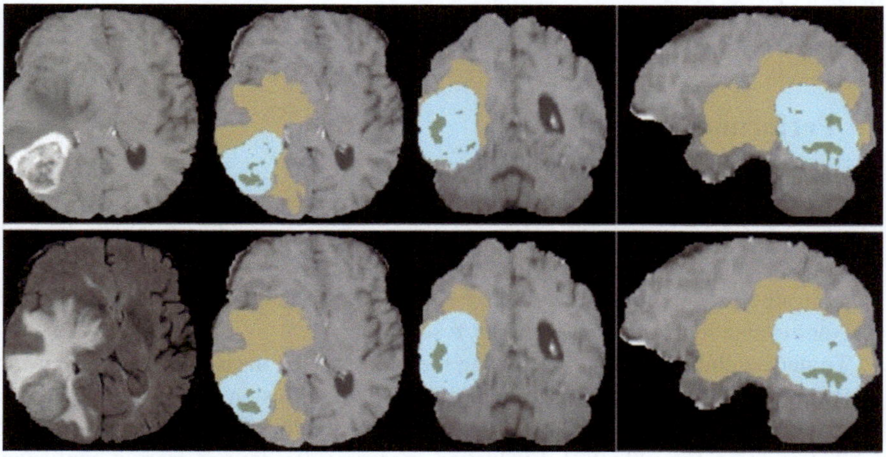

Fig. 19 Samples of recent BraTS datasets

seizures with generalized onsets utilized. During the recording, each patient in this investigation experienced more than one seizure.

The GARD dataset: stands for Gwangju Alzheimer's and Related Dementia The GARD is a Chosun University-owned proprietary dataset that was gathered in Chosun University hospitals. The dataset includes 81 AD participants (39 males, 42 females), 171 health control subjects (83 males, 88 females), 39 mAD (mild cognitive impairment due to AD) patients (25 males, 14 females), and 35 aAD (asymptomatic AD) patients (15 males, 20 females). This dataset was utilized to back up [79]'s conclusions. According to [79], the dataset includes 326 MRI images divided into four categories: 171 patients as normal controls (CNS), 81 Alzheimer's disease dementia (ADD) patients, 35 patients with asymptotic Alzheimer's disease (ADD), and 39 patients with mild Alzheimer's disease (MAD). The average age of the subjects was 70.018 ± 6.074 years old.

The dataset from the University of Chosun for National Dementia Research Center The study presented in [43] had 331 individuals in total. Participants ranged in age from 64 to 85 years old and had a spouse or caretaker who could offer an independent assessment of their cognitive performance. Patients in this dataset were classified into two categories: (1) control normal subjects (CNS) had MMSE scores between 25 and 30, a clinical dementia rating (CDR) of 0, and were neither depressed nor suffering from MCI or dementia. (2) These people had MMSE totals between 14 and 24, a CDR of 0.5 or 1.0, and met the National Institute of Neurological and Communicative Disorders and Stroke (NINCDS) standards for probable Alzheimer's disease. This dataset was collected from Chosun University National Dementia Research Center. An overview of prior research's datasets is illustrated in Table 5.

5 Performance Metrics

Performance metrics are used to assess the effectiveness or quality of the model. These performance measures aid in our comprehension of how well the model handles the supplied data. They also assist measure how well a DL model generalizes on new or previously unexplored data. Table 6 summarizes and defines the relevant evaluation metrics used in our research.

To understand the necessity of using the performance measurements chosen for this study, it is important to take into account the confusion matrix notion. For a prediction task, a confusion matrix summarizes examples of accurate and incorrect predictions (see Fig. 20) [82]. Evaluation measures are derived from a confusion matrix which contains the number of cases classified as positive and negative for each class displayed in the confusion matrix [83].

- The percentage of times a classifier properly predicts the positive class is called the true positive (TP) rate.

Table 5 Overview of datasets in the previous studies

Dataset	Ref	Disease types	Description	Studies
CHB-MIT-EEG	[63]	Epilepsy	24 cases, from 23 patients	[30–32, 34, 35, 37, 38, 59]
The EEG dataset collected by Bonn University	[62]	Epilepsy	5 sets (named from A to E), 100 instances in each set	[33, 36, 37]
UPenn and Mayo Clinic's Seizure Dataset	[67]	Epilepsy	12 cases, 4 dogs and 8 patients (humans)	[31]
The EEG dataset collected by Mount Sinai Hospital for epilepsy patients	[38]	Epilepsy	86 scalps of EEG (sEEG) recordings collected by 28 epilepsy patients	[38]
Alzheimer's dataset	[52]	Alzheimer's disease	6400 MRIs	[27, 41, 45]
OASIS-2	[65]	Alzheimer's disease	150 right-handed themes with age from 60 to 96. With 373 total images	[27, 41]
Alzheimers-Disease-5-Class-Dataset-ADNI	[68]	Alzheimer's disease	1296 MRI files	[27]
ADNI	[80]	Alzheimer's disease	231 images: 4 classes (AD, LMCI, EMCI, NC)	[42]
The Dataset from the University of Chosun for National Dementia Research Center	[NAN]	Alzheimer's disease	331 participants	[43]
The Gwangju Alzheimer's Disease and Related Dementia (GARD)	[NAN]	Alzheimer's disease	326 MRI scans	[44]
Parkinson Disease prediction	[NAN]	Parkinson's disease	Voice records from 42 patients	[46]
Parkinson's dataset	[53]	Parkinson's disease	195 voice records from 31 patients	[47]
Dataset of digitized images from Parkinson's UK Brain Bank	[NAN]	Parkinson's disease	401 participants, 301 of them were classified as having PD. 4154 images (3185 from PD patients, 969 from healthy participants)	[48]
A dataset from the department of neurosurgery from Taiwan General Hospital	[NAN]	Parkinson's disease	12 subjects with PD before and after drug treatment and 12 healthy people	[51]
BRATS 2013	[72]	Brain tumor	25 patients, each one has high-grade gliomas (HGG) and low-grade gliomas (LGG) synthetic images. In addition to other 20 patients, each has images of real HG. Other	[59]

(continued)

Table 5 (continued)

Dataset	Ref	Disease types	Description	Studies
			10 patients, each has LG real images	
BRATS 2016	[81]	Brain tumor	It includes the BRATS 2015 training images that contains 220 images of HHG and 54 images of LGG. In addition, 191 unknown cases for testing	[56]
BRATS 2015	[73]	Brain tumor	It contains 220 MRIs image that HGG and 54 MRIs images that LGG	[56]
BRATS 2017	[75]	Brain tumor	It comprises 285 MRI scans for brain tumor, each scan consists of four different MRI modalities which are T1, T1ce, T2, and Flair.	[56]
Brain Tumor MRI Dataset	[70]	Brain tumor	It contains 7023 MRI images of human brain. This dataset is categorized to four categories: glioma, meningioma, no tumor, and pituitary	[54, 57]

Table 6 Evaluation metrics used in the previous studies

Metric	Equation	Definition
Accuracy	$\frac{TP+TN}{TP+TN+FP+FN}$	Proportion of accurate predictions over the whole number of classes evaluated is represented by the accuracy metric
Specificity	$\frac{TN}{TN+FP}$	Determine the percentage of negatively predicted classes that are correctly predicted
Recall (sensitivity)	$\frac{TP}{TP+FN}$	Used to calculate the fraction of the positive class that is correctly predicted
Precision	$\frac{TP}{TP+FP}$	Determine which of the total anticipated class in a positive class is the accurately classified positive classes
DSC	$\frac{2TP}{2TP+FP+FN}$	The dice coefficient is used in evaluating the accuracy of the area segmentation of the correct classifications of points inside the segmented region
Jaccard	$\frac{TP}{TP+FP+FN}$	A statistic used to assess the similarities and differences between sets of samples is the Jaccard index, also called the Jaccard similarity coefficient

- The percentage of times a classifier correctly predicts a negative class is called its "true negative" (TN).
- The amount of times a classifier incorrectly predicts a negative class as positive is called the false positive rate (FP).
- The amount of times a classifier erroneously predicts a positive class as negative is called the false negative rate (FN rate).

Fig. 20 The structure of the confusion matrix

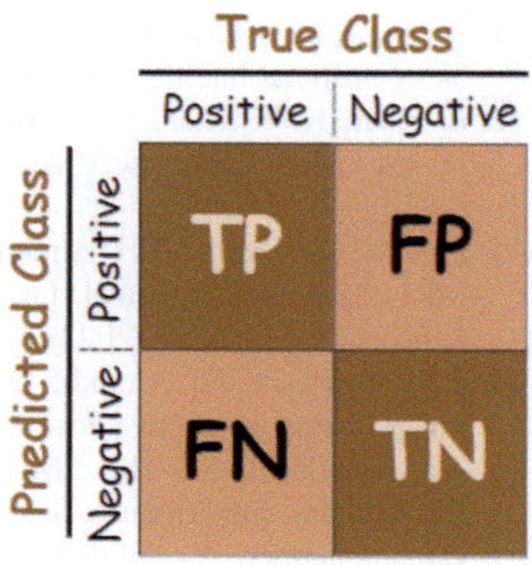

6 Conclusion

In this research, we illustrated the usage of AI techniques in the diagnosis of some of the common brain diseases and disorders, which are brain tumors, epilepsy, dementia especially Alzheimer's disease stage, and Parkinson's disease. Then, popular dataset sources in brain research, brain imaging modalities, and neuropsychological tests are discussed and classified into open-access and private datasets. Finally, the commonly used performance metrics are presented.

References

1. E.A. Buffalo, J.A. Movshon, R.H. Wurtz, From basic brain research to treating human brain disorders. Proc. Natl. Acad. Sci. U S A **116**(52), 26167–26172 (2019)
2. T. Kaufmann, D. van der Meer, N.T. Doan, E. Schwarz, M.J. Lund, I. Agartz, L.T. Westlye, Common brain disorders are associated with heritable patterns of apparent aging of the brain. Nat. Neurosci. (Nature Publishing Group US New York) **22**(10), 1617–1623 (2019)
3. Genetic Brain Disorders: Effective Health Care Program, https://effectivehealthcare.ahrq.gov/health-topics/genetic-brain-disorders, last accessed 30/8/2023
4. Brain diseases, https://www.americanbrainfoundation.org/diseases/, last accessed 30/8/2023
5. Burden of Neurological Conditions, https://www.paho.org/en/enlace/burden-neurological-conditions, last accessed 30/8/2023
6. Essence of AI, https://www.cambridge.org/core/books/abs/cambridge-handbook-of-artificial-intelligence/essence-of-ai/31462D53C85ECABB986BE49C35FEFA6B, last accessed 30/8/2023
7. Artificial Intelligence, https://www.cambridge.org/core/books/abs/cambridge-handbook-of-intelligence/artificial-intelligence/B994B0D29512087BF53979CA9EABC9AB, last accessed 30/8/2023
8. A. Srivastava, The Application & Impact of Artificial Intelligence (AI) On E-Commerce, in *Contemporary Issues in Commerce and Management*, (2021)
9. R. Sharma, Artificial intelligence in agriculture: a review, in *5th International Conference on Intelligent Computing and Control Systems (ICICCS)*, (IEEE, Madurai, India, 2021), pp. 937–942
10. L. Chen, P. Chen, Z. Lin, Artificial intelligence in education: A review. IEEE Access **8**, 75264–75278 (2020)

11. S. Douzi, F.A. AlShahwan, M. Lemoudden, B. El Ouahidi, Hybrid email spam detection model using artificial intelligence. Int. J. Mach. Learn. Comput. **10**(2), 1–7 (2020)
12. P. Kaur, K. Krishan, S.K. Sharma, T. Kanchan, Facial-recognition algorithms: A literature review. Med. Sci. Law (SAGE Publications Sage UK: London, England) **60**(2), 131–139 (2020)
13. H.H.N. Alrashedy, A.F. Almansour, D.M. Ibrahim, M.A.A. Hammoudeh, BrainGAN: brain MRI image generation and classification framework using GAN architectures and CNN models. Sensors **22**(11), 4297 (2022)
14. S.S. Alrumiah, N. Alrebdi, D.M. Ibrahim, Augmenting healthy brain magnetic resonance images using generative adversarial networks. PeerJ Comput. Sci. **9**, e1318 (2023)
15. A. Aouto, T. Huynh-The, J.M. Lee, D.S. Kim, Pose-Based Identification Using Deep Learning for Military Surveillance Systems, in *2019 International Conference on Information and Communication Technology Convergence (ICTC)*, (IEEE, Jeju, Korea (South), 2019), pp. 626–629
16. P. Sweetser, J. Wiles, Current AI in games: A review. Aust. J. Intell. Inf. Process. Syst. (Australian National University, College of Engineering and Computer Science) **8**(1), 24–42 (2002)
17. G. Rong, A. Mendez, E.B. Assi, B. Zhao, M. Sawan, Artificial intelligence in healthcare: review and prediction case studies. Engineering (Elsevier) **6**(3), 291–301 (2020)
18. T.J. Saleem, M.A. Chishti, Exploring the applications of machine learning in healthcare. Int. J. Sensors Wirel. Commun. Control (Bentham Science Publishers) **10**(4), 458–472 (2020)
19. P.-H.C. Chen, Y. Liu, L. Peng, How to develop machine learning models for healthcare. Nat. Mater. (Nature Publishing Group UK London) **18**(5), 410–414 (2019)
20. H. Wang, C. Ma, L. Zhou, A brief review of machine learning and its application, in *2009 International Conference on Information Engineering and Computer Science*, (IEEE, Wuhan, China, 2009), pp. 1–4
21. O.I. Abiodun, A. Jantan, A.E. Omolara, K.V. Dada, N.A. Mohamed, H. Arshad, State-of-the-art in artificial neural network applications: A survey. Heliyon (Elsevier) **4**(11), e00938 (2018)
22. V.S. Dave, K. Dutta, Neural network based models for software effort estimation: a review. Artif. Intell. Rev. (Springer) **42**(2), 295–307 (2014)
23. K. Saravanan, S. Sasithra, Review on classification based on artificial neural networks. Int. J. Ambient Syst. Appl. **2**(4), 11–18 (2014)
24. C. Cao, F. Liu, H. Tan, D. Song, W. Shu, W. Li, Z. Xie, Deep learning and its applications in biomedicine. Genom. Proteom. Bioinform. (Elsevier) **16**(1), 17–32 (2018)
25. H. Chen, O. Engkvist, Y. Wang, M. Olivecrona, T. Blaschke, The rise of deep learning in drug discovery. Drug Discov. Today. (Elsevier) **23**(6), 1241–1250 (2018)
26. P.H. Liu, S.F. Su, M.C. Chen, C.C. Hsiao, Deep learning and its application to general image classification, in *2015 International conference on informative and cybernetics for computational social systems (ICCSS)*, (IEEE, Chengdu, China, 2015), pp. 7–10
27. A. Hashmi, O. Barukab, Dementia classification using deep reinforcement learning for early diagnosis. Appl. Sci. (MDPI) **13**(3), 1464 (2023)
28. A.S. Lundervold, A. Lundervold, An overview of deep learning in medical imaging focusing on MRI. Zeitschrift für Medizinische Physik. (Elsevier) **29**(2), 102–127 (2019)
29. T. Dissanayake, T. Fernando, S. Denman, S. Sridharan, C. Fookes, Deep learning for patient-independent epileptic seizure prediction using scalp EEG signals. IEEE Sens. J (IEEE) **21**(7), 9377–9388 (2021)
30. Z.B. Tariq, A. Iyengar, L. Marcuse, H. Su, B. Yener, Patient-specific seizure prediction using single seizure electroencephalography recording, in *AI for Disease Surveillance and Pandemic Intelligence: Intelligent Disease Detection in Action*, (Springer, 2022), pp. 295–308
31. Z. Zhang, X. Li, F. Geng, K. Huang, A semi-supervised few-shot learning model for epileptic seizure detection, in *43rd Annual International Conference of the IEEE Engineering in Medicine & Biology Society (EMBC)*, (IEEE, Mexico, 2021), pp. 600–603

32. B. Lu, H. Lu, G. Zhou, X. Yin, X. Gu, T. Ni, Multifeature metric learning based on enhanced equidistance embedding for electroencephalogram recognition of epilepsy. Wirel. Commun. Mob. Comput. (Hindawi Limited), 1–11 (2021)
33. I. Ullah, M. Hussain, H. Aboalsamh, An automated system for epilepsy detection using EEG brain signals based on deep learning approach. Expert Syst. Appl. (Elsevier) **107**, 61–71 (2018)
34. S. Hu, J. Liu, R. Yang, Y.N. Wang, A. Wang, K. Li, C. Yang, Exploring the applicability of transfer learning and feature engineering in epilepsy prediction using hybrid transformer model. IEEE Trans. Neural Syst. Rehabil. Eng. **31**, 1321–1332 (2023)
35. H. Daoud, M.A. Bayoumi, Efficient epileptic seizure prediction based on deep learning: IEEE transactions on biomedical circuits and systems. IEEE **13**(5), 804–813 (2019)
36. H.G. Daoud, A.M. Abdelhameed, M. Bayoumi, Automatic epileptic seizure detection based on empirical mode decomposition and deep neural network, in *14th International Colloquium on Signal Processing & Its Applications (CSPA)*, (IEEE, Penang, Malaysia, 2018), pp. 182–186
37. A.M. Abdelhameed, M. Bayoumi, Semi-supervised EEG signals classification system for epileptic seizure detection. IEEE Signal Process. Lett. (IEEE) **26**(12), 1922–1926 (2019)
38. H. Khan, L. Marcuse, M. Fields, K. Swann, B. Yener, Focal onset seizure prediction using convolutional networks. IEEE Trans. Biomed. Eng. (IEEE) **65**(9), 2109–2118 (2017)
39. A. De, A.S. Chowdhury, DTI based Alzheimer's disease classification with rank modulated fusion of CNNs and random forest. Expert Syst. Appl. (Elsevier) **169**, 114338 (2021)
40. H. Nawaz, M. Maqsood, S. Afzal, F. Aadil, I. Mehmood, S. Rho, A deep feature-based real-time system for Alzheimer disease stage detection. Multimed. Tools Appl. (Springer) **80**, 35789–35807 (2021)
41. B.A. Mohammed, E.M. Senan, T.H. Rassem, N.M. Makbol, A.A. Alanazi, Z.G. Al-Mekhlafi, F.A. Ghaleb, Multi-method analysis of medical records and MRI images for early diagnosis of dementia and Alzheimer's disease based on deep learning and hybrid methods. Electronics (Multidisciplinary Digital Publishing Institute) **10**(22), 2860 (2021)
42. S. Korolev, A. Safiullin, M. Belyaev, Y. Dodonova, Residual and plain convolutional neural networks for 3D brain MRI classification, in *14th International Symposium on Biomedical Imaging (ISBI 2017)*, (IEEE, Melbourne, VIC, Australia, 2017), pp. 835–838
43. N.T. Duc, S. Ryu, M.N.I. Qureshi, M. Choi, K.H. Lee, B. Lee, 3D-deep learning based automatic diagnosis of Alzheimer's disease with joint MMSE prediction using resting-state fMRI. Neuroinformatics (Springer) **18**, 71–86 (2020)
44. A. Basher, B.C. Kim, K.H. Lee, H.Y. Jung, Volumetric feature-based Alzheimer's disease diagnosis from sMRI data using a convolutional neural network and a deep neural network. IEEE Access (IEEE) **9**, 29870–29882 (2021)
45. M. Subramoniam, T.R. Aparna, P.R. Anurenjan, K.G. Sreeni, Deep learning-based prediction of Alzheimer's disease from magnetic resonance images, in *Intelligent Vision in Healthcare*, (Springer, 2022), pp. 145–151
46. M. Jyotiyana, N. Kesswani, M. Kumar, A deep learning approach for classification and diagnosis of Parkinson's disease. Soft Comput. (Springer) **26**(18), 9155–9165 (2022)
47. G. Pahuja, T.N. Nagabhushan, A comparative study of existing machine learning approaches for Parkinson's disease detection. IETE J. Res. (Taylor & Francis) **67**(1), 4–14 (2021)
48. B. Pearce, P. Coetzee, D. Rowland, S. Linfoot, D.T. Dexter, D. Gveric, S. Gentleman, *Automatic Sample Segmentation & Detection of Parkinson's Disease Using Synthetic Staining & Deep Learning* (bioRxiv. Cold Spring Harbor Laboratory, 2022), pp. 2022–2008
49. M.B.T. Noor, N.Z. Zenia, M.S. Kaiser, S.A. Mamun, M. Mahmud, Application of deep learning in detecting neurological disorders from magnetic resonance images: a survey on the detection of Alzheimer's disease, Parkinson's disease and schizophrenia. Brain Inform. (Springer) **7**, 1–21 (2020)
50. R. Ullah, Y. Dong, T. Arslan, S. Chandran, A machine learning-based classification method for monitoring Alzheimer's disease using electromagnetic radar data, in *IEEE Transactions on Microwave Theory and Techniques*, (IEEE, 2023), pp. 4012–4026

51. Y.Y. Chen, C.W. Cho, S.H. Lin, H.Y. Lai, Y.C. Lo, S.Y. Chen, S.T. Tsai, A vision-based regression model to evaluate Parkinsonian gait from monocular image sequences. Expert Syst. Appl. (Elsevier) **39**(1), 520–526 (2012)
52. Alzheimer's Dataset (4 class of Images), https://www.kaggle.com/datasets/tourist55/alzheimers-dataset-4-class-of-images, last accessed 30/8/2023
53. Parkinsons Dataset, https://archive.ics.uci.edu/ml/datasets/parkinsons, last accessed 30/8/2023
54. V.P.G.P. Rathi, S. Palani, Brain tumor detection and classification using deep learning classifier on MRI images. Res. J. Appl. Sci. Eng. Technol. (Maxwell Science Publishing) **10**(2), 177–187 (2015)
55. H.R. Almadhoun, S.S. Abu-Naser, Detection of brain tumor using deep learning. Int. J. Acad. Eng. Res. **6**(3), 29–47 (2022)
56. T. Saba, A.S. Mohamed, M. El-Affendi, J. Amin, M. Sharif, Brain tumor detection using fusion of hand crafted and deep learning features. Cogn. Syst. Res. (Elsevier) **59**, 221–230 (2020)
57. M. Siar, M. Teshnehlab, Brain tumor detection using deep neural network and machine learning algorithm, in *9th International Conference on Computer and Knowledge Engineering (ICCKE)*, (IEEE, Mashhad, Iran, 2019), pp. 363–368
58. A.H. Khan, S. Abbas, M.A. Khan, U. Farooq, W.A. Khan, S.Y. Siddiqui, A. Ahmad, Intelligent model for brain tumor identification using deep learning. Appl. Comput. Intell. Soft Comput. (Hindawi Limited) **2022**, 1–10 (2022)
59. S. Sajid, S. Hussain, A. Sarwar, Brain tumor detection and segmentation in MR images using deep learning. Arab. J. Sci. Eng. (Springer) **44**, 9249–9261 (2019)
60. G. Hemanth, M. Janardhan, L. Sujihelen, Design and implementing brain tumor detection using machine learning approach, in *3rd International Conference on Trends in Electronics and Informatics (ICOEI)*, (IEEE, Tirunelveli, India, 2019), pp. 1289–1294
61. A. Ari, D. Hanbay, Deep learning based brain tumor classification and detection system. Turk. J. Electr. Eng. Comput. Sci. **26**(5), 2275–2286 (2018)
62. The Bonn EEG time series download page https://www.upf.edu/web/ntsa/downloads/-/asset_publisher/xvT6E4pczrBw/content/2001-indications-of-nonlinear-deterministic-and-finite-dimensional-structures-in-time-series-of-brain-electrical-activity-dependence-on-recording-regi?inheritRedirect=false&redirect=https%3A%2F%2Fwww.upf.edu%2Fweb%2Fntsa%2Fdownloads%3Fp_p_id%3D101_INSTANCE_xvT6E4pczrBw%26p_p_lifecycle%3D0%26p_p_state%3Dnormal%26p_p_mode%3Dview%26p_p_col_id%3Dcolumn-1%26p_p_col_count%3D1, last accessed 30/8/2023
63. CHB-MIT Scalp EEG Database, https://physionet.org/content/chbmit/1.0.0/, last accessed 30/8/2023
64. CHB-MIT Scalp EEG Database, https://archive.physionet.org/pn6/chbmit/, last accessed 30/8/2023
65. OASIS Brains, https://oasis-brains.org/, last accessed 30/8/2023
66. ADNI, https://adni.loni.usc.edu/, last accessed 30/8/2023
67. UPenn and Mayo Clinic's Seizure Detection Challenge, https://www.kaggle.com/competitions/seizure-detection/data, last accessed 30/8/2023
68. Alzheimers-Disease-5-Class-Dataset-ADNI, https://www.kaggle.com/datasets/madhucharan/alzheimersdisease5classdatasetadni, last accessed 30/8/2023
69. Parkinson Disease Spiral Drawings Using Digitized Graphics Tablet Data Set, https://archive.ics.uci.edu/ml/datasets/Parkinson+Disease+Spiral+Drawings+Using+Digitized+Graphics+tablet, last accessed 30/8/2023
70. Brain Tumor MRI Dataset, https://www.kaggle.com/datasets/masoudnickparvar/brain-tumor-mri-dataset, last accessed 30/8/2023
71. A. Tagaris, D. Kollias, A. Stafylopatis, G. Tagaris, S. Kollias, Machine Learning for Neurodegenerative Disorder Diagnosis—Survey of Practices and Launch of Benchmark Dataset. Int. J. Artif. Intell. Tools (World Scientific) **27**(03), 1850011 (2018)
72. BraTS 2013, https://paperswithcode.com/dataset/brats-2013-1, last accessed 30/8/2023
73. BraTS 2015, https://paperswithcode.com/dataset/brats-2015-1, last accessed 30/8/2023

74. B.H. Menze, A. Jakab, S. Bauer, J. Kalpathy-Cramer, K. Farahani, J. Kirby, K. Van Leemput, The multimodal brain tumor image segmentation benchmark (BRATS). IEEE Trans. Med. Imag. (IEEE) **34**(10), 1993–2024 (2014)
75. BraTS 2017, https://paperswithcode.com/dataset/brats-2017-1, last accessed 30/8/2023
76. BraTS 2020, https://www.kaggle.com/datasets/awsaf49/brats20-dataset-training-validation, last accessed 30/8/2023
77. BraTS 2021, https://www.kaggle.com/datasets/dschettler8845/brats-2021-task1, last accessed 30/8/2023
78. S. Toshkhujaev, K.H. Lee, K.Y. Choi, J.J. Lee, G.R. Kwon, Y. Gupta, R.K. Lama, Classification of Alzheimer's disease and mild cognitive impairment based on cortical and subcortical features from MRI T1 brain images utilizing four different types of datasets. J. Healthc. Eng. (Hindawi) (2020)
79. M. Weiner, P. Aisen, W. Jagust, J.Q. Trojanowki, A.W. Toga, L. Beckett, J.A. Immunotherapy, Acknowledgement List for ADNI Publications
80. BraTS 2016, https://paperswithcode.com/dataset/brats-2016, last accessed 30/8/2023
81. A. Luque, A. Carrasco, A. Martín, A. de Las Heras, The impact of class imbalance in classification performance metrics based on the binary confusion matrix. Pattern Recogn. **91**, 216–231 (2019)
82. M. Hossin, M.N. Sulaiman, A review on evaluation metrics for data classification evaluations. Int. J. Data Min. Knowl. Manag. Process (Academy and Industry Research Collaboration Center (AIRCC)) **5**(2), 01–11 (2015)
83. Confusion Matrix, https://towardsdatascience.com/confusion-matrix-for-your-multi-class-machine-learning-model-ff9aa3bf7826, last accessed 30/8/2023

Empowering Security in Cloud Environment Using Encryption Technique

Mohammad Ali A. Hammoudeh, Aishah Flah, Noura Al Nassar, Wesam Al Rajhi, and Renad Ibrahim

1 Introduction

In today's world, the security of any information has become more vital. Every business corporation, and individual cares deeply about security because they do not want their personal information to be compromised, using sophisticated resources such as data storage and processing power.

The Saudi Arabian Cybersecurity Market is estimated to reach USD9845.26 million by 2026, growing at a CAGR of 17.98 percent. The arrival of COVID-19 boosted the Saudi Arabian Cybersecurity Market since firms allowed workers to work from home to avoid infection. This boosted investment on sophisticated cybersecurity solutions to safeguard data privacy and security. Rapid technological advances adopt sophisticated technologies. Market over the forecast period is shown in Fig. 1. Industry, firm, and geographical distribution are the segments of the Saudi Arabian Cybersecurity Market. The market is divided into cloud and on-premises deployments. During the predicted period, the cloud model is likely to dominate. The increased adoption of cloud technology by enterprises and institutions is boosting demand for cloud technology over the projection period [1].

This study draws inspirations from cloud computing which is a topic in constant evolvements and hybrid cryptography practice. The provision of services via cloud computing involves complex resources such as information storage capacity and processing power. "Cloud" here implies a grouping of data centers that users could access via Internet anywhere in the world. Cloud computing may be restricted to a single business, which is referred to as enterprise clouds, or it can be broadened to

---

M. A. A. Hammoudeh (✉) · A. Flah · N. Al Nassar · W. Al Rajhi · R. Ibrahim
Department of Information Technology, College of Computer, Qassim University, Buraydah, Saudi Arabia
e-mail: maah37@qu.edu.sa; 381212434@qu.edu.sa; 381212840@qu.edu.sa; 382214511@qu.edu.sa; renad.ibrahim@qu.edu.sa

© The Author(s), under exclusive license to Springer Nature Switzerland AG 2024
A. Shaikh et al. (eds.), *Advances in Emerging Information and Communication Technology*, Signals and Communication Technology,
https://doi.org/10.1007/978-3-031-53237-5_25

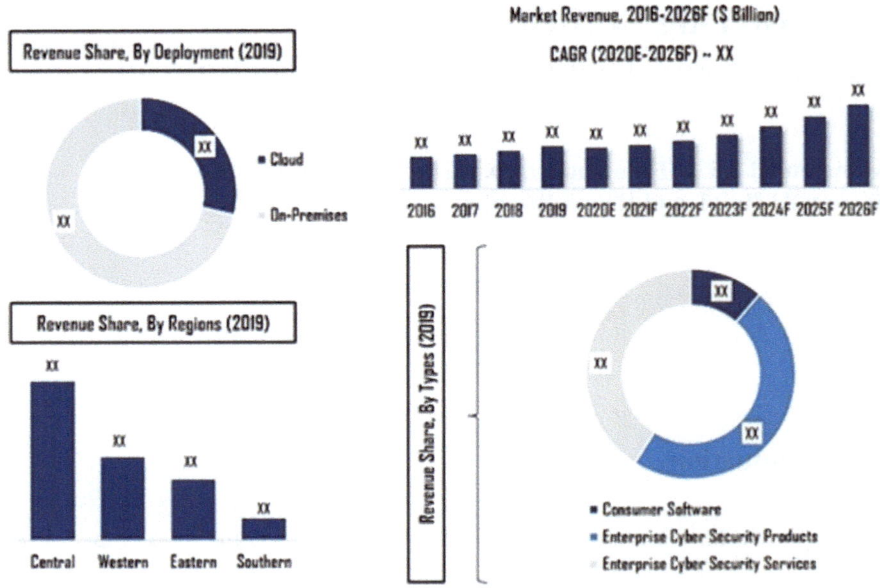

Fig. 1 Saudi Arabia cybersecurity market overview [1]

include several companies, which is referred to as public clouds. Cloud hosting of web applications is also a kind of cloud computing, and it will be used in this project [2]. File storage on the cloud must be safe and secure, which takes us to the topic of hybrid cryptography. Cryptography is a term that refers to a way of securing information via the use of codes to ensure that only authorized individuals have access to that information. The term cryptography refers to a set of rule-based techniques which are used for encryption and decryption of data and information. In cryptography, we strive to accomplish the following goals [3]:

- Confidentiality
- Integrity
- Non-repudiation
- Authentication

As a result, we employ either the symmetric key algorithm or the asymmetric key method. Hybrid cryptography is a way of combining several encryption techniques for making use of its advantages. Moreover, public and private key cryptography assures security as it safeguards the information regarding the two keys. The significance of "Encryption Technique in Cloud Environment" stems from the rise of the related publications number during the past 10 years demonstrated by Fig. 2.

The goal of this research is to combine the hybrid cryptography method with cloud-based files. Creating a textual information repository including pictures, sound, and video data files. The major concept of this effort is to treat security as a

Fig. 2 Number of publications in encryption technique in cloud environment from 2013 to 2022 (May)

significant issue. To ensure security, these files would remain encrypted for recovery using a decryption key.

The following is the proposed work's goal:

(1) Empowering security for data storage based on encryption technique.
(2) Improve data storage confidentiality and integrity by using hybrid technique to perform the security.
(3) Make the system very safe and reliable, with data stored on a cloud server to prevent unwanted access.

As for the scope of the study, it was limited to smaller file forms like texts, photographs, and some tiny audio and video files within the organization. These limitations are put in place to ensure that the allocated limited cloud capacity is not exhausted in the testing and deployment phases. The provision of cloud service is usually accompanied by some significant amount of cloud storage space and a high price.

This paper is organized as follows: security of application is briefly introduced in Sect. 1. Part ii: the related works and methods. Section 3 examines briefly on various topics regarding ciphers, feedback, and resolutions. Chapter 4 includes summary and potential future research.

2 Relevant Concepts and Techniques

This part introduces the relevant concepts and techniques involved in encrypting the cloud domain. In addition, we explain some popular hybrid encryption techniques and how they work.

2.1 Background

Nowadays the need of technology becoming more and more due to the huge number of data and users that need to be managed in a way that help the organization improve. Cloud computing and cryptography are a new orientation that measures how much the country is pioneer in using technology to secure data. It seems that the number of data securing problems has increased, and this was a result of the high request to apply new technologies in the recent years. The kingdom working and aiming for the vision 2030 which digital transformation has a big role in it; our project is one of the many steps to achieve the 2030 vision. Our project is going to securing files using encryption techniques and high language. The government is focusing on solving any problem that obstruct in the way of 2030 Vision.

2.2 Concept and Definitions

Cloud computing has now become a crucial mechanism through which different infrastructure and data service objectives are efficiently achieved at the least budgets, efforts, and yet high scalability levels. It is used in almost all areas of IT. Nevertheless, information security problems still hinder the progress of cloud computing to a certain extent and have to be solved. Security means that data should be saved from danger and vulnerability. It involves multiple risks and weaknesses that ought to be addressed. This approach will comply with the needs associated with the data centers and cloud computing security.

The discovery and production of the algorithms used for encrypting and decrypting data forms are part of cryptography but not so in cryptanalysis where decryption involves analyzing the crypto or rather encrypting system unlike decrypting. Cryptology can be defined as a broad term for the scientific study of secret writing. Cryptology encompasses cryptography and cryptanalysis [4].

Hybrid Encryption—the method whereby public and private keys are combined, in order to ensure effective data transmission by providing higher security. Public key cryptography: the sender generates a private key, encrypts it using such method as, for example, RSA algorithm, and finally, sends an encrypted text including both initial symmetric key and generated one. Using a private key, the sender encodes the cipher after which; the cipher is only readable by the receiver who also possesses another private key [5].

Three most important cryptographic approaches that include symmetrical cryptographic method, asymmetrical cryptographic method, and hash function are usually used for this purpose. Single-key encryption in symmetric technique of encrypting and decrypting data. For instance, some kinds of symmetric encryption include Advanced Encryption Standard (AES), Data Encryption Standard (DES), Blowfish, Carlisle Adams and Stafford Tavares (CAST) algorithm, two fish, International Data Encryption Algorithm (IDEA), and Sec Two different keys may also

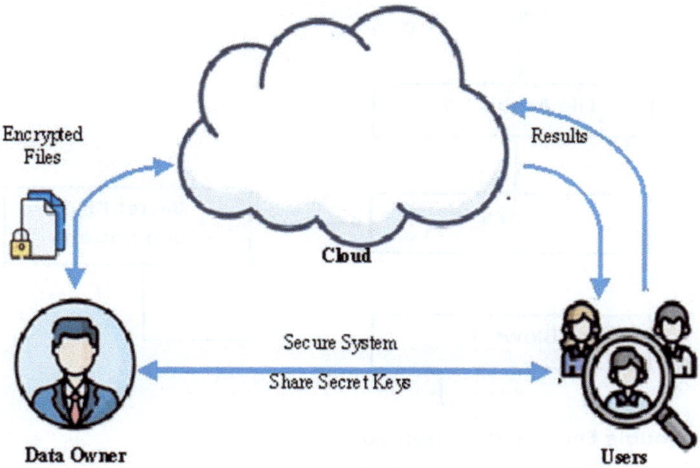

Fig. 3 The work of cryptographic algorithms

be applied in an asymmetric technique, with the public key being used for the encryption and a private key for the decryption or vice versa. Examples include RSA, DSA, Elgama, and ECC. Using of a mathematical transformation, the hash function transforms information into an irreversible "locked state." This is a category that includes message digest of MD5, SHA-1, and SHA-2. Currently, RSA is the most widely employed algorithm for encryption of keys, passwords, and numbers [5].

The architecture of hybrid encryption determines interactions between users, data, and owner and could confirm that they can be executed together. The data owner will encrypt the file with difficult password and after that send it to the user through the cloud. Now the user will be decrypted by the password after that system will send notification to data owner which means the file is opened by the users. The file will be open from a temporary folder as Fig. 3.

2.3 Previous Studies and Related Works

Lately numbers of security problems are increasing, and many ideas come to solve different problems. Several techniques have been proposed them; this is a review some of them below. In other words, cloud computing has a lot of memory space, great processing power, and speedy calculation. Cloud computing is moving to the giant world of corporate components, social networking programs, and individual consumers. Cloud computing security leads us to consider data accessibility, dependability, and reliabilities of information. However, verifying and gaining permission become very critical to access information as cloud is simply an

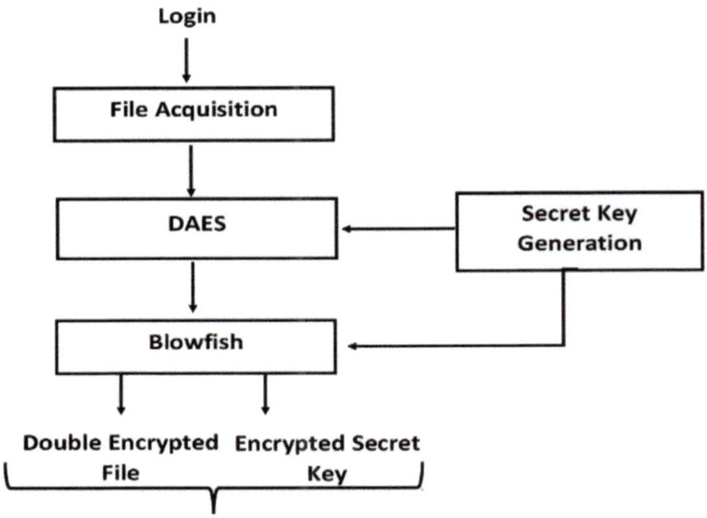

Fig. 4 A proposed hybrid cryptographic model of Salmanetal [6]

amalgamation of real supercomputer speeds all over the world. Security of file encryption under AES algorithm is well-researched. There hasn't been a successful assault against AES yet, but with the rise of cybercrime, attacks such as brute force and algebraic attacks may become conceivable. To enhance data's security, Salma presented a hybrid approach of Blowfish and AES ciphers by Salma et al. Upon a successful login, the file is encrypted with DAES (which is a dynamic version of symmetric AES) and subsequently Blowfish Fig. 4. Blowfish is used to encrypt the secret keys. The cloud is used to store all encrypted data and keys [6].

Shweta and Ashish have proposed a hybrid approach combining transposition and substitution for securing cloud data. A main secret key permutes the input file, as illustrated in Fig. 5. A substitution cipher with a separate secret key is applied to encrypt another layer before uploading it into the cloud. On the other hand, the auditor ensures that secret keys are generated and sent to cloud users [7].

In an effort to develop a new cloud storage security pattern, Fig. 6 illustrates that the original file gets subdivided in three equally divisible parts which are equally encrypted with varying keys. Lastly the encrypted part is assembled and then transported over to the cloud storage server [8].

Some of the research on empowering security in cloud environment using the encryption technique done by many mathematicians, engineer, and city planner has been summarized into Table 1; they have come up with different outcomes as outlined below. Most studies use varying solution approaches; each approach has purpose/motivation/problem, methodology/type, and result and limitation and future direction/recommendation.

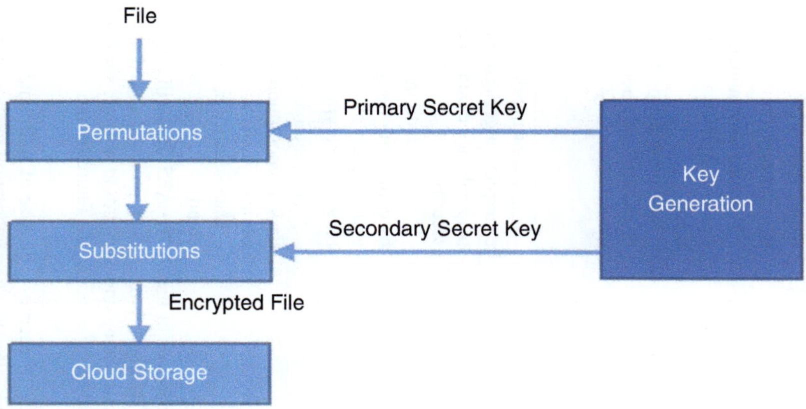

Fig. 5 A framework of hybrid cryptographic model in by Shweta and Ashish [7]

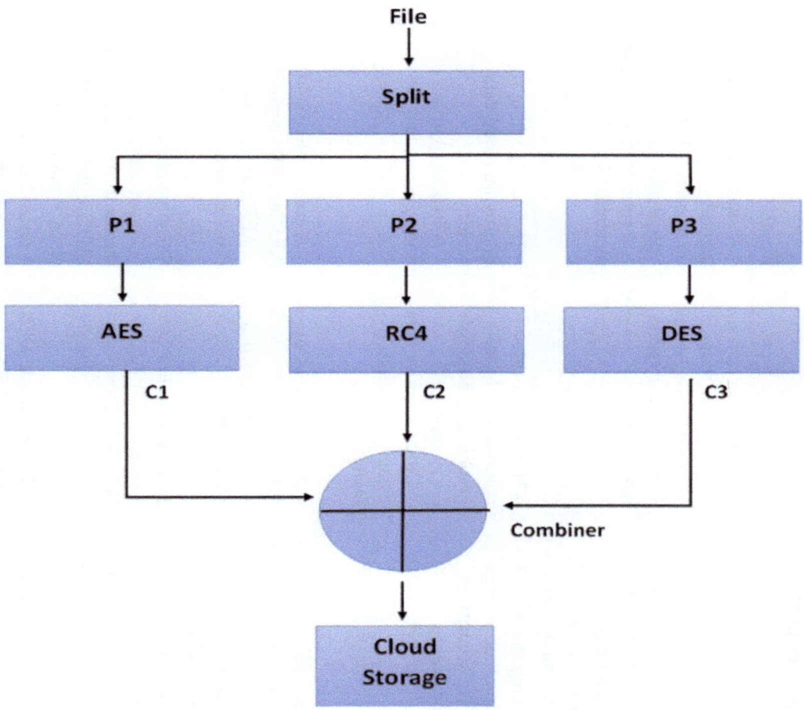

Fig. 6 Hybrid cryptographic model proposed by Shivam et al. [8]

Table 1 Summary of some previous studies

Year	Ref	Purpose/motivation/ problem	Methodology type	Result	Limitation	Future direction Recommendation
2019	[2]	– To give information about hybrid cryptographic algorithms – In closing, this paper will cover a research gap regarding how user authentication and implementation can be addressed	– Hybrid encryption – The simple user-driven, tabular survey versus more detailed interviews	– A study of hybrid cryptography, from late 2015 up to early 2019 – Research gaps in user authentication and implementation identified	– Neglecting user authentication – Improper implementation of hybrid algorithms	– Enhance user authentication in hybrid cryptography – Resolving the problem of insufficiently adopted hybrid algorithms
2019	[3]	– Improving the integrity of data in cloud storage – Ensuring security in cloud computing	– Hybrid algorithm combining. Homographic and blowfish encryption – Cloud security data using encryption method	– Hybrid algorithm for improving cloud data security – A hybrid approach of homographic and blowfish	N/A	– The use of hybrid encryption algorithm can help enhance cloud data security – Achieve better results by combining homographic encryption and blowfish encryption
2020	[4]	– A secure file storage system on the cloud based on hybrid encryption – A comparative study into encryption techniques	– AES and RSA algorithms – AES and blowfish algorithms	– A comparative study of AES and RSA algorithms – AES and blowfish algorithm comparative study	N/A	– A comparative study of various encryption algorithms – Data security in the cloud
2020	[5]	This paper offers innovative scheme of file protection in the cloud based on confidentiality, integrity, and availability of data	Does not explicitly indicate any outcomes or conclusions	Sensitive client data becomes more secure thanks to AES algorithm, MD 5 algorithm, and proxy re-encryption methods	Critical weaknesses of the AES and MD5 algorithms applied for encrypting and generating keys	The AES files will be encrypted and stored on clouds at separate sites on the future course
2021	[9]	– The existing encryption schemes often come with	– These include DES, AES, and PKRSA	– Hybrid cryptography increases data security	– For instance, if encryption keys leak, all existing	– Use of additional encryption methods for

		some limitations and security issues – The hybrid cryptography comes in handy for the purposes of enhancing security and protecting information	– Steganography using LSB for encryption key storage in an image	– It is combinations of AES, DES, and RSA encryption methods	encryption schemes will fail – Therefore, additional security is provided via hybrid cryptography	improved security – Alternative techniques for key storage
2021	[10]	– The key challenge for security in cloud computing – A demand low impact encryption solution	– Modified blowfish algorithm together with ECC algorithm – Verification of data integrity using digital signature	– There is a reduction in execution time against other encryption methods used – The proposed technique features higher throughput than other methods and approaches	N/A	– Carry out validation tests for it in multiple settings, particularly cloud-based computing solutions – Carry out an evaluation on data-heavy solution
2021	[11]	– Encrypt data in the cloud storage – Protect data confidentiality, ensure data integrity and availability	– Encryption algorithms included in AESCCM, AESGMC, and Chacha20poly1305 – The chunks of data are then stored on the cloud	It makes sure the confidentiality, security, and protection of the customers' crucial information from saving client's data on any cloud using the algorithms A/sccm, A/sgmc, chach/20/poly130	N/A	– Improving the security in cloud-based storage – Investigating new data protection algorithms
2021	[12]	– Proposed mechanism: Cloud secure storage mechanism (CSSM) – Purpose: Safeguard privacy and security of cloud-based data and information	– Data dispersion and distributed storage – Hierarchical management and secret sharing	– Mechanism of protection for data at storage level – Acceptable upload/download time for 5G sized file	N/A	– Upload/download speed can be increased to provide an additional optimization over CSSM – Data dispersion and encryption techniques need to be studied further

3 Discussion

In order to develop a new model without repeating the same errors, we must first analyze the models that are already in use and take into account customer needs. We gather data in this phase to assess the user needs, and then we create and construct our model using the most suitable technique.

3.1 Why Hybrid Cryptography?

Cloud Computing: It utilizes high-end resources like databases and processing power. Cloud encompasses several data centers that can be accessed online. Enterprise clouds limit the cloud to a single business, while public clouds broaden them across multiple enterprises. The second kind of cloud computing that will be utilized in this project is web application cloud hosting. So, we finally get to hybrid encryption or cloud file storage that needs to be secure [9, 13].

Cryptography: A technique used for securing information whereby only those that are authorized can decode the set codes and thus access to the information. The rules-based algorithms are used in securing information by using it to encode and decode data as well as information.

To transfer messages in a hard-to-decipher way, there are five fundamental roles of cryptography to be achieved [4, 14, 15]:

(1) Confidentiality: It is also vital to guarantee that there won't be any additional people aside from the targeted receiver that will be able to view the message.
(2) Authentication: Establishing your identity process.
(3) Integrity: This guarantees that the message sent is exactly the same as the original message received and no changes have occurred.
(4) Non-repudiation: A way of showing that the message was sent by the sender.
(5) Key exchange: Different procedures are used in transmitting cryptographic keys from the sender to the receiver.

Various classifications exist for cryptographic algorithms. For the purposes of this research, they will be classified according to the number of keys that are used for encryption and decryption as follows:

- Secret Key Cryptography (SKC): Another type of crypto system is called symmetric method, whereby a single key facilitates the encoding procedure while also allowing for the same key to be used in its decoding operations. Mostly as a source of privacy or secrets.
- Public Key Cryptography (PKC): A type of cryptography makes use of two different keys: one for encryption and another for decryption. This protocol is commonly used in authentication, non-repudiation, and key exchange, among other functions.

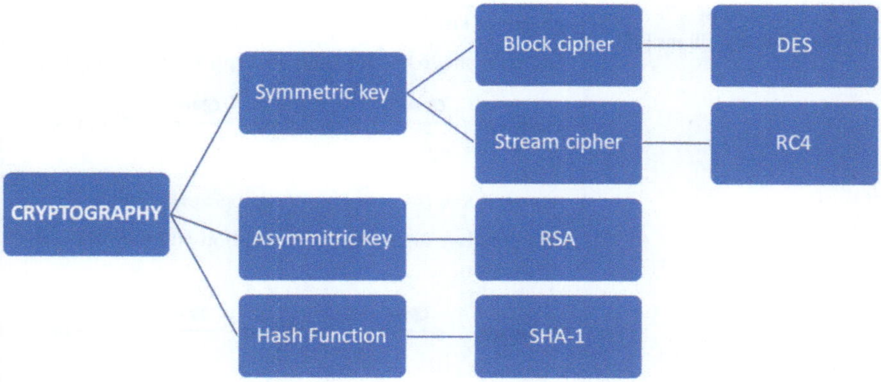

Fig. 7 The types of cryptographic algorithms [8]

- Hash Functions: It uses a mathematical transformation that unalterably encrypts the object's data, resulting in the formation of a digital signature, for main purpose in verification of inputs.

They will also be categorized based on usage type and application. This paper will describe in details the following three types of algorithms (Fig. 7) [8]:

Symmetric Key Algorithms All users of the encrypting system possess the same secret key that may be employed both in encryption and decryption. Symmetric encryption is faster than asymmetric encryption because the sender and receiver share a secret key that is the same. Using the secret key is dangerous because it's a common key for both parties in communication. 128-bit key length should be used for symmetric key encryption in order to avoid or frustrate cracking it. The commonly used encryption methods include AES, RC4, DES, 3DES, RC5, and RC6. AES is the finest of these algorithms. When speed is more important than security, symmetric encryption is utilized. The following are some of the most popular applications for symmetric encryption [10]:

- Banking and financial organizations: Ensures that credit card information and PII are encrypted for completion of transactions.
- Data storage: encrypt data in rest.

Popular Symmetric Algorithms:

(1) Data Encryption Standard (DES).
(2) Triple Data Encryption Symmetric Algorithm (3DES).
(3) Advanced Encryption System (AES).

Asymmetric Key Algorithms Also called public-key cryptography. Encryption and decoding need many keys. It employs two distinct encryption keys; there are two different keys among them—a public key and also a private key. The public key means that anyone can send a message to the sender without restriction. The second

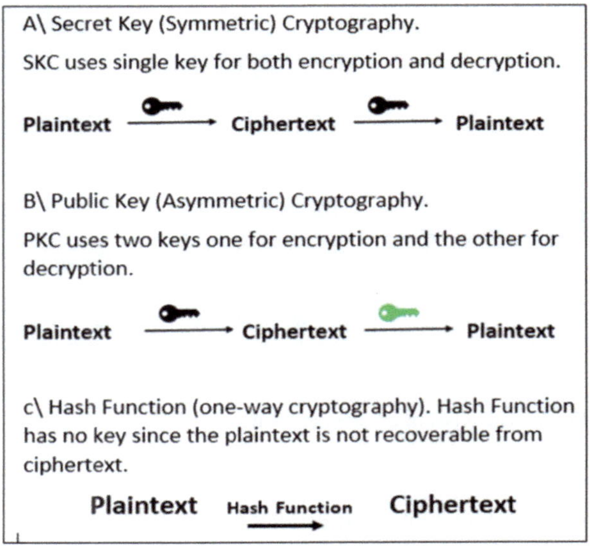

Fig. 8 The work of cryptographic algorithms [8]

private key is only known by the sender. A message is ciphered using a public key, which can then be unscrambled using a private key. A private key can decode an encryption for communication that was done by a public key. It can be disbursed over the Internet as public key security is not required. For information transmission, asymmetric encryption is the best option. Asymmetric is used in client-server model, communicates via asymmetric encryption, ascertains that the server is situated, and creates a certificate with the organization's details in it; both client and server need secured encrypted communication, who returns the certificate. SSL/TLS employs both symmetric and asymmetric encryption techniques [10].

There are no private keys sent in any of the conversations. Encryption and authentication using RSA and PGP to protecting email are examples of algorithms that utilize this concept. When security is more important than speed, asymmetric encryption is utilized as in Fig. 8:

- Digital signature: verification of the identity signature is based on these features.
- Blockchain: verify identity of approval authorized cryptocurrency transaction.
- Public key infrastructure: digital certificate-based authentication for encryption keys.
- Hash functions: known as hash functions or the one-way encryption algorithm that requires no key. Instead, a hash value of set-length is derived from the plaintext such that it is not possible to identify and restore either the contents or the length of the plaintext.

The hash algorithm produces commonly a digital "finger print" that allows to verify whether a file was untouched from third-party interference in the course of transmission or infection as well. Various hash algorithms are used in securing many passwords from different operating systems. Consequently, hash functions, in this case, act as tools that guarantee the file authenticity [8]. Figure 8 shows the work of cryptographic algorithms.

3.2 *Feedback and Statistics*

In this section, Data collection is a process of gathering information from the entire relevant source to gain answers to the study problem and evaluate the result. This study used a questionnaire to get information from the community about a future program for securing data while storing it on cloud environment. Having 68 responses generated with Google Forms, there are also seven specialized questions shown under Fig. 9.

Data analysis is the process of utilizing techniques to represent, explain, and evaluate data, so after analyzing the results, we conclude the following:

(1) 95.7% uses cloud storage.
(2) More than half trust cloud storage.
(3) More than 89% have been shared and uploaded files in data store.
(4) Most of responses carefully choose computational storage by research reviews, security features and trends, cost, and security risk and by a friend respectively as in Fig. 10.
(5) 15.9% of responses has been lost, changed without their approval, or stolen their data in cloud.
(6) Most of responses in Fig. 11 chosen authentication system applied in the cloud environment to protect data, while 37.7% selected encryption techniques.

3.3 *Solution and Recommendation*

After determining requirements and analyzing them, Fig. 12 shows ten-step data flow covered by the system as the following:

Fig. 9 The questionnaire sample

1. **Do you use cloud storage? (iCloud, Drop Box, Google Cloud, Microsoft OneDrive, etc.)**
2. **Do you trust that your data store in the cloud is secure?**
3. **Have you ever shared, uploaded and downloaded a file in the data store?**
4. **How did you choose your cloud provider/service?**
5. **Has your data stored in the cloud been lost, stolen, changed or accessed without your consent?**
6. **If yes, what kind of threat did you face?**
7. **what are the security technologies applied in the cloud environment to protect data?**

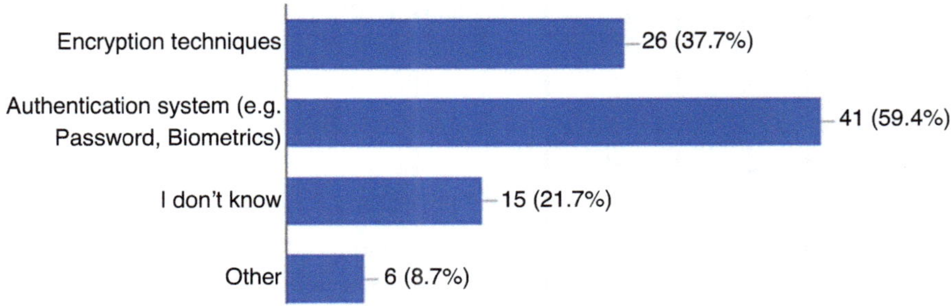

Fig. 10 Security techniques choices

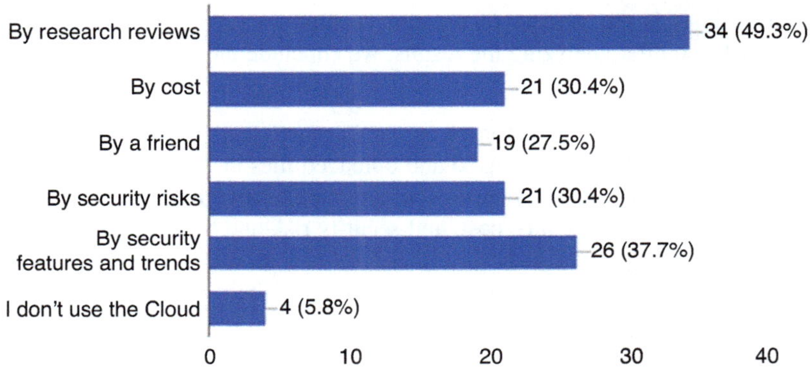

Fig. 11 Security techniques choices

The proposed system for organization starts by the steps below:

- Step1: Sender (User A) must log in to access the system.
- Step2: User will be authorized to use the system.
- Step3: Find the file that need to be encrypted.
- Step4: User will enter the password (16 key), and this is based on AES algorithms that are used in the encryption of the file. RSA encryption for the secret key.
- Step5: User will upload the encrypted file to the cloud for receiver.
- Step6: The system will notify receiver that sender has uploaded the encrypted file to cloud with key.
- Step7: Receiver (User B) must log in to the system (access).
- Step8: User B will be authorized to use the system.
- Step9: Download the file that needs to be decrypted from the cloud.
- Step10: User B (receiver) will enter the password (16 key), and the file is decrypted AES algorithm, and the key is decrypted with RSA algorithm.
- Step11: User B will decrypt the file.
- Step12: System will notify User A that User B has decrypted the file.

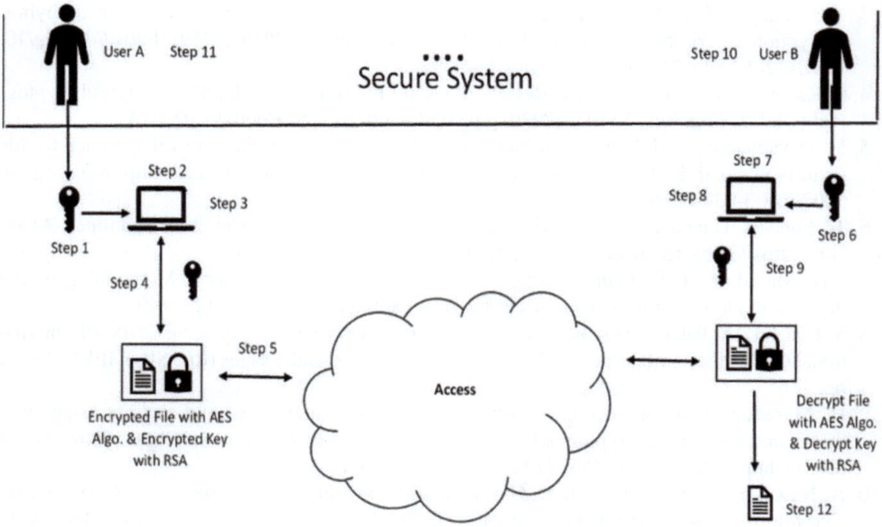

Fig. 12 Recommended approach

4 Conclusion and Future Work

Hybrid Encryption System is one of empowering security tools employing symmetric and asymmetric algorithms for achieving confidentiality of information during its transmitting. To ensure that the system is made more secure, a combination of the RSA algorithm and AES algorithm has been used in its development. Therefore, there are needs for developing more security mechanisms on protecting the attachment documents and files for the email services like Gmail, Yahoo! Mail, and Cloud. Finally, the Hybrid Encryption System can assist users to guarantee that their secret files are not accessed by other people as well as securely deliver to authorized receivers.

The future work is to apply different types of algorithms to detecting attack and improve security.

References

1. Wresearch. (Jul 2021) Saudi Arabia cyber security market (2020–2026). Accessed: 10 May 2022. [Online]. Available: https://www.6wresearch.com/industry-report/saudi-arabia-cyber-security-market-2020-26
2. S.A. Ahmad, A.B. Garko, *Hybrid Cryptography Algorithms in Cloud Computing: A Review* (2019 15th Int. Conf. Electron. Comput. Comput. ICECCO 2019, no. Icecco, 2019). https://doi.org/10.1109/ICECCO48375.2019.9043254

3. K.R. Sajay, S.S. Babu, Y. Vijayalakshmi, Enhancing the security of cloud data using hybrid encryption algorithm. J. Ambient. Intell. Humaniz. Comput. **2018** (2019). https://doi.org/10.1007/s12652-019-01403-1
4. S. Kanatt, P. Talwar, A. Jadhav, Review of secure file storage on cloud using hybrid cryptography. Int. J. Eng. Res. **V9**(02) (2020). https://doi.org/10.17577/ijertv9is020014
5. M. Priyadharsini, I.M. Mary, K. Karuppasamy, F. Margret Sharmila, A novel approach for file security in cloud. EAI/Springer Innov. Commun. Comput., 601–613 (2020). https://doi.org/10.1007/978-3-030-86165-0_50
6. Techopedia. (January 20, 2022) Hybrid encryption. Accessed: 9 May 2022. [Online]. Available: https://www.techopedia.com/definition/1779/hybrid-encryption
7. S.H. Murad, K.H. Rahouma, Hybrid cryptography for cloud security: Methodologies and designs, in *Digital Transformation Technology*, (Springer, 2021), pp. 129–140
8. S. Kaushik, A. Patel, *Secure cloud data using hybrid cryptographic scheme* (2019 4th International Conference on Internet of Things: Smart Innovation and Usages (IoT-SIU). IEEE, 2019), pp. 1–6
9. H. Abroshan, A hybrid encryption solution to improve cloud computing security using symmetric and asymmetric cryptography algorithms. Int. J. Adv. Comput. Sci. Appl. **12**(6), 31–37 (2021). https://doi.org/10.14569/IJACSA.2021.0120604
10. A. Nawal, H. Soni, S. Arewar, V. Gangadhara, *Secure File Storage on Cloud Using Hybrid Cryptography*, vol 5 (Sinhgad Institute of Technology, Lonavala, 2021). https://doi.org/10.48175/IJARSCT-1101
11. S. Gokulraj, P. Ananthi, R. Baby, E. Janani, Secure file storage using hybrid cryptography. SSRN Electron. J. **5**(12), 185–188 (2021). https://doi.org/10.2139/ssrn.3802668
12. H. Song, J. Li, H. Li, A cloud secure storage mechanism based on data dispersion and encryption. IEEE Access **9**, 63745–63751 (2021). https://doi.org/10.1109/ACCESS.2021.3075340
13. S.H. Murad, K.H. Rahouma, Hybrid cryptography for cloud security: Methodologies and designs. Lect. Notes Networks Syst. **224**, 129–140 (2022). https://doi.org/10.1007/978-981-16-2275-5_7
14. R. Adee, H. Mouratidis, A dynamic four-step data security model for data in cloud computing based on cryptography and steganography. Sensors **22**(3), 1–23 (2022). https://doi.org/10.3390/s22031109
15. O.M. Osman, M.E.A. Kanona, M.K. Hassan, A.A.E. Elkhair, K.S. Mohamed, Hybrid multistage framework for data manipulation by combining cryptography and steganography. Bull. Electr. Eng. Inform. **11**(1), 327–335 (2022). https://doi.org/10.11591/eei.v11i1.3451

Convolutional Neural Network Applied to Facial Recognition

Allam Fatima Zohra, Hamami Mitiche Latifa, and Bousbia-Salah Hicham

1 Introduction

The world we live in today is characterized by insecurity and an increase in attacks. Security has become a major concern for both governments and the private sector. In order to combat the risks of crime and terrorism, the use of Facial Recognition Systems is becoming increasingly essential. This security system remains a privilege, especially in public places.

Facial Recognition is based on data and the principles of Artificial Intelligence. Increasingly used because of its practical and effective everyday applications, Facial Recognition is now experiencing dazzling success in terms of surveillance and security. This application motivates researchers in the field of computer vision and remains one of the most studied techniques in recent decades.

As a security measure in protected environments, Facial Recognition analyzes individuals more quickly. The principle is to examine facial features that do not change significantly with age or surgery. This technology has real potential for application in various fields and should constitute a development factor.

In general, the system automatically recognizes a person based on their characteristics. Concretely, this technique makes it possible to identify, authenticate, or verify one's identity. This is the context of our work. The objective is to develop a selfie Facial Recognition System that meets the security and access control needs of highly protected sites.

This chapter is organized as follows: The first section briefly introduces the basics of biometric systems. The second section is devoted to Artificial Neural Networks,

A. F. Zohra (✉) · H. M. Latifa · B.-S. Hicham
Signal and Communication Laboratory, Department of Electronics, National Polytechnic School, Algiers, Algeria
e-mail: fatima_zohra.allam@g.enp.edu.dz; latifa.hamami@g.enp.edu.dz; hicham.bousbia_salah@g.enp.edu.dz

© The Author(s), under exclusive license to Springer Nature Switzerland AG 2024

A. Shaikh et al. (eds.), *Advances in Emerging Information and Communication Technology*, Signals and Communication Technology,
https://doi.org/10.1007/978-3-031-53237-5_26

Machine Learning and Deep Learning. The different Neural Network architectures used in Facial Recognition will be presented, with an emphasis on convolutional networks.

In the third section, we present the system we have developed. We detail the model used for facial recognition, as well as the various stages of image pre-processing and model learning. We also present the tests carried out and the results obtained. Finally, we conclude with a summary of the results obtained and the prospects for improving the recognition rate in biometric systems applied to facial recognition.

2 General Concepts of Biometric Systems

2.1 Principle and Definition

Biometrics is a field of information technology that uses physical and behavioral characteristics to identify and authenticate individuals. These characteristics are collected and stored in recognition systems [1]. Biometrics is an evolving technology that is increasingly used to recognize individuals in various applications. New biometric methods are constantly being developed, requiring constant technological monitoring.

A biometric system has two main processes: enrollment and testing [1, 2].

(a) *Registration Mode* is the first process the user encounters when enrolling in the biometric system. This process involves capturing the user's biometric data, which is then stored in a secure database for later use.
(b) *Recognition Mode (test)* is used when the individual wishes to access the biometric system. The test process compares the biometric data presented by the user with the data stored in the dataset.

There are several types of applications that require user recognition. These applications can be divided into two categories, operating in verification mode or identification mode.

A biometric system consists of the hardware, software, firmware, and network components required to perform the biometric enrollment and comparison processes. The system works by recognizing the shape of the biometric data collected from an individual to be identified via a learning module. After extracting characteristics from this data, the system can make a decision based on the result of the comparison with data stored in a database. In short, a biometric system is a pattern recognition system that uses biometric data to identify individuals [2, 3].

2.2 *Biometric System Modules*

Figure 1 illustrates the general architecture of a biometric system:

Automatic Recognition Systems, whatever the modality used, are based on five main modules [4, 5]:

(a) Acquisition Module is responsible for collecting an individual's biometric data from various devices, such as a camera.
(b) Feature extraction Module is responsible for processing the biometric data collected by the capture module and retaining only the relevant information to create a new representation of the data. This new representation must be specific to each individual and relatively stable, despite the variations that may exist between different samples of the same person.
(c) Matching Module performs an exhaustive comparison between the biometric characteristics extracted from the user and those stored in the system's dataset. It uses matching algorithms to determine the degree of similarity or divergence between the characteristics collected and those stored in the dataset.
(d) Matching Module is used to increase the similarity rate between two objects. In the context of authentication, Matching is performed between the claimed identity model and the biometric data presented by the user.
(e) Decision Module plays a crucial role in the recognition process, by determining whether or not the identity proclaimed corresponds to that of the user. This decision is based on a predefined threshold, which may be difficult to estimate.

3 Artificial Intelligence and Neural Networks Artificial

The use of Artificial Intelligence (*AI*) enables machines to imitate the complex tasks performed by human beings. To achieve this, machines need to be trained to reproduce their own representation of the world using Machine Learning. Machine Learning is a set of methods that enable a machine to learn from examples. Deep

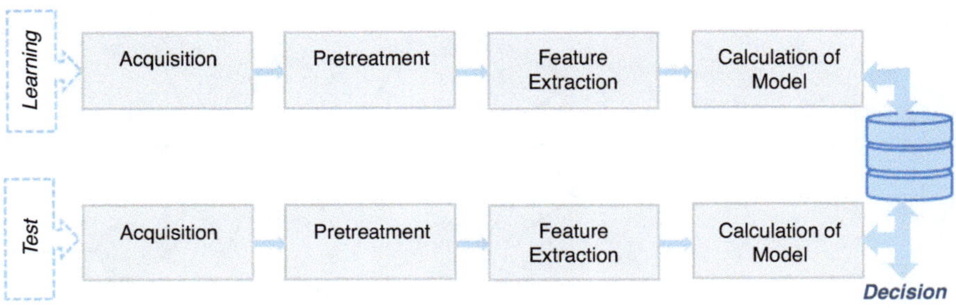

Fig. 1 General architecture of a biometric system

Learning is a subset of machine learning based on Artificial Neural Networks, which mimic the human nervous system [6].

We will present the principles of Neural Networks, Machine Learning, Deep Learning and the different Deep Learning architectures most commonly used.

3.1 Neural Networks

Neural Networks are models of information processing that simulate the functioning of the biological nervous system using interconnected neurons organized in layers. Neural Networks are information processing models that mimic the functioning of the biological nervous system. They are made up of interconnected neurons organized in layers, whose purpose is to perform complex calculations and find a non-linear relationship between numerical data and parameters [7].

Figure 2 shows an example of a Neural Network with two hidden layers [7, 8]:

The main neural network algorithms are summarized below [9–11]:

- *Error Correction Learning Algorithm:* This algorithm consists of finding the optimal weights for correctly classifying a set of inputs. It uses a learning rate and a threshold to update the weights at each iteration until the classification is correct.
- *Gradient Descent Learning Algorithm:* This algorithm is used to minimize the error of a perceptron. There are two types:
  - Stochastic Gradient Descent (SGD), which updates the weights of the perceptron each time a sample is observed.
  - Batch Gradient Descent, which updates the weights of the perceptron after calculating the average error on all the data.

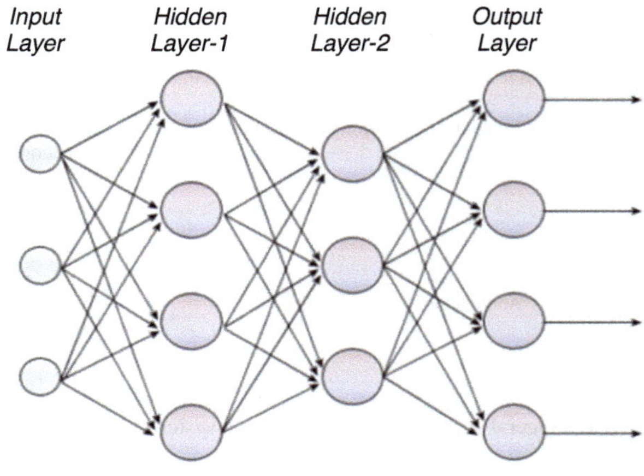

Fig. 2 Neural Network with two hidden layers

- *Multilayer Neural Networks:* This is an Artificial Neural Network model that is organized into several hidden layers. Information flows from the input layer to the output layer in a unidirectional manner. It is a type of direct propagation network.

3.2 Machine Learning

Machine Learning enables a machine to perform complex tasks by learning from data. Its aim is to automatically extract and exploit information in a data set [12].

3.3 Deep Learning

Deep Learning is a Machine Learning method that uses algorithms to learn at multiple levels, at different levels of abstraction. It is capable of extracting features from raw data using linear and non-linear transformations, through multiple layers of processing. This method requires minimal human intervention [13].

3.4 Different Models of the Deep Neural Network

There are several models for a deep neural network. Figure 3 shows the main models [14]:

In our work, we have chosen to use Deep Learning for a number of reasons:

- The failures of traditional AI algorithms in solving different problems.
- The need to manage large quantities of data (big data).
- The ability of Deep Learning to adapt to any type of problem and automatically extract characteristics.

Figure 4 compares Machine Learning and Deep Learning.

In what follows, we will describe the main deep learning networks [14, 15]:

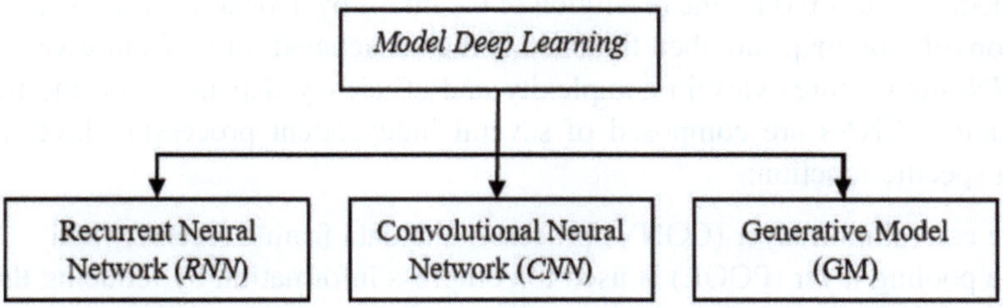

Fig. 3 Different models of Deep Learning

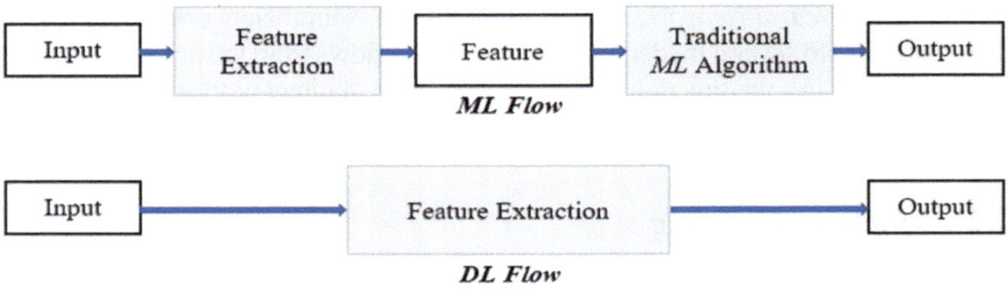

Fig. 4 Comparison between *ML* and *DL*

3.4.1 Recurrent Neural Network (RNN)

A recurrent neural network is a type of neural network in which the connection between neurons forms at least one cycle. There are several types of RNN, such as Jordan networks and Echo State Networks. However, in recent years, the LSTM cell-based neural network has become a common choice for many applications thanks to its high performance.

3.4.2 Generative Model (GM)

Generative models are probabilistic models capable of generating new observations and making predictions using Bayes' law. Different types of generative models are available, such as Boltzmann Machines, Restricted Boltzmann Machines, Deep Belief Networks and Deep Boltzmann Machines.

3.4.3 Convolutional Neural Network (CNN)

Convolutional neural networks are very powerful models for classifying images. They are made up of several convolution layers that perform processing on the input image in the form of a matrix of pixels with a third dimension to represent colours. Filters or convolution kernels create new images called convolution maps. Some intermediate filters reduce the resolution of the image by a local maximum operation. The convolution maps are then flattened and concatenated into a feature vector.

CNN architectures vary in complexity and efficiency depending on the field of application. CNNs are composed of several independent processing layers, each with a specific function:

- The convolution layer (CONV) processes the data from a receiver field
- The pooling layer (POOL) is used to compress information by reducing the size of the intermediate image
- The correction layer (ReLU) applies the activation function

- The Fully Connected layer (FC) is a perception layer
- The Loss layer (LOSS) is responsible for error correction

3.5 *Main CNN Architectures*

We present a general overview of the different existing CNN architectures [16, 17]:

3.5.1 LeNet

LeNet is a seven-level convolution network for number classification. It has been widely used by banks to recognize handwritten numbers on checks scanned at 32x32 pixels. However, this technique is limited by the availability of computing resources, as it requires larger layers to process higher-resolution images.

3.5.2 Alex Net

Alex Net is a convolutional neural network with a similar architecture to LeNet. However, it is deeper and larger, with convolutional layers stacked on top of each other.

3.5.3 VGGNet

VGGNet is a convolution network made up of sixteen convolution layers with a very uniform architecture. It uses only 3x3 convolutions with many filters. It is currently the community's preferred choice for feature extraction from images. Although its weight configuration is publicly available and used in many other applications, VGGNet has 138 million parameters, which can make it difficult to use.

3.5.4 GoogLeNet

GoogLeNet is a convolution network with 22 layers. It uses layers processed in parallel, unlike previous architectures, which had sequential layers. We have used GoogleNet's CNN architecture for facial recognition. In what follows, we will explain this model.

4 Implementation of the Recognition System

We will describe the design of a face recognition system based on the GoogLeNet convolutional neural network, also known as the Inception Module. We have chosen this architecture because it has shown high performance in complex image classification tasks, and in particular in face recognition.

We will begin by describing the architecture of GoogLeNet and the reasons why we chose it for the implementation of the facial recognition system. We will then explain the various stages of data preparation required to train the network, including data collection and data preparation. We will then detail the network training process, describing the parameters and hyper-parameters used, as well as the performance evaluation metrics. We will then present the results of our facial recognition system and discuss the limitations and prospects for improvement.

4.1 Description of the Facial Recognition System

The application block diagram is shown in Fig. 5:

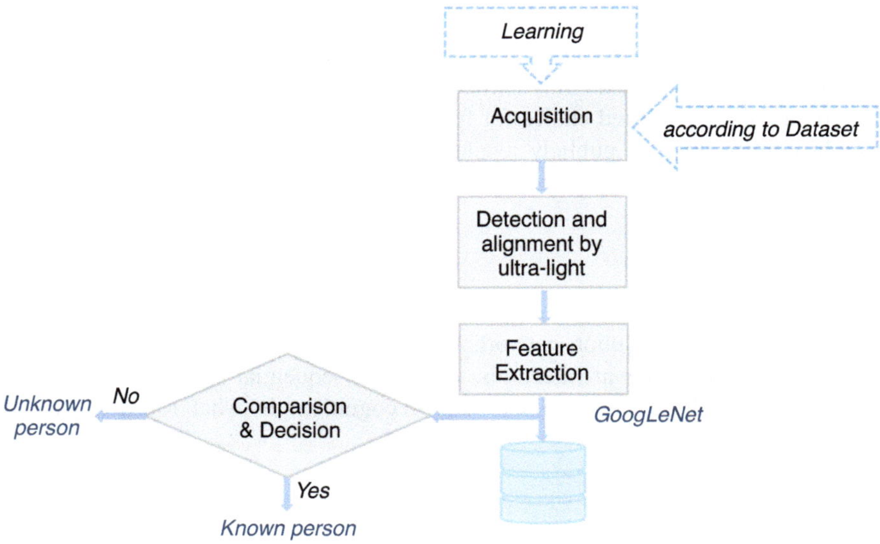

Fig. 5 Block diagram of the developed system

4.2 Development Tool

The implementation of the facial recognition system is based on CNN GoogLeNet, using Deep Network Designer, a visual development environment for creating deep neural networks in Matlab.

Deep Network Designer can be used to design deep neural network architectures for image recognition applications. It is strongly recommended that algorithms be developed in matrix form rather than using "for" loops. It is also recommended to use explicit variable names to improve the readability of the programs.

Deep Network Designer and Matlab's graphical functions allow you to interactively modify the various neural network parameters to suit your application.

4.3 GoogLeNet Architecture

GoogLeNet is known for its complex and innovative structure, which has resulted in greater classification accuracy with less use of computing resources than previous networks. This architecture uses so-called "Inception" modules, which combine convolutions of different sizes to capture image patterns at different scales.

GoogLeNet is still considered one of the most influential models in the field of computer vision and has inspired the development of many other convolutional neural networks. Figure 6 describes the GoogLeNet architecture used in face recognition.

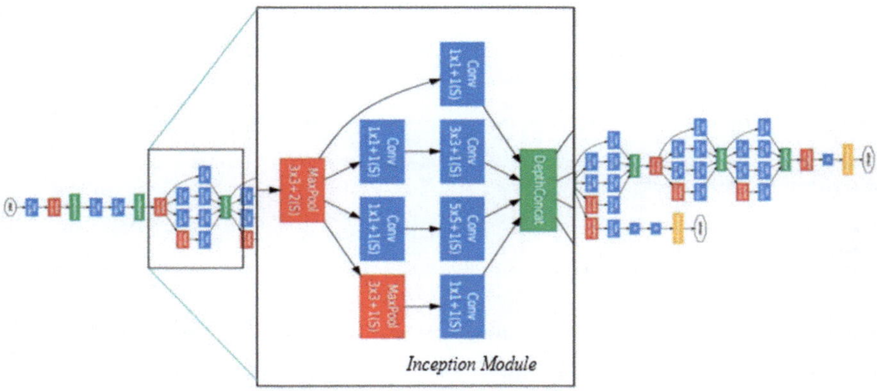

Fig. 6 Architecture of the GoogleNet neural network

4.4 Architecture of GoogLeNet

GoogLeNet is based on the use of inception blocks, which are calculation modules that can process an input with different convolution kernel sizes simultaneously. The GoogLeNet architecture contains 22 layers, with a funnel-shaped architecture that enables a significant reduction in the number of parameters. It is composed of a stack of inception blocks, which include convolution, pooling, and batch normalization layers, as well as output classification.

Each inception block has a complex architecture, with several convolution branches that are concatenated at the end to form the output. These inception blocks allow GoogLeNet to maintain a deep architecture while reducing the number of parameters and improving classification accuracy.

GoogLeNet also uses auxiliary layers to help train and regulate the network. These auxiliary layers have been added to prevent overfitting and to enable faster training.

In GoogleNet, the two layers, "Loss3-Classifier" and "Output," contain information on how to combine the features extracted by the network into class probabilities, loss values, and predicted labels. In order to adapt this network to the image set, we replaced these two layers with new layers adapted to the image set. As mentioned in the code section (Fig. 7).

This program implements a facial recognition system using the GoogLeNet architecture to extract features from training images. The dataset is imported from

```
Input_Layer_Size = net.Layers(1).InputSize;
Layer_Graph = layerGraph(net);

Feature_learner = net.Layers(142);
Output_Classifier = net.Layers(144);

Number_of_Classes = numel(categories(Training_Dataset.Labels));

New_Feature_Learner = fullyConnectedLayer(Number_of_Classes, ...
    'Name','Facial Feature Learner', ...
    'WeightLearnRateFactor', 10,...
    'BiasLearnRateFactor', 10);
New_Classifier_Layer = classificationLayer('Name', 'Face Classifier');

Layer_Graph = replaceLayer(Layer_Graph, Feature_learner.Name, New_Feature_Learner);

Layer_Graph = replaceLayer(Layer_Graph, Output_Classifier.Name, New_Classifier_Layer);
analyzeNetwork(Layer_Graph);

Pixel_Range  = [-30 30];
Scale_Range = [0.9 1.1];
Image_Augmenter = imageDataAugmenter('RandXReflection', true, 'RandXTranslation', Pixel_Range,...
    'RandXScale', Scale_Range, ...
    'RandYScale' , Scale_Range);
Augmented_Training_Image = augmentedImageDatastore(Input_Layer_Size(1:2), Training_Dataset, ...
    'DataAugmentation', Image_Augmenter);
Augmented_Validation_Image= augmentedImageDatastore(Input_Layer_Size(1:2), Validation_Dataset);
```

Fig. 7 Modified GoogleNet Neural Network

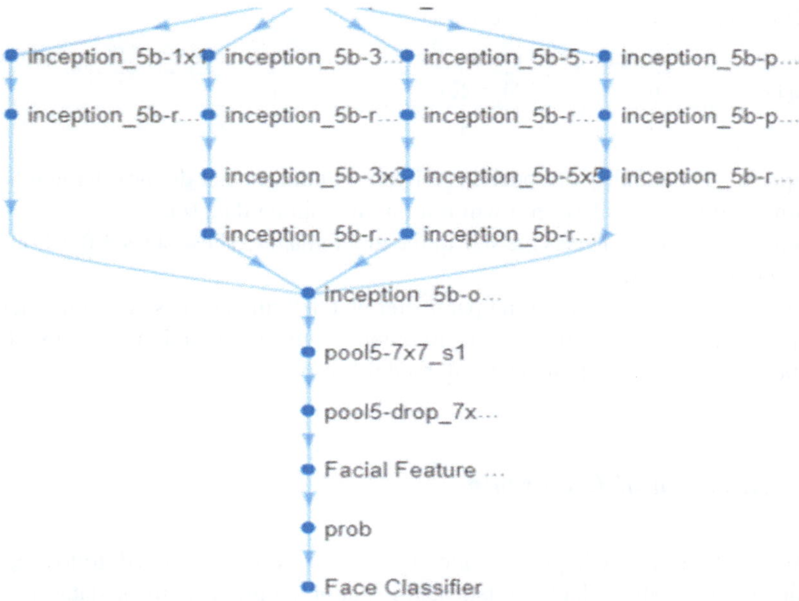

Fig. 8 Visualization of the modified GoogleNet architecture

a "My_Dataset" folder containing subfolders for each class of images. The images are then divided into a training set (70%) and a validation set (30%).

The GoogLeNet architecture is loaded and initialized with its pre-trained weights. The output layer (Classifier Layer) is replaced by a Fully Connected layer for learning new facial features and a classification layer for predicting classes (identities). An image augmentation is applied to the training set to increase the variability of the data and avoid overfitting. The augmentation parameters are specified using the image DataAugmenter.

The model is then trained with a stochastic gradient descent algorithm using specified training options. Options include mini-batch size, number of epochs, initial learning rate, validation frequency, etc. Trained model is then stored in the "netw" variable for later use in predicting face classes for new images.

Figure 8 shows a visualization of the GoogleNet architecture after the last layers have been modified.

4.5 *GoogLeNet Network Settings*

Table 1 shows the various parameters of the GoogleNet CNN model applied in our project.

Table 1 Parameters of the GoogLeNet model

Settings	Epoch	Input images size	Learning rate	Size of minibatch
GoogLeNet	6	224 × 224	3e-4	5

- Epochs: The number of epochs is a hyper-parameter that defines the number of times the learning algorithm will run on the training data set.
- Input image sizes: This is the size of the input images. These sizes differ from one model to another.
- Learning rate: This is a setting parameter that determines the size of the "step" at each iteration while aiming for the minimum of the loss function. It represents the speed at which a machine-learning model learns.

4.6 Results and Discussion

In this section, we will present and discuss the results obtained following the application of GoogLeNet with data augmentation on the face image dataset.

Deep learning algorithms now love large volumes of labeled images. To improve their performance, we use the Data Augmentation technique, which involves transforming all the images before using them for training while preserving their structure.

For each observation, we create several variations. These transformations can be divided into two categories: a set of affine transformations such as horizontal and/or vertical flip, rotation, or non-affine transformations such as variations in brightness and contrast, resizing, etc. In this way, the volume of data is artificially multiplied.

Figure 9 shows the training accuracy and loss curve for GoogLeNet with increasing data. From this figure, we can see that the training accuracy increases across epochs, while the loss decreases over time. The results show that GoogleNet achieved an accuracy rate of 99.8% with an error rate of 0.2%.

5 Conclusion

In conclusion, facial recognition is an evolving technology with many practical and security applications. It is increasingly used to guarantee security in public places and protected environments.

In this chapter, we have implemented an image recognition system based on the CNN architecture, more specifically the GoogLeNet model. The various tests carried out showed satisfactory results in terms of recognition rates, compared with those of other image recognition models. These results are encouraging for the further development of this technology and its application in various fields, including surveillance, medicine, facial recognition, etc.

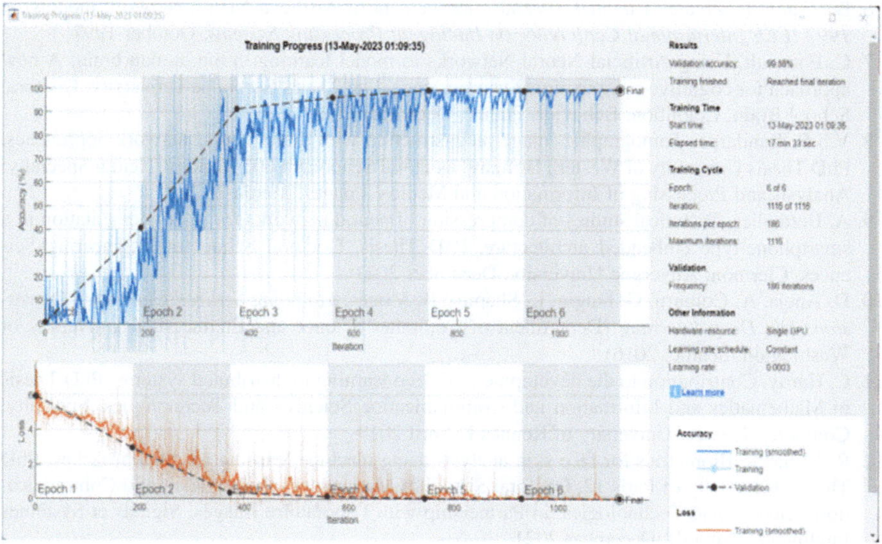

Fig. 9 Learning progression of the GoogleNet network

However, it should be noted that the quality of the results is highly dependent on the quality of the data used to train the model. Therefore, it is important to have a high-quality and large dataset to ensure optimal performance. In addition, the complexity of the CNN architecture can also make it difficult to use for simpler image recognition tasks, where simpler models can be developed.

In conclusion, the use of machine learning and deep neural networks for image recognition is a promising technology that offers numerous possibilities in many fields. It is important to continue to develop and improve this technology in order to get the most out of it and benefit society as a whole.

References

1. N. Khiari-Hili, Multimodal biometrics based on the iris and the face, PhD Theses, University of Tunis El Manar and University of Evry Val d'Essonne Paris-Saclay, Information and Communication Sciences and Technologies, Speciality: Treatment of signal and images, May 2016
2. B. Ammour, Contribution to the development of Biometric Systems based on the face and the iris, PhD Thesis in Sciences, specialty: Electronics, Faculty of Sciences and Technology University Mohammed Seddik Benyahia, December 2018
3. F. Bedad, R. Adjoudj, N. Bousahba, Study of the robustness of a transformation-based multi-biometric template schemes protection. Int. J. Comput. Digit. Syst. **11**(3), 336–344 (2022)
4. A. Aissaoui, Bimodal face recognition by fusion of visual and depth features, PhD Thesis in Computer Science, Fundamental Computer Science Laboratory of Lille (LIFL), June 2014
5. N.T. Deshpande, S. Ravishankar, Face detection and recognition using Viola-Jones algorithm and fusion of PCA and ANN. Adv. Comput. Sci. Technol. **10**(5), 1173–1189 (2017)

6. B. Zhang, L. Zhang, The analysis and improvement of Artificial Neural Network models, in *1997 IEEE International Conference on Intelligent Processing Systems*, October 1997
7. C. Foucault, Using Artificial Neural Networks to model learning in the human brain: A new approach for cognitive computational neuroscience, Phd Thesis Sorbonne University, Doctoral School Brain, Cognition, Behavior, France, December 2020
8. V.S.S. Kandarpa, Tomographic image reconstruction with direct neural network approaches, PhD Thesis University of Western Brittany, Doctoral School No 605 Biology, Health Specialty: Analysis and Processing of Information and Medical Images, March 2022
9. A. Berthelier, Technical studies of compression of neural networks for its implementation in a smartphone-type embedded architecture, PhD Thesis, Doctoral School of Engineering Sciences, Clermont Auvergne University, December 2021
10. D. Aneja, A. Colburn, G. Faigin, L. Shapiro, B. Mones, *Modeling Stylized Character Expressions Via Deep Learning* (Department of Computer Science and Engineering, University of Washington, Seattle, 2016)
11. C. Hardy, Contribution to the development of deep learning in distributed systems, PhD Thesis in Mathematics and Information and Communication Sciences and Technologies, Specialty: Computer Science, University of Rennes I, April 2019
12. R. El Saleh, Biometrics for face skin analysis using machine learning based approaches, PhD Thesis, University of Paris 12, Doctoral School of Mathematics, Information and Communication Sciences and Technologies, in Partnership with Laboratoire Images, Signals et Systèmes Intelligents (Créteil), December 2021
13. A. Caubrière, From signal to concept: Deep neural networks applied to speech understanding, PhD Thesis in Mathematics and Information and Communication Sciences and Technologies Specialty: Computer Science Le Mans University, LIUM, January 2021
14. L. Deng, D. Yu, Deep learning: Methods and applications. Found. Trends Signal Process. **7**(3–4), 197–387 (2014)
15. H. Xiaopeng, L. Jingting, Y. Yang, W. Fan, Reliability verification-based convolutional neural networks for object tracking. IET Image Process. **13**(1), 175–185 (December 2018)
16. M. Swapna, K.S. Yogesh, B.M.G. Prasadh, A survey on face recognition using convolutional neural network, in *Data Engineering and Communication Technology*, (Springer, Singapore, 2020), pp. 649–661
17. M. Swapna, K.S. Yogesh, B.M.G. Prasadh, CNN Architectures: Alex Net, Le Net, VGG, Google Net, Res Net. Int. J. Recent Technol. Eng. (IJRTE) **8**(6), 953 (March 2020)

Power Consumption Analysis for Smarter Robotics Via Industry 4.0 Methods and Technologies

Ryan Samson, Keng Goh, Akshath Sankarraj, Alexandros Gkanatsios, and Hongnian Yu

1 Introduction

The industrial sector accounted for 169 exajoules, representing 38% of the global final energy usage in 2021 [1]. Although a significant portion of energy consumption in the industrial sector is attributed to heavy industry and energy-intensive processes, the World Robotics Report highlights the sustained demand for robotics despite challenges like electronic components shortage and rising energy prices. With over 517 thousand robots installed around factories in 2021, the industry has been experiencing an annual growth rate of 31% year on year. However, the International Federation of Robotics (IFR) report predicts a decline in the growth rate in the future [2].

This paper aims to contribute to three out of five prominent trends identified by the IFR [3], namely, energy efficiency, digital integration, and sustainability focus, through the upgrading or refreshing of existing machinery. By addressing these trends, the work presented in this paper aligns with the evolving needs of the industry and offers insights into leveraging these advancements for improved performance and environmental sustainability.

Industrial sectors consume a high percentage of energy. In this research, our application is on an Industry 3.0 (I3.0) technology, specifically an assembly robot, where Industry 4.0 (I4.0) techniques to harness motor data are used to improve energy efficiency. We showcase that the use of I4.0 methods and technologies has the potential for existing and new manufacturing machinery to operate at increased levels of efficiency.

R. Samson · K. Goh · A. Sankarraj · A. Gkanatsios · H. Yu (✉)
School of Computing, Engineering and the Built Environment, Edinburgh Napier University, Edinburgh, UK
e-mail: Ryan.Samson@napier.ac.uk; K.Goh@napier.ac.uk; 40517059@live.napier.ac.uk; A.Gkanatsios@napier.ac.uk; H.Yu@napier.ac.uk

© The Author(s), under exclusive license to Springer Nature Switzerland AG 2024

A. Shaikh et al. (eds.), *Advances in Emerging Information and Communication Technology*, Signals and Communication Technology,
https://doi.org/10.1007/978-3-031-53237-5_27

With a growing need to reduce the resource waste, industrial environments are actively seeking opportunities to improve their operations, resulting in a reduction in waste and cost. Increasing connection of industrial equipment brought around with the fourth industrial revolution will enable a greater and more valuable communication through the value chain, enabling optimisation [4].

In factories where robots are deployed, the integration of I4.0 technologies such as Internet of Things (IoT), cyber-physical systems (CPS), cloud computing, and data analytics [5–9] has been proving to enhance the capabilities of I3.0 assembly robots. By incorporating these I4.0 technologies, robots gain smart features such as increased connectivity for collaborative capabilities, intelligent decision-making, and advanced perception. This integration enables industrial robots to achieve higher productivity, improved quality, enhanced safety, and greater efficiency in manufacturing environments.

Horizontal integration refers to the seamless coordination and collaboration of different components, processes, or systems within a manufacturing environment. This research uses the opportunity to analyse power consumption of a robot as a platform to discuss vertical and horizontal integration in the context of I4.0. It demonstrates how this data can be turned into information used at various stages in a smart environment, exemplifying the practical application of horizontal integration.

Many articles focused on the energy consumption of industrial robots concentrate on predicting energy consumption through the development of simulation models [10]. In their literature review, Brossog and Bornschlegl [10] discovered the numerous methods to reduce energy consumption in industrial robots. These methods primarily focus on optimising operating parameters, scheduling operations, and developing energy-efficient motion planning. However, they acknowledge that implementing these methods presents challenges due to the technical intricacies of the robotics control systems, especially those that not originally designed to accommodate such functionalities. Consequently, further research and development are necessary to effectively apply these techniques in complex manufacturing systems.

In energy-consuming industrial environments, significant efforts have been made to achieve high levels of energy efficiency. However, traditionally, the design of robotic and automatic systems has not prioritised energy efficiency as an objective [11]. The research question addressed in this study is whether I4.0 technologies and methods can be harnessed to increase the energy efficiency of these machines retrospectively.

Energy-efficient production processes can be realised through the utilisation of I4.0 digital technologies. By incorporating intelligent features, such optimised AI control models have the potential to lower production time while simultaneously decreasing energy consumption [12].

Campo and Calatrava [13] developed and tested an open-source energy monitoring IoT solution in an industrial factory environment. The implementation of this contribution, which is open-source and applicable in an industrial environment not only enables energy monitoring at machine or process level but also serves as a good guide for SMEs seeking to invest in I4.0 practices with limited knowledge and funding.

In a study conducted by Lima and Massote [14], a retrofit approach was presented, integrating I4.0 technologies into legacy machines with a specifical focus on energy monitoring. The significance of this approach lies in its alignment with the principles of sustainability, emphasising the reuse of existing equipment. It raises an important question about the necessity of replacing legacy machines when the retrofit approach effectively allows us to harness the capabilities of I4.0 technologies. By considering this question, we can address Sustainable Development Goals (SDG), including waste reduction and energy conservation that would otherwise be associated with recycling machines. The importance of reusing equipment should be emphasised unless there are compelling reasons to replace old machines. If one of the key drivers for replacement is the incorporation of smart elements to enable superior operational efficiency, the solution of retrofitting with I4.0 technologies becomes even more compelling as it allows for optimised resource utilisation, aligning with SDGs.

With recent articles defining manufacturing sustainability as the conservation of energy [15], efforts towards monitoring and controlling energy usage not only enable cost reduction but also enhance competitiveness by aligning with customers' attitudes and support for sustainability [16]. Implementing measures to monitor and control energy consumption, manufacturing companies can achieve cost savings, demonstrate their commitments to reduce environmental impact, and meet the evolving demands of sustainability-minded customers.

Guerra-Zubiaga and Luong [17] conducted research on the effects of speed, acceleration, temperature, and payload over three joints and joint gearbox of an industrial robot. The results indicated that speed and acceleration account for the majority of power consumption in the first three joints of the robot, starting from the base. The authors highlighted the importance of exploring other factors that contribute to the energy consumption and seeking options to minimise them. Building upon these findings and previous works in this field, our aim is to provide practical implementation data and explore techniques to minimise the energy consumption using technologies associated with the fourth industrial revolution.

This paper explores the potential opportunities for energy efficiency in industrial environments by harnessing I4.0 technology. While integrating I4.0 methods and technologies may be more straightforward in greenfield sites or new developments, this paper specifically focuses on achieving energy efficiency goals through I4.0 using existing legacy equipment. The objective of this study is to demonstrate the feasibility of collecting machine data for monitoring purposes with minimal disruption to plant operations. Additionally it also outlines a path for investigating and implementing I4.0 smart solutions that can effectively improve machine efficiency. The paper is structured as follows: Section 2 presents a I4.0 framework applied in this research. Section 3 introduces the demonstration platform specification and data record architecture. Section 4 presents results and discussion to utilise technologies and methods of I4.0 to improve efficiency. Finally, Sect. 5 concludes the paper.

2 Industry 4.0 Methods and Technologies

The aim of this work is to demonstrate the application of I4.0 methods and technologies to manufacturing equipment that originates from the third industrial revolution. The paper proposes using I4.0 techniques to increase connectivity and integrate CPS and data analytics into the equipment. These techniques are presented within a framework, as depicted in Fig. 1. By leveraging technology advancements in I4.0, the aim is to explore the potential benefits and opportunities for enhancing the performance and capabilities of legacy equipment. This is done in order to better prepare industrial environments for the current and the future landscape of manufacturing.

The initial stage of this project introduces the proposed setup, showing the current operation of the robot (Fig. 1—plant floor). The focus is on creating a closed-loop system that can be connected to cloud infrastructure (Fig. 1—resource planning). This connection allows for real-time information to be transmitted and executed on powerful data analytics platforms.

The setup illustrated in Fig. 1 not only facilitates feedback from data analytical platforms to enhance the system's functionalities mentioned earlier but also provides real-time information from plant floor resources at the enterprise level. This real-time information empowers decision-makers with valuable insights, enabling them to make informed and data-driven decisions to optimise processes, improve efficiency, and drive overall performance in their operations.

The subsequent section of the paper reports on the equipment and experiment configuration, followed by data recording and analysis of the process performed at

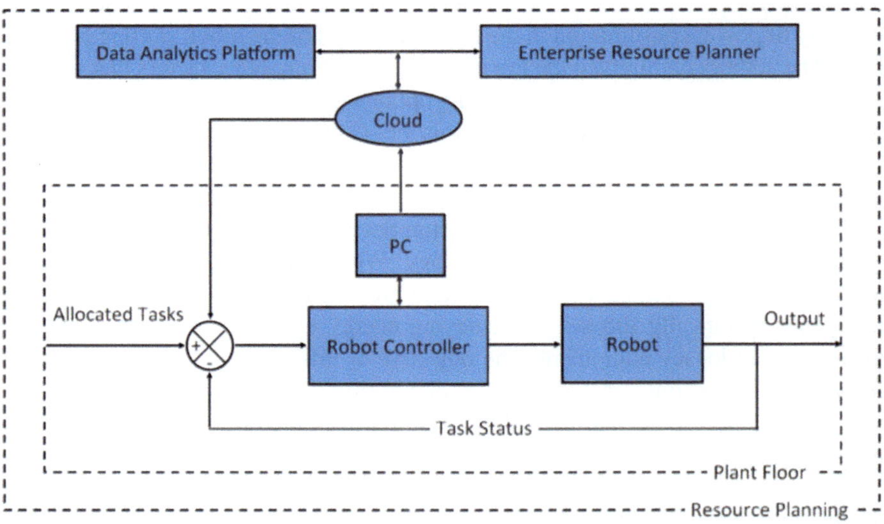

Fig. 1 Data flow from robot to cloud-based data analytics platform and ERP for real-time monitoring and decision-making

the plant floor level. This work represents the early stages of our research and lays the foundations for future application of I4.0 methods and technologies. The objective is to demonstrate the increased system performance and efficiency derived from implementing these approaches.

3 Experiment Configuration

In this section, the technical details regarding hardware and software used for the experiment are presented along with the experimental setup.

3.1 Robot Arm Technical Specifications

The hardware setup includes a Mitsubishi RV-2FB-D industrial assembly robot as shown in Fig. 2a. This robot features a vertical serial joint linkage structure, enabling precise and efficient operations in assembly processes. With 6 degrees of freedom, it enables the robot to operate with a high level of manoeuvrability and flexibility. The platform has a payload capacity of 2kg, making it suitable for various lightweight assembly tasks. The robot exhibits a position repeatability of +/-0.02mm, ensuring consistent and accurate positioning. Equipped with an AC servo system and

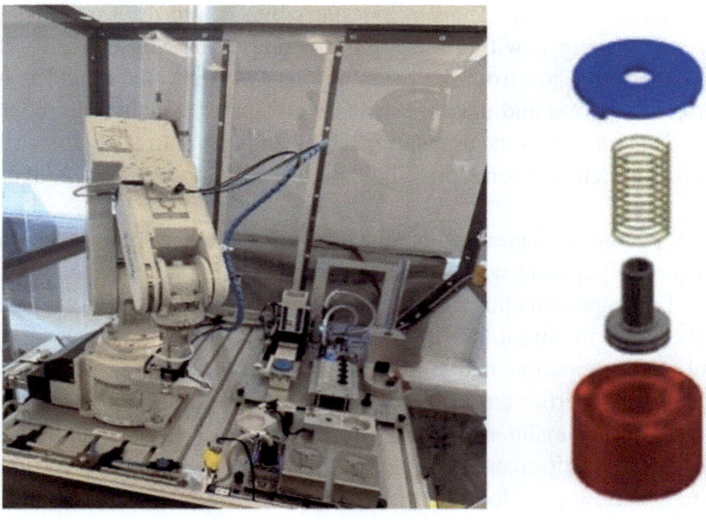

(a) (b)

Fig. 2 Experimental setup, Mitsubishi RV-2FB-D industrial assembly robot (**a**) and exploded view of four component assembly task (**b**) regenerative braking on all axes, the system is designed to enhance energy efficiency. The robot has a maximum reach of 504mm, allowing it to access different areas within the assembly environment. The operational range of each axis is 480, 240, 160, 400, 240, and 720 degrees (J1 to J6). The robot operates at a maximum composite speed of 4955 mm/s, enabling efficient execution of assembly operations.

The assembly process involves four components: a cylinder, piston, spring, and retaining cap, shown in Fig. 2b. The industrial robot is programmed using the original equipment manufacturer (OEM) software, Melfa basic, which is a text-based software language, enabling precise control over the assembly process.

Currently, this robotic arm is situated within an assembly station serving a flexible manufacturing process. The system operates by receiving the individual components, conducting quality checks, and subsequently assembling them using the robotic arm. The assembled pistons are then either stored in a warehouse or sent to the distribution centre for further sorting. This paper aims to demonstrate the potential benefits of full integration with cyber-physical systems (CPS), one of the key technological pillars and building blocks of I4.0, to optimise the manufacturing process by enabling insights to disturbances and varying conditions [18].

3.2 Experimental Setup

During all the experiments, for consistency, the robotic arm follows a pre-programmed trajectory. The process starts by picking up a cylinder from the conveyor and placing it within the robot assembly cell. Subsequently, the robot moves to collect a piston from the pallet and inserts it into the cylinder. Next, the robot retrieves a spring and places it inside the cylinder. Finally, the robot collects the cylinder cap, identifies the orientation, and then fits it onto the cylinder. Once the assembly is complete, the robot picks up the finishes component and delivers it back to the conveyor.

As I4.0 awareness increases, organisations have the option to upgrade their existing legacy equipment with low-cost solutions that enable integration with I4.0 principles. This approach allows organisations to benefit from the advantages of I4.0 without the need to invest in entirely new machines [19]. However, to ensure successful implementation and maximise the potential benefits of the upgrade, various levels of expertise are required due to the nature and complexity of the task.

As parts of robot maintenance, authorised personnel gather operational data to assess the robot's performance and identify any potential faults or unusual behaviour. In this study, we explore the possibility of utilising this existing information locked inside legacy equipment, eliminating the need for additional calibration and setup to avoid costs associated with introducing new devices solely for data capture. The maintenance data includes motor current data for each joint (J1 to J6), as shown in Fig. 3. The users can choose the sample rate depending on the desired resolution of information from the machine.

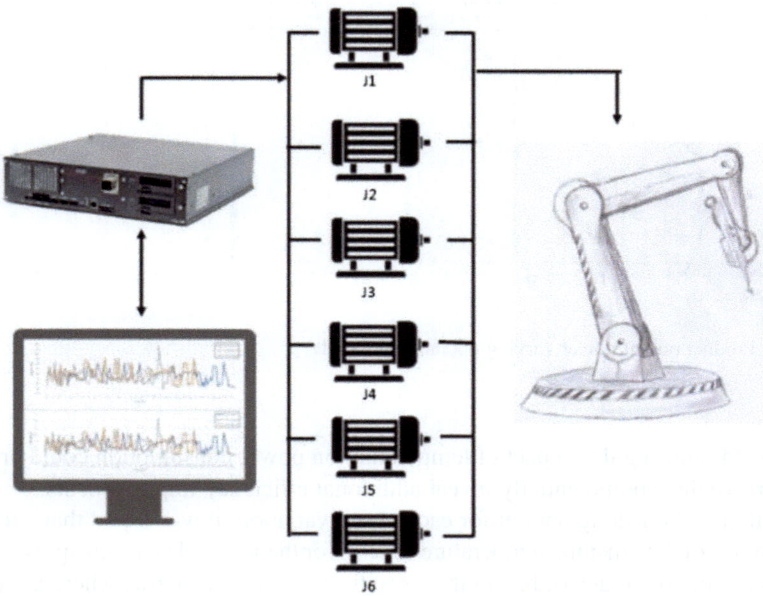

Fig. 3 Block diagram illustrating the information flow and components of the experimental setup, including the six motors, controller, and PC for data monitoring and analysis

To derive meaningful insights from this data, we selected a sample rate of 140Hz. By using the current (in ampere root mean square (ARMS)), motor voltages, and sample period, we calculated the instantaneous power consumption since the previous sample period. This calculation helps in converting the raw data into valuable information about the robot's power usage.

4 Results and Discussion

Motor performance, including power consumption, is known to vary with temperature [20]. To ensure that differences in power consumption were primarily attributed to intentional variations in speed rather than temperature fluctuations, a temperature range of ±3 degrees Celsius was maintained across the three datasets for motor joint J1. Similarly, for motor joint J2, a temperature range of ±3 degrees Celsius was maintained and so on for motor joints J3 to J6. This approach allows for consistent temperature within each dataset for the respective motor joint, ensuring that any observed differences in power consumption were primarily influenced by the intentional variations in speed rather than temperature fluctuations among the different datasets. This approach enhances the reliability and validity of the power consumption analysis in relation to different speeds and weights. However, it also imposes a limitation on the study by restricting the exploration of temperature

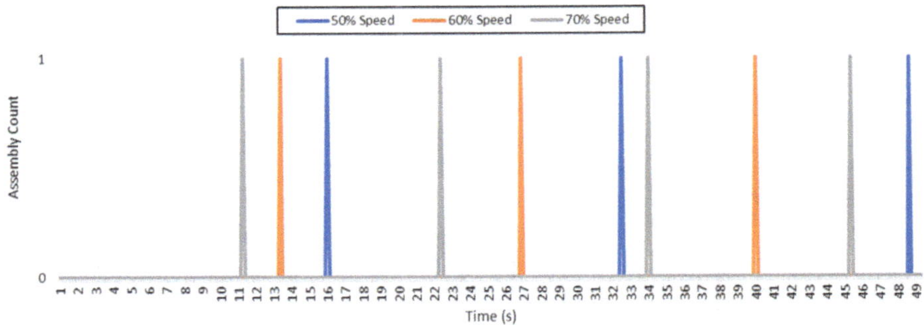

Fig. 4 Product output rate at varying operational speeds

effects. Monitoring the impact of temperature on power consumption could provide further insights and potentially reveal additional efficiency improvements.

Within an hour-long dataset for each speed variation, it was found that a total of 12 minutes of data met the temperature criteria for the joints. The results presented in Figs. 4, 5 and 6 are derived from this specific 12-minute window, where the motor joint temperature conditions were within the desired range.

The study focuses on examining the impact of operational speed on power consumption. One of the primary factors an assembly robot speed would be altered is for the rate of product output. The maximum speed of the robot used in this test was 4966 mm/s; for the purpose of this test, the speed values the robot operates were chosen at 50%, 60%, and 70% of this maximum speed value. Figure 4 displays the data from a counter which outputs a momentary digital signal of 1 when new component has been successfully assembled. The data shows that at 70% speed operation, the time to assemble one component is 10.5s followed by 13.2s (60%) and 16s (50%).

The data presented in Fig. 5 provides insights into the impact of operational speed on power consumption. The variations in instantaneous power consumption are displayed in both watts and joules. By increasing the robot operational speed from 50% to 60%, there is a 16.72% decrease in energy consumption; from 60% to 70% operation speed, there is 11.79% decrease in energy consumption. The findings suggest that due to the higher operational speed and lower operational time, the energy consumed per component assembled decreases with faster robot operation.

Additionally, the study examined the impact of different weights on the energy consumed by the robotic arm. Figure 5 illustrates the recorded data from the robot while operating with varying weighted assemblies, including payloads of 250g, 500g, and 900g. The robot's movements during these tests were consistent with the previous experiments; however, the operational speed and the payload were increased. The bar chart in Fig. 6 provides a comparison of power consumption for different loads and speeds.

The data analysis in this study considered two key elements of the assembly robot task: component weight and operational speed. Since the robot handles

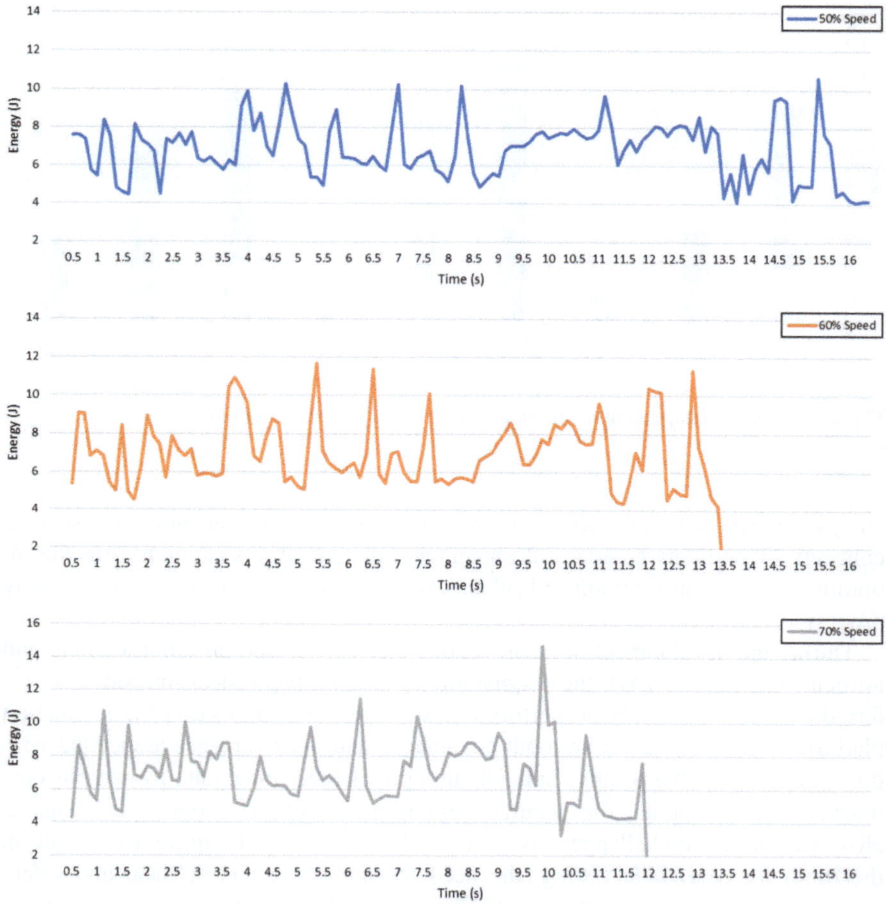

Fig. 5 Power and energy consumption for one product assembly at varying robot operational speeds

components that can have varying weights due to design differences and its operational speed can vary based on task completion time, these two factors were considered in the data analysis process.

The weight aspect in robotics assembly can serve as a valuable smart manufacturing element by incentivising engineers and designers to reduce weight through quantifying energy consumption based on weight margins. By analysing the relationship between weight and energy consumption and regeneration levels during assembly, designers can use this smart tool to aim for optimal weight that maximises efficiency.

I4.0 plays a crucial role in integrating various elements across the product line, connecting designers with manufacturers and others key stakeholders along the product line. This collaboration fosters a harmonious approach towards creating

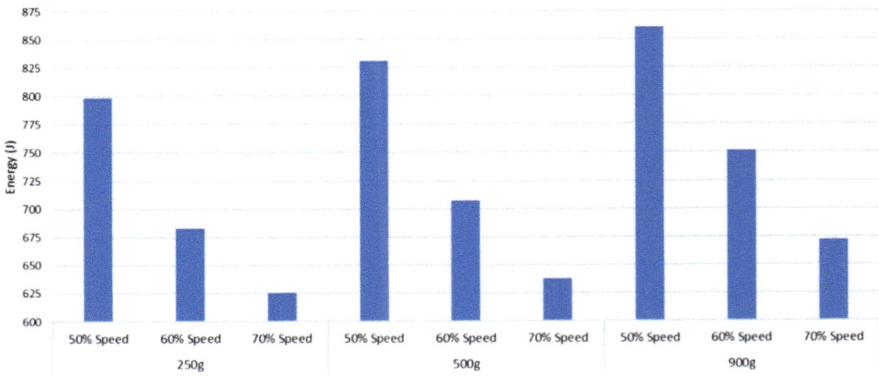

Fig. 6 Energy consumption of the robot assembly at varying speed and loads

energy-efficient solutions. By leveraging the power of I4.0 technologies, such as connectivity and data analytics, designers and manufacturers can work together to optimise energy consumption and enhance overall system efficiency in the assembly process.

During the development of smart robotic platforms based on robot learning and artificial intelligence (AI), the acquisition of data from physical machines is preferred over relying solely on performing tests and data acquisition via simulation platforms. However, there are both advantages and disadvantages associated with this approach. On the positive side, obtaining real operational data from a physical machine allows for a more accurate representation of the robot's performance. However, there are challenges involved, such as the manual requirement to set up the robot for repeatable testing, the need for large amounts of data, and safety concerns related to repeated setup [21]; this is perhaps also the reason that there are many simulation models used to predict the energy consumption of industrial robots [10]. This paper proposes a platform that incorporates I4.0 technologies, aiming to address these challenges, enabling smoother testing of physical robots with AI applications to enable smart robotic platforms.

5 Conclusion

The research presented in this paper focuses on analysing the energy consumption characteristics of an industrial robotic arm during the assembly of a component. The main focus is to highlight the available information at the machine or plant level and demonstrate how an I4.0 framework can be used to plan and deploy optimised processes.

Optimising machine cycle times and achieving good operational equipment effectiveness are important in many industrial environments. With a large push towards resource effectiveness, driven by SDGs, which calls for minimising the consumption of valuable resources, we look towards I4.0 methods and technologies that can help achieve this without negatively impacting production.

The dataset presented in this paper, along with temperature information, can serve as input for an algorithm to enable the robot to adjust its speed and movement for optimal efficiency. While this optimisation can be achieved locally at the machine, the objective of this paper is to provide an example where I4.0 methods and technologies are applied to existing operational technology, contributing to the empirical research on I4.0 and showcasing the development of a system within the I4.0 smart manufacturing domain. This work aims to support industry practitioners in comprehending and implementing the principles of I4.0.

The work complete in this research paper adds to the existing literature for advancing the field of robotics, smart manufacturing, and I4.0. The main contributions of this work includes (1) a practical perspective on understanding the energy requirements and efficiency of robotics in a manufacturing setting. (2) Examining the power consumption of the robotic arm within the context of I4.0, this work demonstrates the application of advanced technologies, data analytics, and intelligent systems to optimise energy usage and enhance the performance of manufacturing processes. (3) This research contributes to the development of smart manufacturing by exploring how power consumption analysis can make the robotic arm smarter. This analysis provides valuable insights into energy optimisation, cost reduction, and efficiency improvement, enabling the advancement of intelligent and interconnected manufacturing systems. (4) This study investigates practical techniques for

industries aiming to adopt I4.0 technologies and improve their manufacturing processes. By understanding the power consumption patterns of the robotic arm, manufacturers can make informed decisions about energy management, resource allocation, and predictive maintenance strategies. The I4.0 methods and technologies investigated helps to make this data, beforehand locked in the machine, now accessible. (5) Finally this study investigates the potential for horizontal and vertical integration. The integration of power consumption analysis and horizontal integration highlights the potential for leveraging energy insights to optimise the coordination and interoperability of interconnected systems within a manufacturing environment, aligning with the principles of I4.0 and further clarifying horizontal integrations.

A flexible manufacturing system's (FMS) overall power consumption can differ based on the system and the production procedures it is intended to support. Robots, conveyor belts, material handling systems, and other pieces of machinery frequently make up FMSs along with other forms of machinery. Each machine uses a different amount of electricity depending on its size, power rating, and usage habits, among other things. In a manufacturing setting, with the vast amount of machinery, the impact of this research acts as building block to visualise the possibilities that can be made by connecting other equipment with the same mindset, creating a fully connected, smart system that has the information, provided from data to be able to make an element more efficient, that could be energy, time, reduced machine failure possibilities.

References

1. EIA, *Industry Sectorial Overview*. 2022; Available from: https://www.iea.org/reports/industry
2. IFR, *World Robotics Report: 'All-Time High' with Half a Million Robots Installed in one Year*. 2022; Available from: https://ifr.org/ifr-press-releases/news/wr-report-all-time-high-with-half-a-million-robots-installed
3. IFR, *Top 5 Robot Trends 2023*. 2023. 16 February 2023; Available from: https://ifr.org/ifr-press-releases/news/top-5-robot-trends-2023#downloads
4. N. Mohamed, J. Al-Jaroodi, S. Lazarova-Molnar, Industry 4.0: Opportunities for enhancing energy efficiency in smart factories, in *2019 IEEE International Systems Conference (SysCon)*, (2019)
5. F. Voigtlander et al., *5G for Robotics: Ultra-Low Latency Control of Distributed Robotic Systems* (IEEE, 2017)
6. S.Y. Wang et al., Cloud-assisted interaction and negotiation of industrial robots for the smart factory. Comput. Electr. Eng. **63**, 66–78 (2017). https://doi.org/10.1016/j.compeleceng.2017.05.025
7. L. Vladareanu et al., Robot digital twin towards industry 4.0, in *21st IFAC World Congress on Automatic Control - Meeting Societal Challenges*, (Electr Network, 2020)
8. Y.T. Tsai et al., Utilization of a reinforcement learning algorithm for the accurate alignment of a robotic arm in a complete soft fabric shoe tongues automation process. J. Manuf. Syst. **56**, 501–513 (2020). https://doi.org/10.1016/j.jmsy.2020.07.001
9. K.H. Tantawi et al., Advances in industrial robotics: From industry 3.0 automation to industry 4.0 collaboration, in *4th Technology Innovation Management and Engineering Science International Conference (TIMES-iCON)*, (Bangkok, Thailand, 2019)
10. M. Brossog, M. Bornschlegl, J. Franke, Reducing the energy consumption of industrial robots in manufacturing systems. Int. J. Adv. Manuf. Technol. **78**, 1315–1328 (2015)
11. G. Carabin, E. Wehrle, R. Vidoni, A review on energy-saving optimization methods for robotic and automatic systems. Robotics **6**(4), 39 (2017). https://doi.org/10.3390/robotics6040039
12. M. Ghobakhloo, M. Fathi, Industry 4.0 and opportunities for energy sustainability. J. Clean. Prod. **295**, 16 (2021). https://doi.org/10.1016/j.jclepro.2021.126427
13. G.D. Campo et al., IoT solution for energy optimization in industry 4.0: Issues of a real-life implementation, in *2018 Global Internet of Things Summit (GIoTS)*, (2018)
14. F. Lima, A.A. Massote, R.F. Maia, IoT energy retrofit and the connection of legacy machines inside the industry 4.0 concept, in *IECON 2019 - 45th Annual Conference of the IEEE Industrial Electronics Society*, (2019)
15. A. Moldavska, T. Welo, The concept of sustainable manufacturing and its definitions: A content-analysis based literature review. J. Clean. Prod. **166**, 744–755 (2017). https://doi.org/10.1016/j.jclepro.2017.08.006
16. M.A. Rosen, H.A. Kishawy, Sustainable manufacturing and design: Concepts, practices and needs. Sustainability **4**(2), 154–174 (2012). https://doi.org/10.3390/su4020154
17. D.A. Guerra-Zubiaga, K.Y. Luong, Energy consumption parameter analysis of industrial robots using design of experiment methodology. Int. J. Sustain. Eng. **14**(5), 996–1005 (2021). https://doi.org/10.1080/19397038.2020.1805040

18. R.A. Rojas, E. Rauch, From a literature review to a conceptual framework of enablers for smart manufacturing control. Int. J. Adv. Manuf. Technol. **104**(1–4), 517–533 (2019). https://doi.org/10.1007/s00170-019-03854-4
19. M. Zambetti et al., Enabling servitization by retrofitting legacy equipment for industry 4.0 applications: Benefits and barriers for OEMs. Procedia Manuf. **48**, 1047–1053 (2020). https://doi.org/10.1016/j.promfg.2020.05.144
20. P.T. Luu et al., Electromagnetic and thermal analysis of permanent-magnet synchronous motors for cooperative robot applications. IEEE Trans. Magn. **56**(3), 1–4 (2020). https://doi.org/10.1109/tmag.2019.2942939
21. Z. Liu et al., Robot learning towards smart robotic manufacturing: A review. Robot. Comput. Integr. Manuf. **77**, 102360 (2022). https://doi.org/10.1016/j.rcim.2022.102360

Enhancing Security Using E-Authentication System

Mohammad Ali A. Hammoudeh, Amjad Ebrahim, Esraa Mohamed, Rawan Almansour, and Renad Ibrahim

1 Introduction

Every time you leave your home, you are probably very careful about locking all the doors and windows and maybe even triggering an intruder alarm if you have one. After all, you can never be too sure. However, why are you particular regarding your computer safety? Does it bother you to create difficult passwords that you should not write down at a public place for other people to see? Even if you did, would not it be possible for someone else to hack into your systems and cause all kinds of damage? Now more and more people are moving their lives online, banks, and e-commerce (shopping) sites, and others must take security more seriously, no longer relying on simple passwords to keep hackers out. The new and upcoming techniques include two-factor authentication that is an Internet version of an evolving combination lock with different codes. How does it work exactly? Let's take a closer look! NIST in the USA has developed a generic electronic authentication model, which is simply an outline for carrying out the authentication process without taking into account jurisdictions and areas of operation. An individual will first apply to a CSP to be enrolled following this paradigm CSP. Before continuing with the transaction, the CSP must verify the applicant's identification. After the CSP confirms the applicant's identity, he or she is granted the position of "subscriber," as well as an authenticator, such as a token, and a credential, which may be in the form of a username as shown in Fig. 1 [1].

Adding an additional layer of authentication, such as two-factor authentication (2FA), to our application makes it considerably more secure by generating a

M. A. A. Hammoudeh (✉) · A. Ebrahim · E. Mohamed · R. Almansour · R. Ibrahim
Department of Information Technology College of Computer, Qassim University, Buraydah, Saudi Arabia
e-mail: maah37@qu.edu.sa; 371216155@qu.edu.sa; 391215706@qu.edu.sa; 372208068@qu.edu.sa; renad.ibrahim@qu.edu.sa

© The Author(s), under exclusive license to Springer Nature Switzerland AG 2024
A. Shaikh et al. (eds.), *Advances in Emerging Information and Communication Technology*, Signals and Communication Technology,
https://doi.org/10.1007/978-3-031-53237-5_28

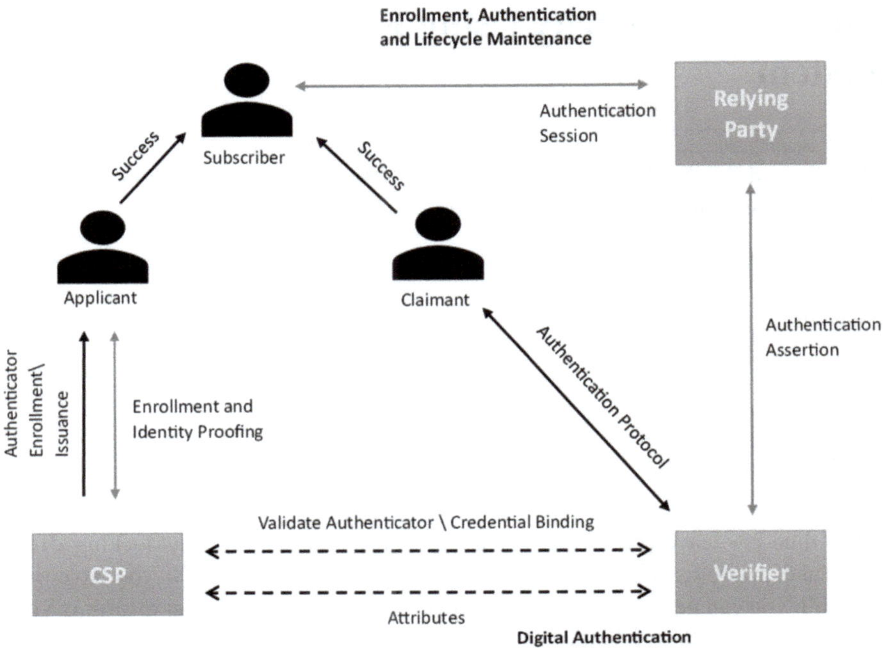

Fig. 1 Digital enrollment and authentication reference process by the American National Institute of Standards and Technology (NIST)

single-use code that is shown in their security token or delivered as a text message to a mobile phone. With 2FA (QR Codes and OTP), you can secure your organization, make users more productive, and drastically reduce the volume of password-reset calls to your help desk. The major objective of the article is to log into the system simply and quickly under the suggested strategy. We assess the security and convenience of use of the proposed scheme, its resistance to password cracking, shoulder-surfing, and other types of intrusion. So, we can list the basic objectives of this study as the following:

1. Assist users in more securely logging into their accounts.
2. Securing accounts from hacking.
3. The password will be hard for him to obtain since shoulder surfing will not permit it.
4. Simple user interface and a strong password approach.

This study explains how to use app to achieve two-factor authentication in E-government system. The suggested solution ensures that services such as online banking and ATM machines are authenticated in a safe and secure way. To start, a user has to fill in some minimum registration data, and then he/she will be registered. Once the registration is done, the user can log in. It starts with authenticating the account through entering the registration's email address and password. Then, if the user's email address and password are verified, he or she proceeds to the next authorization level by selecting either QR Code or OTP.

The paper is organized as follows: the first section discusses the background of E-authentication, the second section displays the recent technologies and techniques used in cybersecurity, the third section rates the user awareness, analyze some related works, and recommend the best solution, and the last section presents conclusion and future work.

2 The Recent Technologies and Techniques

These days security issues are spreading in any field as for instance banks, applications of government agencies, healthcare organizations, armies, etc. Governments are providing standards, giving laws and enforcing them upon corporations and agencies, while nonadherence attracts heavy consequences. Specifically, when discussing security risks across various business domains, a common vulnerability point that affects every situation is passwords. Static passwords are used by most systems nowadays to validate a user's identity. Such passwords, however, raise substantial management security risks. Users are more likely to choose passwords that are simple to guess, use the same password across several accounts, write down or save passwords on their workstations, and so on. Hackers may also steal passwords through shoulder surfing, spying, sniffing, guessing, and other methods. Many correct password schemes have been suggested. Others may be hard to install or might not fit the company's security regulations. Two-factor authentication with device such as token an ATM card has been suggested since they are hard to crack and may solve the problem of password. Two-factor authentication requires the purchase, issuance, and maintenance of tokens or cards as such posing a number of disadvantages. The consumer will lose or steal more than one token for carrying numerous tokens is cumbersome. For many years, mobile phones were primarily seen as devices for placing telephone calls. Nevertheless, thanks to technological innovations and improvements in software, a modern-day phone can be employed for sending text message or checking email and storing contacts, among other things. Mobile connection has also improved. Nowadays, mobile phones incorporate infrared, Bluetooth, 3G, and WLAN as well as the typical GSM link. There are very few people who do not own a mobile phone. Mobile devices' expanding powers are adopted by many mobile banking services. From receiving account balance information through SMS to combining WAP and Java with GPRS, you may transfer money between accounts, trade stocks, and confirm direct payments via the phone's mini browser. Mobile phones may deliver extended new services beyond communication by installing both vendor-specific and third-party apps. As a result, employing a mobile phone as a token will make dealing with numerous two-factor authentication systems simpler for customers while also lowering the cost of making, distributing, and keeping millions of tokens. We propose and construct a full two-factor authentication system that uses mobile phones instead of tokens or cards in this study [1].

2.1 Background

Security measures in computer systems and networks entail prevention of unauthorized data disclosure, theft, physical destruction or damage of hardware, and software or electronically stored data and also safeguard third-party providers' services to user. Because of this, cybersecurity has become one of the major problems in our everyday life. It should be understood that we are not talking only about technical aspects but about everything involving the abuse of opportunities. It was aimed at promoting the security, confidentiality, and integrity of the system.

Electronic authentication is a process for building trust in user identities that are submitted electronically into information systems. The terms electronic authentication or e-authentication can be used interchangeably when one refers to the validation or certification of a person as well as their work. In some instances, it can be used together with an electronic signature to prove that data was not tempered after having been sent by one of its originators. Electronic authentication enables one to verify their authenticity when transacting online so as to reduce identity theft incidences and fraud. There are different types of e-authentications that can be used to verify an individual. It begins with single-factor e-authentications such as password and moves on high-level MFA e-authentications like tokens and biometrics. Security measures in computer systems and networks entail prevention of unauthorized data disclosure, theft, physical destruction or damage of hardware, and software or electronically stored data and also safeguards third-party providers' services to user. Because of this, cybersecurity has become one of the major problems in our everyday life. It should be understood that we are not talking only about technical aspects but about everything involving the abuse of opportunities. It was aimed at promoting the security, confidentiality, and integrity of the system.

Electronic authentication is a process for building trust in user identities that are submitted electronically into information systems. The terms electronic authentication and e-authentication can be used interchangeably when one refers to the validation or certification of a person as well as their work. In some instances, it can be used together with an electronic signature to prove that data was not tempered after having been sent by one of its originators. Electronic authentication enables one to verify their authenticity when transacting online so as to reduce identity theft incidences and fraud. There are different types of e-authentications that can be used to verify an individual. It begins with single-factor e-authentications such as password and moves on high-level MFA e-authentications like tokens and biometrics. Depending on the level of security used, the user may be asked to provide proof of authentication via security tokens, challenges, or a certificate from third-party certificate authority assuring the identity. The prospectus and private filings are often included in the registration. The prospectus is a document that is issued to every investor who acquires a security, while the private filing is information that is sent to the SEC for review. User authentication constitutes one of the essential measures to ensure safe communications and resource-sharing capabilities within an insecure public communication channel. One-time password aims at giving hard time to access of the authorizing resources. However, instead of using password files for conventional authentication purposes, a number of researchers have proposed one-time password schemes based on smart cards, time-synchronized tokens, and SMS mechanisms to reduce vulnerability to tampering and maintenance costs. On the other hand, some of these ideas will be rendered impracticable since there is not enough technologically advanced technology for support as well as infrastructure. Therefore, we could incorporate QR code method together with our one-time password authentication process to overcome these weaknesses. Unlike before, the suggested QR code approach does away with the need for a password verification table and offers a cheaper option that fits many online users who have mobile phones already [2].

2.2 Concepts and Definitions

In this section, we review the most popular E-authentication types which will be presented in the following such as:

SMS Token Authentication: Text messages with a PIN number are the most common form of this check, which is easy to implement for customers and the broader public. As an OTP, the PIN serves as an additional layer of security on top of the conventional username/password combo. Allowing users to authenticate using their mobile devices is a smart move for businesses whose consumers commonly use mobile devices to access their services.

TOTP Stands for Time-Based One-Time Password: A straightforward method using a common unique key and information on the present moment that creates an unforgeable token that expires quickly. For our solution we are only going to use some functions that Google authenticator supports. To utilize TOTP, you need a server with time synchronization, as well as client devices that your visitors may use. In today's world, this is a reasonable assumption.

Hardware Token Authentication: One of the latest and safest authentication mechanisms is the use of a separate hardware token if the client maintains the key. This strategy is more costly but giving out dongles to high-value customers may save money. Business clients are increasingly ready to utilize a hardware token, and acceptance is rising. Except making them mandatory for all but the most valuable and at-risk consumers in banking, insurance, and investing is a bad idea. Mobile users may have to purchase another dongle with an extra USB and/or USB-C port for their mobile devices such as a charger.

Multifactor Authentication: For a long time, it was observed that the technique of hacking is evolving, resulting in an exposure of single-factor authentication to the malwares, replay-themed attacks, offline brute force attacks, key logger Trojan horse viruses, dictionary attacks, shoulder surfing, and other threats. Combining several authentication elements has become increasingly popular to boost security. Multi-factor authentication has become more popular in recent years. Transaction kinds, risk levels, threats, and vulnerabilities may all influence the number of authentication factors necessary.

QR Code: This can be done through QR code (quick response code)-based OTP authentication which obviates the use of the password verification table. Moreover, this approach is cheaper than the former. It can lead to serious health issues among children. The usage of QR codes [3] in mobile phones has a number of benefits, including omnidirectional reading and mistake correcting capabilities.

2.3 Previous Studies

In this section, we present some issues related to E-authentication studies. As shown in Fig. 2, we shall consider a two-factor authentication system for web and mobile applications, so we seek to collect information and studies that help us to solve and reduce secure access problems.

However, most of the present-day authentication schemes use static word-based passwords to authenticate and identify users. However, the use of password raised serious questions but turned out to be traditional. As users do not have strong password management experience, they tend to use weak password, use one

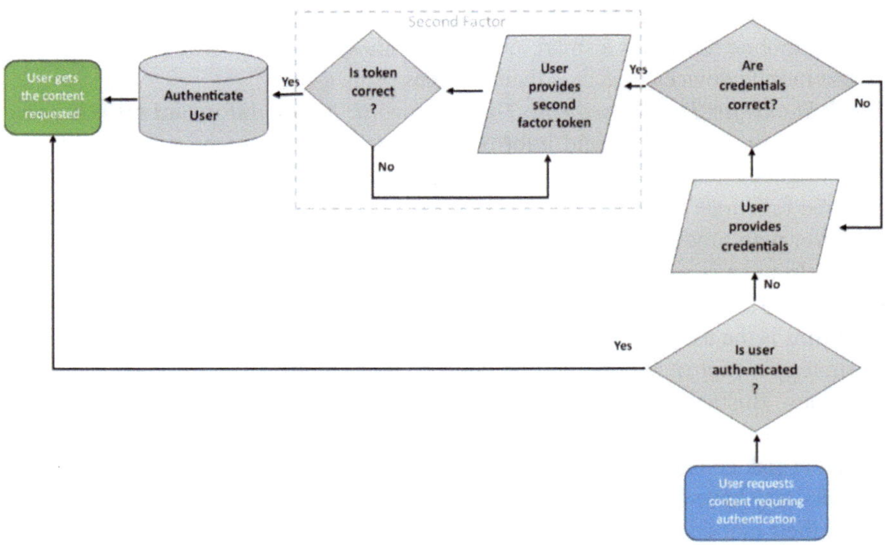

Fig. 2 Two-factor authentication flow [4]

password for many accounts, write it on personal devices and note papers. In addition, different ways for unauthorized users include shoulder surfing, snooping, sniffing, guessing, man-in-the-middle attack, and phishing. SMS-based OTP one-time password is arguably the most popular two-factor authentication. This kind of one-time password is involved in a simple procedure, such as entering an OTP after logging in with a username and password or authorizing a transaction using this OTP A paper presents an architecture for construction of OTP-based two-factor authentication that acts as OTP as a service along others services offered by the cloud-like software as a service, infrastructure as a service, hardware as a service, and others. This paper also states that offering OTP as a secure and private cloud service can assist enterprises to thwart attacks with two-factor authentication. However, many security flaws are associated with sending SMS OTP from the cloud, such as SIM swapping, malware for intercepting and forwarding 292 D. T. Irfe and V. K. Anand's message, and social engineering attacks.

Secondly, currently SMS-based OTP is a less secured two-factor authentication as it is highly dependent upon secure delivery of SMS message which ultimately depends on secure cellular networks thereby exposing it to numerous cyber criminals' attacks. Authors suggested that in paper there is feasible solution which could deal with the possible problem that may arise on SMS-based OTP in the mobile devices. This paper aims at improving the SMS-based one-time password two-factor authentication through dealing with various problems among them: SMS cost, during each login request SMS delay due to the network reasons among others; international roaming, traveling overseas; and SMS security. Further, another author suggests a remedy for use of mobile phones to generate one-time passwords. This is a suggestion which seeks to protect gate of services like e-banking and ATM systems. Mobile phone operates as the software token that produces a one-time password for the user's end. The temporary one-time password which is generated based on the uniqueness of a combination of the user as well as the device which generates it is valid for up to 30 minutes. The authors in addition proposed a secondary backup mechanism of SMS-based authentication procedure.

Paper [5] has also suggested time-based one-time password (OTP) for preventing the vulnerability of authentication in claim token method provided in private blockchain which is protected by enforcement two-factor authentication. Ticketing systems can also be applied in transport sector based on OTP-based authentication [6]. The other one refers to two-factor verification based on the use of QR code. The paper [7] asserts that OTP-based 2FA can be compromised via man-in-the-middle attack, phishing, and also spoofing attack, and as its substitute IMEI authentication technique is proposed wherein a unique encrypted IMEI of the phone is displayed in the two-factor authentication consists of QR code as the first attribute, while mobile phones take the second part; on that note, a novel technique for two-factor authentication entails using the real GPS position of the users as documented in [3]. S-Mbank paper [8] offers other mechanisms that replace the use of one-time password with the contact less smart card in order to avoid attackers exploiting an unencrypted message transmitted to user's mobile phone and eliminate vulnerability of the imposter using un

Some of them can be seen in Table 1 regarding enhanced security via e-authentication system, a two-factor authentication system. Mostly, research uses various solution approaches – each with a purpose, methodology, result, and limitation as well as recommendation or future direction.

3 Discussion

3.1 Data Collection

Data collection is a process of gathering information from the entire relevant source to gain answers to the study problem and evaluate the result. This study collected data through an online questionnaire that sought to establish how much users are aware of web and mobile application security challenges. It has 46 responses collected by Google forms, which included in Fig. 3 15 specialized questions:

3.2 Data Analysis

Data analysis is the process of utilizing techniques to represent, explain, and evaluate data. Following the publication of this questionnaire, we got 46 responses, and the questionnaires start from inquiring about their gender, out of which 78.3% were female (21.7%) and male (87%) aged 87%, all within the age range of 20–30 years. Most of the respondents have bachelor with a percentage of 76.1%, 15.2% are secondary school, and the rest have higher education. Figure 4 shows the variety of occupation.

Figure 5 presents the security knowledge level of the respondents

The responses about the type of online system the respondents used to use comes varies between different fields as shown in Fig. 6.

Almost half of respondents using smart phones to access the online systems, laptop comes in the second order with 29%, and tablet and desktop have 12% and 8%, respectively. More than half respondents do not trust the accessing to their online accounts. Figure 7 shows the ways that respondents managed their passwords.

Many of respondents believe that the security is important, 65.2% of responses were not attacked, and the rest of the responses have ordered the attacks that threaten respondents as follows: accessing by someone without their approval, changing sittings without their approval, stealing data, and finally losing files.

So, almost half of the respondents agreed with strengthening the current system. Lastly, more than half of respondents agreed with effective security techniques for E-authentication system by using two-factor (QR codes OTP) authentication as shown in Fig. 8.

Table 1 Literature review

Ref	Year	Purpose/motivation/ problem	Methodology	Result	Limitation	Future direction and recommendation
[5]	2021	Five different types of 2fa systems should be evaluated for their reliability	– SMS – TOTP – Progenerated-codes – Push – U2F security keys	– As a whole, people found that tested 2FA was highly rated – Most of them indicated a keen interest in utilizing 2FA for a majority of their online accounts	– It is also a study of a group of educated, young university students with better technological orientation than that of their parents – Participants did not necessarily create the 2FA themselves	– Conduct more research regarding employing 2FA processes – Improvement of 2FA setup process research
[7]	2019	Mobile application with an authentication program based on a two-factor authentication model	The developed model is based on two types of two-factor authentication: download a software-based authenticator onto your smartphone, receiving temporary login credentials in the form of a text message, while submitting only general information	On basis of a certain algorithm and according to a validity period for one instance authentication session, a server forms a temporary or one-time password	The present paper contains vulnerability analysis of information leaks and development of the application-authenticator algorithm	This paper does not compare the suggested authentication method with other existed two-factor authentications in order to measure its performance. When designing such app, one would have to ensure that it runs on all operational system platforms that are available today
[8]	2019	In this case, 2FYSH provides more secure process of tokenized two steps for password replacement	– Two tokens: mobile phone and NFC card – Protocol based on public and private key challenge response	– Majority of those sampled for usability testing opted for 2FYSH – Provides an appropriate security solution for confidentiality needs	N/A	– However, there is a method to 2fysch that allows for diverse form of tokens – Some researchers could carry out biometrics' validity studies of 2FYSH

(continued)

Table 1 (continued)

Ref	Year	Purpose/motivation/ problem	Methodology	Result	Limitation	Future direction and recommendation
[6]	2020	Threat models and countermeasures in the two-factor authentication schemes proposed for mobile money	– A computerized literature search was conducted on several scientific databases – A critique about a total of 97 articles related to this	– Threat models in the 2FA scheme for mobile money were categorized into five: privacy, authentication, confidentiality, integrity, and availability issues with information services – The countermeasures include cryptographic functionality and personal identification	– Security of mobile money issues in new 2fa schemes – This kind of authentication is based exclusively on PIN and SIM	– Some improvements need to be made in the current 2FA mechanism for mobile money – Such countermeasures include cryptographic functions and/or biometrics means of identification
[9]	2022	Implementation of a decentralized and privacy-preserving scheme for the public auditing of cloud storage	– Randomized challenge generation using blockchain – Privacy protection utilizing zero knowledge proof for user	– Public auditing scheme that is decentralized and privacy preserving – This includes effectiveness of security analysis and performance review	– Challenge message should not solely be based on user/auditor – Auditor should provide incontrovertible evidence	– Other auditing spheres using a blockchain in the future – It also enables better efficiency and scales up the process suggested
[10]	2020	The files in a blockchain-based decentralized cloud storage are secure; they get encrypted for safety and stored across other interconnected peers on it	– Blockchain for decentralized cloud storage – File encryption and storing under IPFS protocol	– Blockchain decentralized secure data storage – Data availability and optimal utilization of capacity	N/A	– Further research on scalability and optimized performance – Compatibility with other blockchain compatible applications and devices
[11]	2021	Briefly, covering some fundamental issues about security of web applications along with enough	In this document, diverse vulnerable scanners, feedback, as well as solutions are discussed, to have	This paper presents an overview of architecture structure as it relates to web-based applications,	N/A	In the future this work will be extended by looking at the data in

		explanation of other concepts and practices related in security assessment	better understanding of vulnerability in web-based application	application software, database and its communication system		more detail using graphical approaches
[12]	2021	In order to improve security, this paper suggests two-way authentication through the use of QR codes and Android phones	– A proposal for a QR code based two-factor user login system – Smartphones are used by students as another authentication factor	N/A	Vulnerable OTP mechanisms (MITM, phishing, spoofing)	– To curb QR code tampering, put an additional safety measure – Carrying out user-testing for improving convenience and practicality
[13]	2022	Phishing with a twist of safety through QR-based authentication	N/A	These tests included some quality, accuracy, suitability, and security tests that succeeded	– Limited number of test users – Small observational test data	Add more security features in identifying and verifying

Fig. 3 The conducted questionnaire

1. What is your gender?
2. Your age
3. What is your nationality?
4. Level of education
5. Occupation
6. Assess your knowledge in Security.
7. What are the types of Online System you use?
8. Specify your Daily Used Device
9. Do you trust that accessing to your accounts online is secure?
10. Are you managing your passwords as the following?
11. Do you think Security is important in Web/Mobile Application?
12. Have you ever been attack through the internet?
13. If yes, what type of threaten you have faced?
14. Do you agree that your current system in (government, organization…) is strong enough to protect your privacy from illegal use?
15. In your opinion, what is/are the effective security techniques for the E-Authentication System?

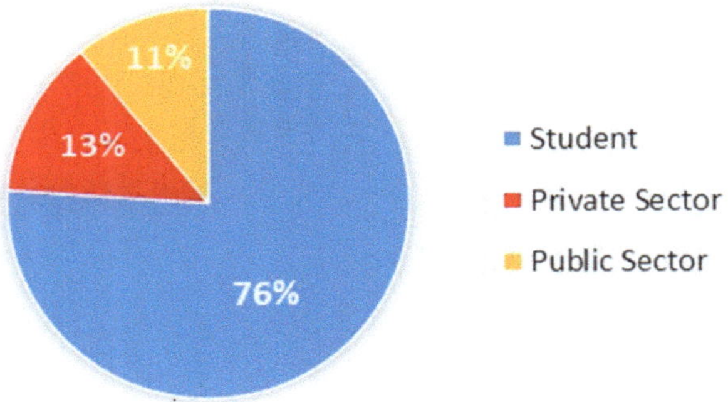

Fig. 4 Answer of occupation

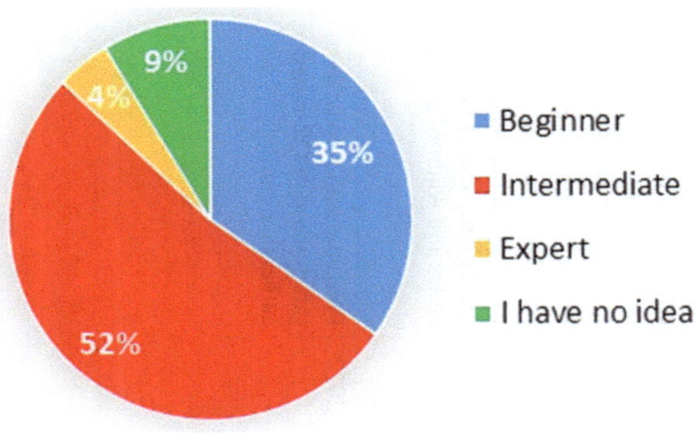

Fig. 5 Answer of user's knowledge about security

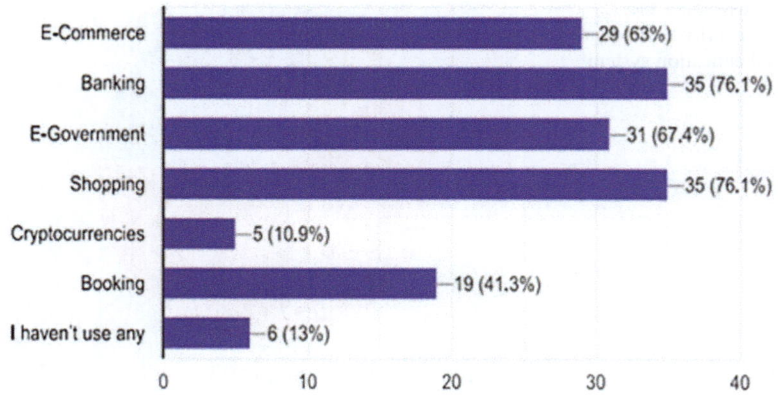

Fig. 6 Answers of the types of the used online system

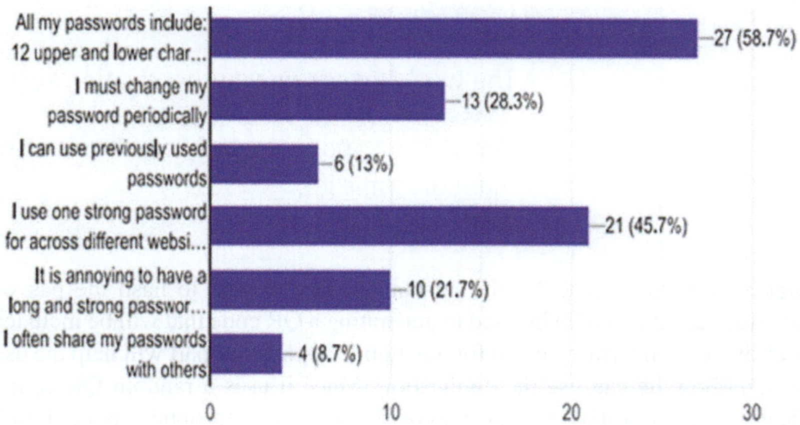

Fig. 7 Answers of the ways of managing the password

3.3 Recommendation and Solution

After completing the requirements, Fig. 9 shows the flowchart of the proposed system, where the user can access by getting the authentication using two factors: OTP and QR code. The user has three trials to get the authentication for each of them unless the access will be rejected.

The desktop websites and mobile applications are two items that are required for this study. Username and OTP are the two types of user input that the desktop is meant to accept. The site will first request the user's username and then followed by a corresponding password. Server-side scripting, therefore, is necessary for the system. It will also generate a random key and pull the password from the database

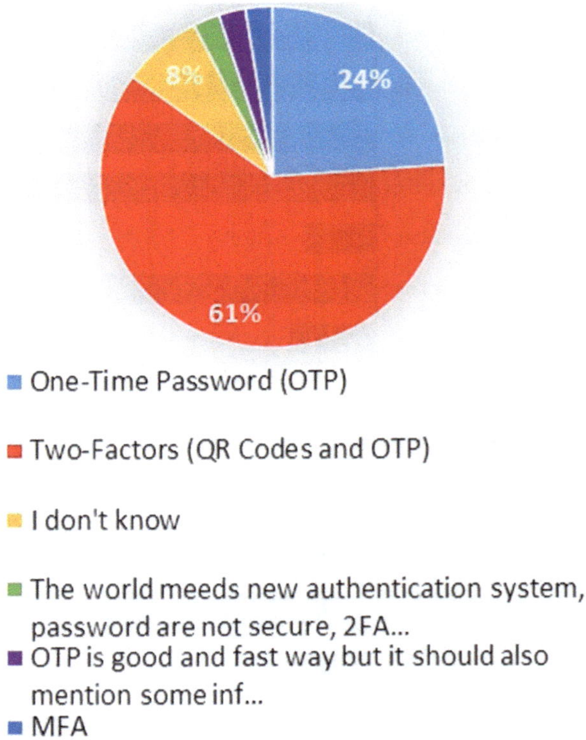

Fig. 8 Answers of the effective security techniques for E-authentication system

through server-side script. A selected random key is used to hash the password. Server-side scripting would be used in generating a QR code that will be included on the website. User interface design for the mobile application part will help the user to understand how he can use the application. Since it uses a random QR code, the application has to be able to take photos for the phone to obtain its random key. Another way is to take a picture of the QR code and then use software in one's phone to obtain its secret keys. Afterward, a program will again ask for the user's password, using it to generate an OTP. In response to the problem with the current system, whenever a user logs in to the system to produce an OTP, the server should generate different random keys every time. This, however, prevents the attackers from determining the actual password. The login page should be restricted to three attempts (logins) for a user's password to minimize chance of success for brute force attacks [12]. So, in these subsections, we will illustrate what we use and each step-in detail as the following: (1) a smart phone is used to scan QR code from website and then generate OTP to put it in website. (2) Laptop: the system also requires a server to run authentication services. Instead of setting up the system with a real server, use your laptop as a virtual server. The laptop is also allowed you to browse website and log in.

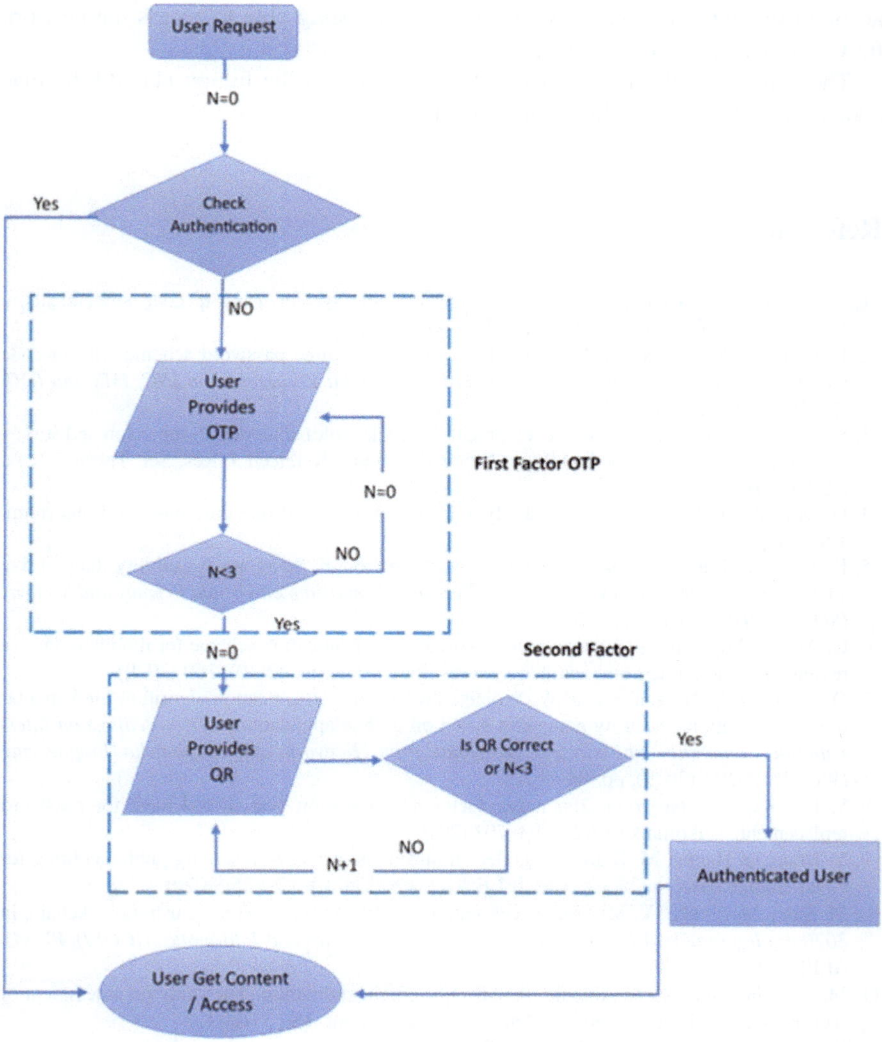

Fig. 9 The solution tool structure

4 Conclusion and Future Work

The effort has been a major success in preventing attack, where the attackers would need to guess the keys and hack the system, which takes a long time. The system also uses another type of the security that is two-factor authentication, and this involves the use of the real password with the OTP so that for the system to work the two have to match. The ability to create OTPs without an Internet connection is another big accomplishment that helps to stop hackers from being able to access the network flow and extract the real password.

There are several services that will be provided in the future: (1) add the mac address. (2) Use addition factor authentication.

References

1. A.A. Azeez, A. Ajibola, A secured medical booking system using qr-code authentication. Int. J. Sci. Res. Comput. Sci. Eng. Inf. Technol. (2019)
2. K.-C. Liao, W.-H. Lee, M.-H. Sung, T.-C. Lin, A one-time password scheme with qr-code based on mobile phone, in *2009 Fifth International Joint Conference on INC, IMS and IDC*, (IEEE, 2009), pp. 2069–2071
3. S.K. Dora, P. Dubey, Software development life cycle (sdlc) analytical comparison and survey on traditional and agile methodology. National Monthly Refereed J. Res. Sci. Technol. **2**(8), 22–30 (2013)
4. D. M'Raihi, S. Machani, M. Pei, J. Rydell, Totp: time-based one-time password algorithm. Tech. Rep. (2011)
5. K. Reese, T. Smith, J. Dutson, J. Armknecht, J. Cameron, K. Seamons, A usability study of five {Two-Factor} authentication methods, in *Fifteenth Symposium on Usable Privacy and Security (SOUPS 2019)*, (2019), pp. 357–370
6. G. Ali, M. Ally Dida, A. Elikana Sam, Two-factor authentication scheme for mobile money: a review of threat models and countermeasures. Future Internet **12**(10), 160 (2020)
7. O. Ussatova, S. Nyssanbayeva, W. Wójcik, Two-factor authentication algorithm implementation with additional security parameter based on mobile application, in *Proceedings on International Conference on Wireless Communication, Network and Multimedia Engineering (WCNME2019)*, (2019), pp. 84–86
8. S. Pranata, H.T. Nugroho, 2fysh: two-factor authentication you should have for password replacement. Telkomnika **17**(2), 693–702 (2019)
9. Y. Miao, Q. Huang, M. Xiao, H. Li, Decentralized and privacy-preserving public auditing for cloud storage based on blockchain. IEEE Access **8**, 139813–139826 (2020)
10. M. Shah, M. Shaikh, V. Mishra, G. Tuscano, Decentralized cloud storage using blockchain, in *2020 4th International Conference on Trends in Electronics and Informatics (ICOEI)(48184)*, (IEEE, 2020), pp. 384–389
11. M.A.A. Hammoudeh, A. Alobaid, A. Alwabli, F. Alabdulmunim, The study on assessment of security web applications. Int. J. Interact. Mobile Technol. **15**(23) (2021)
12. B. Rodrigues, A. Chaudhari, S. More, Two factor verification using qr-code: a unique authentication system for android smartphone users, in *2016 2nd International Conference on Contemporary Computing and Informatics (IC3I)*, (IEEE, 2016), pp. 457–462
13. M.T.R.M. Teja, K. Praveen, Prevention of phishing attacks using qr code safe authentication, in *Inventive Computation and Information Technologies: Proceedings of ICICIT 2021*, vol. 336, (2022), p. 361

Toward Extensible Low-Code Development Platforms

Gregory Popov, Joan Lu, and Vladimir Vishnyakov

1 Introduction

Low-code development (LCD) is being positively recognized both by the industry and academia [1, 2]. It offers development platforms (LCDP) for a vast range of application domains: business services (e.g., mobile, web, healthcare services), IoT (e.g., Node-RED, ArduBlock, LabView), education and design (Scratch, 3DGrasshopper), aeronautics [1], chatbots [3], and machine learning (e.g., Google's AutoML, RapidMiner). Because of the wider target audience, lower complexity, and a much higher level of automation, it is believed that LCDPs will fare better than their predecessors, namely, CASE and MDE tools [3]. This is also reflected in a Gartner report, which predicts that "by 2024 low-code development will be responsible for more than 65% of application development activity" [4]. However, existing platforms are not reusable: most of them can solve a particular problem, such as learning [5], website creation [6], or microcontroller programming [7]. A better solution is needed to take the challenge.

The purpose of this research is to present a review of modern technologies and identify advanced state-of-the-art technologies for the creation of extensible, open [8], and open-source LCDP. The rest of the paper is organized as follows: Sect. 2 reviews the most popular platforms to identify limitations inherent to LCDP in general. In Sect. 3 we research existing technologies, selecting the essential LCDP components to overcome identified limitations in a new platform. The selection process is performed, by answering the following research questions:

G. Popov (✉) · J. Lu · V. Vishnyakov
Computer Science, University of Huddersfield, Huddersfield, UK
https://www.linkedin.com/in/h-popov/
e-mail: j.lu@hud.ac.uk; https://pure.hud.ac.uk/en/persons/joan-lu;
V.Vishnyakov@hud.ac.uk; https://pure.hud.ac.uk/en/persons/vladimir-vishnyakov

© The Author(s), under exclusive license to Springer Nature Switzerland AG 2024
A. Shaikh et al. (eds.), *Advances in Emerging Information and Communication Technology*, Signals and Communication Technology,
https://doi.org/10.1007/978-3-031-53237-5_29

RQ1. Rich text editing support: what are available methods and technologies?
RQ2. Rich diagram editing support: how to define visual representation of the programming language?
RQ3. IDE platform selection: what could be the basis for an extendable platform?

In Sect. 4, we discuss the means to combine textual and visual representations of the DSL synchronously, within a set of selected technologies. Finally, we conclude this article with achievements and suggestions for future works.

2 Background: Problems of Modern LCDPs

The modern world of LCDPs is diverse. Recent literature reviews [1, 3, 9–13] reported more than 60 distinct tools but not limited. This paper selected the most relevant platforms for closer review, including industry leaders [4], enterprise platforms with high publication trends [1], and open-source platforms mentioned in publications [8, 14–16]. The 11 platforms that were reviewed are listed below, and their limitations are presented in Table 1 and Fig. 1:

Table 1 Problems addressing frequency in LCDPs

Modern LCDP problem	LCDPs, dealing with it		$LCDPs_{dealing}$/ $LCDPs_{total}$
	Fully	*Partially*	
Ignored integration API standards	OutSystems, Salesforce	NodeRED	2+1/11
Proprietary languages interoperability	Mendix	RESTsec, OPP	1+2/11
Collaborative development	Appian, Microsoft PowerApps, Mendix		3/11
Architecture extensibility		OutSystems, NodeRED	0+2/11
Hidden generated code	NodeRED, OLP, RESTsec		3/11

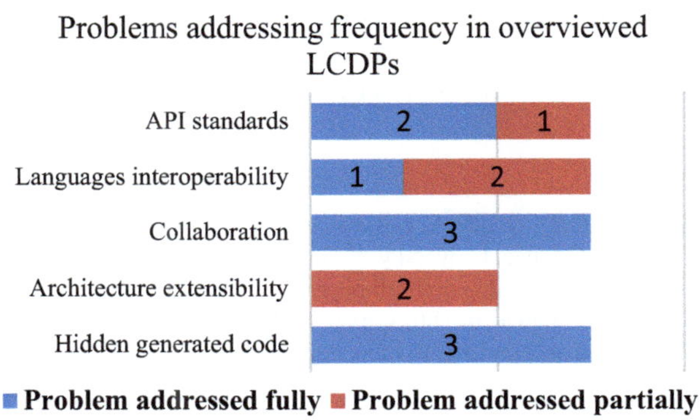

Fig. 1 Problems addressing frequency in overviewed LCDPs

- OutSystems [17–21]
- Mendix [19, 21–23]
- Salesforce [19, 24, 25]
- Zoho Creator [19, 23, 26]
- Appian [19, 23, 27]
- NodeRED [28, 29]
- Microsoft PowerApps [20, 21, 23]
- Google AppSheet [19, 30]
- Open Low-code Platform (OLP) [8, 31]
- Object-Process Programming (OPP) [15, 32]
- RESTsec [16, 33]

As indicated in Fig. 1, the least addressed problem in LCDPs is platform extensibility. It is an essential problem: if it is resolved, many other problems are becoming much easier to deal with [34].

3 Extensible LCDP Technology Overview

Based on the above background study and literature review, the questions proposed in Sect. 1 will be answered as follows.

3.1 RQ1. Methods and Technologies for Rich Text Editing

Nowadays, Language Server Protocol (LSP) is a de facto standard for enabling rich text editing for languages [35]. It is a common practice to develop a language server per programming language to use it in various IDEs [36, 37]. It facilitates the reusability of rich language editing, such as error highlighting, autocomplete, or navigation by name across multiple source code editors. Also, it implements a "single responsibility" pattern, decoupling UI editing-related features from the supporting logic (e.g., error-checking rules, the composition of autocomplete hints, or file-by-name locations) [35, 36].

There are technologies called *language engineering tools* [38] available to aid LS development. They automate the creation of most LS features and provide a framework to ease LS customization. Hence, a language engineering tool should be one of the core platform technologies.

To find available mature candidates, the Google search engine was used instead of popular scientific publication platforms. The search was conducted twice with the following queries: "language engineering tools" and "generate language server." Then the links from the first ten pages were gathered and filtered based on the Google page description and date updated (considering webpages updated after 2005). This yields eight potentially relevant results, among which five were

Table 2. Language engineering tools

Eclipse XText	IntelliJ MPS	Langium
Java	Java	TypeScript
🔥 ANTLRWorks for checking and optimizing language grammar	🔥 Advertised as efficient	🔥 Outperform XText
🔥 Grammar definition is the only input to generate basic LS	🔥 More language features out-of-the-box	🔥 Supports cross-language references
❄ Scale poorly: additional manual work is required for decent UX	🔥 Rich toolset to experiment and enhance the design of developed DSLs	🔥 2 grammar definition syntaxes: XText-like and TS-like
❄ Is based on EMF, which adds not always necessary complexity and computation overhead	❄ Exhibits vendor lock-in, since it is not compatible with a standalone LS, but serves as an entire IDE +language platform	🔥 Bleeding-fast Chevrotain parser with ALL(*) support
	❄ Poor diagram editor capabilities: only as an alternative view in the projection editor	❄ No language grammar debug tool developed

discarded as outdated, non-mature, or irrelevant. The three remaining candidates are the following:

1. *Eclipse XText*: Java framework is based on EMF and uses a popular and sophisticated ANTLR parser under the hood. Many academic- and even industry-level languages are written with it [39, 40].
2. *TypeFox Langium*: TypeScript library, recently built and inspired by XText, but with no bound to EMF. Features elaborated AST definition: it is possible to define TypeScript interfaces of AST nodes inside a grammar, alongside XText-like parsing rules [41–43]. Language parser supports the most powerful ALL(*) parsing algorithm [43, 44].
3. *JetBrains MPS*: Language engineering workbench, providing versatile tools to design a language, and its projections (text, tables, math formulas, graphs) [45]. While offering more IDE features out of the box, than XText or Langium, such as code transformation through intentions, refactoring actions, or language testing [46, 47], it lacks rich diagram editing support [45].

The comparison summary is listed in Table 2.

3.2 RQ2. Technologies and Approaches for Rich Diagram Editing

Rodriguez-Echeverria et al. [36] suggest several approaches for graphical LSP design and list up-to-date custom solutions, such as Obeo GSP [48] and Typefox Sprotty LS wrapper [49, 50] as the only alternatives relying on nonproprietary protocols. Nevertheless, nowadays Obeo GSP seems to have been abandoned (the last source code update happened in 2018), while the Typefox initiative has grown into the Graphical Language Server Platform (GLSP).

GLSP is an open standard under Eclipse Foundation [51] designed to handle diagram editing following the same architectural pattern as broadly adopted LSP [52]. This technology is available for multiple platforms (Java/Node.js) and with arbitrary source model (as a JSON file, a database, or even a standalone model server) [52, 53]. Eclipse GLSP comes with a modern web library for rendering diagrams rendering called Sprotty, which suggests high performance and SVG support [49, 54]. All these features make it a perfect candidate for open-source LCDP.

3.3 RQ3. IDE Platform Selection

Most of the modern IDEs (such as IntelliJ IDEA, VSCode, Eclipse) provide a way to extend or even customize them to be suitable for different frameworks, languages, etc. This can help to reuse existing rich IDE functionality instead of rewriting it from scratch. To choose the best platform, an Internet search has been run, looking for the IDE platforms that:

- are open-source
- support diagrams editing
- are potential long-term market players

There were five platform candidates selected for comparison. Their features are presented in Table 3, and a conclusive Euler diagram, indicating Theia as a winner, is presented in Fig. 2.

4 Integration of Textual and Visual Representations

Designing a low-code platform, where development can be performed equally with textual or graphical means, does not only imply that models of the visual languages should be stored in text files but also that these text files should be easy to read and modify on their own. This introduces *a challenge of textual DSL integration with its visual representation.*

Table 3 Comparison of IDE platform candidates

Name	Features
VSCode [54–56]	🔥 Custom WebView support 🔥 Most popular extensions mechanism on the market 🔥 Existing integration with GLSP ❄ Customization is limited with external API for extensions ❄ Builds and extension marketplace are close-sourced
IntelliJ IDEA [57]	🔥 The most popular IDE (for Java) 🔥 Language plugin development support ❄ Language plugin provides only textual language support natively
Eclipse GEF [15, 58]	🔥 Support the creation of LCDP on top 🔥 Allows more customization than Sirius ❄ Involves more work, being a low-level framework
Sirius [59–61]	🔥 Aimed specifically at graphical modeling workbench creation 🔥 Rich diagrams support (synchronization, validation, etc.) 🔥 Provides a DSL for modeling tool configuration 🔥 Web IDE 🔥 Supports models interpreting ❄ Constrained with eclipse modeling framework, since Sirius is based on it
Theia [62, 63]	🔥 Aimed specifically at custom IDE creation 🔥 Web IDE 🔥 Custom WebView support 🔥 Integrates with GLSP out-of-the-box 🔥 Compatible with VSCode extensions 🔥 Cloud-native, which makes it a long-term player ❄ May require a lot of development effort to customize the platform to one's needs

At the last EclipseCon, presenting the release of GLSP v1.0, Philip Langer mentioned an idea to use LS as a source model provider for GLSP [64]. This synergy can bring the following advantages:

- Comprehendible format of textual source code (unlike EMF.cloud Model Server, which stores source models in XML [65]).
- Real-time bidirectional synchronization between source code and its visual representation. This allows the user to edit the same aspect of a program model in the most convenient way for him. Whenever a diagram is modified – its changes are propagated to the textual code (source model) employing GLSP SourceModel storage [52], while backward propagation can be implemented by following the overall architecture of EMF.cloud Model Server [66].

Fig. 2 IDE selection

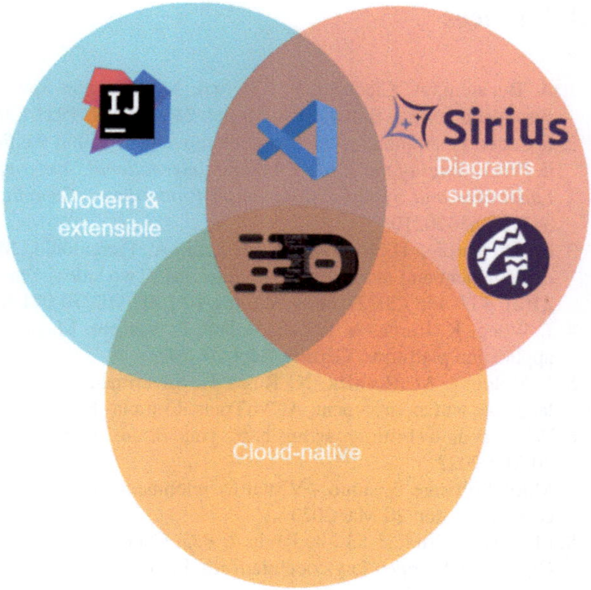

- Increased diagrams expressiveness. Diagrams will be able to augment validated expressions in textual language [67].

The problem with such integration, though, is its high complexity [68].

5 Conclusion and Future Suggestions

In this paper the following contributions are made:

1. We highlight the lack of platform extensibility within the modern LCDPs' limitations.
2. We overview existing technologies, answer research questions, and select the best candidates to be used for an extensible open-source LCDP, namely, Langium as a language engineering tool, GLSP as a diagram editing framework, and Theia as an IDE platform.
3. We recognize the challenge of synchronization between textual code and visual diagram.

We suggest the integration of LS and GLSP as one of the future works on the path toward extensible open-source LCDP. Another suggested future work is the development of an architectural design of the platform itself and its extensions, to augment different DSLs, at the same time falling into Theia plugins paradigm.

References

1. A. Bucaioni, A. Cicchetti, F. Ciccozzi, Modelling in low-code development: a multi-vocal systematic review. Softw. Syst. Model. **21**(5), 1959–1981 (2022)
2. J. Cabot, Positioning of the low-code movement within the field of model-driven engineering, in *Proceedings of the 23rd ACM/IEEE International Conference on Model Driven Engineering Languages and Systems: Companion Proceedings, Canada*, (2020). https://doi.org/10.1145/3417990.3420210
3. D. Di Ruscio, D. Kolovos, J. de Lara, A. Pierantonio, M. Tisi, M. Wimmer, Low-code development and model-driven engineering: Two sides of the same coin? Softw. Syst. Model. **21**(2), 437–446 (2022). https://doi.org/10.1007/s10270-021-00970-2
4. J. Wong, K. Iijima, A. Leow, A. Jain, P. Vincent, Magic Quadrant for enterprise low-code application platforms. Gartner, Sept 20, 2021
5. J. Maloney, M. Resnick, N. Rusk, B. Silverman, E. Eastmond, The scratch programming language and environment. ACM Trans. Comput. Educ. (TOCE) **10**(4), 1–15 (2010)
6. WiX, Your website, your business, your future. Available: https://www.wix.com. Accessed 20 Mar 2023.
7. Mitov Software, Visuino – Visual Development for Arduino. Available: https://www.visuino.com/. Accessed 20 Mar 2023
8. M.A.A. da Cruz, H.T.L. de Paula, B.P.G. Caputo, S.B. Mafra, P. Lorenz, J.J.P.C. Rodrigues, Olp—a restful open low-code platform. Future Internet **13**(10), 249 (2021). https://doi.org/10.3390/fi13100249
9. F. Ihirwe, D. Di Ruscio, S. Mazzini, P. Pierini, A. Pierantonio, Low-code engineering for Internet of Things: a state of research, presented at the *23rd ACM/IEEE International Conference on Model Driven Engineering Languages and Systems, MODELS-C 2020*, 2020, [Online]. https://doi.org/10.1145/3417990.3420208
10. M. Shamsujjoha, J. Grundy, L. Li, H. Khalajzadeh, Q. Lu, Developing mobile applications via model driven development: a systematic literature review. Inf. Softw. Technol. **140**(106693) (2021). https://doi.org/10.1016/j.infsof.2021.106693
11. I. Khodnenko, S.V. Ivanov, K. Prokofiev, A. Lantseva, A Lightweight Visual Programming tool for Machine Learning and Data Manipulation, in *Presented at the 2020 International Conference on Computational Science and Computational Intelligence (CSCI), Las Vegas, NV, USA*, (2020), pp. 981–985. https://doi.org/10.1109/CSCI51800.2020.00182
12. P.M. Gomes, M.A. Brito, Low-code development platforms: a descriptive study, in *Presented at the 2022 17th Iberian Conference on Information Systems and Technologies (CISTI), Madrid, Spain*, (2022), pp. 1–4. https://doi.org/10.23919/CISTI54924.2022.9820354
13. Y. Luo, P. Liang, C. Wang, M. Shahin, J. Zhan, Characteristics and challenges of low-code development: the practitioners perspective, in *Presented at the International Symposium on Empirical Software Engineering and Measurement*, (2021., [Online]). https://doi.org/10.1145/3475716.3475782
14. M. Altendeitering, S. Schimmler, Data-flow programming for smart homes and other smart spaces, in *Presented at the 2020 IEEE Symposium on Visual Languages and Human-Centric Computing (VL/HCC), Dunedin, New Zealand*, (2020). https://doi.org/10.1109/VL/HCC50065.2020.9127268
15. A. Bibliowicz, Object-Process Programming: A Visual Programming Language for Complex Systems Design and Implementation. Ph.D. dissertation, Fac. of Ind. Eng. and Manage., Technion – Israel Inst. of Technol., Haifa, Tevet, Israel (2017)
16. C. Zolotas, K.C. Chatzidimitriou, A.L. Symeonidis, RESTsec: a low-code platform for generating secure by design enterprise services. Enterprise Inf. Syst. **12**(8–9), 1007–1033 (2018). https://doi.org/10.1080/17517575.2018.1462403
17. Outsystems, High-Performance Low-Code for App Development. Available: https://www.outsystems.com/. Accessed 9 Nov 2022

18. H. Lourenço, C. Ferreira, J.C. Seco, OSTRICH – a type-safe template language for low-code development, in *Presented at the 2021 ACM/IEEE 24th International Conference on Model Driven Engineering Languages and Systems (MODELS), Fukuoka, Japan*, (2021), pp. 216–226. https://doi.org/10.1109/MODELS50736.2021.00030
19. A. Sahay, A. Indamutsa, D. Di Ruscio, A. Pierantonio, Supporting the understanding and comparison of low-code development platforms, in *Presented at the 2020 46th Euromicro Conference on Software Engineering and Advanced Applications (SEAA), Portorož, Slovenia*, (2020), pp. 171–178. https://doi.org/10.1109/SEAA51224.2020.00036
20. K. Talesra, G.S. Nagaraja, Low-Code platform for application development. Int. J. Appl. Eng. Res. **16**(5), 346–351 (2021) Available: https://www.ripublication.com/ijaer21/ijaerv16n5_02.pdf
21. F. Gürcan, G. Taentzer, Using Microsoft PowerApps, Mendix and OutSystems in two development scenarios: an experience report, in *Presented at the 2021 ACM/IEEE International Conference on Model Driven Engineering Languages and Systems Companion (MODELS-C), Fukuoka, Japan*, (2021), pp. 67–72. https://doi.org/10.1109/MODELS-C53483.2021.00017
22. Mendix, The Mendix Low-Code Platform. Available: https://www.mendix.com/platform/. Accessed 3 Nov 2022
23. U. Frank, P. Maier, A. Bock, Low code platforms: promises, concepts and prospects: a comparative study of ten systems. ICB Res. Rep. **70** (2021). https://doi.org/10.17185/duepublico/75244
24. Mulesoft, Anypoint Platform. Available: https://www.mulesoft.com/platform/enterprise-integration. Accessed 8 Nov 2022
25. Salesforce, Salesforce Platform Solutions (2022). Available: https://www.salesforce.com/products/platform/solutions/. Accessed 8 Nov 2022
26. Zoho, Zoho Creator: Best Low-Code Custom Application Development Platform. Available: https://www.zoho.com/creator/index.html. Accessed 7 Nov 2022
27. Appian, Appian: Low-Code Platform for Workflow, Automation & Process Mining. Available: https://appian.com/. Accessed 7 Nov 2022.
28. N. O'Leary, D. Conway-Jones, Node-RED, Mar. 02, 2023. Available: https://github.com/node-red/node-red. Accessed 16 Mar 2023
29. NodeRED, OpenJS Foundation & Contributors, Node-RED. Available: https://nodered.org. Accessed 7 Nov 2022
30. Google, Get started with AppSheet. Available: https://support.google.com/appsheet/answer/11581986?visit_id=638034330140875778-561841910&rd=1. Accessed 14 Nov 2022
31. H. T. L. de Paula, B. P. G. Caputo, heitor-lassarote/iolp, July 29, 2021. Available: https://github.com/heitor-lassarote/iolp. Accessed 4 Nov 2022.
32. A. Bibliowicz, vainolo/Object-Process-Programming, Feb. 27, 2017. Available: https://github.com/vainolo/Object-Process-Programming. Accessed 4 Nov 2022
33. C. Zolotas, M. Orlić, T. Diamantopoulos, L. Gašpar, K. Chatzidimitriou, s-case/mde, Dec. 09, 2016. Available: https://github.com/s-case/mde. Accessed 24 Nov 2022
34. A. López, O. Casquero, E. Estévez, A. Armentia, D. Orive, M. Marcos, An industrial agent-based customizable platform for I4.0 manufacturing systems. Comput. Ind. **146**(C) (2023). https://doi.org/10.1016/j.compind.2023.103859
35. learn.microsoft, Language Server Protocol. Available: http://learn.microsoft.com/en-us/visualstudio/extensibility/language-server-protocol. Accessed 9 Nov 2022.
36. R. Rodriguez-Echeverria, J.L.C. Izquierdo, M. Wimmer, J. Cabot, Towards a language server protocol infrastructure for graphical modeling, in *Presented at the 21th ACM/IEEE International Conference on Model Driven Engineering Languages and Systems, Copenhagen, Denmark*, (2018), pp. 370–380. https://doi.org/10.1145/3239372.3239383
37. Langserver, A community-driven source of knowledge for Language Server Protocol implementations. Available: https://langserver.org/. Accessed 9 Nov 2022.

38. G. Kardas, F. Ciccozzi, L. Iovino, Introduction to the special issue on methods, tools and languages for model-driven engineering and low-code development. J. Comput. Lang. **74**, 101190 (2023). https://doi.org/10.1016/j.cola.2022.101190
39. eclipse.org, Xtext – Language Engineering Made Easy! Available: https://www.eclipse.org/Xtext/. Accessed 17 Mar 2023
40. L. Lemiengre, S. Zarnekow, Scaling Xtext, Presented at the *EclipseCon 2022*, 2022, [Online]. Available: https://www.eclipsecon.org/2022/sessions/scaling-xtext. Accessed 17 Mar 2023
41. langium, Langium. Available: https://langium.org. Accessed 17 Mar 2023
42. M. Spönemann, Textual and Graphical Languages for the Cloud Era, Nov. 04, 2022. Available: https://www.typefox.io/blog/textual-graphical-languages-cloud-era. Accessed 17 Mar 2023
43. M. Spönemann, Langium 1.0: A Mature Language Toolkit, Dec. 16, 2022. Available: https://www.typefox.io/blog/langium-1.0-a-mature-language-toolkit. Accessed 17 Mar 2023
44. M. Sujew, ALL(*) Lookahead in Langium, Dec. 08, 2022. Available: https://www.typefox.io/blog/allstar-lookahead. Accessed 20 Mar 2023
45. JetBrains, MPS: The Domain-Specific Language Creator by JetBrains, 2023. Available: https://www.jetbrains.com/mps/. Accessed 20 Mar 2023
46. JetBrains, MPS User's Guide | MPS, 2023. Available: https://www.jetbrains.com/help/mps/mps-user-s-guide.html. Accessed 20 Mar 2023
47. M. Felderer, Experimentation in DSL Engineering – Integrated Tool Environment, June 22, 2017. Available: https://modeling-languages.com/experimentation-in-dsl-engineering/. Accessed 20 Mar 2023
48. S. Bégaudeau, M. Bats, P. Langer, A. Pupier, ObeoNetwork/GraphicalServerProtocol, Mar. 23, 2018. Available: https://github.com/ObeoNetwork/GraphicalServerProtocol. Accessed 9 Nov 2022
49. J. Köhnlein, Sprotty – A Web-based Diagramming Framework, June 08, 2017. Available: https://www.typefox.io/blog/sprotty-a-web-based-diagramming-framework/. Accessed 9 Nov 2022
50. J. Koehnlein et al., eclipse-sprotty/sprotty-server, May 12, 2022. Available: https://github.com/eclipse-sprotty/sprotty-server. Accessed 20 Mar 2022
51. N. Doschek, E. Neufeld, C. Letavernier, J. Helming, S. Graband, and O. Lessenich, Eclipse GLSP™, June 30, 2022. Available: https://projects.eclipse.org/projects/ecd.glsp. Accessed 9 Nov 2022
52. Eclipse Foundation, Documentation – Eclipse Graphical Language Server Platform." Available: https://www.eclipse.org/glsp/documentation/. Accessed 17 Nov 2022
53. T. Ortmayr et al., eclipse-glsp/glsp, Mar. 09, 2023. Available: https://github.com/eclipse-glsp/glsp. Accessed 20 Mar 2023
54. J. Köhnlein, Beyond LSP: Getting Your Language into Theia and VS Code, Presented at the *EclipseCon*, 2020, [Online]. Available: https://www.eclipsecon.org/2020/sessions/beyond-lsp-getting-your-language-theia-and-vs-code
55. GitHub, microsoft/vscode, Nov. 17, 2022. Available: https://github.com/microsoft/vscode. Accessed 17 Nov 2022
56. J. Helming, M. Koegel, P. Langer, Using VS Code extensions in Eclipse Theia and Che, Oct. 31, 2019. Available: https://eclipsesource.com/blogs/2019/10/31/using-vs-code-extensions-in-eclipse-theia-and-che/. Accessed 24 Nov 2022
57. JetBrains, Custom Language Support Tutorial, Sept. 27, 2022. Available: https://plugins.jetbrains.com/docs/intellij/custom-language-support-tutorial.html. Accessed 10 Nov 2022
58. A. Nyssen, GEF. Available: https://www.eclipse.org/gef/. Accessed 9 Nov 2022
59. eclipse.org, Sirius, 2022. Available: https://www.eclipse.org/sirius/. Accessed 9 Nov 2022
60. V. Vujovic, M. Maksimovic, B. Perisic, Comparative analysis of DSM graphical editor frameworks: Graphiti vs. Sirius, Presented at the *Erk 2014,* Portoroz, Slovenia, 2014
61. A. Richard, T. Giraudet, Langium + Sirius Web = Heart," presented at the *EclipseCon 2022*, Oct. 26, 2022, [Online]. Available: https://www.eclipsecon.org/2022/sessions/langium-sirius-web

62. Theia-ide, Theia – Cloud and Desktop IDE platform. Available: https://theia-ide.org/. Accessed 11 Nov 2022
63. J. Helming, Getting started with eclipse theia, Presented at the *Cloud IDE Days 2022,* [Online], Apr 28, 2022.
64. P. Langer, Diagram editors with GLSP: Why flexibility is key, Presented at the *EclipseCon 2022,* 2022, [Online]. Available: https://www.eclipsecon.org/2022/sessions/diagram-editors-glsp-why-flexibility-key
65. Eclipse, EMF.cloud, 2022. Available: https://www.eclipse.org/emfcloud/#. Accessed 23 Nov 2022
66. M. Koegel, P. Langer, Building web-based tools with Theia and Che, Presented at the *EclipseCon Europe 2019,* Ludwigsburg, Germany, 2019. Available: https://www.eclipsecon.org/europe2019/sessions/building-web-based-tools-theia-and-che
67. P. Langer, Embedded LSP support in GLSP diagrams, Mar. 15, 2021. Available: https://github.com/eclipse-glsp/glsp/issues/192
68. J. Koehnlein, M. Spönemann, Domain-specific languages in the cloud – with eclipse technologies, Presented at the *EclipseCon 2018,* Ludwigsburg, Germany, Oct. 24, 2018. Available: https://www.eclipsecon.org/europe2018/sessions/domain-specific-languages-cloud-%E2%80%93-eclipse-technologies

Bitcoin Prediction Analysis Using Deep Learning Techniques

Muhammad Muneeb, Noman Islam, Mana Saleh Al Reshan, Mohammed Hamdi, Hani Alshahrani, and Safeeullah Soomro

1 Introduction

Bitcoin prediction analysis is now becoming an important research area for the researchers due to its popularity among people. It lacks a physical form, and it is not backed up by any government or agencies or any public body [1]. Their transactions are made through blockchain system. Although the bitcoin price follows a random walk [2] as it is gaining more popularity, many giant companies are now joined with bitcoin [3], and it is expected that it will expand more and more in the future. Many giant companies like JP Morgan have even announced that they have planned to launch their own cryptocurrencies on February 2019 [4]. Revolut in Europe also received European Banking License in December 2018. Furthermore they are also trading in digital currency.

M. Muneeb
Data Science, NED University of Engineering and Technology, Karachi, Pakistan

N. Islam
Karachi Institute of Economics and Technology, Karachi, Pakistan

M. S. Al Reshan (✉)
Department of Information Symens, College of Computer Science and Information Systems, Najran University, Najran, Saudi Arabia
e-mail: msalreshan@nu.edu.sa

M. Hamdi · H. Alshahrani
Department of Computer Science, College of Computer Science and Information Systems, Najran University, Najran, Saudi Arabia
e-mail: mahamdi@nu.edu.sa; hmalshahran@nu.edu.sa

S. Soomro
American National University and University of Fairfax, Vienna, VA, USA
e-mail: ssoomro@an.edu

© The Author(s), under exclusive license to Springer Nature Switzerland AG 2024

A. Shaikh et al. (eds.), *Advances in Emerging Information and Communication Technology*, Signals and Communication Technology,
https://doi.org/10.1007/978-3-031-53237-5_30

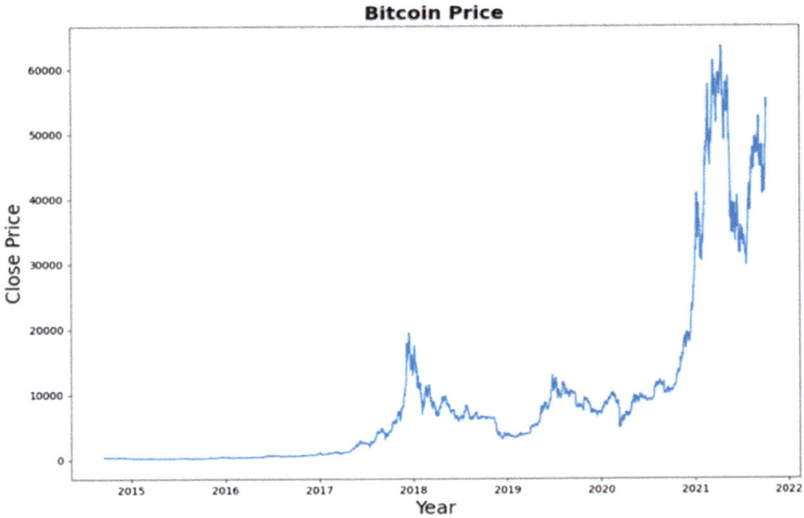

Fig. 1 Bitcoin price over the years

In the past, many questions have been raised about bitcoin such as about the value of bitcoin, the market factor, the area of financial success, and its value that implies the trust of the stakeholders on cryptocurrency [5]. According to the economist, investors don't consider bitcoin as a currency but an opportunity for investment such as the Internet stocks of the previous century [6]. Below Fig. 1 can be seen through which popularity and increase and decrease of bitcoin in yearly basis can be analyzed from the start to till now.

In the past, different deep learning algorithms have been analyzed and investigated for the bitcoin price prediction. But most of them employed very few deep learning algorithms such as recurrent neural networks (RNN), convolutional neural network (CNN), and pre-trained models such as ResNet, etc. These models only addressed basic regression problems in which their model predicts next bitcoin prices on the basis of previous data. They didn't work on classification problem to predict whether the price will go up or down on the basis of previous data [7]. There are some algorithms like decision trees, naïve Bayes, support vector machines (SVM), etc. that are conventional machine learning algorithms that are mostly used sentiment analysis to find the polarity like neutral, positive, and negative, but deep learning algorithms are more popular and are mostly preferred by the researchers for higher clarification.

This paper has used some popular deep learning techniques, in which we have done comparative analysis between different algorithms. Deep learning is a part of machine learning family which is inspired by the function and the structure of the brain which is called artificial neural networks (ANN). In this work, following techniques were used. Long short-term memory network (LSTM) is a type of recurrent neural network (RNN), and it is capable of learning order dependencies in sequence prediction problems. A Convolutional neural network (CNN) is a deep learning algorithm which takes input image and assigns importance to various objects/aspects in the image and get able to differentiate one form the other. Deep hybrid learning is an approach that it uses to combine different neural networks.

In the next section, the first related work is discussed which is followed by the methodology and then results. The paper concludes with limitations and discussions on future work.

2 Related Work

There has been a great deal of research on the prediction of bitcoin price using machine learning. However, the problem is at its infancy, and not much success has been observed, especially using deep learning techniques. As per the researches done in the past, more than 600 papers have been published on the bitcoin. Bitcoin is relatively a new technology and can be regarded as most expensive cryptocurrency in the world. This work investigated different work done on prediction of bitcoin the bitcoin that is reported below.

Dennys and Mallqui [8] have used various innovative techniques for the selection of attributes/features for price prediction. They have then applied different machine learning algorithms. Support vector machine (SVM) and artificial neural network (ANN) are included among them [9]. employed the bitcoin blockchain information as attributes for price prediction. This includes features like number of wallets, block mining difficulty, etc. These features were then used (considering only those that are highly correlated to the price) to finally develop a prediction model. They developed different regression models based on linear regression, random forest [10], and gradient boosting [11].

In another research, [12] considered various factors like S&P 500, EURO STOXX 50, DOW 30, NASDAQ, and exchange rate between major fiat currencies. They investigated various prediction models such as Bayesian neural network [13], linear regression, and support vector regression (SVR) [14]. The authors concluded that Bayesian neural network is better than the other two algorithms.

Ji et al. [15] investigated different techniques and concluded that LSTM outperformed other models. Also, for classification problems, DNN slightly outperformed the other models [16]. have used various machine learning algorithms for the predictions of bitcoin price in an efficient manner [17]. also attempted the prediction problem for other popular cryptocurrencies [18]. implemented different machine learning algorithms to predict exchange rate of price variation, by using recurrent neural network (RNN) and generalized regression neural network (GRNN) to get accurate prediction [19]. reported an accuracy of around 53% by using LSTM. Velankar et al. [20] addressed the prediction problem by using the Bayesian regression and generalized linear model.

This paper explores the following two models: LSTM and hybrid deep learning models, and investigates the performance of these algorithms on bitcoin price prediction.

3 Methodology

This section discusses the methodology used for prediction of bitcoin.

3.1 *Dataset and Feature Engineering*

Feature engineering is a very important step of any machine learning pipeline. It involves the extracting useful features from the data such that it can be used for further exploration, data visualization, analytics and prediction, etc.

It is one of the most important steps of machine learning, because it can affect the accuracy in the prediction. Suppose, if we left any null values in any feature from the dataset, and it is one of the most important feature which is need to be in proper format, and we neglect the null values; the whole analysis can go wrong, if that feature is highly dependent on the output variable. Furthermore, it is also important to remove unnecessary features not required in the dataset or you can say that it is not highly dependent to the output variable and will not impact the result, and then it is highly recommended to remove that feature from the dataset. In our dataset we have done the same. The dataset that is used for the research is taken by using yfinance library in python which is provided by Yahoo Finance which provides financial news, data, and commentary including stock quotes, press releases financial reports, etc. The data that we are getting is real-time data.

For our work we have excluded all the features having correlation less than 0.95 and taken all the features whose correlation is greater than 0.95. The features that we have selected are Open, High, Low, and Close. The features that are removed are due to having less correlation coefficient is less than 0.95. The features that are removed are Volume and Adj Close. Below tables show the dataset detail for the selected features. Table 1 shows the summary of statistics used for the experiment. Table 2 shows the correlation between features. Table 3 shows the details of the featureset.

Figure 2 shows the Spearman rank correlation coefficient matrix between different features of Bitcoin.

Table 1 Summary statistic used for experiment

	Open	High	Low	Close
Count	2586.000000	2586.000000	2586.000000	2586.000000
Mean	9461.390072	9730.210401	9174.024105	9484.257866
Std	13884.982767	14301.827478	13427.314028	13980.803466
Min	176.897003	211.731003	171.509995	178.102997
25%	577.754990	582.823242	573.438751	577.567490
50%	5515.076172	5647.827148	5353.012939	5540.485107
75%	9699.704834	9908.871582	9507.834717	9700.673340
Max	64280.031250	64863.097656	63811.187500	63896.273438

Table 2 Correlation between features

	Open	High	Low	Close
Open	1.000000	0.999518	0.999369	0.999025
High	0.999518	1.000000	0.999108	0.999474
Low	0.999369	0.999108	1.000000	0.999543
Close	0.999025	0.999474	0.999543	1.000000

Table 3 Details of the selected features from the dataset

Column	Details of the columns
Datetime	Current date time with in that 1 minute time
Open	The open price of bitcoin on that 1 minute time
High	The highest price of bitcoin on that 1 minute time
Low	The lowest price of bitcoin on that 1 minute time
Close	The close price of bitcoin on that 1 minute time

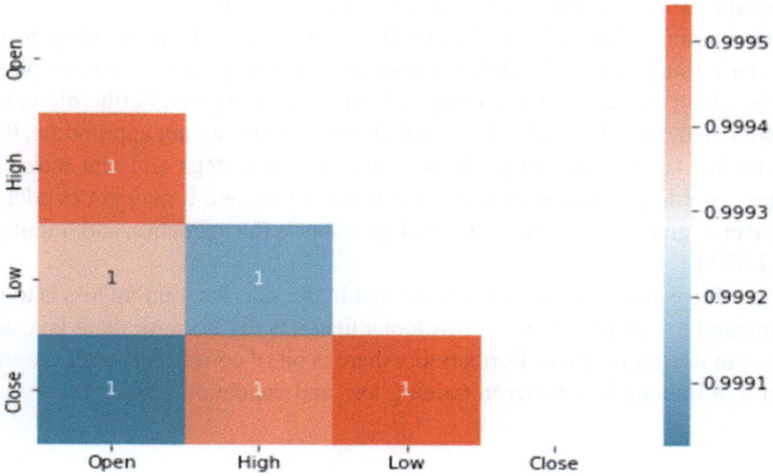

Fig. 2 Spearman's correlation coefficient of features

4 Experiments and Results

In our experiment we first normalized the data values in our dataset, and the purpose for doing this normalization is to change the values in our dataset from numeric value to a common scale without disturbing the range of values, and for this purpose we have used MinMaxScaler which is provided by sklearn library in python. Below Table 4 shows the difference before and after normalization.

After normalizing the data, we then split the data into training data and testing data, to use training data to train our model and the test data for comparison for our prediction. It is one of the most crucial step as we need to assure that our prediction make sense because we cannot use the same data for training our model as it can do overfitting. For splitting purpose we have created our own method that returns x_train, y_train, x_test, and y_test.

4.1 Experiment 1: Using LSTM

After splitting the data, we started to build our model, and the first model that we built was LSTM in which we were given 50 epochs (epoch is called an iteration), ADAM as an optimizer and mean square root as loss function. Below is the LSTM model summary shown in Fig. 3.

After compilation, we use fit method in which we set validation split as 0.1 for our analysis in which order of information depends in our analysis. If we change the order, then our results will not make any sense. Furthermore, training of the model can also be done without GPU as the amount of data is less and due to having simple architecture. These models can take hours or days to train.

After training of model, we then evaluate the results of our training, to know about our training loss and validation loss, and for this purpose we have plotted a graph in which we can see the training loss and validation loss. Furthermore, in this step, after analyzing the result, if we think that the results are not appropriate, then in that case we have choice to go back in the previous steps and can make some changes accordingly. In our case the results seem fine, and it may not be ideal, but they are enough for our purpose. Below Fig. 4 shows the validation and training loss using LSTM.

In the above graph, we can clearly see that in the start the training loss is too high as compared to validation loss which looks fine actually because there is a sudden decrease in the training loss. Furthermore there is bit of consistency after the sudden decrease of training loss between training loss and validation loss.

Table 4 Actual and normalized values

Actual: 320.19299316	Normalized: 0.00215894
Actual: 314.24899292	Normalized: 0.00206862
Actual: 315.03201294	Normalized: 0.00208052

Fig. 3 Summary of LSTM model

```
Model: "sequential_8"
_________________________________________________
Layer (type)                              Output
Shape                   Param #
=================================================
lstm_10 (LSTM)                            (None,
64)                     16896
_________________________________________________
leaky_re_lu_10 (LeakyReLU)                (None,
64)                     0
_________________________________________________
dropout_10 (Dropout)                      (None,
64)                     0
_________________________________________________
dense_10 (Dense)                          (None,
1)                      65
=================================================
Total params: 16,961
Trainable params: 16,961
Non-trainable params: 0
_________________________________________________
```

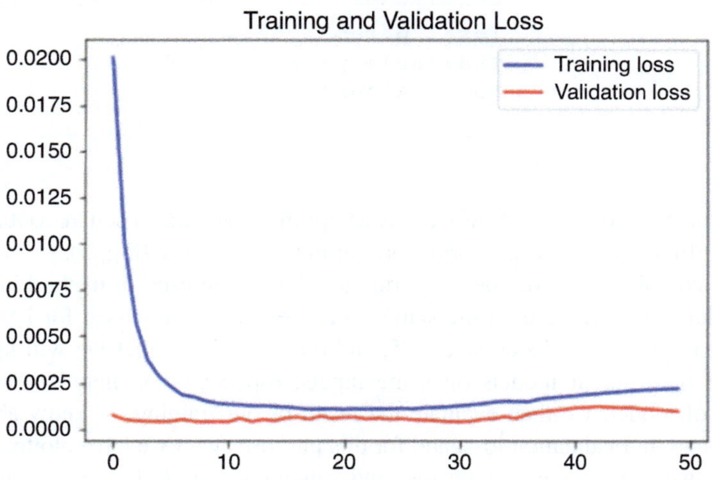

Fig. 4 Training and validation loss with LSTM

4.2 Experiment 2: Deep Hybrid Learning Model

The second model that we built was deep hybrid learning model in which we used the combination of CNN and LSTM, in which the first layer was of CNN and the second and third layers were of LSTM. In this model we have also given 50 epochs

Fig. 5 Deep hybrid model

```
Model: "sequential_11"
_________________________________________________________________
Layer (type)                                            Output
Shape                      Param #
=================================================================
conv1d_11 (Conv1D)                                      (None,
None, 64)                  384
_________________________________________________________________
lstm_17 (LSTM)                                          (None,
None, 64)                  33024
_________________________________________________________________
lstm 18 (LSTM)                                          (None,
None, 64)                  33024
_________________________________________________________________
leaky re lu 1 (LeakyReLU)                               (None,
None, 64)                  0

dropout_1 (Dropout)                                     (None,
None, 64)                  0

dense_8 (Dense)                                         (None,
None, 1)                   65
=================================================================
Total params: 66,497
Trainable params: 66,497
Non-trainable params: 0
_________________________________________________
```

(epoch is called an iteration), ADAM as an optimizer and mean square root as a loss function. Below is the deep hybrid learning model summary (Fig. 5).

After compilation of our deep hybrid model, we then used fit method in to train our model in which we used the same parameter as we have used for LSTM like validation split = 0.1, batch size = 5, and epoch = 50 so that we will know the behavior of different models on same inputs, especially, on iteration, and after training of model, we then evaluate the result of our training, to know about our training loss and validation loss, and for this purpose again we have plotted a graph in which we can see the training loss and validation loss. Below Fig. 6 shows the validation and training loss using the deep hybrid model, with first layer of CNN and second and third layers of LSTM.

In the above graph, we can clearly see that in the start the training loss is too high as compared to validation loss, and after that there is a sudden decrease in the training loss, then there is a sudden increase in the training loss, and then we can see the consistency between training loss and validation loss. But if we compare it to our previous model which was LSTM, we can clearly say that LSTM is performing far better than our deep learning model because there is very less training loss in LSTM.

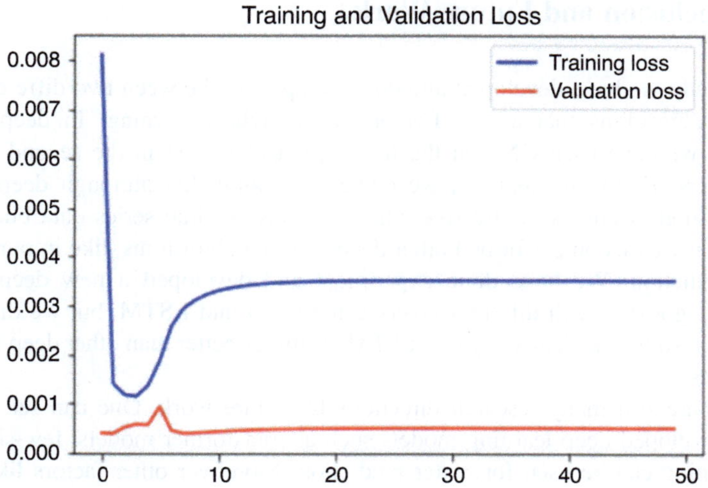

Fig. 6 Training and validation loss with deep hybrid model

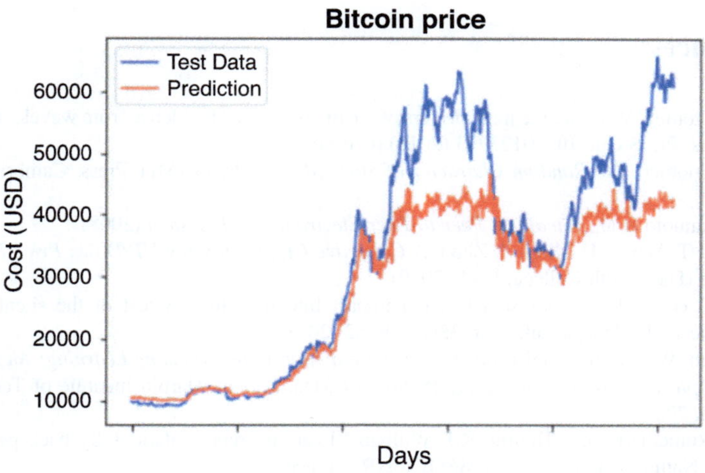

Fig. 7 Using LSTM for prediction

Therefore we have selected LSTM for our prediction, and below Fig. 7 can show the prediction results using LSTM.

In the above we can clearly say that the results of LSTM are pretty good and LSTM can be used for time series data and bitcoin data for analysis and prediction.

5 Conclusion and Future Work

In this study we have developed and done comparison between two different deep learning algorithms that are LSTM and deep hybrid learning. In deep hybrid learning, we have used CNN in the first layer and LSTM in the second and the third layers. From our analysis, we come to a point that although deep hybrid learning models are more effective when it comes to time series data but LSTM still in many cases outperformed other deep learning algorithms, like in our case of price prediction. We have done experiment and developed a new deep hybrid learning algorithm with different layers using CNN and LSTM, but we didn't get expected results that clearly say that LSTM is still far better than other deep learning algorithms.

There are still many research directions for future work. One can see if other newly developed deep learning models such as transformer models, few-shot/one-shot learning can be used for better prediction. Moreover other factors like open, high, and low attributes can also be considered to get more accuracy of the prediction models.

References

1. L. Kristoufek, What are the main drivers of the bitcoin price? Evedence from wavelet coherence analysis. PLoS One **10**, e0123923 (2015) [CrossReference]
2. P.H. Cootner, *The Random Character of Stock Market Prices* (MIT Press, Cambridge, MA, 1964)
3. S. Nakamoto et al., *Bitcoin: A Peer-to-Peer Electronic Cash System* (2008)
4. C. Bai, T. White, L. Xiao, Z. Zhou, *A Collective Cryptocurrency UP/Down Price Prediction Engine* (Dartmouth College, USA, 2019)
5. F. Mai et al., How does social media impact bitcoin value? A test of the silent majority hypothesis. J. Manage. Inf. Syst. **35**(1), 19–52 (2018)
6. Z. Chen, W.S. ChunhongLi, *Bitcoin Price Prediction Using Machine Learning: An Approach to Sample Dimension Engineering* (School of Management, Harbin Institute of Technology, Harbin, 2020)
7. D.E. Rumelhart, G.E. Hinton, R.J. Williams, Learning representations by back propagating errors. Nature **323**, 533–536 (1986) [CrossReference]
8. C.A. Dennys, R.A.F. Mallqui, Predicting the direction, maximum, minimum and closing prices of daily bitcoin exchange rate using machine learning techniques. Int. J. Soft Comput. (IJSC), 596–606 (2018)
9. M. Saad, A. Mohaisen, Towards characterizing blockchain based cryptocurrencies for highly accurate predictions, in *Proceedings of the IEEE INFOCOM 2018—IEEE Conference on Computer Communications Workshops (INFOCOM WKSHIPS)*, Honolulu, HI, USA, 16 April 2018, pp. 704–709
10. T.K. Ho, Random decision forests, in *Processing of the Third International Conference on Document Analysis and Recognition*, Montreal, QC, Canada, 14–16 August 1995; vol. 1, pp. 278–282
11. Y. Freund, R.E. Schapire, A decision-theoretic generalization of on-line learning and an application to boosting. J. Comput. Syst. Sci. **55**, 199–139 (1997) [CrossReference]

12. H. Jang, J. Lee, An empirical study on modeling and prediction of bitcoin prices with Bayesian neural networks based on blockchain information. IEEE Access **6**, 5427–5437 (2018)
13. C.M. Bishop, *Pattern Recognition and Machine Learning* (Springer, Berlin/Heidelberg, 2006)
14. C. Cortes, V. Vapnik, Support Vector networks. Mach. Learn. **20**, 273–297 (1995) [CrossReference]
15. S. Ji, J. Kim, H. Im, *A Comparitive Study of Bitcoin Price Prediction Using Deep Learning* (Department of Computer Science, Kangwon National University, Chuncheon-si, Gangwon-do 24341, Korea, 2019)
16. T. Phaladisailoed, T. Numnonda, Machine learning models comparison for bitcoin price prediction, in *10th International Conference on Information Technology and Electrical Engineering (ICITEE)* (IEEE, 2018), pp. 506–511
17. J. Abraham, D. Higdon, J. Nelson, J. Ibarra, Cryptocurrencies price prediction using tweet volumes and sentiment analysis. SMU Data Sci. Rev. **1**(3), 1. SMU (2018)
18. S. Lahmiri, S. Bekiros, *Cryptocurrency Forecasting with Deep Learning Chaotic Neural Networks* (Elsevier, 2019), pp. 35–40
19. S. McNally, J. Roche, S. Caton, Predicting the price of bitcoin using machine learning, in *26th Euromicro International Conference on Parallel, Distributed and Neural-Based Processing (PDP)*, 2018
20. S. Velankar, S. Valecha, S. Maji, Bitcoin price prediction using machine learning, in *20th International Conference on Advance Communications Technology (ICACT)*, Chuncheon-si, Korea, 11–14 February 2018

A Blockchain-Based Renewable Energy Authenticated Marketplace: BEAM of Flexibility

Mutiullah Shaikh, Sundas Munir, Uffe Kock Wiil, and Amina Shaikh

1 Introduction

Smart grids must be capable of promptly adapting to dynamic shifts in energy supply, enabling users to adjust their electricity consumption patterns effectively. The widespread implementation of renewable energy sources (RES), such as electrical vehicles (EVs), photovoltaic cells (i.e., solar panels), and wind turbines, introduces new challenges to managing smart grids [1]. The prosumers who own RES can offer excess electricity in exchange for their desired compensation. Grid operators, including transmission system operators (TSOs) and distributed system operators (DSOs), can request this excess electricity (i.e., flexibility) to manage the peak demand and congestion in the electricity grid [2]. This market mechanism is called the local flexibility market (LFM). Utilizing LFM can facilitate monitoring energy movement, incentivize alterations in prosumers' energy consumption and production patterns, attain equilibrium in local energy generation and consumption, and optimize the distribution of electricity flows. To achieve this, a microgrid is typically linked to the nearby distribution power system, allowing it to function on a grid-connected island mode based on the balance between energy generation and consumption. The deployment of these systems, distinguished by their capacity to

M. Shaikh (✉) · U. K. Wiil
The Maersk Mc-Kinney Moller Institute, University of Southern Denmark, Odense, Denmark
e-mail: mutsh@mmmi.sdu.dk; ukwiil@mmmi.sdu.dk

S. Munir
Halmstad University, Halmstad, Sweden
e-mail: sundas.munir@hh.se

A. Shaikh
Kulliyyah of Information and Communication Technology, International Islamic University Malaysia, Kuala Lumpur, Malaysia
e-mail: amina.a@live.iium.edu.my

© The Author(s), under exclusive license to Springer Nature Switzerland AG 2024
A. Shaikh et al. (eds.), *Advances in Emerging Information and Communication Technology*, Signals and Communication Technology,
https://doi.org/10.1007/978-3-031-53237-5_31

efficiently manage numerous loads and generation sources while accommodating bidirectional operational methods, has the potential to promote the assimilation of distributed and renewable energy resources into the power distribution network. Moreover, within the contemporary Energy Internet (EI) framework, the rise of mass producers who concurrently generate and consume energy (i.e., prosumers) has introduced a novel energy trading paradigm [3]. Nevertheless, managing energy trading through centralized means entails substantial costs and necessitates a sophisticated communication infrastructure. Hence, a compelling research and development objective exists to explore avenues for optimizing power systems and markets.

Demand and supply of electricity are often balanced through demand response (DR) programs that pay large energy consumers who agree to reduce their energy demand during electricity grid stress [4]. The adoption of DR programs encounters obstacles stemming from prosumers' limited awareness and safeguarding the privacy of their energy data. These complexities necessitate the creation of decentralized strategies and the engaged involvement of prosumers and assets not owned by the grid. Consequently, an imperative exists for developing privacy-preserving solutions that reconcile the benefits of decentralization with the protection of prosumer data privacy. Hence, establishing a distributed, trustworthy, efficient trading mechanism becomes imperative.

The decentralization of these programs, facilitated by promising technologies like blockchain, holds the potential for substantial benefits [5]. Blockchain represents a decentralized data processing innovation wherein all network participants share data distribution and storage responsibility [6]. In recent years, the widespread proliferation of blockchain technology has prompted numerous energy departments worldwide to integrate it into smart grid applications, encompassing energy trading, electric vehicle management, power grid security, and microgrid operations. By combining blockchain with cryptography, smart grids can attain secure real-time metering and communication, thereby alleviating the considerable transmission costs associated with traditional models. The adoption of blockchain in electric energy trading holds the potential to mitigate energy losses and bolster the real-time reliability and security of the trading system. This approach facilitates peer-to-peer and trust-based power and financial exchange between producers and consumers, eliminating the need for intermediaries. These advantages encompass enhanced transparency, increased openness, improved control, and heightened prosumer engagement. Nonetheless, while transparency is generally a favorable attribute in the energy domain, it's important to recognize that prosumers' energy data is highly sensitive and deeply personal.

1.1 Problem Statement

As the integration of RES continues to expand, the effective utilization of flexibility becomes increasingly crucial. This flexibility, originating primarily from decentralized prosumers who both consume and generate energy, plays a pivotal role in preserving grid stability, particularly considering the intermittent nature of renewables. However, achieving a seamless exchange of this flexibility between prosumers and grid operators remains a significant challenge. This challenge remains unresolved partially due to insufficient marketplaces to verify the provision. Another challenge is the prosumers' concern for their privacy and rules and regulations, such as GDPR in Europe. Therefore, there is a need for a robust digital platform acting as an intermediary to enable seamless and secure energy flexibility exchange between prosumers and grid operators.

2 Methodology

Resolving the above challenges requires addressing the following research question.

RQ. What is the minimum set of information that must be set up for prosumers and grid operators while considering the concerns of trust, privacy, and verifiability?

To address this question, our work is dedicated to establishing a decentralized renewable energy platform, culminating in a marketplace where prosumers and grid operators can seamlessly trade energy flexibility. Given these considerations, we propose a novel system designed for energy trading within decentralized grids featuring numerous prosumers to exchange feasibility to gird operators with due privacy, trust, and verification without involving any third party. This system, referred to as blockchain-based renewable energy automated marketplace (BEAM), is built around a streamlined private permissioned blockchain. It incorporates intelligent agent-free trading support and infrastructure features that significantly lower the barriers to public participation. Our system design is strategically fortified with dMIDs to foster stakeholder trust, emphasizing privacy, verifiability, reliability, and commitment.

Furthermore, the BEAM platform is founded on the concept of a shared grid and a common carrier/producer, serving as both the medium of exchange and the supplier of last resort. Upon its implementation, BEAM is expected to offer cost-effective operations and scalability, courtesy of using a private permissioned distributed ledger, aligning with previous work. Moreover, BEAM ensures transparency for all system stakeholders and enables regulatory authorities to oversee transactions directly. As such, we hypothesize that this platform can serve as a pioneering framework for exchanging power transaction data within present and future energy systems.

2.1 Aims and Objectives

The aims and objectives of the BEAM platform are as follows:

1. To develop a real-time data system on blockchain that caters to the needs of both prosumers and grid operators, focusing on privacy, verifiability, trustworthiness, and immutability.
2. To devise a flexible framework for handling data generated by decentralized energy assets, enabling the efficient management of flexibility offers and demands within the energy marketplace.
3. To integrate smart contract-based flexibility offers into existing energy systems to enhance data security, transparency, and reliability, with a particular focus on facilitating trustworthy energy transactions.

2.2 Proposed Solution

Our proposed scheme's conceptualization and significance are as follows:

Blockchain Integration. Blockchain's decentralized ledger offers transparency, enabling all stakeholders to view and verify transactions and data in real time. This transparency reduces the need for intermediaries, fostering trust and facilitating decentralized energy trading.

Renewable Energy Verification. Blockchain's capacity to track the origin and generation of renewable energy ensures authenticity and compliance with sustainability standards, boosting confidence in renewable energy claims.

Data Access and Verification. Our system accesses comprehensive information encompassing transaction and master data derived from the prosumer's flexibility offer. The master data of provisions consists of information from dMIDs. Table 1 shows information typically contained in a dMID. This information contains concrete details about the source of provision, which ensures grid operators get verifiable flexibility. The transaction data refers to the actual provision, such as the amount and time of excess electricity provided as flexibility be prosumers.

Zero-Knowledge Proofs (ZKP). We employ ZKP to address the trade-off between transparency and verifiability inherent in blockchain technology. This cryptographic method verifies without revealing sensitive data, thereby safeguarding prosumers' privacy. This rigorous verification process ensures verifiability and trustworthiness in fulfilling chosen commitments.

By combining dMIDs, blockchain technology, and ZKP, our proposed solution offers a robust, privacy-preserving, and verifiable platform that empowers prosumers to actively participate in energy markets while maintaining the integrity of their sensitive data.

Table 1 Information typically contained in a dMID

Information	Details
Public key	A cryptographic key is used for encryption, authentication, and secure communication
Private key	A cryptographic key is kept secret and known only to the entity or device to which it belongs. It is mathematically related to the corresponding public key and is used for various security purposes, including encryption, decryption, and digital signatures
Digital certificate	A digital certificate is a digitally signed document issued by a trusted Certificate Authority (CA). It binds a public key to a specific entity, device, or machine, thereby providing a verifiable link between the entity and its public key
Revocation info.	Information about how the certificate can be revoked if compromised or no longer valid. This includes details about Certificate Revocation Lists (CRLs) or Online Certificate Status Protocol (OCSP) responders
Subject information	Information about the machine or device that the digital certificate belongs to. This may include the machine's name, IP address, and other relevant identifying details
Issuer information	Information about the entity that issued the digital certificate, typically a Certificate Authority (CA). This includes the CA's name and digital signature
Validity period	The certificate will have a defined validity period during which it is considered valid. After this period, it may need to be renewed or replaced
Digital signature	A digital signature from a trusted Certificate Authority (CA) that vouches for the authenticity and validity of the machine's identity
Key usage	Information about how the public key associated with the digital machine identity can be used. This can include details such as whether it can be used for encryption, digital signatures, or both
Extended key usage	This extension specifies the purposes for which the public key can be used. For example, it might indicate that the key is only for server authentication

3 Related Work

Integrating renewable energy sources into the electricity grid presents opportunities and challenges for electricity system operators (TSOs/DSOs). Several of these challenges have been addressed through blockchain-based solutions. Particularly, the study [7] introduces a prototype for a decentralized power trading system, leveraging distributed ledger technology. That work is most similar to ours, with the main difference in the kinds of stakeholders involved and our increased focus on privacy and verifiability. That platform incorporates power companies, energy producers, and consumers as network participants who are not part of our platform. Our platform can be considered more focused and niche as it is intended for trustworthy and verifiable interaction between prosumers and grid operators. Another similar approach is a decentralized energy flexibility marketplace that enables small-scale prosumers to engage in peer-to-peer (P2P) flexibility trading [8]. That approach introduces the energy flexibility tokens to represent the tradable asset, utilizing smart contracts that empower prosumers to register their flexibility bids and offers as token-based orders during market sessions. That work is also similar in spirit to ours, with the main difference in our adoption of dMIDs to ensure the verifiability of provision details. Additionally, contrary to our work, that approach employs a decentralized flexibility matching mechanism, employing a greedy heuristic and bipartite graph approach to minimize energy flexibility. Another (similar in focus to ours) decentralized P2P energy marketplace is built on blockchain technology, specifically using the private and permissioned blockchain Hyperledger Fabric (HF) and its smart contracts for energy trading settlements [9]. A distinctive factor in that marketplace is that it includes a particular regulator acting as a governmental representative overseeing marketplace operation. The primary purpose of the regulator is generating guarantees of origins (GOs), which enable prosumers to guarantee the buyer that the energy was generated with a RES. In this way, the suggested P2P marketplace can address the governance issues needed in electricity marketplaces. Our approach does not incorporate GOs, as verified information from dMIDs guarantees that excess energy was produced using a certain RES. Thus, our approach eliminates the need for GOs for prosumers.

Orthogonal to our work in this paper, some studies focus on managing the inherent variability and intermittent nature of renewable electricity generation. Particularly, a flexibility market platform [10] powered by blockchain technology emerged as a promising solution. That work uses blockchain-based smart contracts and ledger technology to facilitate congestion management and TSO-DSO coordination. In similar effort to [10], a decentralized solution has been proposed for administering demand response programs within the context of smart grids [11]. That approach integrates grid components with a blockchain framework to manage the energy consumption information collected from Internet of Things smart metering devices. In the proposed framework, smart contracts programmatically define the expected energy flexibility at the level of each prosumer, the associated rewards or penalties, and the rules for balancing the energy demand with the energy production at the grid level. To ensure the security and dependability of trading within smart grids, an energy trading system anchored on a consortium blockchain framework was proposed [12]. That system strategically incorporates encryption and digital signatures to safeguard data confidentiality and integrity. Furthermore, that platform autonomously executes tailored procedures in response to user and power company requirements via smart contracts, delivering mutual benefits to both users and power companies responsible for maintaining nodes.

Challenges concerning privacy have also been addressed. Particularly, Khan and Khan [13] introduced an approach involving a multi-authority attribute-based signature scheme. This scheme was integrated with blockchain technology to establish a privacy-preserving framework for recording energy trading transactions. In that framework, the private key used in the signature algorithm is collaboratively generated by multiple institutions. However, it's noteworthy that within that framework, the computational overhead associated with the signature algorithm is considerable, and the signature size scales linearly with the number of attributes. In similar efforts, another privacy-focused energy trading clearing model was proposed [14]. That model employs interoperability techniques to create a blockchain network capable of concurrently handling the iterative quotation negotiation process on multiple sub-chains, thereby enhancing scalability and transaction speed. Laszka et al. [15] investigated security, privacy, and resilience challenges in microgrid energy trading. That platform ensures producer anonymity, incorporates a privacy protection mechanism using an anonymous mixer, and devises a hybrid solution that combines a smart contract-based validator with an external optimizer to securely and efficiently clear quotations.

Another set of challenges in this domain pertains to the growing number of renewable energy facilities leading to transaction volume that can overwhelm blockchain capabilities. In this effort, the study [16] addresses scalability, data size, and privacy challenges in blockchain-based renewable energy trading platforms. That work introduces two innovative approaches: one involves managing electricity production resources (EPRs) through an extended UTXO token, while the other employs data aggregation with evidentiality using ZKP to consolidate smart meter records.

In addition to blockchain-based solutions, several surveys and exploratory studies have also been performed to investigate the strengths and weaknesses of such solutions. Specifically, Farrokh A. Rahimi [17] reviewed that the electric industry is undergoing major shifts due to new technologies, widespread distributed energy resources (DERs), and informed prosumers. These changes affect both utility operations, as they grapple with managing grid-edge devices with limited control, and business models, with customer-owned DERs reducing net energy sales while increasing accommodation costs. Another survey [18] on integrating blockchain technology with smart grids examines the need for blockchain across various aspects of smart grids. That survey discusses current solutions' challenges and highlights integration frameworks and techniques. Ya Nan Cao et al. [19] present and compare blockchain-driven solutions for tackling issues related to privacy safeguarding, identity verification, data consolidation, and electricity pricing within the realms of data gathering and energy trading in smart grids. Though some centralization is needed for oversight, blockchain automates P2P energy trading while preserving privacy. Desen Kirli et al. [20] reviewed that blockchain-based smart contracts are increasingly recognized as pivotal in reshaping the energy sector toward decentralization and efficiency. This comprehensive review highlights their potential applications, outlines challenges, and provides valuable guidance for researchers and practitioners seeking to harness this technology for more transparent and transactive energy markets.

4 Technical Overview of the BEAM Platform

Figure 1 shows the high-level overview of the BEAM platform.

The platform (as shown in the center of Fig. 1) has the following components:

A decentralized network of nodes (i.e., blockchain platform). The platform consists of a private and permissioned decentralized network of nodes. Each node runs an open-source Ethereum client, written in Go, that implements proof of authority consensus mechanisms.

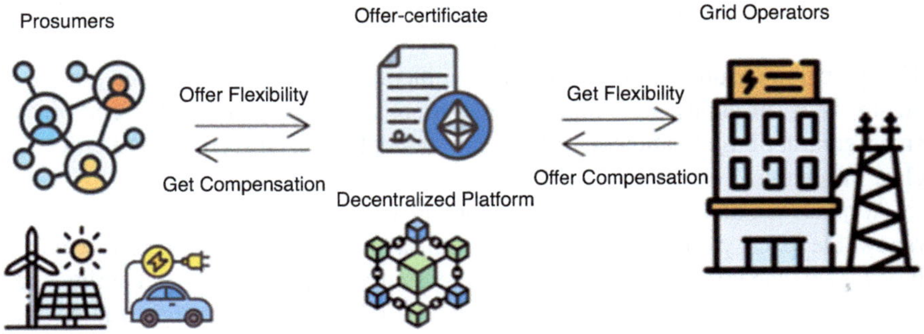

Fig. 1 A high-level overview of the BEAM platform

Offer-certificates. Every 15 min, a digital certificate is generated, representing units and times of excess electricity offered by prosumers. These digital certificates of the offer are written as smart contracts that can be stored and executed on the blockchain. The offer-certificates (smart contracts) are written in Solidity language and contain the following information:

- Information from prosumers – The information from the prosumer includes the following:
  - *A unique ID of the offer:* An offer-certificate includes a unique ID representing the offer. This ID is similar to the unique address of the smart contract that represents the offer.
  - *ID of the prosumer:* An offer-certificate includes a unique ID of the prosumer who is the generator (or owner) of the offer. The ID of the prosumer is similar to the address of the smart contract owner, who holds administrative rights to the contract. For instance, a prosumer can submit a transaction to destroy (or delete) the offer-certificate if they are no longer interested in offering the flexibility.
  - *Status of validity*: An offer-certificate is only valid for a certain period, after which the offer is no longer available.
  - Compensation: The prosumers set the amount they want in exchange for the flexibility they offer. This compensation is set as the amount requested by the prosumer in the offer-certificate.
  - *Confirmation of offer request:* The prosumer can accept or reject the request or bid submitted by grid operators for the flexibility they offer.
  - *Verified master data:* An offer-certificate contains verified (through ZKProofs) data of RES so that grid operators can be assured of the offer's legitimacy and prosumers can be assured of their privacy concerns.
  - *Transaction data (actual provision):* An offer-certificate contains provision data representing the amount of electricity a prosumer offers.
- Information from grid operators: The information from the grid operators includes the following:

- *Confirmation of request:* Similar to the confirmation from prosumers, the grid operators can directly interact with offer-certificates to submit the confirmation of request (in the form of a transaction) for the flexibility outlined in the offer-certificate.

5 Workflow and Implementation

In this section, we present the workflow and implementation of the BEAM platform.

5.1 Workflow of the BEAM Platform

Once the prosumers sign the consent form, they need to create a wallet account. The grid operators also need wallet accounts to interact with the offer-certificates. As shown in Fig. 2, the verifier and provision oracles provide external information from RES to offer-certificates. An offer-certificate itself does not contain any private information in plain text. The transactions submitted to offer-certificate are visible on the blockchain to anyone; therefore, the verifier oracle uses ZKP to provide verified information in the form of yes or no answers to the queries interesting to grid operators. Additionally, the provision oracle first calculates cryptographic hash values of the amount and time of provision and then provides this data to

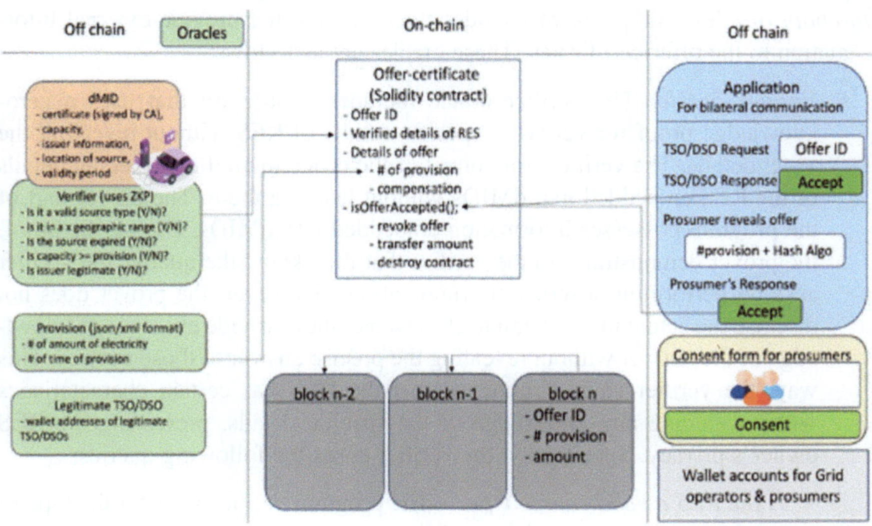

Fig. 2 Components of the BEAM platform

offer-certificate. Once a grid operator is interested in flexibility offered through an offer-certificate, they can communicate with the prosumer off-chain through an application. The off-chain application allows grid operators and prosumer to communicate their terms of flexibility. It is hosted on the same network of nodes managing the blockchain. The grid operators can request a certain provision by using its Offer ID in the application and request the prosumers to share the exact details of the provision. The prosumers can reveal this information by exchanging the key and hash algorithm that was used to calculate hash values of provision by the provision oracle. Once both parties agree on the provision details (i.e., verified details of master data, amount and time of provision, and compensation for provision), they can submit a transaction to the offer-certificate confirming the offer. Then the offer-certificate revokes the offer, ensuring that it can no longer be availed by other grid operators, and transfers the compensation to prosumers. The offer-certificate also self-destructs so that it can no longer be referred to again.

5.2 Implementation of the BEAM Platform

The BEAM platform uses various components, as shown in Fig. 2. Following are the implementation-level details of these components.

Consent form for prosumers. The prosumers sign a consent form agreeing to share relevant information and offer flexibility on the platform. Only the prosumers who sign this form can interact with the platform and its other components.

Wallet accounts. The platform includes wallet applications for grid operators and prosumers to interact with the offer-certificate executing on the blockchain.

Off-chain oracles. The platform includes three oracles that provide external information to the offer-certificates. These oracles are as follows:

- *Verifier oracle:* The verifier oracle includes a program that uses a zero-knowledge proof for verifying specific details of RES without revealing the exact details. The verifier component of the program confirms specific details of the RES embedded into dMIDs without learning them. The prover part of the program possesses information embedded into dMIDs. During the proof, the prover demonstrates to the verifier that they know the queries the verifier asks by performing a series of computations. However, the prover does not disclose the actual information itself. Instead, they provide evidence of knowledge of the answer without revealing the precise answer to those queries. This way, the verifier gains confidence that the RES has certain characteristics without gaining any knowledge of the specific details, preserving the RES owner's privacy. Specifically, the verifier poses the following queries:
  - *Is the RES a valid source type?* - The program verifies whether the type of the RES matches one of the valid sources included in the query specification. It does not reveal information about the exact source type.

- *Is the RES located within a certain geographic range?* - The program verifies whether the location of RES matches with one of the locations included in the specification of the query without revealing the exact location of RES.
- *Is the RES expired?* - The RES may have a certain expiry date; for instance, the battery of an EV may expire within a few years after the manufacturing date. The program verifies whether the battery has expired without revealing the exact expiry date of the battery stored in dMIDs.
- *Is the capacity of RES more than the provision?* - When a prosumer offers a provision, the grid operators may need to verify whether the RES can produce the offered electricity. The verifier queries this information, and the prover proves it without revealing the exact capacity of the RES.
- *Is the issuer of RES a legitimate entity?* - Not all issuers of the RES may be trustworthy. Grid operators may need to ensure whether the issuer of RES is a legitimate entity. The program helps verify this information without revealing the exact details of the issuer.

- *Provision oracle:* The provision oracle gathers the amount of electricity, and the time of provision is generated from the RES every 15 min. This program calculates the cryptographic hash value of the amount and time of provision before it is provided to the offer-certificate.
- *Legitimate grid operator oracle:* The offer-certificate only accepts transactions from verified grid operators; therefore, the BEAM platform contains an oracle that provides an updated list of legitimate grid operators.

Off-chain Application. The prosumers and grid operators use a specific application to communicate off-chain regarding the flexibility offered in the offer-certificate. The application is hosted on the private network of the blockchain and offers the following features:

- *Sign-in from grid operators and prosumers:* Grid operators and prosumers need to sign into this application because it allows only verified users to interact with it. For this reason, it takes information from "legitimate grid operator" oracle.
- *Offer request from grid operators*: As shown in Fig. 2, the grid operators can provide a unique ID of the offer-certificate to refer to the specific flexibility offer. This action shows that a certain grid operator is requesting a certain offer.
- *Offer reveal from prosumers:* The prosumers can choose to reveal the offer's details requested by grid operators. Since the provision oracle calculates the hash value of the offer, the prosumers provide the key and hash algorithm, which can be used to reveal the actual information of the flexibility.
- *Offer confirmation from grid operators and prosumers:* After revealing the information, both parties, grid operators and prosumers, can choose to accept or reject the offer through this application.

Offer-certificate. An offer-certificate is generated every 15 min, with flexibility being offered by the prosumers. The offer-certificates are written as Solidity smart contracts and stored and executed on the blockchain (see Algorithm 1 below).

Algorithm 1: Offer-certificate

Interface Oracle_GOP_Service
Function fetchLegitimateGridOperators
Input: -
Output: address[] memory ligitimate_grid_operators
End
Interface Oracle_Verified_dMIDs
Function fetchVerifiedResults
Input: bytes proof, bytes inputs
Output: bool[5] memory results
End
Interface Oracle_Transaction_Data
Function getTransactioData
Input: -
Output: bytes timeOfProvision, bytes amountOfprovision
End
State (set from prosumers): address offer_ID, address prosumer, bool validityOfOffer, uint256 amountOFOffer
State (set from grid operators): bool hasGOPAcceptedOffer
Oracle (legitimate grid operators): Oracle_GOP_Service oracle_GOP
Oracle (verified master data): Oracle_Verified_dMIDs verified_dMIDs
Oracle (transaction data): Oracle_Transaction_Data transactionData
Constructor
Input: address _oracleGOPAddress, address _oracleVerifierAddress, address _oracleTransactionAddress
Begin
verified_dMIDs = Oracle_Verified_dMIDs(_oracleVerifierAddress);
transactionData = Oracle_Transaction_Data(_oracleTransactionAddress);
oracle_GOP = Oracle_GOP_Service(_oracleGOPAddress);
End
Function confirmOfferGOP
Input: bool value
Output: -
Modifiers: isOfferValid, onlyLigitimateGridOperator
Begin:
hasGOPAcceptedOffer = value;
End
Function confirmOfferProsumer
Input: bool value
Output: -

(continued)

Algorithm 1: (continued)

```
Modifiers: isOfferValid, onlyProsumer
Begin:
    Require hasGOPAcceptedOffer;
    validityOfOffer = false;
    payable(prosumer).transfer(amountOFOffer);
    selfdestruct(payable(prosumer));
End
```

Algorithm 2: Oracle to fetch verified information from dMIDs

```
Function fetchVerifiedResults
Input: bytes proof, bytes inputs
Output: bool[5] memory results
Begin:
    results = zkSnarkVerify(proof, inputs);
End
```

Algorithm 3: Oracle to fetch transaction data from (dMIDs of) RES

```
Function getTransactioData
Input: -
Output: bytes timeOfProvision, bytes amountOfprovision
Begin:
    transaction_data = json.load(meterData.json);
    provision = transaction_data['provision'];
    timeOfProv = transaction_data['time'];
    amountOfprovision = Web3.keccak(text=str(provision));
    timeOfProvision = Web3.keccak(text=str(timeOfProv));
End
```

The offer-certificates include the following details:

- *Unique offer ID:* Every offer has a unique ID, which is referred to by the grid operators when they request flexibility represented by an offer. On the implementation level, the unique address of the smart contract representing the offer-certificate is the offer ID.
- *Information from verifier oracle:* The verifier oracle provides the verified data from RES that offer flexibility. Algorithm 2 shows the structure of this oracle. The returned information from this oracle is only true or false answers to the queries posed by the verifier part of the program.
- *Information from provision oracle:* The offer-certificate takes cryptographic hash values of the amount of provision and time of provision from the provision oracle. Algorithm 3 outlines the structure of this oracle.
- *Transactions confirming the offer:* The offer-certificates allow only a legitimate grid operator and its owner (the prosumer) to confirm the offer. Since the list of

Algorithm 4: Oracle to fetch legitimate grid operators

```
Function fetchLegitimateGridOperators
Input: -
Output: address[] memory ligitimate_grid_operators
Begin:
   ligitimate_grid_operators = parseApiResponse();
End
```

legitimate grid operators is data that resides outside the blockchain, an oracle uses relevant APIs to fetch this information in real-time and provides it to the offer-certificate. Algorithm 4 outlines a function in this oracle that returns a list of legitimate grid operators. Once an offer is requested by a grid operator and approved by the owner of the offer-certificate, it allows the automatic transfer of compensation from grid operators to the prosumer (owner of the offer-certificate) and revokes the offer (as shown in Algorithm 1).

6 Discussion

Lastly, we discuss system provisions to reflect on our research question. The key features of the BEAM platform guarantee trust, privacy, and financial incentives for prosumers and reliability, security, and operational benefits for grid operators. The specific system provisions for prosumers and grid operators are as follows:

System provisions for prosumers

1. The system guarantees trust by allowing consumers to agree directly without needing third-party intermediaries.
2. Transparency is increased by using this decentralized approach, and it reduces reliance on central entities.
3. The system also safeguards data privacy by implementing zero-knowledge proofs (ZKP) and hash commitments, a cryptographic technique that verifies information without exposing real data, thus protecting confidential information.
4. Smart contracts offer a mechanism for the automated, clear, and secure execution of agreements, with prosumers receiving financial incentives for participating in the renewable energy system.

System provisions for grid operators

1. The system provides a secure platform for participants and dMIDs to verify the authenticity of their digital identities, which are stored on the blockchain. This eliminates the risk of double spending in energy trading, as each transaction is recorded on the blockchain.

2. Grid operators can take advantage of the ability to revoke trading offers, which increases their flexibility and adaptability to changing grid conditions and unforeseen circumstances.
3. Utilizing machine identities allows for more efficient and rapid trading in the energy market, leading to increased liquidity in the grid.
4. With improved visibility into consumers' energy production and consumption patterns, grid operators can better manage their load and respond to demand fluctuations.

The utilization of dMID in a RES provides a range of benefits; however, certain limitations and challenges must be considered:

1. Implementing and managing machine identities for a wide range of devices and users can be complex and require many resources. As the system increases in size, there may be problems with network congestion, transaction throughput, and data storage. Additionally, blockchain-based systems that rely on machine identity management may be energy-intensive, potentially offsetting the environmental benefits of a renewable energy system, particularly if energy consumption is high on the blockchain network.
2. Integrating machine identity, ZKP, and other cryptographic technologies may require a high level of understanding, implementation, and maintenance expertise.
3. Zero-knowledge proof and hash commitments can increase privacy but also reduce transparency. It is important to balance transparency and privacy to comply with regulations and avoid misuse.
4. Interoperability of the new system with existing energy systems and infrastructure may be challenging.
5. The introduction of new technologies can bring about a variety of legal and regulatory issues. It is essential to ensure that all transactions are legally recognized, data ownership is respected, and liability in the event of a dispute is addressed. Furthermore, blockchain technology is not immune to security risks, as smart contracts and blockchain networks can be vulnerable to hacking, errors, and other breaches. Moreover, it is essential to educate users, providers, and other stakeholders on the advantages and limitations of the new system.

In conclusion, dMIDs can have a significant impact on renewable energy systems. However, it is important to consider the limitations and challenges when planning and implementing the technology.

Acknowledgment The BEAM platform was built based on the use case scenario provided by Fraunhofer FIT. We express our gratitude to Prof. Dr. Jens Strüker, Dr. Marc-Fabian Körner, and Till Zwede for their invaluable support and mentorship throughout the development process. Additionally, we extend our thanks to Prof. Dr. Dr. h.c. Roman Beck for providing valuable feedback on the use case report. We also acknowledge Gabriela Mariutac for her valuable assistance with the use case.

References

1. F. D'Ettorre, M. Banaei, R. Ebrahimy, S.A. Pourmousavi, E.M.V. Blomgren, J. Kowalski, et al., Exploiting demand-side flexibility: state-of-the-art, open issues and social perspective. Renew. Sustain. Energy Rev. **165**, 112605 (2022)
2. I. Lampropoulos, T. Alskaif, J. Blom, W. van Sark, A framework for the provision of flexibility services at the transmission and distribution levels through aggregator companies. Sustain. Energy Grids Netw. **17**, 100187 (2019)
3. J. Parra-Domínguez, E. Sánchez, Á. Ordóñez, The prosumer: a systematic review of the new paradigm in energy and sustainable development. Sustainability **15**(13), 10552 (2023)
4. S.J. Darby, Demand response and smart technology in theory and practice: customer experiences and system actors. Energy Policy **143**, 111573 (2020)
5. C.D. Pop, M. Antal, T. Cioara, I. Anghel, I. Salomie, Blockchain and demand response: zero-knowledge proofs for energy transactions privacy. Sensors **20**(19), 5678 (2020)
6. T.R. Gadekallu, T. Huynh-The, W. Wang, G. Yenduri, P. Ranaweera, Q.V. Pham, D.B. da Costa, M. Liyanage, Blockchain for the metaverse: A review. arXiv preprint arXiv:2203.09738 (2022)
7. D.B. Gajić, V.B. Petrović, N. Horvat, D. Dragan, A. Stanisavljević, V. Katić, J. Popović, A distributed ledger-based automated marketplace for the decentralized trading of renewable energy in smart grids. Energies **15**(6), 2121 (2022)
8. C. Antal, T. Cioara, M. Antal, V. Mihailescu, D. Mitrea, I. Anghel, et al., Blockchain based decentralized local energy flexibility market. Energy Rep. **7**, 5269–5288 (2021)
9. R.V. Tkachuk, D. Ilie, R. Robert, V. Kebande, K. Tutschku, Towards efficient privacy and trust in decentralized blockchain-based peer-to-peer renewable energy marketplace. Sustain. Energy Grids Netw. **35**, 101146 (2023)
10. V. Mladenov, V. Chobanov, G.C. Seritan, R.F. Porumb, B.A. Enache, V. Vita, et al., A flexibility market platform for electricity system operators using blockchain technology. Energies **15**(2), 539 (2022)
11. C. Pop, T. Cioara, M. Antal, I. Anghel, I. Salomie, M. Bertoncini, Blockchain based decentralized management of demand response programs in smart energy grids. Sensors **18**(1), 162 (2018)
12. D. Zheng, K. Deng, Y. Zhang, J. Zhao, X. Zheng, X. Ma, Smart grid power trading based on consortium blockchain in Internet of Things, in *International Conference on Algorithms and Architectures for Parallel Processing*, (Springer International Publishing, Cham, 2018), pp. 453–459
13. S. Khan, R. Khan, Multiple authorities attribute-based verification mechanism for Blockchain mircogrid transactions. Energies **11**(5), 1154 (2018)
14. B. Wang, S. Zhao, Y. Li, C. Wu, J. Tan, H. Li, K. Yukita, Design of a privacy-preserving decentralized energy trading scheme in blockchain network environment. Int. J. Electr. Power Energy Syst. **125**, 106465 (2021)
15. A. Laszka, S. Eisele, A. Dubey, G. Karsai, K. Kvaternik, TRANSAX: A blockchain-based decentralized forward-trading energy exchanged for transactive microgrids, in *2018 IEEE 24th International Conference on Parallel and Distributed Systems (ICPADS)*, (IEEE, 2018), pp. 918–927
16. T. Miyamae, F. Kozakura, M. Nakamura, S. Zhang, S. Hua, B. Pi, M. Morinaga, ZGridBC: zero-knowledge proof based scalable and private blockchain platform for smart grid, in *2021 IEEE International Conference on Blockchain and Cryptocurrency (ICBC)*, (IEEE, 2021), pp. 1–3
17. F.A. Rahimi, S. Mokhtari, Distribution management system for the grid of the future: a transactive system compensating for the rise in distributed energy resources. IEEE Electrific. Mag. **6**(2), 84–94 (2018)

18. Y. Guo, Z. Wan, X. Cheng, When blockchain meets smart grids: a comprehensive survey. High-Confidence Comput. **2**(2), 100059 (2022)
19. Y.N. Cao, Y. Wang, Y. Ding, Z. Guo, Q. Wu, H. Liang, Blockchain-empowered security and privacy protection technologies for smart grid. Comput. Stand. Inter. **85**, 103708 (2022)
20. D. Kirli, B. Couraud, V. Robu, M. Salgado-Bravo, S. Norbu, M. Andoni, et al., Smart contracts in energy systems: a systematic review of fundamental approaches and implementations. Renew. Sustain. Energy Rev. **158**, 112013 (2022)

Correction to: Contribution of Sustainable and Responsible Finance: Issues and Perspectives

Fatima Bellaali, Abdelhamid El Bouhadi, Mohammed Zaryouhi, and Najah Bouchao

Correction to:
Chapter 6 in: A. Shaikh et al. (eds.), ***Advances in Emerging Information and Communication Technology*****, Signals and Communication Technology,**
https://doi.org/10.1007/978-3-031-53237-5_6

The original version of the book was inadvertently published with an error. The chapter author's name [Abdelhamid El Bouhadi] was misspelled in the initially published version of this chapter as [Abdelhamid Elbouhadi], which has now been corrected. The book has been updated with the change.

The updated version of this chapter can be found at
https://doi.org/10.1007/978-3-031-53237-5_6

© The Author(s), under exclusive license to Springer Nature Switzerland AG 2024

A. Shaikh et al. (eds.), *Advances in Emerging Information and Communication Technology*, Signals and Communication Technology,
https://doi.org/10.1007/978-3-031-53237-5_32

Index

© The Editor(s) (if applicable) and The Author(s), under exclusive license to Springer Nature Switzerland AG 2024
A. Shaikh et al. (eds.), *Advances in Emerging Information and Communication Technology*, Signals and Communication Technology, https://doi.org/10.1007/978-3-031-53237-5

MIX
Papier aus verantwortungsvollen Quellen
Paper from responsible sources
FSC® C105338

If you have any concerns about our products,
you can contact us on
ProductSafety@springernature.com

In case Publisher is established outside the EU,
the EU authorized representative is:
Springer Nature Customer Service Center GmbH
Europaplatz 3, 69115 Heidelberg, Germany

Printed by Libri Plureos GmbH
in Hamburg, Germany